									VIIIA 18
									2 **He** 4.002602 — Helium
					IIIA 13	IVA 14	VA 15	VIA 16	VIIA 17	

IIIA 13	IVA 14	VA 15	VIA 16	VIIA 17	VIIIA 18
Boron 5 **B** 10.811	Carbon 6 **C** 12.011	Nitrogen 7 **N** 14.0067	Oxygen 8 **O** 15.9994	Fluorine 9 **F** 18.998403	Neon 10 **Ne** 20.179
Aluminium 13 **Al** 26.98154	Silicon 14 **Si** 28.0855	Phosphorus 15 **P** 30.97376	Sulfur 16 **S** 32.066	Chlorine 17 **Cl** 35.453	Argon 18 **Ar** 39.948

10	IB 11	IIB 12						
28 **Ni** 58.69	29 **Cu** 63.546	30 **Zn** 65.38	31 **Ga** 69.723	32 **Ge** 72.59	33 **As** 74.9216	34 **Se** 78.96	35 **Br** 79.904	36 **Kr** 83.80
46 **Pd** 106.42	Silver 47 **Ag** 107.8682	48 **Cd** 112.41	49 **In** 114.82	50 **Sn** 118.710	51 **Sb** 121.75	52 **Te** 127.60	53 **I** 126.9045	54 **Xe** 131.29
78 **Pt** 195.08	Gold 79 **Au** 196.9665	Mercury 80 **Hg** 200.59	81 **Tl** 204.383	82 **Pb** 207.2	83 **Bi** 208.9804	84 **Po** (209)	85 **At** (210)	86 **Rn** (222)

63 **Eu** 151.96	64 **Gd** 157.25	65 **Tb** 158.9254	66 **Dy** 162.50	67 **Ho** 164.9304	68 **Er** 167.26	69 **Tm** 168.9342	70 **Yb** 173.04	71 **Lu** 174.967
95 **Am** (243)	96 **Cm** (247)	97 **Bk** (247)	98 **Cf** (251)	99 **Es** (252)	100 **Fm** (257)	101 **Md** (258)	102 **No** (259)	103 **Lr** (260)

Basic Concepts of Chemistry

Basic Concepts of Chemistry

Fourth Edition

Alan Sherman

Sharon Sherman

Leonard Russikoff

Middlesex County College
Edison, New Jersey

Houghton Mifflin Company Boston
Dallas / Geneva, Illinois / Palo Alto / Princeton, New Jersey

We dedicate this text to Robert and Michael and their grandparents

Cover photograph by Ralph Mercer

Printed in the U.S.A.

Library of Congress Catalog Card Number: 87-80579

ISBN: 0-395-38127-4

CDEFGHIJ-D-99876543210

Contents Overview

Contents

Preface

Basic Concepts of Chemistry, Fourth Edition, is intended for use in the one-quarter, one-semester, or two-quarter introductory and preparatory chemistry course. Traditionally, students enrolled in these courses have had limited backgrounds in science and mathematics. However, professors tell us that their students read, enjoy, and learn from our text. For more than a decade, users have continued to comment positively on our student-oriented approach, which has been our trademark in the previous three editions. To promote student learning, we have maintained the conversational, approachable writing style in the fourth edition. We have also added several new features to help stimulate student interest.

Readability

Using short paragraphs, concise sentences, and a conversational tone, we make the material easier for students to read. Abstract concepts are made more concrete by the use of numerous applications of chemistry to everyday life. In addition, line drawings, diagrams, and photographs that clarify and emphasize important points are frequently placed in the margin next to the corresponding material. Examples of research and discoveries are blended with historical excerpts so that students can gain a sense of how chemistry evolves. An expanded feature of this edition is the presentation of *Career Sketches*, which describe careers related to chemistry.

With the help of many excellent reviewers, we have fine-tuned the textbook so that concepts and definitions are presented simply and accurately. In the fourth edition we take great care to present explanations an' details clearly, while still maintaining complete topical coverage. Important concepts are discussed thoroughly. We believe that this approach is essential in preparing students for future courses.

Problem-solving

Our experience indicates that problem-solving in chemistry continues to be an area in which some students encounter difficulty. For this reason, our problem-solving program has been further expanded in the fourth edition. The number of worked-out examples has been increased, and numerous solved problems are explained step-by-step throughout the text. As in the previous edition, practice exercises follow the worked-out examples so that students can immediately test their understanding of the subject matter.

End-of-chapter problems are divided into two groups, self-test exercises and extra exercises. Self-test exercises appear in matched pairs and are keyed to the specific learning goals listed at the beginning of each chapter. These problems help reinforce the learning of individual skills. Extra exercises also help students review skills. These exercises, however, are in a quiz-like format.

New to the fourth edition are cumulative review problems. They appear after every three or four chapters and are designed to test the student's knowledge of each set of chapters. The cumulative reviews help students reinforce knowledge of material learned earlier in the course. Students may find the reviews also to be useful in studying for tests, midterms, and final examinations.

As in previous editions, a section on mathematics (Supplement A) is included at the end of the book. This section can be treated as a reference or used at the beginning of a course as a basic skills review.

Topical Coverage

Basic chemical principles are thoroughly covered. In the fourth edition, the chapter on acids and bases now follows the chapter on solutions; a complete section on acid-base titrations has been added; and the subject matter has been thoroughly revised and updated. Several chapters have been extensively rewritten. We continue to offer topics such as kinetics and equilibrium, the quantum mechanical model of the atom, orbitals and probability, oxidation-reduction reactions, coordinate covalent bonding, specific heat, ionization and dissociation, pH, colligative properties, radio-isotopes and medicine, and rules for writing and naming chemical compounds. Environmental chemistry is also emphasized, in keeping with its global importance.

In addition to the thorough coverage, we present material in a way that allows instructors considerable flexibility. Topics are organized into discrete sections within chapters to enable instructors to include or omit topics when making reading assignments.

Study Aids

Our text has been planned to suit the backgrounds and needs of a wide variety of students. Each chapter begins with a series of learning goals. Throughout each chapter the learning goals appear in the margin next to the discussion of the corresponding material. A summary at the end of each chapter—in narrative rather than outline form—reviews the most important points discussed in the chapter. The Glossary defines important terms and concepts for the student's easy reference.

New to the fourth edition is an end-of-text supplement covering study skills for chemistry students. Concrete suggestions to help students study more efficiently are offered. The supplement provides tips for learning the material and a discussion of college life in general.

Teaching/Learning Aids

Basic Concepts of Chemistry, Fourth Edition, is part of a complete instructional package for introductory or preparatory chemistry courses. Our *Laboratory Experiments for Basic Chemistry*, Fourth Edition, contains 25 experiments and 2 laboratory exercises. New to the fourth edition of the laboratory manual is a chemical hazards guide, which clearly discusses potential hazards and stresses safety in the laboratory. Once again, *Laboratory Experiments* includes the copyrighted card game CHEM-DECK. This popular game teaches students how to write and name chemical compounds while playing card games that are similar to gin rummy and poker.

The *Instructor's Manual with Test Bank* provides coverage of both the textbook and the laboratory manual. It includes a 500-question test bank and a broad selection of final examination questions. Each lab experiment description lists the quantities of chemicals that a class of 24 students needs to conduct the experiment.

New to the fourth edition is the *Solutions Manual*, which contains worked-out solutions to the more than 2,000 problems in the textbook. Also new to this edition is a set of 52 two-color transparencies, which duplicate some of the figures and tables found in the textbook.

As with previous editions, the fourth edition of the textbook is accompanied by a *Study Guide* by James Braun of Clayton State College in Morrow, Georgia. This guide offers further explanation and additional exercises for students who need more practice. Every chapter in the textbook is supplemented by a *Study Guide* chapter of additional worked-out examples and extra practice exercises for the student to solve. The opportunity for further study is offered in special review sections that cover groups o chapters.

Acknowledgments

We thank the people who thoroughly reviewed and commented on portions of the manuscript so that we could continue to improve its quality. Without their time, effort, and helpful suggestions, we could not have forged ahead so easily. These people are Anthony Barone, Shoreline Community College; John M. Goodenow, Lawrence Institute of Technology; Irmgard Karle, Queensborough Community College; Robert Kowerski, College of San Mateo; Arnold Loebel, Merritt College; Fred Redmore, Highland Community College; Walter Rothaug, Bergen Community College; Walter Weibrecht, University of Massachusetts–Harbor Campus; John Weyh, Western Washington University; and Joseph Wijnen, Hunter College.

Once again, we thank the friends and relatives who have supported us and helped us through the many hours of rigorous work that were required to make this fourth edition possible. Professors Dominic Macchia, John Murray, Charles Oxman, Robert Smith, Virginia Filardi, Linda Christopher, Barbara Drescher, and Booker Alexander—all of Middlesex County College—were a tremendous help. Our thanks also go to Lee Walling of the East Brunswick Public Library and to the staff of the North Brunswick Public Library, who helped us with some of our research. We would also like to express our appreciation to Dave Wittmer and the staff at the North Brunswick Post Office, who moved our many Express Mail packages to Boston so efficiently. Thanks also go to Dr. Ethan Glickman, Dr. Jack Kurlansik, and the staff at the Park Dental Group; Max Sapirstein; Gertrude Sapirstein; Pamela Kance; Paula Russikoff; Marlene Heller; and Stanley Kokowicz.

Finally, we'd like to thank our sons, Robert and Michael, for their continued cooperation and assistance. We look forward to the time when they are both old enough to study chemistry, and we hope that they, too, will marvel at its wonders.

To the Student

In our increasingly complex, technological society, a knowledge of chemistry is more important than ever. Events such as the Chernobyl nuclear plant accident in April, 1986, and the toxic chemical leak in Bhopal, India, in December, 1984, focused global attention on the effects that science and technology have on all of us. And who can forget the space shuttle tragedy of January, 1986? As citizens of the worldwide community, we continue to hope for cures for cancer, heart disease, acquired immune deficiency syndrome (AIDS), and other illnesses. We look to the future for scientific discoveries and new applications that will improve the quality of life on our planet.

By studying chemistry, you will comprehend more about these and other problems. As your knowledge increases, you will better understand the ever-changing world.

The field of chemistry constantly grows. Every day scientists make discoveries and devise new applications for many of these discoveries. Today a great deal of research in chemistry and physics focuses on superconductivity. In the past, superconductivity was shown to occur in certain materials when they were cooled to extremely low temperatures. Current research reveals that some compounds show superconductivity at near room temperature conditions. Applications of basic research in superconductivity could bring us incredible energy savings, not to mention practical electric cars, magnetically levitated trains, safer nuclear reactors operating on fusion—not fission—and smaller, more powerful computers. Imagine the positive impact such developments would have on society.

As you begin your study of chemistry, you'll open the door to many exciting career opportunities in fields such as basic research, biochemistry, medicine, dentistry, nursing, pharmaceutical chemistry, engineering, an environmental chemistry, to name just a few. Throughout the text we describe chemistry-related careers, and in this edition you will find descriptions of more than 30 careers, including new high-technology fields of employment. We have prepared an extensive study skills supplement (located at the end of the text) to help you study more efficiently and to attain success.

As you study basic chemistry, keep an open mind toward the subject. Then you'll acquire a feeling for the discipline and for the people who devote their lives to finding new knowledge about the nature of the universe. We hope our textbook will help you achieve your career goals and guide you as you grow intellectually.

Alan Sherman
Sharon Sherman
Leonard Russikoff

1 *The Origins of Chemistry:*
Where It All Began

Learning Goals

After you've studied this chapter, you should be able to:

1. Discuss the beginnings of scientific thought.
2. Name the substances that Greek philosophers thought were basic to the composition of the earth.
3. Discuss the role of the Egyptians in the development of chemistry.
4. Describe the major goal of the alchemists.
5. Explain how Robert Boyle's book laid the foundations of modern chemistry.
6. Name the two main branches of chemistry.
7. Discuss some of the positive and some of the negative aspects of modern chemistry.
8. Explain the importance of modern-day chemical research.

Introduction

Human beings are curious. They have always wondered about what they are and why they are here. The desire to understand their environment has made people search for explanations, principles, and laws, and modern chemistry is one result of that search.

Chemistry is the science that deals with matter and the changes it undergoes. Chemistry also focuses on energy and how it changes when matter is transformed. Chemists seek to understand how all matter behaves and to discover the principles governing this behavior.

Chemists ask fundamental questions and look for solutions in their attempt to unlock the secrets of nature. Many of their questions have been answered, and this information has helped to build the body of knowledge that we call science.

Many of the basic concepts of chemistry have been incorporated into different fields of study. Physicists, biologists, engineers, medical and dental professionals, and nutritionists are just some of the professionals whose fields require chemical knowledge.

In this chapter we first look at the beginnings of chemistry and then briefly touch upon some areas of modern chemical research.

1.1 History

Learning Goal 1
Beginnings of scientific thought

Learning Goal 2
Influence of Greek philosophy on chemistry

Learning Goal 3
Egyptians' role in the development of chemistry

Some early alchemical symbols
(Courtesy Fisher Collection, Fisher Scientific Co., Pittsburgh, PA)

The earliest attempts to explain natural phenomena led to fanciful inventions—to myths and fantasies—but not to understanding. Around 600 B.C., a group of Greek philosophers became dissatisfied with these myths, which explained little. Stimulated by social and cultural change as well as curiosity, they began to ask questions about the world around them. They answered these questions by constructing lists of logical possibilities. Thus Greek philosophy was an attempt to discover the basic truths of nature by thinking things through, rather than by running laboratory experiments. The Greek philosophers did this so thoroughly and so brilliantly that the years between 600 and 400 B.C. are called the "golden age of philosophy."

Some of the Greek philosophers believed they could find a single substance from which everything else was made. A philosopher named Thales believed that this substance was water, but another named Anaximenes thought it was air. Some earlier philosophers believed the universe was composed of four elements—earth, air, fire, and water. The Greek philosopher Empedocles took this belief a step further, maintaining that these four elements combine in different proportions to make up all the objects in the universe.

During this period, the Greek philosophers laid the foundation for one of our main ideas about the universe. Leucippus (about 440 B.C.) and Democritus (about 420 B.C.) were trying to determine whether there was such a thing as a smallest particle of matter. In doing so, they established the idea of the atom, a particle so tiny that it could not be seen. At that time there was no way to test whether atoms really existed, and more than 2,000 years passed before scientists proved that they do exist.

While the Greeks were studying philosophy and mathematics, the Egyptians were practicing the art of chemistry. They were mining and purifying the metals gold, silver, and copper. They were making embalming fluids and dyes. They called this art *khemia*, and it flourished until the seventh century A.D., when it was taken over by the Arabs. The Egyptian word *khemia* became the Arabic word *alkhemia* and then the English word *alchemy*. Today our version of the word is used to mean everything that happened in chemistry between A.D. 300 and A.D. 1600.

· · · · · · · · · · · · ·

Learning Goal 4
Alchemists' major goals

A major goal of the alchemists was to transmute (convert) "base metals" into gold. That is, they wanted to transform less desirable elements such as lead and iron into the element gold. The ancient Arabic emperors employed many alchemists for this purpose, which, of course, was never accomplished.

The alchemists also tried to find the "philosopher's stone" (a supposed cure for all diseases) and the "elixir of life" (which would prolong life indefinitely). Unfortunately they failed in both attempts, but they did have some lucky accidents. They discovered acetic acid, nitric acid, and ethyl alcohol, as well as many other substances used by chemists today.

· · · · · · · · · · · · ·

Learning Goal 5
Boyle's influence on the development of modern chemistry

The modern age of chemistry dawned in 1661 with the publication of the book *The Sceptical Chymist*, written by Robert Boyle, an English chemist, physicist, and theologian. Boyle was "skeptical" because he was not willing to take the word of the ancient Greeks and alchemists as truth, especially about the elements that make up the world. Instead Boyle believed that scientists must start from basic principles, and he realized that every theory had to be proved by experiment. His new and innovative scientific approach was to change the whole course of chemistry.

Table 1.1 chronologically outlines some of the major contributions in chemistry throughout history.

1.2 Chemistry Is a Diverse Field

· · · · · · · · · · · · ·

Learning Goal 6
Branches of chemistry

During the 1700s and early 1800s, most chemists believed that there were two main branches of chemistry: organic and inorganic. Organic substances were thought to have been derived from living or once-living organisms; inorganic substances were said to have originated from nonliving materials. Sugars, fats, and oils were classified as organic chemicals. Salt and iron were classified as inorganic chemicals.

These two branches of chemistry still exist today, but the rules governing their classification have changed. Chemists now classify organic substances as those containing the element carbon. Inorganic substances are generally those that are composed of elements other than carbon. (Some inorganic substances are exceptions to this rule and do contain carbon.) Sugars, fats, and oils contain the element carbon and are indeed classified as organic substances. Salt and iron do not contain carbon, and both fall into the category of inorganic substances.

Other branches of chemistry include nuclear chemistry, biochemistry, pharmaceutical chemistry, analytical chemistry, environmental chemistry, and physical chemistry.

Robert Boyle
(Reprinted with permission from Torchbearers of Chemistry, by Henry Monmouth Smith, © Academic Press, Inc., 1949)

1.3 How Chemistry Affects Our World

Drawing upon such natural resources as the air, forests, oceans, mines, wells, and farms around us, the chemical industry produces more than

Table 1.1
Time Line of Chemistry

600 B.C.	Thales	Water is the main form of matter
546 B.C.	Anaximenes	Air is the main form of matter
450 B.C.	Empedocles	Idea that the four elements combine in different proportions
420 B.C.	Democritus	Idea of the atom
400 A.D.	Ko Hung	Attempts to find the elixir of life
1000	Avicenna	*Book of the Remedy*
1330	Bonus	*Introduction to the Arts of Alchemy*
1500		*Little Book of Distillation*
1620	Van Helmont	Foundations of chemical physiology
1625	Glauber	Contributions to practical chemistry
1661	Boyle	*The Sceptical Chymist*
1766	Cavendish	Discovery of hydrogen
1775	Lavoisier	Discovery of the composition of air
1787	Lavoisier & Berthollet	A system of naming chemicals
1800	Proust	Law of Definite Composition
1800	Dalton	Proposes an atomic theory
1820	Berzelius	Modern symbols for elements
1829	Döbereiner	Law of Triads
1860	Bunsen and Kirchhoff	Spectroscopic analysis
1869	Mendeleev	Periodic Law
1874	Zeidler	Discovery of DDT
1874	Van't Hoff and Le Bel	Foundations of stereochemistry
1886	Goldstein	Naming of cathode rays
1897	Thomson	Proposes a structure of the atom
1905	Einstein	$E = mc^2$ Matter–energy relationship
1908	Gelmo	Discovers sulfanilamide
1911	Rutherford	Proposes the nuclear atom

50,000 different chemicals. These chemicals are used to manufacture a variety of products and to provide raw materials for other industries to use in producing their goods. It is by this process that the items required to sustain our basic human needs are made available to us.

· · · · · · · · · · · ·
Learning Goal 7
Positive and negative aspects of chemistry

Because of chemistry, many of our foods are processed so that they remain fresh and free of harmful toxins for a longer period of time. Chemists are hard at work seeking and researching ways to alleviate the world food shortage. Thousands of drugs to help us treat disease have be-

Table 1.1
Time Line of Chemistry (*continued*)

1913	Bohr	Proposes energy levels in atoms
1922	Banting, Best, & Macleod	Discover insulin
1928	Fleming	Discovers penicillin
1932	Urey	Discovers deuterium
1942	Fermi	First atomic pile
1945		First atomic bomb; nuclear fission
1950	Pauling	Helical shape of polypeptides
1952		First hydrogen bomb; nuclear fusion
1953	Watson and Crick	Structure of DNA
1958	Townes and Schawlow	Develop laser beam
1969		Discovery of first complex organic interstellar molecule—formaldehyde
1970	Ghiorso	Synthesis of element number 105
1973	Cohen, Chang, Boyer, and Helling	Initiation of recombinant DNA studies
1974	Seaborg and Ghiorso	Synthesis of element 106
1979	A. Crewe	First color motion pictures of individual atoms
1982	Munzenberg and Armbruster	Synthesis of element 109
1984	Gallico	Development of test-tube skin
1985		Genetically engineered human insulin and human growth hormone become commercially available
1985	Diana	Development of compounds that kill cold viruses
1986		Nearly 800 genes mapped using techniques of genetic engineering
1987		New class of high-temperature superconducting oxides with major implications for use in electronics discovered by researchers at the University of Houston and at AT&T Bell Laboratories in Murray Hill, NJ

Taken in part from the *Mallinckrodt Outline of the History of Chemistry*, Mallinckrodt Chemical Works, St. Louis, 1961.

come available through the application of medical knowledge in the chemical and pharmaceutical industries. Just a few of the chemically based products we use are plastics, cleansing agents, paper products, textiles, hardware, machinery, building materials, dyes and inks, fertilizers, and paints.

At the same time, many thorny environmental problems have been created by the chemical industry. Chemists, engineers, and environmental scientists are now working together to devise safe and efficient means

Insecticides are sprayed over wide regions in order to kill pests that damage vegetation and crops.
(USDA photo from Acme)

of destroying the toxic products that are generated along with the desired products of industrial activity. As long as funds are made available and the industry behaves responsibly, this important endeavor can progress.

1.4 And the Quest to Uncover New Knowledge Continues

Today nearly every part of life is touched by the science of chemistry. Studies of the chemistry of natural and synthetic substances have led the way to the production of fuels, drugs, explosives, food, and clothing. The applications of chemistry surround us. The following examples show how current chemical research is forging ahead to benefit humankind.

Understanding Cystic Fibrosis

Learning Goal 8
Importance of chemistry today

The most common fatal genetic disease among Caucasians is cystic fibrosis. While geneticists have uncovered the chromosomal markers for the disease, biochemists have learned that those suffering from cystic fibrosis have a defect in lung cell function.

Cystic fibrosis affects the pancreas, respiratory system, and sweat glands. It usually begins in infancy and is characterized by chronic lung infection, improper functioning of the pancreas, and susceptibility to hot weather and excessive heat.

VIIIA
18

				IIIA 13	IVA 14	VA 15	VIA 16	VIIA 17	2 He 4.002602

Helium

	10	IB 11	IIB 12	5 B 10.811	6 C 12.011	7 N 14.0067	8 O 15.9994	9 F 18.998403	10 Ne 20.179
				13 Al 26.98154	14 Si 28.0855	15 P 30.97376	16 S 32.066	17 Cl 35.453	18 Ar 39.948
28 Ni 58.69	29 Cu 63.546	30 Zn 65.38	31 Ga 69.723	32 Ge 72.59	33 As 74.9216	34 Se 78.96	35 Br 79.904	36 Kr 83.80
46 Pd 106.42	47 Ag 107.8682	48 Cd 112.41	49 In 114.82	50 Sn 118.710	51 Sb 121.75	52 Te 127.60	53 I 126.9045	54 Xe 131.29
78 Pt 195.08	79 Au 196.9665	80 Hg 200.59	81 Tl 204.383	82 Pb 207.2	83 Bi 208.9804	84 Po (209)	85 At (210)	86 Rn (222)

(handwritten labels: Boron, Carbon, Nitrogen, Oxygen, Fluorine, Neon, Aluminum, Silicon, Phosphorus, Sulfur, Chlorine, Argon, Silver, Gold, Mercury)

63 Eu 151.96	64 Gd 157.25	65 Tb 158.9254	66 Dy 162.50	67 Ho 164.9304	68 Er 167.26	69 Tm 168.9342	70 Yb 173.04	71 Lu 174.967
95 Am (243)	96 Cm (247)	97 Bk (247)	98 Cf (251)	99 Es (252)	100 Fm (257)	101 Md (258)	102 No (259)	103 Lr (260)

Basic Concepts of Chemistry

Basic Concepts of Chemistry

Fourth Edition

Alan Sherman

Sharon Sherman

Leonard Russikoff

Middlesex County College
Edison, New Jersey

Houghton Mifflin Company Boston
Dallas / Geneva, Illinois / Palo Alto / Princeton, New Jersey

*We dedicate this text to Robert and Michael and
their grandparents*

Cover photograph by Ralph Mercer

Printed in the U.S.A.

Library of Congress Catalog Card Number: 87-80579

ISBN: 0-395-38127-4

CDEFGHIJ-D-99876543210

Contents Overview

Contents

Preface

Basic Concepts of Chemistry, Fourth Edition, is intended for use in the one-quarter, one-semester, or two-quarter introductory and preparatory chemistry course. Traditionally, students enrolled in these courses have had limited backgrounds in science and mathematics. However, professors tell us that their students read, enjoy, and learn from our text. For more than a decade, users have continued to comment positively on our student-oriented approach, which has been our trademark in the previous three editions. To promote student learning, we have maintained the conversational, approachable writing style in the fourth edition. We have also added several new features to help stimulate student interest.

Readability

Using short paragraphs, concise sentences, and a conversational tone, we make the material easier for students to read. Abstract concepts are made more concrete by the use of numerous applications of chemistry to everyday life. In addition, line drawings, diagrams, and photographs that clarify and emphasize important points are frequently placed in the margin next to the corresponding material. Examples of research and discoveries are blended with historical excerpts so that students can gain a sense of how chemistry evolves. An expanded feature of this edition is the presentation of *Career Sketches*, which describe careers related to chemistry.

With the help of many excellent reviewers, we have fine-tuned the textbook so that concepts and definitions are presented simply and accurately. In the fourth edition we take great care to present explanations and details clearly, while still maintaining complete topical coverage. Important concepts are discussed thoroughly. We believe that this approach is essential in preparing students for future courses.

Problem-solving

Our experience indicates that problem-solving in chemistry continues to be an area in which some students encounter difficulty. For this reason, our problem-solving program has been further expanded in the fourth edition. The number of worked-out examples has been increased, and numerous solved problems are explained step-by-step throughout the text. As in the previous edition, practice exercises follow the worked-out examples so that students can immediately test their understanding of the subject matter.

End-of-chapter problems are divided into two groups, self-test exercises and extra exercises. Self-test exercises appear in matched pairs and are keyed to the specific learning goals listed at the beginning of each chapter. These problems help reinforce the learning of individual skills. Extra exercises also help students review skills. These exercises, however, are in a quiz-like format.

New to the fourth edition are cumulative review problems. They appear after every three or four chapters and are designed to test the student's knowledge of each set of chapters. The cumulative reviews help students reinforce knowledge of material learned earlier in the course. Students may find the reviews also to be useful in studying for tests, midterms, and final examinations.

As in previous editions, a section on mathematics (Supplement A) is included at the end of the book. This section can be treated as a reference or used at the beginning of a course as a basic skills review.

Topical Coverage

Basic chemical principles are thoroughly covered. In the fourth edition, the chapter on acids and bases now follows the chapter on solutions; a complete section on acid-base titrations has been added; and the subject matter has been thoroughly revised and updated. Several chapters have been extensively rewritten. We continue to offer topics such as kinetics and equilibrium, the quantum mechanical model of the atom, orbitals and probability, oxidation-reduction reactions, coordinate covalent bonding, specific heat, ionization and dissociation, pH, colligative properties, radioisotopes and medicine, and rules for writing and naming chemical compounds. Environmental chemistry is also emphasized, in keeping with its global importance.

In addition to the thorough coverage, we present material in a way that allows instructors considerable flexibility. Topics are organized into discrete sections within chapters to enable instructors to include or omit topics when making reading assignments.

Study Aids

Our text has been planned to suit the backgrounds and needs of a wide variety of students. Each chapter begins with a series of learning goals. Throughout each chapter the learning goals appear in the margin next to the discussion of the corresponding material. A summary at the end of each chapter—in narrative rather than outline form—reviews the most important points discussed in the chapter. The Glossary defines important terms and concepts for the student's easy reference.

New to the fourth edition is an end-of-text supplement covering study skills for chemistry students. Concrete suggestions to help students study more efficiently are offered. The supplement provides tips for learning the material and a discussion of college life in general.

Teaching/Learning Aids

Basic Concepts of Chemistry, Fourth Edition, is part of a complete instructional package for introductory or preparatory chemistry courses. Our *Laboratory Experiments for Basic Chemistry*, Fourth Edition, contains 25 experiments and 2 laboratory exercises. New to the fourth edition of the laboratory manual is a chemical hazards guide, which clearly discusses potential hazards and stresses safety in the laboratory. Once again, *Laboratory Experiments* includes the copyrighted card game CHEM-DECK. This popular game teaches students how to write and name chemical compounds while playing card games that are similar to gin rummy and poker.

The *Instructor's Manual with Test Bank* provides coverage of both the textbook and the laboratory manual. It includes a 500-question test bank and a broad selection of final examination questions. Each lab experiment description lists the quantities of chemicals that a class of 24 students needs to conduct the experiment.

New to the fourth edition is the *Solutions Manual*, which contains worked-out solutions to the more than 2,000 problems in the textbook. Also new to this edition is a set of 52 two-color transparencies, which duplicate some of the figures and tables found in the textbook.

As with previous editions, the fourth edition of the textbook is accompanied by a *Study Guide* by James Braun of Clayton State College in Morrow, Georgia. This guide offers further explanation and additional exercises for students who need more practice. Every chapter in the textbook is supplemented by a *Study Guide* chapter of additional worked-out examples and extra practice exercises for the student to solve. The opportunity for further study is offered in special review sections that cover groups of chapters.

Acknowledgments

We thank the people who thoroughly reviewed and commented on portions of the manuscript so that we could continue to improve its quality. Without their time, effort, and helpful suggestions, we could not have forged ahead so easily. These people are Anthony Barone, Shoreline Community College; John M. Goodenow, Lawrence Institute of Technology; Irmgard Karle, Queensborough Community College; Robert Kowerski, College of San Mateo; Arnold Loebel, Merritt College; Fred Redmore, Highland Community College; Walter Rothaug, Bergen Community College; Walter Weibrecht, University of Massachusetts–Harbor Campus; John Weyh, Western Washington University; and Joseph Wijnen, Hunter College.

Once again, we thank the friends and relatives who have supported us and helped us through the many hours of rigorous work that were required to make this fourth edition possible. Professors Dominic Macchia, John Murray, Charles Oxman, Robert Smith, Virginia Filardi, Linda Christopher, Barbara Drescher, and Booker Alexander—all of Middlesex County College—were a tremendous help. Our thanks also go to Lee Walling of the East Brunswick Public Library and to the staff of the North Brunswick Public Library, who helped us with some of our research. We would also like to express our appreciation to Dave Wittmer and the staff at the North Brunswick Post Office, who moved our many Express Mail packages to Boston so efficiently. Thanks also go to Dr. Ethan Glickman, Dr. Jack Kurlansik, and the staff at the Park Dental Group; Max Sapirstein; Gertrude Sapirstein; Pamela Kance; Paula Russikoff; Marlene Heller; and Stanley Kokowicz.

Finally, we'd like to thank our sons, Robert and Michael, for their continued cooperation and assistance. We look forward to the time when they are both old enough to study chemistry, and we hope that they, too, will marvel at its wonders.

To the Student

In our increasingly complex, technological society, a knowledge of chemistry is more important than ever. Events such as the Chernobyl nuclear plant accident in April, 1986, and the toxic chemical leak in Bhopal, India, in December, 1984, focused global attention on the effects that science and technology have on all of us. And who can forget the space shuttle tragedy of January, 1986? As citizens of the worldwide community, we continue to hope for cures for cancer, heart disease, acquired immune deficiency syndrome (AIDS), and other illnesses. We look to the future for scientific discoveries and new applications that will improve the quality of life on our planet.

By studying chemistry, you will comprehend more about these and other problems. As your knowledge increases, you will better understand the ever-changing world.

The field of chemistry constantly grows. Every day scientists make discoveries and devise new applications for many of these discoveries. Today a great deal of research in chemistry and physics focuses on superconductivity. In the past, superconductivity was shown to occur in certain materials when they were cooled to extremely low temperatures. Current research reveals that some compounds show superconductivity at near room temperature conditions. Applications of basic research in superconductivity could bring us incredible energy savings, not to mention practical electric cars, magnetically levitated trains, safer nuclear reactors operating on fusion—not fission—and smaller, more powerful computers. Imagine the positive impact such developments would have on society.

As you begin your study of chemistry, you'll open the door to many exciting career opportunities in fields such as basic research, biochemistry, medicine, dentistry, nursing, pharmaceutical chemistry, engineering, and environmental chemistry, to name just a few. Throughout the text we describe chemistry-related careers, and in this edition you will find descriptions of more than 30 careers, including new high-technology fields of employment. We have prepared an extensive study skills supplement (located at the end of the text) to help you study more efficiently and to attain success.

As you study basic chemistry, keep an open mind toward the subject. Then you'll acquire a feeling for the discipline and for the people who devote their lives to finding new knowledge about the nature of the universe. We hope our textbook will help you achieve your career goals and guide you as you grow intellectually.

Alan Sherman
Sharon Sherman
Leonard Russikoff

1 The Origins of Chemistry: Where It All Began

Learning Goals

After you've studied this chapter, you should be able to:

1. Discuss the beginnings of scientific thought.
2. Name the substances that Greek philosophers thought were basic to the composition of the earth.
3. Discuss the role of the Egyptians in the development of chemistry.
4. Describe the major goal of the alchemists.
5. Explain how Robert Boyle's book laid the foundations of modern chemistry.
6. Name the two main branches of chemistry.
7. Discuss some of the positive and some of the negative aspects of modern chemistry.
8. Explain the importance of modern-day chemical research.

Introduction

Human beings are curious. They have always wondered about what they are and why they are here. The desire to understand their environment has made people search for explanations, principles, and laws, and modern chemistry is one result of that search.

Chemistry is the science that deals with matter and the changes it undergoes. Chemistry also focuses on energy and how it changes when matter is transformed. Chemists seek to understand how all matter behaves and to discover the principles governing this behavior.

Chemists ask fundamental questions and look for solutions in their attempt to unlock the secrets of nature. Many of their questions have been answered, and this information has helped to build the body of knowledge that we call science.

Many of the basic concepts of chemistry have been incorporated into different fields of study. Physicists, biologists, engineers, medical and dental professionals, and nutritionists are just some of the professionals whose fields require chemical knowledge.

In this chapter we first look at the beginnings of chemistry and then briefly touch upon some areas of modern chemical research.

1.1 History

Learning Goal 1
Beginnings of scientific thought

Learning Goal 2
Influence of Greek philosophy on chemistry

Learning Goal 3
Egyptians' role in the development of chemistry

Some early alchemical symbols
(Courtesy Fisher Collection, Fisher Scientific Co., Pittsburgh, PA)

The earliest attempts to explain natural phenomena led to fanciful inventions—to myths and fantasies—but not to understanding. Around 600 B.C., a group of Greek philosophers became dissatisfied with these myths, which explained little. Stimulated by social and cultural change as well as curiosity, they began to ask questions about the world around them. They answered these questions by constructing lists of logical possibilities. Thus Greek philosophy was an attempt to discover the basic truths of nature by thinking things through, rather than by running laboratory experiments. The Greek philosophers did this so thoroughly and so brilliantly that the years between 600 and 400 B.C. are called the "golden age of philosophy."

Some of the Greek philosophers believed they could find a single substance from which everything else was made. A philosopher named Thales believed that this substance was water, but another named Anaximenes thought it was air. Some earlier philosophers believed the universe was composed of four elements—earth, air, fire, and water. The Greek philosopher Empedocles took this belief a step further, maintaining that these four elements combine in different proportions to make up all the objects in the universe.

During this period, the Greek philosophers laid the foundation for one of our main ideas about the universe. Leucippus (about 440 B.C.) and Democritus (about 420 B.C.) were trying to determine whether there was such a thing as a smallest particle of matter. In doing so, they established the idea of the atom, a particle so tiny that it could not be seen. At that time there was no way to test whether atoms really existed, and more than 2,000 years passed before scientists proved that they do exist.

While the Greeks were studying philosophy and mathematics, the Egyptians were practicing the art of chemistry. They were mining and purifying the metals gold, silver, and copper. They were making embalming fluids and dyes. They called this art *khemia,* and it flourished until the seventh century A.D., when it was taken over by the Arabs. The Egyptian word *khemia* became the Arabic word *alkhemia* and then the English word *alchemy.* Today our version of the word is used to mean everything that happened in chemistry between A.D. 300 and A.D. 1600.

• • • • • • • • • • • •
Learning Goal 4
Alchemists' major goals

A major goal of the alchemists was to transmute (convert) "base metals" into gold. That is, they wanted to transform less desirable elements such as lead and iron into the element gold. The ancient Arabic emperors employed many alchemists for this purpose, which, of course, was never accomplished.

The alchemists also tried to find the "philosopher's stone" (a supposed cure for all diseases) and the "elixir of life" (which would prolong life indefinitely). Unfortunately they failed in both attempts, but they did have some lucky accidents. They discovered acetic acid, nitric acid, and ethyl alcohol, as well as many other substances used by chemists today.

• • • • • • • • • • • •
Learning Goal 5
Boyle's influence on the development of modern chemistry

The modern age of chemistry dawned in 1661 with the publication of the book *The Sceptical Chymist*, written by Robert Boyle, an English chemist, physicist, and theologian. Boyle was "skeptical" because he was not willing to take the word of the ancient Greeks and alchemists as truth, especially about the elements that make up the world. Instead Boyle believed that scientists must start from basic principles, and he realized that every theory had to be proved by experiment. His new and innovative scientific approach was to change the whole course of chemistry.

Table 1.1 chronologically outlines some of the major contributions in chemistry throughout history.

1.2 Chemistry Is a Diverse Field

• • • • • • • • • • • •
Learning Goal 6
Branches of chemistry

During the 1700s and early 1800s, most chemists believed that there were two main branches of chemistry: organic and inorganic. Organic substances were thought to have been derived from living or once-living organisms; inorganic substances were said to have originated from nonliving materials. Sugars, fats, and oils were classified as organic chemicals. Salt and iron were classified as inorganic chemicals.

These two branches of chemistry still exist today, but the rules governing their classification have changed. Chemists now classify organic substances as those containing the element carbon. Inorganic substances are generally those that are composed of elements other than carbon. (Some inorganic substances are exceptions to this rule and do contain carbon.) Sugars, fats, and oils contain the element carbon and are indeed classified as organic substances. Salt and iron do not contain carbon, and both fall into the category of inorganic substances.

Other branches of chemistry include nuclear chemistry, biochemistry, pharmaceutical chemistry, analytical chemistry, environmental chemistry, and physical chemistry.

Robert Boyle
(Reprinted with permission from Torchbearers of Chemistry, by Henry Monmouth Smith, © Academic Press, Inc., 1949)

1.3 How Chemistry Affects Our World

Drawing upon such natural resources as the air, forests, oceans, mines, wells, and farms around us, the chemical industry produces more than

Table 1.1
Time Line of Chemistry

600 B.C.	Thales	Water is the main form of matter
546 B.C.	Anaximenes	Air is the main form of matter
450 B.C.	Empedocles	Idea that the four elements combine in different proportions
420 B.C.	Democritus	Idea of the atom
400 A.D.	Ko Hung	Attempts to find the elixir of life
1000	Avicenna	*Book of the Remedy*
1330	Bonus	*Introduction to the Arts of Alchemy*
1500		*Little Book of Distillation*
1620	Van Helmont	Foundations of chemical physiology
1625	Glauber	Contributions to practical chemistry
1661	Boyle	*The Sceptical Chymist*
1766	Cavendish	Discovery of hydrogen
1775	Lavoisier	Discovery of the composition of air
1787	Lavoisier & Berthollet	A system of naming chemicals
1800	Proust	Law of Definite Composition
1800	Dalton	Proposes an atomic theory
1820	Berzelius	Modern symbols for elements
1829	Döbereiner	Law of Triads
1860	Bunsen and Kirchhoff	Spectroscopic analysis
1869	Mendeleev	Periodic Law
1874	Zeidler	Discovery of DDT
1874	Van't Hoff and Le Bel	Foundations of stereochemistry
1886	Goldstein	Naming of cathode rays
1897	Thomson	Proposes a structure of the atom
1905	Einstein	$E = mc^2$ Matter–energy relationship
1908	Gelmo	Discovers sulfanilamide
1911	Rutherford	Proposes the nuclear atom

50,000 different chemicals. These chemicals are used to manufacture a variety of products and to provide raw materials for other industries to use in producing their goods. It is by this process that the items required to sustain our basic human needs are made available to us.

Because of chemistry, many of our foods are processed so that they remain fresh and free of harmful toxins for a longer period of time. Chemists are hard at work seeking and researching ways to alleviate the world food shortage. Thousands of drugs to help us treat disease have be-

Learning Goal 7
Positive and negative aspects of chemistry

Table 1.1
Time Line of Chemistry (*continued*)

1913	Bohr	Proposes energy levels in atoms
1922	Banting, Best, & Macleod	Discover insulin
1928	Fleming	Discovers penicillin
1932	Urey	Discovers deuterium
1942	Fermi	First atomic pile
1945		First atomic bomb; nuclear fission
1950	Pauling	Helical shape of polypeptides
1952		First hydrogen bomb; nuclear fusion
1953	Watson and Crick	Structure of DNA
1958	Townes and Schawlow	Develop laser beam
1969		Discovery of first complex organic interstellar molecule—formaldehyde
1970	Ghiorso	Synthesis of element number 105
1973	Cohen, Chang, Boyer, and Helling	Initiation of recombinant DNA studies
1974	Seaborg and Ghiorso	Synthesis of element 106
1979	A. Crewe	First color motion pictures of individual atoms
1982	Munzenberg and Armbruster	Synthesis of element 109
1984	Gallico	Development of test-tube skin
1985		Genetically engineered human insulin and human growth hormone become commercially available
1985	Diana	Development of compounds that kill cold viruses
1986		Nearly 800 genes mapped using techniques of genetic engineering
1987		New class of high-temperature superconducting oxides with major implications for use in electronics discovered by researchers at the University of Houston and at AT&T Bell Laboratories in Murray Hill, NJ

Taken in part from the *Mallinckrodt Outline of the History of Chemistry*, Mallinckrodt Chemical Works, St. Louis, 1961.

come available through the application of medical knowledge in the chemical and pharmaceutical industries. Just a few of the chemically based products we use are plastics, cleansing agents, paper products, textiles, hardware, machinery, building materials, dyes and inks, fertilizers, and paints.

At the same time, many thorny environmental problems have been created by the chemical industry. Chemists, engineers, and environmental scientists are now working together to devise safe and efficient means

Insecticides are sprayed over wide regions in order to kill pests that damage vegetation and crops.
(USDA photo from Acme)

of destroying the toxic products that are generated along with the desired products of industrial activity. As long as funds are made available and the industry behaves responsibly, this important endeavor can progress.

1.4 And the Quest to Uncover New Knowledge Continues

Today nearly every part of life is touched by the science of chemistry. Studies of the chemistry of natural and synthetic substances have led the way to the production of fuels, drugs, explosives, food, and clothing. The applications of chemistry surround us. The following examples show how current chemical research is forging ahead to benefit humankind.

Understanding Cystic Fibrosis

Learning Goal 8
Importance of chemistry today

The most common fatal genetic disease among Caucasians is cystic fibrosis. While geneticists have uncovered the chromosomal markers for the disease, biochemists have learned that those suffering from cystic fibrosis have a defect in lung cell function.

Cystic fibrosis affects the pancreas, respiratory system, and sweat glands. It usually begins in infancy and is characterized by chronic lung infection, improper functioning of the pancreas, and susceptibility to hot weather and excessive heat.

Several years ago at the University of California at Riverside, researcher Paul Quinton discovered that sweat gland cells of cystic fibrosis sufferers do not allow the passage of chloride ions, which are substances found in the blood. After this discovery, research centered on finding the problem with the channels through which these chloride ions pass. At four major universities across the United States, researchers recently learned that the problem lies not with the channels themselves, but with the cells that control the channels.

In a normal lung, chloride ions cause water to move into the lining of the lung from the surrounding tissues. This water thins the mucus normally found in the lung, allowing proper lung function to take place. For those afflicted with cystic fibrosis, the mucus in the lung is thick and sticky, and the lung cannot function normally, because sufficient water does not enter the lung to thin the mucus. Current therapy includes pounding the chest to loosen lung secretions and administering antibiotics to combat frequent infections. It is hoped that these recent discoveries about the chemistry involved will lead to a more successful treatment of the disease.

A New Coating for Protein-Containing Drugs

A group of Ohio researchers has devised a new way to package protein-containing drugs. When perfected, these drugs will free diabetics from their daily insulin injections and provide a way to administer some painkillers, certain vaccines, and new contraceptives by mouth.

When protein-containing drugs without the new coating are ingested, the enzymes and high acid levels in the stomach, followed by the alkaline conditions in the small intestine, break down the proteins in the drugs before they can enter the bloodstream and get to work. Because of this, therapeutic proteins, such as insulin, must be taken by injection. The new protective coating cannot be broken down by the digestive system and therefore keeps the drug intact until it is ready to do its job. The next several years will see continued research aimed at enabling those who require insulin or any other protein-containing drug to receive treatment by mouth rather than by injection.

Chemistry is truly a powerful tool. It can create a better future or, if misdirected, can lead to destruction. A knowledge of chemistry and chemical principles is therefore extremely important in helping us to understand ourselves and our world.

Summary

Scientific thought had its beginnings in the contributions of the Greek philosophers of 600 to 400 B.C. The earliest chemistry was practiced at about that same time by the Egyptians. Later the art of chemistry—or alchemy—was taken over and extended by the Arabs. "Alchemy" now includes all of chemistry from the fourth through the sixteenth centuries A.D. The alchemists made several important discoveries, although they were primarily interested in transmuting base metals into gold. Modern chemistry began with the work of Robert Boyle in the mid-seventeenth century. Boyle believed that theories must be founded on basic principles and must be supported by experiment.

Modern chemistry and its applications affect us in a variety of ways as we eat or sleep, work or play. Present chemical research is directed toward the solution of problems in physiology, in medicine, and in the environment, among many other areas.

. . . .

Career Sketches*

Chemistry not only touches our daily lives but also is a part of many occupations. At the end of each chapter, we have included brief descriptions—career sketches—of some of these occupations. We hope you will enjoy reading about the various jobs that require an understanding of chemistry. You might even find that you want to pursue one of these occupations as your educational goal. To find out more about any of the careers described, ask at your college counseling office.

. . . .

Career Sketch

Nursing Occupations

More than half of all health care workers are employed in the nursing occupations. They include registered nurses (RN's) licensed practical nurses (LPN's), nursing aides, orderlies, and attendants. Those who work in nursing need physical stamina for the long hours that are required. Some nursing tasks are unpleasant, as well. Nurses must be able to follow directions and work under close supervision. They must be patient, understanding, and emotionally stable.

*Career sketch material in this book was adapted from two U.S. Department of Labor publications: *Occupational Outlook Handbook* and *Dictionary of Occupational Titles*. In addition, information about the latest high-tech careers was obtained from the *Encyclopedia of Careers*, Volume 6 (Chicago, Ill.: J.G. Ferguson Publishing Co., 1984).

More than half of all health
care workers are employed in
the nursing occupations.
San Pedro Peninsula Hospital, California

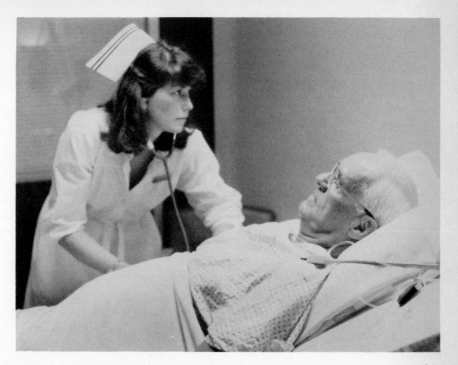

RN's: There are over one million registered nurses working in the United States (more than one-third are employed part-time). They comfort patients, assess patient behavior, and administer treatment and prescribed medication. They record patient progress and prepare equipment. In addition, some RN's assist in surgery and in examinations. They also supervise other personnel. There are specialties in cardiology, obstetrics, rehabilitation, surgery, and psychiatry. Registered nurses can enroll in a two-year associate's degree program, a two- or three-year diploma program, or a four-year baccalaureate program, and are licensed in the state in which they practice.

LPN's: Licensed practical nurses make up 20% of those employed in nursing occupations and administer 80% of the bedside nursing care in the United States. Their supervisors are usually registered nurses, physicians, or dentists. They observe patients and record information about them, take their temperature and blood pressure, measure respiration, assist with personal hygiene, give injections and medication, and change dressings. They work in specialties such as pediatrics, obstetrics, and hemodialysis. The various locations where they work include operating rooms, recovery rooms, day care facilities, physician's offices, private homes, schools, camps, and industry. They take a 12- to 18-month course of study and must pass a written examination.

Nursing aides, orderlies, and attendants: These people make beds, carry meals, assist patients, give baths, and set up equipment. There are over one million such workers in the United States today. These positions require a high school education plus on-the-job training and some classroom training.

Career Sketch

Biomedical Engineer

Biomedical engineers are highly trained scientists who employ the principles of engineering and the principles of bio-behavioral science to conduct research on the biology of humans and other animals. In addition to developing new facts and theories, they test and modify existing theories on life systems. They also design life-support apparatus.

Working with computers, they use their knowledge of graphics and related technologies to develop mathematical models that simulate bio-behavioral systems, and provide data for the measurement of life systems. They design artificial organs, ultrasonic imagery devices, heart pacemakers, and surgical lasers. They also develop artificial limbs to help people with physical deformities or injury.

The minimum requirement is a baccalaureate in engineering, although many biomedical engineers go on to earn Ph.D.'s.

Dr. William Kolff, biomedical engineer and director of the Institute for Biomedical Engineering and Division of Artificial Organs at the University of Utah, developed the first artificial kidney.
University of Utah

Self-Test Exercises

The self-test exercises are arranged in matched pairs. The first exercise of each pair appears in the left column; the exercises whose numbers are in color are answered in the back of the book. Answers to exercises in the right column are not given. Difficult problems are marked with asterisks.

Learning Goal 1: Beginnings of Scientific Thought

1. Discuss the evolution of scientific thought from the golden age of philosophy to the birth of modern chemistry.

2. What years are called "the golden age of philosophy"? Why are those years considered the beginning of scientific thought?

Learning Goal 2: Influence of Greek Philosophy on Chemistry

3. List the four elements that Empedocles thought made up the world.

4. What did Thales think the world was composed of?

5. What contribution did the Greek philosopher Democritus make to modern-day society?

6. Who established the concept of the atom?

7. If Empedocles had been right and the world were indeed made up of four elements, what two elements would compose paper?

8. If the world were made up of Empedocles's four elements, what elements would compose vegetation?

Learning Goal 3: Egyptians' Role in the Development of Chemistry

9. List some of the contributions of the Egyptians to the development of chemistry.

10. True or False: The Egyptians mined and purified the metals gold, silver, and antimony.

11. While the Greeks were studying philosophy and mathematics, the Egyptians were already practicing the art of chemistry, which they called _____ .

12. Until what time in history did the art of *khemia* flourish?

Learning Goal 4: Alchemists' Major Goals

13. What was the aim of chemistry from A.D. 300 to A.D. 1600?

14. Name the metals that the alchemists tried to turn into gold.

15. The people who tried to turn lead into gold were known as _____ .

16. What is meant by the term *elixir of life*?

Learning Goal 5: Boyle's Influence on the Development of Modern Chemistry

17. What was the message behind Robert Boyle's book, *The Sceptical Chymist?*

18. How do you think Boyle might have set about to prove that water is not an element and that it can be broken down into simpler substances?

Learning Goal 6: Branches of Chemistry

19. Name an organic chemical and an inorganic chemical.

20. Explain by what criterion chemicals are now classified as organic or inorganic.

Learning Goal 7: Positive and Negative Aspects of Chemistry

21. Choose one application of chemistry that is positive and one that is negative and discuss the impact of each on society today.

22. Weigh the benefits that chemical research and applications yield to society against the problems caused by the chemical industry. What legislative measures have been taken (or are pending) to solve some of these problems?

Learning Goal 8: Importance of Chemistry Today

23. Make a list of the industries that are based on the science of chemistry.

24. Explain why the mucus of an individual afflicted with cystic fibrosis is thick.

25. From the "Time Line of Chemistry" (pages 4–5), choose those events in the history of chemistry that you believe have had the greatest social importance. Give reasons for your choices.

26. Why are protein-containing drugs ineffective when taken orally unless they are made with a protective coating?

27. If you could extend the "Time Line of Chemistry" 50 years into the future, what would you like to see happen? What advances in research do you believe will take place?

28. Can an educated public, better equipped with scientific knowledge, more easily assess the effects of science on society? Explain your answer.

Extra Exercises*

29. With what is the science of chemistry concerned?

30. According to Empedocles, how would each of the following be classified?
(*a*) Copper (*b*) Rain (*c*) Lightning
(*d*) Nitrogen gas

31. Explain in a few sentences how you believe chemistry has affected life in the context of one topic, such as food, clothing, or medicine.

32. Discuss the contributions of alchemy to the science of chemistry.

33. List four ways in which chemistry has had a positive, healthy effect on life in your home. Explain each.

*The answers to Extra Exercises, when they are short, are found at the back of the book.

2 *Systems of Measurement*

Learning Goals

After you've studied this chapter, you should be able to:

1. Find the area of any square, rectangle, circle, or triangle.
2. Find the volume of any cube, other rectangular solid, cylinder, or sphere.
3. Convert units of mass, length, and volume within the metric system using the factor-unit method.
4. Convert from a metric unit to the corresponding English unit using the factor-unit method.
5. Convert from an English unit to the corresponding metric unit using the factor-unit method.
6. Distinguish between the mass of an object and its weight.
7. Calculate the density, mass, or volume of an object when you are given the other two.
8. Find the number of significant figures in a measurement.
9. Do calculations using the rules for significant figures.
10. Write numbers in scientific notation.
11. Convert temperatures from the Celsius to the Fahrenheit scale, and vice versa.

Introduction

Measurements are descriptions of the dimensions of things. Today, we take our systems of measurement for granted. When we buy a quart of milk, four square yards of cloth, or five pounds of sugar, the salesperson knows exactly how much we want. And we know exactly how much we are going to get.

In early England, if you wanted to buy one foot of wooden planking, you couldn't be sure how long a piece you would get. A foot was taken to be the length of the king's foot, so every time there was a new king the length of the foot changed. And, because every country had its own king, the length of the foot varied from country to country. This varying unit of length must have caused many problems in both domestic and international trade. Eventually a single, fixed unit of length was adopted, and standardized measurement began. In this chapter we discuss the use of such standard units to describe various properties that are important in chemistry.

2.1 Area and Volume

Learning Goal 1
Areas of geometric figures

Learning Goal 2
Volumes of geometric figures

Suppose you want to know the area of a tennis court. A flat rectangular surface like a tennis court has just *two* dimensions: length and width, abbreviated *l* and *w*. As Table 2.1 shows, the area of a rectangle is equal to its length *l* times its width *w*. So the area of the tennis court would be *l* × *w*.

Now suppose you want to find the volume of a cereal carton. A geometric figure like a carton (a rectangular prism) has three dimensions: length *l*, width *w*, and height *h*. As Table 2.1 and Figure 2.1 show, the volume of your carton is *l* × *w* × *h*.

Area is a measure of the extent of a surface. It is a two-dimensional measure that is always stated in squared units such as square feet (ft^2). **Volume** is a measure of the capacity of an object. It is a three-dimensional measure that is given in cubed units such as cubic meters (m^3).

Figure 2.1

Formulas for calculating the volume of a cereal carton and a soup can

Volume = length × width × height
$V = l \times w \times h$

Volume = π × (radius)² × height
$V = \pi \times r^2 \times h$
$\pi = 3.14$

Table 2.1
Areas and Volumes of Various Geometric Figures

Areas (Note that areas are always given in squared units: in², ft², and so on.)

1. Square

Area = side × side
$A = s \times s$

2. Rectangle

Area = length × width
$A = l \times w$

3. Circle

Area = π × (radius)²
$A = \pi \times r^2$

4. Triangle

Area = $\frac{1}{2}$ × base × height
$A = \frac{1}{2} \times b \times h$

Volumes (Note that volumes are always given in cubed units: in³, ft³, and so on.)

5. Cube

$V = s \times s \times s$

6. Carton
   (rectangular solid)

$V = l \times w \times h$

7. Cylinder

$V = \pi \times r^2 \times h$

8. Sphere

$V = \frac{4}{3}\pi \times r^3$

Table 2.1 lists formulas for the areas and volumes of several geometric figures. Example 2.1, which follows, shows you how to use them. Then, in Practice Exercise 2.1, you are asked to apply a volume formula on your own. You should take the time to work each practice exercise as you come to it. It will help you learn and apply what you have just read, and it will serve as a quick check of your understanding of each topic.

Example 2.1

The soup can shown in Figure 2.1 has a height of 4.00 inches and a radius of 1.00 inch. What is the volume of the can?

Solution

A soup can has the shape of a cylinder. Table 2.1 gives the formula for the volume of a cylinder as $V = \pi r^2 h$. First record what is given in the problem.

Given: $\pi = 3.14$, $r = 1.00$ inch, $h = 4.00$ inches

Then substitute these numbers into the formula and solve for V.

$$V = \pi r^2 h$$
$$= (3.14)(1.00 \text{ in.})^2 (4.00 \text{ in.}) = 12.56 \text{ in}^3$$

The can holds 12.56 cubic inches of soup, which is about two servings.

Practice Exercise 2.1   A cylinder has a height of 6.0 inches and a radius of 4.0 inches. What is its volume?

2.2 The English System of Measurement

The English system of units, as used in the United States today, has a built-in problem. It is not an orderly system and is therefore difficult to use. We will look at three units in the English system and then discuss this inherent difficulty.

In the English system, the *foot* is the unit of length and is divided into 12 smaller units called *inches*. The inch can be used to measure short distances, and the foot can be used to measure longer distances. To measure still longer distances there is the *yard* (which equals 3 feet) and the *mile* (which equals 5,280 feet).

The unit of weight in the English system is the *pound*, and the pound is divided into 16 smaller units called *ounces*. To measure larger weights, the *ton* may be used. The ton is equal to 2,000 pounds.

Table 2.2
The English System of Measurement (As Used in the United States)

Length	Weight	Volume
12 inches = 1 foot	16 ounces = 1 pound	16 fluid ounces = 1 pint
3 feet = 1 yard	2,000 pounds = 1 ton	2 pints = 1 quart
5,280 feet = 1 mile		4 quarts = 1 gallon

In the English system as used in the United States today, the unit of liquid volume is the *quart*. To measure smaller quantities, the *fluid ounce* is used (32 fluid ounces equal 1 quart). Other units of liquid volume are the *pint* (2 pints equal 1 quart) and the gallon (4 quarts equal 1 gallon).

Table 2.2 lists these units. If you examine that table closely, the problem will be evident: There is no systematic relationship among units used to measure the same property. Consider the length unit. To convert inches to yards, it is necessary to know that 12 inches equal 1 foot and that 3 feet equal 1 yard. You cannot easily move from one unit to the other. The units for weight and volume are just as inconvenient to convert. Such problems led to the development of the metric system of measurement.

2.3 The Metric System of Measurement

In the late 1700s, the French decided to change their own system of measurement. To replace it they developed a logical and orderly system called the **metric system.** The advantages of this system led to its adoption in most countries of the world and in all branches of science. The British held out until 1965 and then began a changeover to the metric system.

In countries using the metric system, almost everything is measured in metric units—distances between cities, the weight of a loaf of bread, the size of a sheet of plywood. Nearly all countries have adopted the metric system. In 1975 the United States Congress passed a bill establishing a policy of voluntary conversion to the metric system and creating the U.S. Metric Board. A program of gradual conversion was the goal of this Board, whose funding ended in 1982. The Office of Metric Programs, established by the U.S. Department of Commerce, now has the job of promoting the increased use of metric units of measurement by business and industry.

The metric system consists of (1) a set of standard units of measurement for distance, weight, volume, and so on, and (2) a set of prefixes

Table 2.3
Basic SI (Metric) Units

Quantity	Unit	Symbol
Length	meter	m
Mass	kilogram	kg
Time	second	s
Electric current	ampere	A
Temperature	kelvin	K
Light intensity	candela	cd
Amount of substance	mole	mol

that are used to express larger or smaller multiples of these units. All the prefixes designate factors of 10 that multiply or divide every unit. The official name of the metric system is *Système International d'Unités* (International System of Units), or the SI system.

In the SI system there are seven basic units. Table 2.3 lists them. Other units, called *derived units*, are composed of combinations of basic units. (For example, the unit for speed, meters per second, is derived from the basic units *meters* and *second*.) The basic units were all carefully defined. Occasionally one of the definitions is modified to make it more precise or more useful. For example, the meter was first defined as one ten-millionth of the distance between the North Pole and the Equator, along a meridian of the earth. Later, to ensure that 1 meter meant the same thing everywhere, it was redefined as the distance between two scratches on a

Table 2.4
Metric Prefixes

Prefix	Symbol	Multiplier
nano	n	0.000000001
micro	μ (Greek mu)	0.000001
milli	m	0.001
centi	c	0.01
deci	d	0.1
deka	da	10
hecto	h	100
kilo	k	1,000
mega	M	1,000,000

platinum bar that is kept at exactly the freezing point of water (zero degrees on the Celsius temperature scale) in a vault outside Paris. Still later, as more exacting measuring instruments were developed, the meter was defined as 1,650,763.73 wavelengths in vacuum of the orange-red line of the spectrum of krypton-86. You will learn more about this in Chapter 5.

Table 2.4 lists the more commonly used metric prefixes. To use one, simply "tack it on" to the front of a metric unit. That gives a related unit that is a multiple of the original unit. For example, the prefix *kilo* has a multiplier of 1,000, so a kilometer is equal to 1,000 meters. The prefix *centi* has a multiplier 0.01, so a centigram is equal to 0.01 gram. Conversely, there are 100 centigrams in a gram. The examples and practice exercises that follow show how easy it is to make such conversions.

. . . .

Learning Goal 3
Conversion of units within the metric system

Example 2.2

Change 40 meters to centimeters.

Solution

In this solution and in future examples, a method called the *factor-unit method*, or *dimensional analysis* (see Supplement A, especially Examples A.5 through A.9) will be used. With this approach the problem is set up in the following manner.

$$\text{Quantity wanted} = \text{quantity given} \times \text{factor unit}$$

The idea is that when you multiply the *quantity given* by the proper *factor unit*, some of the units cancel to give the desired quantity. Let us see how this works in the example.

$$\text{Quantity wanted} = \text{quantity given} \times \text{factor unit}$$

$$\text{centimeters} = \text{meters} \times \frac{\text{centimeters}}{\text{meter}}$$

Note that the factor unit expresses a relationship between the *quantity wanted* and the *quantity given*. The factor unit is written in such a way that the given units cancel when you multiply the quantity given times the factor unit. Then you're left with the *quantity wanted:*

$$\text{centimeters} = \cancel{\text{meters}} \times \frac{\text{centimeters}}{\cancel{\text{meter}}}$$

$$\text{centimeters} = \text{centimeters}$$

Put in the numbers.

$$? \text{ centimeters} = 40 \ \cancel{\text{meters}} \times \frac{100 \text{ centimeters}}{1 \ \cancel{\text{meter}}} = 4,000 \text{ centimeters}$$

There are 4,000 centimeters in 40 meters.

If you wonder where the term

$$\frac{100 \text{ centimeters}}{1 \text{ meter}}$$

came from, look back at Table 2.4 and you'll see that *centi* means 0.01. This means that there are 100 centimeters in 1 meter. Simply write this as

$$\frac{100 \text{ centimeters}}{1 \text{ meter}}$$

which reads "100 centimeters per meter"; the division line means *per*.

Practice Exercise 2.2   Change 35 meters to centimeters.

Of course any factor unit can be expressed in two ways. For example, the relationship 1 meter = 100 centimeters can be expressed as

$$\frac{100 \text{ centimeters}}{1 \text{ meter}} \quad \text{or} \quad \frac{1 \text{ meter}}{100 \text{ centimeters}}$$

You choose the factor unit that will make the proper terms cancel.

Example 2.3

Change 430 milligrams to grams.

Solution

Again we use the factor-unit method. (This time we will use abbreviations.)

$$g = \cancel{mg} \times \frac{g}{\cancel{mg}}$$

$$? \text{ g} = 430 \, \cancel{mg} \times \frac{1g}{1,000 \, \cancel{mg}} = 0.430 \text{ g}$$

There is 0.430 g in 430 mg.
   If you wonder where

$$\frac{1 \text{ g}}{1,000 \text{ mg}}$$

was found, look back again at Table 2.4: you'll see that *milli* means 0.001, so 1 g equals 1,000 mg. This means there is

$$\frac{1 \text{ g}}{1,000 \text{ mg}} \qquad \text{(which reads 1 g per 1,000 mg)}$$

Practice Exercise 2.3   Change 350 mg to grams.

Example 2.4

How many centimeters are there in 8 meters?

Solution

$$\text{cm} = \cancel{\text{m}} \times \frac{\text{cm}}{\cancel{\text{m}}}$$

$$? \text{ cm} = 8 \; \cancel{\text{m}} \times \frac{100 \text{ cm}}{1 \; \cancel{\text{m}}} = 800 \text{ cm}$$

There are 800 cm in 8 m.

Practice Exercise 2.4   How many centimeters are there in 18 meters?

Example 2.5

Convert 580 millimeters to centimeters.

Solution

$$\text{cm} = \cancel{\text{mm}} \times \frac{\text{cm}}{\cancel{\text{mm}}}$$

$$? \text{ cm} = 580 \; \cancel{\text{mm}} \times \frac{1 \text{ cm}}{10 \; \cancel{\text{mm}}} = 58 \text{ cm}$$

There are 58 cm in 580 mm.
   If you wonder where the term

$$\frac{1 \text{ cm}}{10 \text{ mm}}$$

came from, look at a metric ruler, and you'll see that there are 10 mm in 1 cm.

Practice Exercise 2.5   Convert 660 mm to centimeters.

Example 2.6

Convert 75 millimeters to its corresponding length in   (*a*) centimeters,   (*b*) meters,   (*c*) kilometers.

Solution

(a) $? \text{ cm} = 75 \; \cancel{\text{mm}} \times \frac{1 \text{ cm}}{10 \; \cancel{\text{mm}}} = 7.5 \text{ cm}$

(b) $? \text{ m} = 7.5 \; \cancel{\text{cm}} \times \frac{1 \text{ m}}{100 \; \cancel{\text{cm}}} = 0.075 \text{ m}$

(c) $? \text{ km} = 0.075 \text{ m} \times \dfrac{1 \text{ km}}{1000 \text{ m}} = 0.000075 \text{ km}$

Practice Exercise 2.6   Convert 55 millimeters to the corresponding length in   (*a*) centimeters,   (*b*) meters,   (*c*) kilometers.

Example 2.7

Convert 2.3 kilograms to the corresponding mass in   (*a*) grams, (*b*) decigrams,   (*c*) centigrams,   (*d*) milligrams.

Solution

(a) $? \text{ g} = 2.3 \text{ kg} \times \dfrac{1000 \text{ g}}{1 \text{ kg}} = 2300 \text{ g}$

(b) $? \text{ dg} = 2300 \text{ g} \times \dfrac{10 \text{ dg}}{1 \text{ g}} = 23{,}000 \text{ dg}$

(c) $? \text{ cg} = 23{,}000 \text{ dg} \times \dfrac{10 \text{ cg}}{1 \text{ dg}} = 230{,}000 \text{ cg}$

(d) $? \text{ mg} = 230{,}000 \text{ cg} \times \dfrac{10 \text{ mg}}{1 \text{ cg}} = 2{,}300{,}000 \text{ mg}$

Practice Exercise 2.7   Convert 4.4 kilograms to the corresponding mass in   (*a*) grams,   (*b*) decigrams,   (*c*) centigrams,   (*d*) milligrams.

You should now realize that conversion between units of the metric system is not difficult. If you want to convert metric units into English units, consult Table 2.5. (More precise conversion factors can be found in Supplement B, Table B.5.)

Learning Goal 4
Conversion of metric to
English units

Although it will always be convenient to be able to convert between systems, it is even more important to be able to associate metric measurements with commonly used items. Start now to "think metric," for only in this way will the metric system become meaningful to you. The following example offers a starting point.

Example 2.8

A new car is described in an advertisement as having an overall length of 5.50 meters and an overall width of 1.50 meters. Find the dimensions of the car in feet.

Solution
Consulting Table 2.5, we see that to convert meters to feet it is necessary to multiply by 3.28. In other words, 1 meter = 3.28 feet. There-

fore the length and width can be determined as follows:

$$\text{Length:} \qquad ? \text{ ft} = 5.50 \text{ m} \times \frac{3.28 \text{ ft}}{1 \text{ m}} = 18.0 \text{ ft}$$

$$\text{Width:} \qquad ? \text{ ft} = 1.50 \text{ m} \times \frac{3.28 \text{ ft}}{1 \text{ m}} = 4.92 \text{ ft}$$

Practice Exercise 2.8   The dimensions of a new American car are as follows: overall length, 4.20 meters; overall width, 1.25 meters. Is this a compact car? Find its dimensions in feet.

Learning Goal 5
Conversation of English to metric units

Before we work the next example, we need to discuss the SI unit of volume. This unit is a derived unit that has the dimension (length)³—for example, m^3 or cm^3. However, before the SI system was developed, the metric unit of volume was the liter: One liter (1 L) is equal in volume to 1,000 cm^3, and 1 mL = 1 cm^3. For some measurements the liter is simpler to use than the SI units, and many people continue to use it in such cases. In this book, we shall use whichever unit best fits a particular situation. In the next example we work with liters, which are universally used in selling gasoline.

Table 2.5
Conversion of Units

To convert	into	multiply by
Length		
inches	centimeters	2.54 cm/in.
centimeters	inches	0.39 in./cm
feet	meters	0.30 m/ft
meters	feet	3.28 ft/m
Weight (mass; see Section 2.4)		
ounces	grams	28.35 g/oz
grams	ounces	0.035 oz/g
pounds	grams	454 g/lb
grams	pounds	0.0022 lb/g
Volume		
liters	quarts	1.06 qt/liter
quarts	liters	0.946 liter/qt

Example 2.9

An American family visiting Mexico wants to fill its car with 15 gallons of gasoline. Mexico uses the metric system. How many liters of gasoline should the visitors ask the attendant to put in their tank?

Solution

We know that 15 gallons of gasoline are the same as 60 quarts (4 quarts = 1 gallon). Consulting Table 2.5, we see that to convert quarts to liters we multiply by 0.946 liter/quart. Therefore we may solve the problem as follows:

$$? \text{ liters} = 60 \; \cancel{\text{quarts}} \times \frac{0.946 \text{ liter}}{1 \; \cancel{\text{quart}}} = 57 \text{ liters}$$

Practice Exercise 2.9   To conserve water, the Simon family installed a reducing valve and a water-metering device in their shower. If Mr. Simon uses 20 gallons of water in his shower, and Mrs. Simon uses 18 gallons in hers, how many liters of water do the Simons use?

It is valuable to know that a liter is just a little more than a quart. A meter is slightly longer than a yard (it is 1 yard and 3.375 inches).

When you are dealing with weights, it is helpful to remember that a nickel weighs about 5 g.

If you can keep these three approximations in mind, you'll find it easier to work in the metric system.

2.4 Mass and Weight

Learning Goal 6
Mass and weight

The mass of an object and its weight are often thought of as being the same, so the words *mass* and *weight* are frequently used interchangeably. This is incorrect, because, by definition, the two terms have different meanings. **Mass** is *the quantity of matter in an object,* whereas **weight** is *the gravitational force that attracts an object.*

The definition of mass implies that your body's *mass* is constant no matter where you are. Your body's *weight,* on the other hand, varies from planet to planet and even varies slightly at different places on the earth. An astronaut who weighs 180 pounds on earth weighs only 30 pounds on the moon (which has only one-sixth of the gravitational pull of the earth) and has no weight at all in outer space. But *the astronaut's mass is the same in all places.* (We can define weight mathematically as

$$\text{Weight} = \text{mass} \times \text{gravity} \qquad \text{or} \qquad W = m \times g$$

Figure 2.2
Three cubes of equal volume

Iron cube   Glass cube   Oak wood cube

The force of gravity varies from planet to planet, so the weight of an object must vary too.)

However, because all of our weighing will be done here on earth, and because the weights of objects vary by only small amounts at different places on the earth, we can use the terms *mass* and *weight* interchangeably without introducing too much error into our calculations.

2.5 Density

Learning Goal 7
Calculation of density, mass, and volume

Which is heavier: glass, iron, or wood from an oak tree? Naturally it depends on the size of each piece. However, what if all three were the same size? In other words, what if we had cubes of glass, iron, and oak wood, each with a volume of 1 cm³? (See Figure 2.2.) Suppose we weighed each cube to determine which was the heaviest. We would find that 1 cm³ of iron weighs 7.9 g, 1 cm³ of glass weighs 2.4 g, and 1 cm³ of oak wood weighs 0.6 g. We would conclude that, *for a particular volume*, iron has the greatest mass.

The concept of density enables us to express this relationship conveniently. **Density** can be defined as *the mass per unit volume of a substance or object*. The density of a substance can be determined by using the formula

$$D = \frac{m}{V}$$

where D is the substance's density, m is its mass, and V is its volume. If the mass is measured in grams and the volume in cubic centimeters, then the unit of density is

$$\frac{grams}{cubic\ centimeters} \quad or \quad \frac{g}{cm^3}$$

(See Supplement A for a review of solving algebraic word equations.)

Example 2.10

A block of iron that is 5.0 cm long, 3.0 cm high, and 4.0 cm wide weighs 474 g. What is the density of iron?

Solution

We first calculate the volume of the block.

$$\text{Volume} = \text{length} \times \text{width} \times \text{height}$$

$$V = 5.0 \text{ cm} \times 4.0 \text{ cm} \times 3.0 \text{ cm} = 60 \text{ cm}^3$$

We now know that $V = 60 \text{ cm}^3$ and $m = 474$ g, so all we have to do is solve the density formula for D.

$$D = \frac{m}{V} = \frac{474 \text{ g}}{60 \text{ cm}^3} = 7.9 \frac{\text{g}}{\text{cm}^3}$$

Practice Exercise 2.10   A block of aluminum is 2.0 cm long, 3.0 cm high, and 5.0 cm wide, and it weighs 81.0 g. What is the density of the aluminum?

Example 2.11

Suppose you are told that 400 g of alcohol occupy a volume of 500 mL. What is the density of alcohol?

Solution

In this example we are given the mass and the volume of the substance whose density we are asked to find. Note that the volume of the alcohol is given in milliliters (mL). But *one milliliter is equal to one cubic centimeter*. That is, 1 mL = 0.001 L = 1 cm³. So it is okay to interchange the units mL and cm³. Now let's solve this problem.

$$D = \frac{m}{V} = \frac{400 \text{ g}}{500 \text{ cm}^3} = 0.8 \frac{\text{g}}{\text{cm}^3}$$

Practice Exercise 2.11   Suppose you are told that 880 g of a clear, colorless liquid occupy a volume of 110 mL. What is the density of this liquid?

Example 2.12

Determine which is denser, carbon tetrachloride or chloroform, from the following information.
(a) 16 g of carbon tetrachloride occupy a volume of 10 mL.
(b) 30 g of chloroform occupy a volume of 20 mL.

Solution

We simply calculate the density of each liquid and see which is greater:

$$D\text{(carbon tetrachloride)} = \frac{m}{V} = \frac{16 \text{ g}}{10 \text{ cm}^3} = 1.6 \frac{\text{g}}{\text{cm}^3}$$

$$D\text{(chloroform)} = \frac{m}{V} = \frac{30 \text{ g}}{20 \text{ cm}^3} = 1.5 \frac{\text{g}}{\text{cm}^3}$$

The carbon tetrachloride is the denser liquid.

Practice Exercise 2.12   Determine which is denser, liquid A or liquid B, given the following information.
(a) 55.0 g of liquid A occupy a volume of 10.0 mL.
(b) 25.0 g of liquid B occupy a volume of 12.5 mL.

The preceding calculations are straightforward uses of the density formula. But what if you were given the density of a material and its mass? Could you calculate its volume? Here's how to do it.

Example 2.13

The density of alcohol is 0.8 g/cm³. Calculate the volume of 1.6 kg of alcohol.

Solution
First we have to solve the density formula

$$D = \frac{m}{V}$$

for V. To do so, we multiply both sides of the equation by V. This gives us $D \times V = m$. Now we divide both sides of the equation by D. This gives us the formula we want:

$$V = \frac{m}{D}$$

We are given that $D = 0.8$ g/cm³ and $m = 1.6$ kg, or 1,600 g. We substitute these numbers into the formula and solve for V.

$$V = \frac{m}{D} = \frac{1600 \text{ g}}{0.8 \text{ g/cm}^3} = 2,000 \text{ cm}^3 \text{ (or 2,000 mL)}$$

Practice Exercise 2.13   The density of chloroform is 1.5 g/cm³. Calculate the volume of 2.0 kg of chloroform.

Figure 2.3

(a) We can measure the temperature of a liquid using a thermometer calibrated only in degrees. (b) We can also measure the temperature of a liquid using a thermometer calibrated in tenths of degrees.

2.6 Significant Figures

Here is an important fact about density: Most substances *expand when heated*. Therefore, when a certain weight of a substance is hot, it occupies a larger volume than it does when it is cool. This means that the density of the substance decreases as it is warmed. Think about this for a moment, and make sure you understand why it is true. Most densities reported in chemical references, such as the *Handbook of Chemistry and Physics*, are the densities at 20°C, which is about room temperature.

Suppose that, in an experiment, you are measuring the temperature of a liquid. The thermometer you are using is calibrated, or marked off, only in whole degrees. Imagine that the mercury in the thermometer is halfway between 34 and 35 degrees (Figure 2.3a). You can then estimate that the temperature of the liquid is 34.5 degrees.

Now you pick up another thermometer, which is calibrated in tenths of degrees (Figure 2.3b). When you read this thermometer, it looks as though the temperature of the liquid is 34.55 degrees. It seems that the two thermometers differ in sensitivity, but which measurement is correct?

To answer this question, we consider the number of significant figures in each measurement. **Significant figures** are *digits that express information that is reasonably reliable. The number of significant figures equals the number of digits written, including the last digit even though its value is uncertain.* For example, the temperature 34.5 degrees given by the first thermometer is known to be correct to three significant figures. These significant

Learning Goal 8
Significant figures

Figure 2.4
Measurement of a piece of glass

19.45 cm

15.3 cm

14.24 cm

18.74 cm

figures are the 3, the 4, and the 5. (Expressing the measurement as 34.50 degrees would mean that the temperature is known to four significant figures—the zero on the right is significant.)

The temperature 34.55 degrees given by the second thermometer shows a measurement with confidence to four significant figures. It may not be more accurate than the first measurement, but it is more *precise*. In general, the more significant figures there are, the more precise the measurement.

Actually, neither of the two temperature measurements may be accurate. (They certainly wouldn't be accurate if the temperature of the liquid were actually, say, 47.51 degrees.) Thus *accuracy* involves closeness to the actual dimension. *Precision* is related to the detail with which a measurement is known (expressed by the number of significant figures.)

A measuring *instrument* is considered accurate if it consistently provides measurements that are close to the actual amount. It is considered to have good precision if it consistently provides the same measurement when it is used to measure the same amount. To see the difference, consider several darts being thrown at a target. If the darts land very close together, the results exhibit precision. If all the darts land on the bull's eye, they exhibit accuracy as well.

Learning Goal 9
Calculations with significant figures

Another experiment will show how significant figures play an important role in measurement. Suppose that we have to measure a certain piece of glass to find its perimeter (we do this by measuring the four edges of the glass and adding the lengths). We have two rulers, one calibrated in centimeters and the other calibrated in millimeters. We set out to measure each edge of the glass with the more precise ruler (the one calibrated in millimeters). However, when we get to the fourth side of the glass, we absent-mindedly pick up the less precise ruler (the one calibrated in centimeters). Looking at the glass, shown in Figure 2.4, we can see that three edges are measured to two decimal places and that the fourth is measured to one decimal place. What can we say about the perimeter of the glass? Is it 67.73 cm or 67.7 cm? The answer is 67.7 cm. We can report only what we know for sure—and in this case that is the answer to one decimal place. *Our least precise measurement determines the number of significant figures in our result.*

Whenever we *add or subtract* measured quantities, we must report the results in terms of the least precise measurement. We do this by rounding off to the least number of significant figures. Our result must have no more decimal places than our least precise quantity.

To round off a number, we drop one or more digits at the right end of the number and, if necessary, adjust the rightmost digit that we keep. In some cases, dropped digits must be replaced with zeros. The rules we shall use in rounding numbers are as follows.

Rule 1 ▶ If the leftmost dropped digit is smaller than 5, simply drop the digits. Replace dropped digits with zeros as necessary to maintain the magnitude of the rounded number. Thus 7.431 rounded to two significant figures is 7.4. And 7,431 rounded to two significant figures is 7,400.

Rule 2 ▶ If the leftmost dropped digit is 5 or greater, increase the last retained digit by 1. Replace dropped digits with zeros as necessary to maintain the magnitude of the rounded number. Thus 93.56 rounded to three significant figures is 93.6. And 9,356 rounded to three significant figures is 9,360.

Example 2.14

Add 18.7444 and 13, and report the result to the appropriate number of significant figures.

Solution

The 13 is the less precise measurement; therefore we must round the sum to a whole number.

$$\begin{array}{r} 18.7444 \\ +13 \\ \hline 31.7444 \end{array}$$ Round off to 32 (whole number).

Practice Exercise 2.14   Add the numbers 12.4432 and 15. Provide the appropriate number of significant figures in your answer.

Example 2.15

Subtract 0.12 from 48.743, and use the appropriate number of significant figures in your answer.

Solution

The 0.12 is the less precise number, because we know it to only two decimal places. Therefore we subtract and round off to two decimal places.

$$\begin{array}{r} 48.743 \\ -\ 0.12 \\ \hline 48.623 \end{array} \qquad \text{Round off to 48.62 (two decimal places).}$$

Practice Exercise 2.15   Subtract 1.23 from 54.667, giving the appropriate number of significant figures in your answer.

Here is a rounding rule that is useful in *multiplying and dividing* measured quantities.

Step 1
Count the number of significant figures in each of the quantities to be multiplied or divided.

Step 2
Report the result to the least number of significant figures determined in step 1. Round as required.

Example 2.16

What is the area of a square whose side is measured as 1.5 cm?

Solution
The area of a square is equal to the length of its side squared, or

$$A = s \times s = 1.5 \text{ cm} \times 1.5 \text{ cm} = 2.25 \text{ cm}^2$$

Round off to 2.3 cm^2. (We report to only two significant figures, because the measurement has only two significant figures.)

Practice Exercise 2.16   Calculate the area of a square whose side is measured as 2.5 cm.

Example 2.17

Divide 20.8 by 4, and give the result to the proper number of significant figures.

Solution

The result must be rounded to one significant figure because of the one-digit divisor 4.

$$\frac{20.8}{4} = 5.2 \qquad \text{Round off to 5.}$$

Practice Exercise 2.17   Divide 48.2 by 4, giving the result to the proper number of significant figures.

Example 2.18

Multiply 20.8 by 4.1, and report the result to the proper number of significant figures.

Solution

The result must be rounded to two significant figures.

$$20.8 \times 4.1 = 85.28 \qquad \text{Round off to 85.}$$

Practice Exercise 2.18   Multiply 25.5 by 3.2, giving the result to the proper number of significant figures.

Another problem arises when we use significant figures: What do we do about zeros? Are they significant figures, or not? Here are some helpful rules. Read them carefully, work through Example 2.19, and then do Practice Exercise 2.19.

Rule 1 ▶ Zeros *between* nonzero digits are significant:

4.004 has four significant figures

Rule 2 ▶ Zeros to the *left* of nonzero digits are *not* significant, because these zeros show only the position of the decimal point:

0.00254 has three significant figures

0.0146 has three significant figures

0.06 has one significant figure

Rule 3 ▶ Zeros that fall at the *end* of a number are not significant unless they are marked as significant. If a zero does indicate the number's precision, we can mark it as significant by *placing a line over it.*

84,000 has two significant figures

84,$\overline{0}$00 has three significant figures

84,000.0 has six significant figures

In the last example, it really isn't necessary to put a line over each zero. The number, as it is written, shows that you have measured to one decimal place and found it to be zero. So all the zeros must be significant.

Example 2.19

In this example, we tell you how many significant figures there are in different numbers. In Practice Exercise 2.19, *you* are asked to tell *us*.

Solution

(a) 0.00087 has two significant figures.
(b) 1.004 has four significant figures.
(c) 873.005 has six significant figures.
(d) 9.00000 has six significant figures.
(e) 320,000 has two significant figures.
(f) 18$\overline{0}$,000 has four significant figures.
(g) 180,000.0 has seven significant figures.
(h) 2,$\overline{0}$00 has four significant figures.

Practice Exercise 2.19   Find the number of significant figures in each of the following numbers.

(a) 0.0023
(b) 5.025
(c) 123.456

(d) 5,$\overline{000}$
(e) 5,$\overline{0}$00
(f) 12.000

2.7 Scientific Notation: Powers of 10

....

Learning Goal 10
Scientific notation

In the physical sciences, we often deal with numbers that are *very* large or *very* small. Numbers like 100,000,000,000 (a hundred billion) or 0.0000008 (eight ten-millionths) arise frequently, and they are troublesome to work with in calculations. There is a shorthand method of writing such numbers, based on powers of 10. The number 100,000,000,000 can be written as 1×10^{11}, and 0.0000008 as 8×10^{-7}. In the first example, we moved the decimal point eleven places to the left:

$$1\,0\,0,\,0\,0\,0,\,0\,0\,0,\,0\,0\,0.\ = 1 \times 10^{11}$$

Moving the decimal point to the left corresponds to multiplying by a positive power of 10. In the second example, we moved the decimal point seven places to the right.

$$0.\ 0\ 0\ 0\ 0\ 0\ 8 = 8 \times 10^{-7}$$

Moving the decimal point to the right corresponds to multiplying by a negative power of 10. In both cases, the power that 10 is raised to is called the **exponent** of the **base number** 10. (If this causes you to hesitate, turn to Section A.3 in Supplement A and see that explanation, plus Examples A.1 through A.4.) We have

$$10^{11} \longleftarrow \quad \text{Exponent}$$
$$\underset{\uparrow}{\phantom{10^{11}}} \longleftarrow \quad \text{Base number}$$

Here's how to use the shorthand method: To express a number in **scientific notation,** write the number with only one significant figure to the left of the decimal point. Multiply it by 10 raised to the number of places you moved the decimal point—positive if it was moved to the left, and negative if it was moved to the right.

$$3,\ 8\ 0\ 0 = 3.8 \times 10^3$$

$$0.\ 0\ 0\ 6\ 2\ 5 = 6.25 \times 10^{-3}$$

$$1\ 0\ 0,\ 0\ 0\ 0,\ 0\ 0\ 0 = 1 \times 10^8$$

$$0.\ 0\ 0\ 0\ 0\ 0\ 1 = 1 \times 10^{-7}$$

The number of significant figures is often clarified by using scientific notation. For example, if we know the number 500 to three significant figures, we write 5.00×10^2. If we know the number 500 to only one significant figure, we write 5×10^2. Writing 5.00×10^2 is the same as writing $5\overline{00}$ (three significant figures). Writing 5×10^2 is the same as writing 500 (one significant figure).

With a little practice, you'll soon be using scientific notation as easily as you use the more standard notation, but do not deny yourself the practice you need!

Example 2.20

Write the following numbers in scientific notation, showing the number of significant figures requested.
(a) $5,\overline{0}00$ (two significant figures)
(b) $48,\overline{0}00$ (three significant figures)
(c) 4,090,000 (three significant figures)
(d) 0.000087 (two significant figures)

Solution

(a) 5, $\overline{0}$ 0 0 . = 5.0×10^3 (two significant figures)

(b) 4 8, $\overline{0}$ 0 0 . = 4.80×10^4 (three significant figures)

(c) 4, 0 9 0, 0 0 0 . = 4.09×10^6 (three significant figures)

(d) 0 . 0 0 0 0 8 7 = 8.7×10^{-5} (two significant figures)

Practice Exercise 2.20   Write the following numbers in scientific notation, showing the number of significant figures requested.
(a) 4,200 (two significant figures)
(b) 56,$\overline{0}$00 (three significant figures)
(c) 6,023,000 (four significant figures)
(d) 0.00123 (three significant figures)

2.8 Temperature Scales and Heat

When we heat a substance, we add a quantity of heat to that substance.
We can then use a thermometer to measure the temperature of the substance. The thermometer measures the *intensity* of the heat; it tells us
nothing about the quantity of heat that has entered the substance. (We
shall discuss heat quantity in Chapter 10.)

Three different temperature scales are commonly used in measuring
heat intensity. Two of these—the Fahrenheit and Celsius scales—are in
general use. The third, the Kelvin scale, is used mainly by scientists.

The Fahrenheit scale was devised by Gabriel Daniel Fahrenheit, a German scientist, in 1724. On this scale (Figure 2.5), the freezing point of
pure water is at 32 degrees (32°F), and the boiling point of water is at 212
degrees (212°F). There are thus 180 Fahrenheit degrees between the
freezing point and the boiling point of water.

The Celsius scale was devised in 1742 by Anders Celsius, a Swedish astronomer. His objective was to develop an easier-to-use temperature
scale; he did so by assigning a nice, round 100 Celsius degrees between
the freezing and boiling points of pure water. On the Celsius scale (Figure
2.5), the freezing point of water is at zero degrees (0°C), and the boiling
point of water is at 100 degrees (100°C). (The Celsius scale is also sometimes referred to as the centigrade scale.)

The Kelvin scale is an *absolute* temperature scale. That is, its zero point
(0 K) is at absolute zero, the lowest possible temperature theoretically at-
tainable. The divisions of the Kelvin scale are the same size as Celsius degrees, but they are called kelvins (abbreviated K) rather than degrees.
Chapter 11 discusses the Kelvin scale in more detail.

Figure 2.5

A comparison of Fahrenheit
and Celsius scales

°C

100°C — 212°F

37°C — 98.6°F

20°C — 68°F

0°C — 32°F

−40°C — −40°F

°F

2.9 Converting Celsius and Fahrenheit Degrees

. . . .
Learning Goal 11
Celsius and Fahrenheit
temperature conversions

On the Celsius scale, there are 100 divisions between the freezing point and the boiling point of water. On the Fahrenheit scale, there are 180 divisions between these two points. Therefore 100 Celsius degrees cover the same range as 180 Fahrenheit degrees, so that 1 Celsius degree = 1.8 Fahrenheit degrees. Moreover, 0°C is equivalent to 32°F. Formulas for converting from a temperature on one scale to a temperature on the other are based on these facts. The formulas are

$$°F = (1.8 \times °C) + 32 \quad \text{and} \quad °C = \frac{°F - 32}{1.8}$$

Let us use these formulas to convert from Fahrenheit to Celsius degrees and from Celsius to Fahrenheit degrees.

Example 2.21

Convert 122°F to degrees Celsius.

Solution
Substitute 122°F into the conversion formula for changing °F to °C.

$$°C = \frac{°F - 32}{1.8} = \frac{122 - 32}{1.8} = \frac{90}{1.8} = 50$$

$$122°F = 50°C$$

Practice Exercise 2.21   Convert 244°F to degrees Celsius.

Example 2.22

Convert 100°C to degrees Fahrenheit.

Solution
Substitute 100°C into the conversion formula for changing °C to °F.

$$°F = (1.8 \times °C) + 32 = (1.8 \times 100) + 32 = 180 + 32 = 212$$

$$100°C = 212°F$$

Practice Exercise 2.22   Convert 150°C to degrees Fahrenheit.

Summary

The metric (or SI) system is a system of measurement used in most parts of the world and in all scientific work. It consists of seven basic units and a number of derived units, along with a set of prefixes that indicate multiples of these units. The unit of length is the meter, the unit of mass is the kilogram, and the unit of volume is a derived unit (cubic centimeters or, sometimes, liters). The corresponding units in the English system, as used in the United States, are the foot (length), pound (mass), and quart (volume). The factor-unit method may be used to convert from unit to unit within either system, or from system to system. Mass is a measure of the quantity of matter, and weight is a measure of the gravitational force on an object. The density of an object is its mass divided by its volume. Temperatures, which indicate heat intensity, are measured on the Fahrenheit, Celsius, or Kelvin temperature scale and may be converted from one scale to another.

Significant figures are digits that convey the precision of our information. The results of measurements and computations should always be rounded to the appropriate number of significant figures. Scientific notation, in powers of 10, may be used to indicate significant figures; it also simplifies computations involving very large or very small numbers.

Career Sketch

Chemical Engineer

Chemical engineers plan, design, and construct chemical plants and equipment. They do research to develop and improve the processes by which large quantities of chemicals are produced for industry and commerce. They often set up pilot plants to test the processes they devise. They work with such materials as plastics, detergents, textile fibers, drugs, and synthetic rubbers.

Chemical engineering is so diversified and complex that chemical engineers often specialize in a particular chemical process, such as oxidation or polymerization. Others specialize in a particular area, such as pollution control, or in the production of a specific substance, such as a plastic or paint.

In the United States today, there are 55,000 chemical engineers working in the chemical and petrochemical industry and 10,000 in higher education, government, research, and consulting. When working in areas that affect the public welfare, chemical engineers must obtain a license or

An example of one of the many different industrial settings in which chemical engineers work is this chemical pilot plant.
(IBM)

be registered. This requires that they earn a degree, have four years of experience, and be awarded a passing grade on a licensing examination.

Self-Test Exercises

The self-test exercises are arranged in matched pairs. The first exercise of each pair appears in the left column; the exercises whose numbers are in color are answered in the back of the book. Answers to exercises in the right column are not given. Difficult problems are marked with asterisks.

Learning Goal 1: Areas of Geometric Figures

1. Find the area of a rectangular room that measures 6.0 m by 3.0 m. Use the proper number of significant figures in reporting your answer.

2. Find the area of a rectangular room that measures 5.0 m by 4.0 m. Use the correct number of significant figures in reporting your answer.

3. Determine the area of this page in (a) square centimeters, (b) square inches.

4. Find the area of this page in (a) square meters and (b) square feet.

5. Find the area occupied by a rectangular house that measures $3\overline{0}$ m by $2\overline{0}$ m.

6. Find the area of a table that measures 1.5 meters by 2.0 meters.

Learning Goal 2: Volumes of Geometric Figures

7. Find the volume of a milk carton that measures 5.00 cm long, 6.00 cm wide, and 15.0 cm high.

8. Find the volume of a box that measures 5.0 m long, 4.0 m wide, and 2.0 m high.

9. Determine the volume of a cylinder-shaped can that measures 60.0 cm high and has a radius of 15.0 cm.

10. Determine the volume of a metal cylinder that is 25 cm high and has a radius of 3.0 cm.

11. A rectangular solid—a gasoline can—has the following dimensions: length, 18.0 cm; width, 10.0 cm; and height, 25.0 cm. Determine the volume of this gasoline can.

12. There are 100 cm in 1 m. How many cubic centimeters (cm^3) are there in 2 cubic meters (2 m^3)?

13. Determine the volume of a cylinder-shaped soup can that measures 15.0 cm high and has a radius of 6.00 cm.

14. Find the volume of a cylinder-shaped jar of applesauce that measures $1\overline{0}$ cm high and has a radius of 5.0 cm.

15. There are 12 inches in one foot. From this information, calculate the number of cubic inches in one cubic foot.

16. A cookie jar has the dimensions $4\overline{0}$ cm by 25 cm by 15 cm. Calculate its volume.

Learning Goal 3: Conversion of Units Within the Metric System

17. Convert 0.15 kg to (a) grams, (b) decigrams, (c) milligrams.

18. Convert 0.84 kg to (a) grams, (b) decigrams, (c) milligrams.

19. Convert 3.1 m to (a) decimeters, (b) centimeters, (c) millimeters.

20. Convert 35,000 mm to (a) centimeters, (b) decimeters, (c) meters, (d) kilometers.

21. Convert 149 mm to (a) centimeters, (b) meters, (c) kilometers.

22. Convert 5.5 m to (a) decimeters, (b) centimeters, (c) millimeters.

23. Convert 7.850 m to (a) millimeters, (b) centimeters, (c) decimeters, (d) kilometers.

24. Convert 125 mm to (a) centimeters, (b) meters, (c) kilometers.

25. Convert 2.5 km to (a) meters, (b) decimeters, (c) centimeters.

26. Convert 1.234 m to (a) millimeters, (b) centimeters, (c) decimeters, (d) kilometers.

27. Convert $56,78\overline{0}$ mg to (a) centigrams, (b) decigrams, (c) grams, (d) kilograms.

28. Convert 7.5 km to (a) meters, (b) decimeters, (c) centimeters.

29. Convert $3,5\overline{00}$ mL to (a) liters, (b) deciliters.

30. Convert $15\overline{00}$ mL to (a) liters, (b) deciliters.

31. There are $1\overline{00}$ cm in one meter. From this information, calculate the number of cubic centimeters (cm^3) in one cubic meter (m^3).

32. Convert $25,55\overline{0}$ mg to (a) centigrams, (b) decigrams, (c) grams, (d) kilograms.

Learning Goal 4: Conversion of Metric to English Units

33. Express 25.0 m in   (*a*) feet,   (*b*) inches.

35. Express 1.2 liters in   (*a*) gallons,   (*b*) quarts.

37. Express 5.00 g in   (*a*) pounds,   (*b*) ounces.

39. Without looking at a conversion table, determine the number of cubic feet in a cubic meter.

41. Express $10\overline{0}$ m in   (*a*) yards,   (*b*) feet.

43. There are 3.28 ft in one meter (1.00 m). From this information, calculate the number of square feet (ft^2) in one square meter (m^2).

34. Express 100.0 m in   (*a*) feet,   (*b*) inches.

36. Express 10.0 L in   (*a*) gallons,   (*b*) quarts.

38. Express 100.0 kg in   (*a*) pounds,   (*b*) tons.

40. Express $25\overline{0}$ m in   (*a*) yards,   (*b*) feet.

42. Express 2.50 g in pounds.

44. Express 50.0 L in gallons.

Learning Goal 5: Conversion of English to Metric Units

45. Convert your height from feet and inches to meters.

47. Convert your weight from pounds to kilograms.

49. Express $1\overline{0}$ ft in   (*a*) meters,   (*b*) centimeters, (*c*) millimeters.

51. Express 6.00 ft in   (*a*) meters,   (*b*) centimeters.

53. Express $10\overline{0}$ yd in   (*a*) meters,   (*b*) centimeters.

55. Express 5.00 gal in   (*a*) liters,   (*b*) milliliters.

46. Express $2\overline{0}$ gal in   (*a*) liters,   (*b*) milliliters.

48. Express 100.0 lb in   (*a*) kilograms,   (*b*) grams.

50. Express 5.0 ft in   (*a*) meters,   (*b*) centimeters.

52. Express 25 ft in   (*a*) meters,   (*b*) centimeters.

54. Express 15 gal in   (*a*) liters,   (*b*) milliliters.

56. Express $22\overline{0}$ lb in   (*a*) kilograms,   (*b*) grams.

Learning Goals 6 and 7: Calculation of Density, Mass, and Volume

57. Determine the density of this textbook. You'll have to weigh the book and determine its volume. Report your answer in grams per cubic centimeter.

59. A cube is 5.00 cm on each side and weighs 600.0 g. What is its density?

61. A block of aluminum with a density of 2.7 g/cm^3 weighs 549 g. What is the volume of the block?

63. The element barium, which is a soft, silvery-white metal, has a density of 3.5 g/cm^3. What would be the mass of a rectangular block of barium with the dimensions 1.0 cm $\times$ 3.0 cm $\times$ 5.0 cm?

58. A cube is 4.00 cm on each side and weighs 250.0 g. What is its density?

60. A block of aluminum with a density of 2.70 g/cm^3 weighs 274.5 g. What is the volume of the block?

62. The element barium has a density of 3.5 g/cm^3. What would be the mass of a rectangular block of barium with the dimensions 2.0 cm $\times$ 3.0 cm $\times$ 4.0 cm?

64. Calculate the mass of a cube that has a density of 2.50 g/cm^3 and is 1.5 cm on each side. Report your result to the proper number of significant figures.

*65. A spherical balloon is filled with helium, a gaseous element. Helium has a density of 0.177 g/liter. The balloon has a radius of 3.0 cm. What is the mass of the helium in the balloon? (*Hint:* Volume of a sphere = $\frac{4}{3}\pi r^3$.)

*67. A 1.00-liter container holds: a block of cesium that is 1.00 cm × 2.00 cm × 3.00 cm and whose density is 1.90 g/cm³; 14.0 g of iron, whose density is 7.86 g/cm³; and 0.500 liter of mercury, whose density is 13.6 g/cm³. The rest of the container is filled with air, whose density is 1.18×10^{-3} g/cm³. Calculate the *average* density of the contents of the container. *Hint:* To calculate average density, use the formula

$$\text{Average density} = \frac{m_a + m_b + \cdots}{V_a + V_b + \cdots}$$

*69. A fish tank whose dimensions are 1.50 ft × 1.00 ft × 0.500 ft contains six fish. Fish one and two weigh 2.00 g each, fish three weighs 2.50 g, and the combined weight of fish four, five, and six is 9.80 g. The bottom of the fish tank is covered with gravel, which occupies one-eighth of the volume of the tank. The density of this gravel is 3.00 g/cm³. The rest of the fish tank is filled with water (density 1.00 g/cm³). Assuming that the volume occupied by the fish is negligible, calculate the average density of the contents of the fish tank.

71. Calculate the mass of a cube that has a density of 1.61 g/cm³ and is 4.1 cm on each side. Express your answer to the proper number of significant figures.

*73. Suppose you are drinking a chocolate milk shake. Its average density is 2.00 g/cm³. The volume of the container that holds it is $25\overline{0}$ cm³. The milk shake consists of 90.0% milk products and 10.0% chocolate powder by volume. Assuming that the density of the milk products is 1.50 g/cm³, calculate the amount of chocolate powder that was used to make the drink.

75. A cube measures 3.50 cm on each side and has a mass of $40\overline{0}$ g. What is its density?

77. Gold has a density of 19.3 g/cm³. A certain block of gold is a rectangular solid that measures 30.0 cm by 5.00 cm by 10.0 cm. What is its mass?

66. A cube has a density of 0.250 g/cm³ and a mass of 15.0 g. What is the volume of this cube?

68. A small metal sphere weighs 75.0 g. The sphere is placed in a graduated cylinder containing water. The water level in the cylinder changes from 10.0 cm³ to 20.0 cm³ when the sphere is submerged. What is the density of the sphere?

70. Determine the density of a rectangular solid that has the dimensions 8.0 cm × 2.0 cm × 3.0 cm and a mass of 192.0 g.

72. A balloon contains a gaseous compound. The radius of the balloon is 1.06 cm, and the mass of the gaseous compound in the balloon is 9.80 g. What is the density of the gaseous element?

74. Determine the length of a metallic cube that has a density of 10.5 g/cm³ and a mass of 672 g.

76. The element copper has a density of 8.92 g/cm³. What is the mass of a cylinder-shaped piece of copper that has a radius of 2.00 cm and a height of 10.00 cm?

78. An empty graduated cylinder weighs 80.00 g. The empty cylinder is filled with exactly 30.0 cm³ of a liquid and weighed again. The cylinder with the liquid in it weighs 488 g. What is the density of the liquid? If you are told that this liquid is an element, can you determine what element it is?

79. The density of lead is 11.3 g/cm^3. What is the volume of a chunk of lead that weighs $25\overline{0}$ g?

80. A small stone that appears to be an irregularly shaped diamond is weighed and found to have a mass of 21.06 g (about 105 carats if it is indeed a diamond). The stone is placed in a graduated cylinder that initially has 10.0 mL of water in it. When the stone is immersed in the water, the water level rises to 16.0 cm^3. What is the density of the stone? Could the stone be a diamond? (*Hint:* Check the *Handbook of Chemistry and Physics.*)

81. A small rock weighs 55.0 g. The rock is placed in a graduated cylinder containing water. The water level in the cylinder changes from 25.0 cm^3 to 40.0 cm^3 when the rock is submerged. What is the density of the rock?

82. A piece of copper that has a mass of 120.0 g is melted and poured into some liquid silver that has a mass of 200.0 g. The two metals are mixed and allowed to solidify. What is the average density of this metal alloy? [*Hint:* D(copper) = 8.92 g/cm^3 and D(silver) = 10.5 g/cm^3.]

*83. An object in the shape of a sphere—for example, a hollow globe—is filled with ethyl alcohol. The sphere has a radius of 18.0 cm, and ethyl alcohol has a density of 0.800 g/cm^3. What is the mass of ethyl alcohol in the sphere? (*Hint:* Volume of a sphere = $\frac{4}{3}\pi r^3$.)

84. Determine the mass in grams of one gallon of gasoline if the density of gasoline at 25°C is 0.56 g/cm^3.

85. A cube has a density of 0.500 g/cm^3 and a mass of 25.0 g. What is the volume of this cube?

86. What weight of lead [D(lead) = 11.3 g/cm^3] occupies the same volume as $10\overline{0}$ g of aluminum [D(aluminum) = 2.70 g/cm^3]?

Learning Goal 8: Significant Figures

87. How many significant figures are there in each of the following numbers?
(a) 4,000.0   (b) 4,000   (c) 0.808   (d) 35.000
(e) 101,010.0   (f) 2×10^2 (g) 1.000×10^3
(h) 48.1×10^5 (i) $4,\overline{0}00$ (j) $4,\overline{00}0$

88. How many significant figures are there in each of the following numbers?
(a) 4250.0   (b) 0.509   (c) 3×10^2 (d) $5,\underline{0}00$
(e) 32.1×10^3 (f) 0.00034   (g) $402,\overline{0}00$ (h) 90

89. Determine the number of significant figures in each of the following numbers.
(a) $3,\overline{0}00,000$ (b) $3,0\overline{0}0,000$ (c) 3×10^6
(d) 3.00×10^6 (e) 3.000000×10^6
(f) 0.0000305   (g) 0.100054
(h) 6.720×10^{-8} (i) 305,075   (j) 35.00

90. Determine the number of significant figures in each of the following numbers:
(a) $4,\overline{0}00,000$ (b) $4,0\overline{0}0,000$ (c) 2×10^5
(d) 0.00123   (e) 4.02×10^{-5} (f) 23.00
(g) 2×10^{-1} (h) 750

Learning Goal 9: Calculations with Significant Figures

91. Find the perimeter of a triangle whose sides are 4.23 cm, 4.198 cm, and 4 cm. How many significant figures are there in your answer?

92. Find the perimeter of a triangle whose sides are 1.23 cm, 5.006 cm, and 3 cm. How many significant figures are there in your answer?

93. Find the area of a square whose sides are 1.3 cm. How many significant figures do you have in your answer?

94. Find the area of a square whose sides are 4.2 cm. How many significant figures are there in your answer?

95. Find the perimeter of a four-sided object whose sides are 7.382 cm, 3.95 cm, 5.4342 cm, and 3.83 cm. (Use the proper number of significant figures.)

96. Find the area of a rectangle whose sides are 10.12 cm and 10.25 cm. (Use the proper number of significant figures.)

97. Find the area of a rectangle whose sides are 20.62 cm and 10.4 cm. How many significant figures are there in your answer?

98. Find the area of a square whose sides are 2.1 cm each. (Use the proper number of significant figures.)

99. State the number of significant figures in each number given in Exercise 105.

100. Find the perimeter of a square whose sides are 2.1 cm each. (Use the proper number of significant figures.)

Learning Goal 10: Scientific Notation

101. Express the following numbers in scientific notation.
(a) 8,000,000 (b) 0.002 (c) 0.00004
(d) 905,000 (e) 0.00207 (f) 305,000,000

102. Express the following numbers in scientific notation.
(a) 900,000 (b) 45,000 (c) 2,970
(d) 2,546,000 (e) 0.00006 (f) 0.0122
(g) 0.056 (h) 20.0040

103. Express the following numbers in scientific notation.
(a) 10,581 (b) 0.00205 (c) 1,000,000 (d) 802

104. Express the following numbers in scientific notation.
(a) 45,$\overline{000}$,000 (b) 4$\overline{00}$ (c) 4$\overline{0}$0
(d) 400.0 (e) 425,$\overline{000}$

105. Express the following numbers in scientific notation.
(a) 850,000,000 (b) 0.00000607 (c) 6,308,000
(d) 0.06005 (e) 5$\overline{0}$0 (f) 5$\overline{0}$0 (g) 50$\overline{0}$
(h) 500.0 (i) 23,$\overline{000}$,000 (j) 0.0000000930

106. Express the following numbers in scientific notation.
(a) 0.123 (b) 0.006 (c) 0.00601

Learning Goal 11: Celsius and Fahrenheit Temperature Conversions

107. Convert 28°F to degrees Celsius.

108. Convert 54.0°F to degrees Celsius.

109. Convert −40°F to degrees Celsius. What is unusual about this temperature?

110. Convert −20.0°F to degrees Celsius.

111. Convert 20°C to degrees Fahrenheit.

112. Convert 23.0°C to degrees Fahrenheit.

113. Change each of the following temperatures from degrees Fahrenheit to degrees Celsius.
(a) 50°F (b) −94°F (c) 419°F (d) −130°F

114. Convert −20.0°C to degrees Fahrenheit.

115. Change each of the following temperatures from degrees Celsius to degrees Fahrenheit.
(a) 95.0°C (b) −80.0°C (c) 80.0°C (d) 210°C

116. Convert 15.0°C to degrees Fahrenheit.

Extra Exercises

117. Would a carefully weighed object weigh the same in Death Valley as it would on the top of Mount McKinley?

118. In filling out an application form in a company that uses the metric system, a job applicant recorded his height as 4.00 m and his weight as 200.0 kg. Did he fill the application out correctly?

119. An art dealer needs to know the volume of a piece of irregularly shaped sculpture. A chemist friend tells the dealer to submerge the object in water and measure the amount of water displaced by it. The dealer finds that 48 g of water are displaced. If the density of the water is 1.0 g/mL, what is the volume of the art object?

120. A chemistry student decides to cook a pizza in the lab. The instructions call for cooking the pizza for 10 minutes at 425°F. The oven dial, however, is set in degrees Celsius. At what temperature should the dial be set so that the pizza will be done in 10 minutes?

121. Choose a room where you live and measure its dimensions in English units; then convert each dimension to metric units.

122. Express the following relationships in scientific notation.
(a) _____ g = 1 kg   (b) _____ kg = 1 g
(c) _____ μg = 1 g   (d) _____ mg = 1 g

123. Assuming that the following numbers are from experimental measurements, calculate the answer, and state it to the proper number of significant figures.

$$\frac{(3.12 \times 10^6)(8.123 \times 10^{-4})}{3.1}$$

124. A temperature is measured as 16.0°F. What is this temperature in degrees Celsius?

125. Change the following temperatures to °F.
(a) $16\overline{°}$C   (b) $20\overline{0}$°C

126. Change the following temperatures to °C.
(a) $30\overline{0}$°F   (b) $-15\overline{0}$°F

127. The 1976 world's record for the hundred-meter dash was 9.90 seconds. What would this speed be in miles per hour?

3

Matter and Energy, Atoms and Molecules

Learning Goals

After you've studied this chapter, you should be able to:

1. Explain what is meant by the scientific method.

2. Explain the Law of Conservation of Mass and Energy.

3. Describe the difference between homogeneous and heterogeneous matter, between mixtures and compounds, and between compounds and elements.

4. Explain the difference between physical and chemical properties.

5. Describe the difference between an atom and a molecule.

6. Explain the Law of Definite Composition (or Definite Proportions).

7. Explain the terms atomic mass, atomic weight, molecular weight, gram-molecular weight, mole, and Avogadro's number.

8. Calculate the number of moles in a sample of an element when you are given the weight of the sample.

9. Calculate the weight, in grams, of a sample when you are given the number of moles.

10. Calculate the number of atoms of an element present in a sample when you are given the weight of the sample.

11. Calculate the empirical formula for a compound when you are given its percentage composition.

12. Determine the molecular weight of a compound when you are given the formula for the molecule.

13. Figure out the molecular weight, number of molecules, or number of moles in a sample of a certain weight.

14. Determine the molecular formula of a compound from its empirical formula and molecular weight.

15. Calculate the percentage composition of a compound when you are given its formula.

Introduction

This chapter is really the beginning of your study in chemistry. We start with discussions of the most elementary concepts, those of matter and energy. Then, in this chapter and later chapters, we build on and extend these concepts. As we do this, we shall be discussing the results of centuries of scientific research—the theories and laws of modern chemistry. These theories and laws are sometimes presented to students as though each resulted from a quick flash of insight on the part of some scientist. Actually they are the fruit of years—and sometimes decades or centuries—of hard work by many people.

3.1 The Scientific Method

Chemistry is an experimental science that is concerned with the behavior of matter. Much of the body of chemical knowledge consists of abstract concepts and ideas. Without application, these concepts and ideas would have little impact on society. Chemical principles are applied for the benefit of society through technology. Useful products are developed by the union of basic science and applied technology.

Learning Goal 1
The scientific method

Over the past 200 years, science and technology have moved forward at a rapid pace. Ideas and applications of these ideas are developed through carefully planned experimentation, in which researchers adhere to what is called the scientific method. The scientific method is composed of a series of logical steps that allow researchers to approach a problem and try to come up with solutions in the most effective way possible. It is generally thought of as having four parts:

1. Observation and classification. Scientists begin their research by carefully observing natural phenomena. They carry out experiments, which are observations of natural events in a controlled setting. This allows results to be duplicated and rational conclusions to be reached. The data the scientists collect are analyzed, and the facts that emerge are classified.

2. Generalization. Once observations are made and experiments carried out, the researcher seeks generalizations, regularities, or patterns in the results. If such a generalization, pattern, or regularity consistently emerges, is basic, and can be communicated in a concise statement or a mathematical equation, the statement or equation is called a law.

3. Hypothesis. Researchers try to find reasons and explanations for the generalizations, patterns, and regularities they discover. A hypothesis expresses a tentative explanation of a generalization that has been stated. Further experiments then test the validity of the hypothesis.

4. Theory. The new experiments are carried out to test the hypothesis. If they support it without exception, the hypothesis becomes a theory. A theory is a tested model that explains some basic phenomenon of nature. It cannot be proven to be absolutely correct. As further research is performed to test the theory, it may be modified or a better theory may be developed.

The scientific method represents a systematic means of doing research. There are times when discoveries are made by accident, but most knowledge has been gained via careful, planned experimentation. In your study of chemistry you will examine the knowledge and understanding that researchers using the scientific method have uncovered.

3.2 Matter and Energy

We begin with the two things that describe the entire universe: *matter* and *energy*. **Matter** is *anything that occupies space and has mass*. That includes trees, clothing, water, air, people, minerals, and many other things. Matter shows up in a wide variety of forms.

Energy is *the ability to perform work*. And, like matter, energy is found in a number of forms. Heat is one form of energy, and light is another. There are also chemical, electrical, and mechanical forms of energy. And energy can change from one form to another. In fact, matter can also change form or change into energy, and energy can change into matter.

3.3 Law of Conservation of Mass and Energy

Learning Goal 2
Law of Conservation of Mass and Energy

The brilliant French chemist–physicist Antoine Lavoisier performed experiments involving matter. In one of these experiments, he heated a measured amount of tin and found that part of it changed to a powder. He also found that the *product* (powder plus tin) weighed *more* than the original piece of tin. To find out more about the added weight, he began heating metals in sealed glass jars. (The jars had air in them, of course.) He first measured the mass of his starting materials (*reactants*). Then, when the reaction was finished (that is, when the metal in the sealed jar stopped changing to powder), he carefully measured the mass of the resulting products. In every such reaction the mass of the reactants (the oxygen from the air in the jar plus the original metal) equaled the mass of the products (the remaining metal plus the powder). Today we know that the reaction stopped when all the oxygen in the sealed jar had combined with the metal to form the powder. Lavoisier then came to the correct conclusion that *matter is neither created nor destroyed*. It just changes from one form to another (Figure 3.1). Lavoisier's oxygen and metal changed form when they combined into the powder.

Figure 3.1

An experiment like Lavoisier's. An experimenter puts a test tube containing a lead nitrate solution into a flask containing a potassium chromate solution. The experimenter weighs the flask and contents, then turns the flask upside down to mix the two solutions. A chemical reaction takes place, producing a yellow solid. The experimenter weighs the flask and contents again and finds no change in weight.

Potassium chromate solution (yellow solution)
Lead nitrate solution (colorless)

Invert flask to mix solutions

Lead chromate plus potassium nitrate solution

Yellow solid

Weight of flask plus contents equals 120 g

Before reaction

Weight of flask plus contents equals 120 g

After reaction

In the 1840s, more than half a century after Lavoisier, three scientists—the Englishman James Joule, and the Germans Julius von Mayer and Hermann von Helmholtz—did various experiments. They showed that in a reaction, *energy is neither created nor destroyed*; it, too, just changes from one form to another.

An auto engine provides a good example of how one form of energy is converted to a different form. *Electrical* energy from the battery generates a spark that contains *heat* energy. The heat ignites the gasoline–air mixture, which explodes, transforming chemical energy into heat and *mechanical* energy. The mechanical energy causes the pistons to rise and fall, rotating the engine crankshaft and moving the car.

At the same time, in the same engine, matter is changing from one form to another: When the gasoline explodes and burns, it combines with oxygen in the cylinders to form carbon dioxide and water vapor. (Unfortunately, carbon monoxide and other dangerous gases may also be formed. This is one of the major causes of air pollution, as you will see in Chapter 17.)

To appreciate the significance of these facts, think of the universe as a giant chemical reactor or system. At any given time there are certain amounts of matter and energy present, and the matter has a certain mass. Matter is always changing from one form to another, and so is energy. Besides that, matter is changing to energy and energy to matter. But *the sum of all the matter (or mass) and energy in the universe always remains the same*. This repeated observation is a law, called the **Law of Conservation of Mass and Energy.**

3.4 Potential Energy and Kinetic Energy

Which do you think has more energy, a metal cylinder held 1 foot above the ground or an identical cylinder held 5 feet above the ground? If you dropped them on your foot, you would know immediately that the cylinder with more energy was the one that was 5 feet above the ground. But where does this energy come from?

Work had to be done to raise the two cylinders to their respective heights—to draw them up against the pull of gravity. And energy was needed to do that work. The energy used to lift each cylinder was "stored" in each cylinder. The higher the cylinder was lifted, the more energy was stored in it—due to its position. *Energy that is stored in an object by virtue of its position* is called **potential energy.**

If we drop the cylinders, they fall toward the ground. As they do so, they lose potential energy because they lose height. But now they are moving; their potential energy is converted to "energy of motion." The more potential energy they lose, the more energy of motion they acquire. *The energy that an object possesses by virtue of its motion* is called **kinetic energy.** The conversion of potential energy to kinetic energy is a very common phenomenon. It is observed in a wide variety of processes, from downhill skiing to the generation of hydroelectric power.

3.5 The States of Matter

Matter may exist in any of the three physical states: solid, liquid, and gas.

A **solid** has a definite shape and volume that it tends to maintain under normal conditions. The particles composing a solid stick rigidly to one another. Solids most commonly occur in the **crystalline** form, which means they have a fixed, regularly repeating, symmetrical internal structure. Diamonds, salt, and quartz are examples of crystalline solids. A few solids, such as glass and paraffin, do not have a well-defined crystalline structure, although they do have a definite shape and volume. Such solids are called **amorphous,** which means they have no definite internal structure or form.

A **liquid** does not have its own shape but takes the shape of the container in which it is placed. Its particles cohere firmly, but not rigidly, so the particles of a liquid have a great deal of mobility while maintaining close contact with one another.

A **gas** has no fixed shape or volume and eventually spreads out to fill its container. Its particles move independently of one another. Compared with those of a liquid or a solid, the particles of a gas are quite far apart. Unlike solids and liquids, which cannot be compressed very much at all, gases can be both compressed and expanded.

Often referred to as the fourth state of matter, **plasma** is *a form of matter composed of electrically charged atomic particles.* Many objects found in the earth's outer atmosphere, as well as many celestial bodies found in space (such as the sun and stars), consist of plasma. A plasma can be created by heating a gas to extremely high temperatures or by passing a current through it. A plasma responds to a magnetic field and conducts electricity well.

Matter can also be classified as either heterogeneous or homogeneous. **Heterogeneous** matter is made up of different parts with different properties. **Homogeneous** matter is made of parts with the same properties

Learning Goal 3
Types of matter

Figure 3.2
Classification of matter

or the same consistency throughout. These categories can be broken down still further (Figure 3.2).

3.6 Physical and Chemical Properties

Learning Goal 4
Physical and chemical
properties

Matter—whether it is solid, liquid, or gas—possesses two kinds of properties: physical and chemical. **Chemical properties** stem from the ability of a substance to change to a new substance (react) that has different properties, usually in the presence of some other material. For example, iron reacts with oxygen to produce iron (III) oxide (rust). This is an example of a *chemical change*. The ability to enter into this combination is a chemical property of both iron and oxygen. The new substance formed (iron (III) oxide) has its own chemical properties, which are different from the properties of the original substances (iron and oxygen).

   Physical properties are qualities such as state, color, odor, taste, hardness, boiling point, and melting point. These properties do not depend on the ability of the substance to react and form new substances. The melting of an ice cube (a change of state) is an example of a *physical change*. The resulting substance is unchanged chemically, but its physical form is different.

3.7 Mixtures

Learning Goal 3
Types of matter

A **mixture** is a combination of two or more substances in variable proportion. The substances in a mixture can usually be separated by physical means. In other words, we can separate them without making any chemical changes.

   Mixtures may be heterogeneous or homogeneous. In a *heterogeneous mixture* the different parts have different properties. In a *homogeneous mixture* every part is exactly like every other part. A mixture of salt and sand, for instance, is heterogeneous, because it is made up of two substances, *each of which keeps its own unique properties*. You can separate a salt–sand mixture by physical means, such as putting it in water. The water dissolves the salt, and you can remove the sand by filtering (Figure 3.3). Then you can recover the salt by heating the salt-and-water mixture until the water evaporates.

Figure 3.3
Separating sand from salt
water

3.8 Solutions

Solutions are homogeneous mixtures. That is, they are uniform in composition. Every part of a solution is exactly like every other part. The salt-and-water mixture described in Section 3.7 is a solution. Even if we add more water to the solution, it will still be homogeneous, because the salt particles will continue to be distributed evenly throughout the solution. We would get the same result if we added more salt to the solution. However, we couldn't do this indefinitely. Homogeneity would end when the solution reached *saturation* (the point at which no more salt could dissolve in the limited amount of water). We will discuss this further in Chapter 15.

3.9 Elements

Learning Goal 3
Types of matter

An **element** is *a substance that cannot be broken down into simpler substances, with different properties, by physical or chemical means.* The elements are the basic building blocks of all matter. There are now 109 known elements. Each has its own unique set of physical and chemical properties. (The elements are tabulated on the inside front cover of this book, along with their chemical symbols—a shorthand notation for their names.)

The elements can be classified into three groups: *metals, metalloids,* and *nonmetals.*

Examples of metallic elements are sodium (which has the symbol Na), calcium (Ca), iron (Fe), cobalt (Co), and silver (Ag). These elements are all classified as metals because they have certain properties in common. They have luster (in other words, they are shiny), they conduct electricity well, they conduct heat well, and they are malleable (can be pounded into sheets) and ductile (can be drawn into wires).

Some examples of *non*metals are chlorine, which has the symbol Cl (note that the second letter of this symbol is a lower-case "el" and not the figure *one*), oxygen (O), carbon (C), and sulfur (S). Again, these elements are classified as nonmetals because they have certain properties in common. They don't shine, they don't conduct electricity well, they don't conduct heat well, and they are neither malleable nor ductile.

The metalloids fall halfway between metals and nonmetals. Metalloids have some properties like those of metals and other properties like those of nonmetals. Some examples are arsenic (As), germanium (Ge), and silicon (Si). These particular metalloids are used in manufacturing transistors and other *semiconductor devices.*

3.10 Atoms

Learning Goal 5
Difference between atom and molecule

Suppose we had a chunk of some element, say gold, and were able to divide it again and again, into smaller and smaller chunks. Eventually we could get a particle that could not be divided any further without losing its properties. This particle would be an atom of gold. An **atom** is *the smallest particle of an element that enters into chemical reactions.*

The atom is the ultimate particle that makes up the elements. Gold is composed of gold atoms, iron of iron atoms, and neon of neon atoms. These atoms are so small that billions of them are needed to make a speck large enough to be seen with a microscope. In 1970, Albert Crewe and his staff at the University of Chicago's Enrico Fermi Institute took the first black-and-white pictures of single atoms, using a special type of electron microscope. In late 1978, Crewe and his staff took the first time-lapse moving pictures of individual uranium atoms.

Uranium single atoms and microcrystals obtained from the solution of uranyl acetate
(Courtesy Albert V. Crewe)

3.11 Compounds

A **compound** is *a substance that is made up of two or more elements, chemically combined in a definite proportion by mass or weight.* Unlike mixtures, compounds have a definite composition. Water, for instance, is made up of hydrogen and oxygen in the ratio of 11.1% hydrogen to 88.9% oxygen by weight. No matter what the source of the water, it is always composed of hydrogen and oxygen in this ratio. This idea, that *every compound is com-*

posed of elements in a certain fixed proportion, is called the **Law of Definite Composition** (or the Law of Definite Proportions). It was first proposed by the French chemist Joseph Proust in about 1800.

The properties of a compound need not be similar to the properties of the elements that compose it. For example, water is a liquid, whereas hydrogen and oxygen are both gases. When two or more elements form a compound, they truly form a new substance.

Compounds can be broken apart into elements only by chemical means—unlike mixtures, which can be broken down by physical means. More than 6 million compounds have been reported to date, and millions more may be discovered. Some compounds you may be familiar with are sodium chloride (table salt), which is composed of the elements sodium and chlorine, and sucrose (cane sugar), which is composed of the elements carbon, hydrogen, and oxygen.

3.12 Molecules

We have seen what happens when a chunk of an element is continually divided: We eventually get down to a single atom. What happens when we keep dividing a chunk of a compound? Suppose we do so with the compound sugar. As we continue to divide a sugar grain, we eventually reach a small particle that can't be divided any further without losing the physical and chemical properties of sugar. This ultimate particle of a compound, *the smallest particle that retains the properties of the compound,* is called a **molecule.** Like atoms, molecules are extremely small, but with the aid of an electron microscope we can observe some of the very large and more complex molecules.

Molecules, as you might have guessed, are made up of two or more atoms. They may be composed of different kinds of atoms (for instance, water contains hydrogen and oxygen) or the same kinds of atoms (for instance, a molecule of chlorine gas contains two atoms of chlorine). (See Figure 3.4.) The Law of Definite Composition states that the atoms in a compound are combined in definite proportions by weight. We can see in Figure 3.4 that they are also combined in definite proportions by number. For example, water molecules always contain two hydrogen atoms for every oxygen atom.

Figure 3.4
Molecules of water and chlorine

O

H   H

Water

Cl   Cl

Chlorine

3.13 Symbols and Formulas of Elements and Compounds

We have already mentioned the chemical symbols—the shorthand for the names of the elements. In some cases, as in the symbol O (capital "oh") for oxygen, the chemical symbol is the first letter of the elements's name, capitalized. Often, though, the symbol for an element contains two letters. In these cases only the first letter is capitalized; the second letter is never capitalized. For instance, the symbol for neon is Ne and the symbol for cobalt is Co. (Be careful of this: CO does not represent cobalt, but a combination of the elements carbon and oxygen, which is a compound.) Some symbols come from the Latin names of the elements: Iron is Fe, from the Latin *ferrum*, and lead is Pb, from the Latin *plumbum*.

Because compounds are composed of elements, we can use the chemical symbols as a shorthand for compounds too. We use the symbols to write the chemical formula, which shows the elements that compose the compound. For example, sodium chloride (table salt) contains one atom of sodium (Na) and one atom of chlorine (Cl). The chemical formula for sodium chloride is NaCl. Water is another example: The water molecule contains two atoms of hydrogen (H) and one atom of oxygen, so we write it as H_2O. The number 2 in the formula indicates that there are two atoms of hydrogen in the molecule; note that it is written as a *subscript*. A molecule of ethyl alcohol contains two atoms of carbon, six of hydrogen, and one of oxygen. It is written as follows:

3.14 Atomic Mass or Weight

· · · · · · · · · · · · · ·
Learning Goal 7
Atomic and molecular weight,
Avogadro's number

Suppose we want to find the relative masses of the atoms of the various elements. Suppose also that we have a double-pan balance that can weigh a single atom. We begin by assigning an arbitrary mass to one of the elements. Let's say we decide to assign a mass of 1 unit to hydrogen. (This is what chemists did originally, because hydrogen was known to be the lightest element even before the relative atomic masses were determined.)

Then, to find the relative mass of, say, carbon, we place an atom of carbon on one pan. We place hydrogen atoms on the other pan, one by one, until the pans are exactly balanced. We find that it takes 12 hydrogen atoms to balance 1 carbon atom, so we assign a relative atomic mass of 12 to carbon. In the same way for an oxygen atom, we find that it takes 16 hydrogen atoms to balance 1 oxygen atom. So we assign a relative mass of 16 to oxygen. By doing this experiment for all the other elements, we can determine all the masses relative to hydrogen (which is assigned mass 1).

Amedeo Avogadro
(Reprinted with permission from Torchbearers of Chemistry, by Henry Monmouth Smith, © Academic Press, Inc., 1949)

Unfortunately this kind of balance has never existed, and more elaborate means had to be developed to find the relative atomic masses of the elements. But the logic we used here is the same as that used by the many scientists who determined the relative masses. (The atomic mass scale has been revised. The present scale is based on a particular type of carbon atom, called carbon-12. This carbon is assigned the value of 12 atomic mass units.) The periodic table (inside the front cover) lists the relative atomic masses of the elements below their symbols. From now on, instead of "relative atomic mass," we will use the simpler term **atomic mass** or **atomic weight.**

3.15 Gram-Atomic Weight and the Mole

The **gram-atomic weight** of an element is *its atomic mass expressed in grams.* For example, the atomic mass of gold is 197, so gold's gram-atomic weight is 197 g. And the gram-atomic weight of carbon is 12 g.

One gram-atomic weight of any element contains the same number of atoms as one gram-atomic weight of every other element. For example, 197 g of gold contain the same number of atoms as 12 g of carbon and the same number of atoms as 1 g of hydrogen. In about 1870, scientists discovered that this number is 6.02×10^{23}. It is called *Avogadro's number,* in honor of the scientist whose thinking led to its discovery.

Later scientists took this idea one step further and defined 6.02×10^{23} atoms as **1 mole** of atoms. Thus 1 mole is equal to 6.02×10^{23}; it is simply a number of items. We could just as easily talk about 1 mole of people, 1 mole of apples, or 1 mole of dollars. (By the way, if we had 1 mole of dollar bills, we would have $602,000,000,000,000,000,000,000. If we distributed our money equally among the 5 billion people on the earth, each person would get more than 120 trillion dollars:

$$\frac{\$602,000,000,000,000,000,000,000}{5,000,000,000} = 1.20 \times 10^{14}$$

A person could spend a million dollars every day, each day of the year, and never run out of money for 330,000 years.)

A mole of hydrogen atoms weighs 1 g; a mole of oxygen atoms weighs 16 g; a mole of carbon atoms weighs 12 g; and a mole of uranium atoms weighs 238 g. (To simplify the mathematics, we are rounding off the atomic masses found in the periodic table to the nearest whole numbers. We will do this often.) Figure 3.5 shows the weights of a mole of some other kinds of atoms.

The mole is very important in chemistry. It gives the chemist a convenient way to describe a large number of atoms or molecules and to relate this number of atoms or molecules to a weight (usually in grams). In fact, it has replaced the gram-atomic weight, which is seldom used today. But you do have to be able to convert from moles of atoms of an element to

Learning Goal 8
Moles from grams of an element

Figure 3.5
One mole of atoms of some common substances

Pure silver spoon

1 mole of silver (Ag) atoms weighs 108 g.

Lead pipe

1 mole of lead (Pb) atoms weighs 207 g.

Neon gas (colorless)

1 mole of neon (Ne) atoms weighs 20.2 g.

weight in grams, and vice versa. This is like converting inches to feet and feet to inches.

Example 3.1

Do the following conversions: (*a*) 60 in. = ? ft   (*b*) 8.0 ft = ? in.

Solution

(a) There are 12 in. in 1 ft, so

$$? \text{ ft} = 60 \text{ in.} \times \frac{1 \text{ ft}}{12 \text{ in.}} = 5.0 \text{ ft}$$

(b) In the same way,

$$? \text{ in.} = 8.0 \text{ ft} \times \frac{12 \text{ in.}}{1 \text{ ft}} = 96 \text{ in.}$$

Practice Exercise 3.1   Do the following conversions: (*a*) 72.0 in. = ? ft
(*b*) 10.0 ft = ? in.

Example 3.2

(a) How many moles of oxygen atoms weigh $8\bar{0}$ g?
(b) What is the weight in grams of 0.5 mole of oxygen atoms?

Solution

(a) We use the factor-unit method just as in Chapter 2. First we write

$$\text{moles of O atoms} = \text{g} \times \frac{\text{moles of O atoms}}{\text{g}}$$

Then, because the atomic weight of oxygen is 16,

$$? \text{ moles of O atoms} = 8\bar{0} \text{ g} \times \frac{1 \text{ mole of O atoms}}{16 \text{ g}}$$

$$= 5.0 \text{ moles of O atoms}$$

(b) Again we use the atomic weight of oxygen. Our unit factor is

$$\text{g} = \overline{\text{moles of O atoms}} \times \frac{\text{g}}{\overline{\text{moles of O atoms}}}$$

$$?\ \text{g} = 0.5\ \overline{\text{mole of O atoms}} \times \frac{16\ \text{g}}{1\ \overline{\text{mole of O atoms}}} = 8\ \text{g}$$

Practice Exercise 3.2   (*a*) How many moles of oxygen atoms weigh 16$\overline{0}$ grams?   (*b*) What is the weight in grams of 2.00 moles of oxygen atoms?

Example 3.3

Find the number of moles of atoms in the following elements.
(*a*) 46 g of Na   (*b*) 5.4 g of Al   (*c*) 0.14 g of N   (*d*) 23.8 g of U

Solution
We have to look up the atomic weight of each element in the periodic table inside the front cover.
(a) The atomic weight of Na is 23. Therefore,

$$?\ \text{moles of Na atoms} = 46\ \cancel{\text{g}} \times \frac{1\ \text{mole of Na atoms}}{23\ \cancel{\text{g}}}$$

$$= 2.0\ \text{moles of Na atoms}$$

(b) The atomic weight of Al is 27. Therefore,

$$?\ \text{moles of Al atoms} = 5.4\ \cancel{\text{g}} \times \frac{1\ \text{mole of Al atoms}}{27\ \cancel{\text{g}}}$$

$$= 0.20\ \text{mole of Al atoms}$$

(c) The atomic weight of N is 14. Therefore,

$$?\ \text{moles of N atoms} = 0.14\ \cancel{\text{g}} \times \frac{1\ \text{mole of N atoms}}{14\ \cancel{\text{g}}}$$

$$= 0.010\ \text{mole of N atoms}$$

(d) The atomic weight of U is 238. Therefore,

$$?\ \text{moles of U atoms} = 23.8\ \cancel{\text{g}} \times \frac{1\ \text{mole of U atoms}}{238\ \cancel{\text{g}}}$$

$$= 0.100\ \text{mole of U atoms}$$

Practice Exercise 3.3   Find the number of moles of atoms in each of the following elements.
(*a*) 11.50 g of Na   (*b*) 2.70 g of Al   (*c*) 28$\overline{0}$ g of N   (*d*) 0.0238 g of U

Example 3.4

Find the weight in grams of each of the following.
(a) 0.20 mole of Zn atoms   (b) 4.0 moles of Br atoms
(c) 1.5 moles of Ca atoms

Solution

We have to look up the atomic weight of each element in the periodic table.
(a) The atomic weight of Zn is 65. Therefore,

Learning Goal 9
Weight, in grams, of a
sample, given the moles

$$? \text{ g} = 0.20 \text{ mole of Zn atoms} \times \frac{65 \text{ g}}{1 \text{ mole of Zn atoms}} = 13 \text{ g}$$

(b) The atomic weight of Br is $8\bar{0}$. Therefore,

$$? \text{ g} = 4.0 \text{ moles of Br atoms} \times \frac{8\bar{0} \text{ g}}{1 \text{ mole of Br atoms}} = 320 \text{ g}$$

(c) The atomic weight of Ca is $4\bar{0}$. Therefore,

$$? \text{ g} = 1.5 \text{ moles of Ca atoms} \times \frac{4\bar{0} \text{ g}}{1 \text{ mole of Ca atoms}} = 6\bar{0} \text{ g}$$

Practice Exercise 3.4   Find the weight in grams of each of the following.
(a) 15.0 moles of Zn atoms
(b) 0.200 mole of Br atoms
(c) 3.00 moles of Ca atoms

Example 3.5

How many *atoms* of each element are there in Example 3.4?

Solution

We know that 1 mole of atoms of *any* element is 6.02×10^{23} atoms of that element; in other words, there are

Learning Goal 10
Number of atoms of an
element from grams of
sample

$$\frac{6.02 \times 10^{23} \text{ atoms}}{1 \text{ mole of atoms}}$$

for any element. But for this example we will round off Avogadro's number to 6.0×10^{23}.
(a) For the zinc,

$$? \text{ Zn atoms} = 0.20 \text{ mole of Zn atoms} \times \frac{6.0 \times 10^{23} \text{ Zn atoms}}{1 \text{ mole of Zn atoms}}$$

$$= 1.2 \times 10^{23} \text{ Zn atoms}$$

(b) For the bromine,

$$? \text{ Br atoms} = 4.0 \; \cancel{\text{moles of Br atoms}} \times \frac{6.0 \times 10^{23} \text{ Br atoms}}{1 \; \cancel{\text{mole of Br atoms}}}$$

$$= 24 \times 10^{23} \text{ Br atoms}$$

(c) For the calcium,

$$? \text{ Ca atoms} = 1.5 \; \cancel{\text{moles of Ca atoms}} \times \frac{6.0 \times 10^{23} \text{ Ca atoms}}{1 \; \cancel{\text{mole of Ca atoms}}}$$

$$= 9.0 \times 10^{23} \text{ Ca atoms}$$

Practice Exercise 3.5   How many atoms of each element are there in Practice Exercise 3.4?

3.16 Empirical Formulas

Learning Goal 11
Formula from percentage composition

Chemists prepare hundreds of new compounds in their search for substances that may be beneficial in medicine, agriculture, industry, and the home. A first step in determining the nature of such a new or unknown compound is to obtain its empirical formula. The **empirical formula** of a compound is *the simplest whole-number ratio of the atoms that make up a compound.* For example, the empirical formula of water is H_2O. The subscripts indicate that the ratio of hydrogen to oxygen atoms in this molecule is two to one, often written 2:1. (The lack of a subscript on the O is taken to mean a subscript of 1.)

To find the empirical formula of a compound, the chemist measures the percentage by weight of each element in the compound. From the percentage of each element, the chemist then determines (1) the number of moles of atoms of each element in 100 g of the compound, and (2) the ratio of the moles of atoms. (We take 100 g of compound merely as a convenience.)

Example 3.6

Determine the empirical formula of a compound whose composition is 50.05% S and 49.95% O by weight.

Solution

If we had 100 g of the compound, 50.05 g would be sulfur and 49.95 g would be oxygen. All we have to do now is convert these weights to moles of atoms and then find their whole-number ratio.

$$? \text{ moles of S atoms} = 50.05 \cancel{g} \times \frac{1 \text{ mole of S atoms}}{32.1 \cancel{g}}$$

$$= 1.56 \text{ moles of S atoms}$$

$$? \text{ moles of O atoms} = 49.95 \cancel{g} \times \frac{1 \text{ mole of O atoms}}{16.0 \cancel{g}}$$

$$= 3.12 \text{ moles of O atoms}$$

Therefore the formula may be written as $S_{1.56}O_{3.12}$, but this formula does not have whole-number subscripts. One way to obtain a formula with whole-number subscripts is to *divide* all the subscripts by the *smallest* subscript. This gives us

$$S_{1.56/1.56}O_{3.12/1.56} \quad \text{or} \quad SO_2$$

This is an empirical formula, because it has the lowest possible ratio of whole-number subscripts. (The compound is called sulfur dioxide.)

Practice Exercise 3.6   Determine the empirical formula of a compound whose composition is 88.9% oxygen and 11.1% hydrogen by weight.

Example 3.7

Find the empirical formula of a compound whose composition is 3.1% H, 31.5% P, and 65.4% O by weight.

Solution

In 100 g of this compound there are 3.1 g of H, 31.5 g of P, and 65.4 g of O. Therefore there are

$$? \text{ moles of H atoms} = 3.1 \cancel{g} \times \frac{1 \text{ mole of H atoms}}{1.0 \cancel{g}}$$

$$= 3.1 \text{ moles of H atoms}$$

$$? \text{ moles of P atoms} = 31.5 \cancel{g} \times \frac{1 \text{ mole of P atoms}}{31.0 \cancel{g}}$$

$$= 1.02 \text{ moles of P atoms}$$

$$? \text{ moles of O atoms} = 65.4 \cancel{g} \times \frac{1 \text{ mole of O atoms}}{16.0 \cancel{g}}$$

$$= 4.09 \text{ moles of O atoms}$$

The formula may be written $H_{3.1}P_{1.02}O_{4.09}$, but this does not have whole-number subscripts. Therefore we divide each subscript by the smallest subscript to obtain

$$H_{3.1/1.02}P_{1.02/1.02}O_{4.09/1.02} \quad \text{or} \quad H_3PO_4$$

Practice Exercise 3.7   Find the empirical formula of a compound whose composition is 51.9% Cr and 48.1% S by weight.

Example 3.8

Determine the empirical formula of a compound that has 23.8% C, 5.9% H, and 70.3% Cl by weight.

Solution

In 100 g of this compound there are 23.8 g of C, 5.9 g of H, and 70.3 g of Cl. Therefore there are

$$? \text{ moles of C atoms} = 23.8 \text{ g} \times \frac{1 \text{ mole of C atoms}}{12.0 \text{ g}}$$

$$= 1.98 \text{ moles of C atoms}$$

$$? \text{ moles of H atoms} = 5.9 \text{ g} \times \frac{1 \text{ mole of H atoms}}{1.0 \text{ g}}$$

$$= 5.9 \text{ moles of H atoms}$$

$$? \text{ moles of Cl atoms} = 70.3 \text{ g} \times \frac{1 \text{ mole of Cl atoms}}{35.5 \text{ g}}$$

$$= 1.98 \text{ moles of Cl atoms}$$

One possible formula is $C_{1.98}H_{5.9}Cl_{1.98}$, but this does not have whole-number subscipts. Again we divide by the smallest subscript to obtain

$$C_{1.98/1.98}H_{5.9/1.98}Cl_{1.98/1.98} \quad \text{or} \quad CH_3Cl.$$

Practice Exercise 3.8   Determine the empirical formula of a compound whose composition is 41.5% Zn, 17.8% N, and 40.7% O by weight.

After you divide the subscripts in determining an empirical formula, you may round subscripts to the nearest whole number if they are within 0.1 of the whole number. Otherwise you must search for a factor that will give you whole numbers. For example, suppose a compound contains 68.4% Cr and 31.6% O by weight. This means that 100 g of the compound contains 68.4 g (or 1.32 moles) of Cr and 31.6 g (or 1.98 moles) of O. Therefore the formula is

$$Cr_{1.32}O_{1.98}$$

Following our usual procedure, we divide by the smallest number of moles (the smallest subscript) and obtain

$$Cr_{1.32/1.32}O_{1.98/1.32} \quad \text{or} \quad CrO_{1.5}$$

The formula $CrO_{1.5}$ does not have *whole*-number subscripts. What do we do now? Should we simply round off the 1.5 to the number 2 and make the empirical formula CrO_2? The answer is NO! We may round off only when the number is within 0.1 of the whole number. Therefore we must search for a factor that will give us a whole-number ratio. When we look carefully at $Cr_1O_{1.5}$, we see that, if we multiply each subscript by 2, we can get whole-number subscripts. $Cr_1O_{1.5}$ becomes Cr_2O_3, and the empirical formula is Cr_2O_3.

3.17 Molecular Weights

Learning Goal 12
Molecular weight of a
compound from the formula

The **molecular weight** of a compound is *the sum of the atomic weights of all the atoms that make up a molecule of the compound.* For example, the molecule P_2O_5 has two phosphorus atoms and five oxygen atoms. The atomic weight of each P is 31, and the atomic weight of each O is 16. Therefore the molecular weight of the compound is

$$(2 \times 31) + (5 \times 16) = 62 + 80 = 142$$

(Check this calculation and see whether you get the same answer. If not, you may have forgotten that, in an equation like this, multiplication and division are done *before* addition and subtraction.)

An EPA cleanup crew wears self-contained breathing apparatus and protective clothing while cleaning up hazardous waste. Many hazardous wastes are compounds that have high molecular weights.
(Steve Delaney/EPA)

Example 3.9

Find the molecular weights of the following compounds.
(*a*) H_2O (*b*) NaCl   (*c*) $Ca(OH)_2$ (*d*) $Zn_3(PO_4)_2$

Note: In a chemical formula, parentheses followed by a subscript mean that everything inside the parentheses is multiplied by the subscript. For example, in one molecule of $Ca(OH)_2$ there is one Ca atom plus two O atoms and two H atoms.

Solution

We must find the atomic weight of each element in the periodic table and then sum up the weights of all the atoms in each compound.

(a) The atomic weight of H is 1, and the atomic weight of O is 16.

$$\text{Molecular weight of } H_2O = (2 \times 1) + (1 \times 16)$$
$$= 2 + 16 = 18$$

(b) The atomic weight of Na is 23, and the atomic weight of Cl is 35.

$$\text{Molecular weight of NaCl} = (1 \times 23) + (1 \times 35)$$
$$= 23 + 35 = 58$$

(c) The atomic weight of Ca is 40.

$$\text{Molecular weight of } Ca(OH)_2 = (1 \times 40) + (2 \times 16) + (2 \times 1)$$
$$= 40 + 32 + 2 = 74$$

(d) The atomic weight of Zn is 65, and the atomic weight of P is 31.

$$\text{Molecular weight of } Zn_3(PO_4)_2 = (3 \times 65) + (2 \times 31) + (8 \times 16)$$
$$= 195 + 62 + 128 = 385$$

Practice Exercise 3.9   Find the molecular weights of the following compounds: (*a*) Na_2CO_3 (*b*) $CoCl_2$

Learning Goal 7
Gram-molecular weight

. . . .

Learning Goal 13
Molecular weight, number of molecules, or number of moles in a sample

The molecular weight of a compound in grams is called its **gram-molecular weight.** It is the weight of a collection of 6.02×10^{23} molecules (a mole of molecules). Therefore one gram-molecular weight of a compound contains one mole of molecules. For example, the molecular weight of water (H_2O) is 18. So the weight of 1 mole of water molecules is 18 g (Figure 3.6). It is important to be able to convert the number of grams of a compound into the number of moles of molecules, and vice versa.

Figure 3.6
One mole of molecules of some common substances

1 mole of water (H_2O) molecules weighs 18 g.

1 mole of glucose sugar $(C_6H_{12}O_6)$ molecules weighs 180 g.

1 mole of TNT $(C_7H_5N_3O_6)$ molecules weighs 227 g and can make a lot of noise.

1 *Scalopus aquaticus* alias 1 mole—weighs varying amounts and can make very little noise.

Example 3.10

Find the number of moles of molecules in each of the following:
(a) 32 g of CH_4 (b) 0.32 g of CH_4 (c) 81 g of H_2O
(d) 37 g of $Ca(OH)_2$

Solution

We first find the molecular weight of each compound and then use factor units.

(a) The molecular weight of CH_4 is 16. Therefore we have

$$? \text{ moles of } CH_4 \text{ molecules} = 32 \text{ g} \times \frac{1 \text{ mole of } CH_4 \text{ molecules}}{16 \text{ g}}$$

$$= 2.0 \text{ moles of } CH_4 \text{ molecules}$$

(b) The molecular weight of CH_4 is 16. Therefore we have

$$? \text{ moles of } CH_4 \text{ molecules} = 0.32 \text{ g} \times \frac{1 \text{ mole of } CH_4 \text{ molecules}}{16 \text{ g}}$$

$$= 0.020 \text{ mole of } CH_4 \text{ molecules}$$

(c) The molecular weight of H_2O is 18. Therefore

$$? \text{ moles of } H_2O \text{ molecules} = 81 \text{ g} \times \frac{1 \text{ mole of } H_2O \text{ molecules}}{18 \text{ g}}$$

$$= 4.5 \text{ moles of } H_2O \text{ molecules}$$

(d) The molecular weight of $Ca(OH)_2$ is 74. Therefore

$$? \text{ moles of } Ca(OH)_2 \text{ molecules}$$

$$= 37 \text{ g} \times \frac{1 \text{ mole of } Ca(OH)_2 \text{ molecules}}{74 \text{ g}}$$

$$= 0.50 \text{ mole of } Ca(OH)_2 \text{ molecules}$$

Practice Exercise 3.10   Find the number of moles of molecules in each of the following: (*a*) 9$\overline{0}$ grams of H_2O (*b*) 0.016 gram CH_4

Example 3.11

In Example 3.10, how many *molecules* are present in each sample?

Solution

To obtain our factor unit, we recall that there are 6.0×10^{23} molecules in 1 mole of molecules.
(a) For the first sample of CH_4,

$$? \text{ molecules of } CH_4 = 2.0 \text{ moles} \times \frac{6.0 \times 10^{23} \text{ molecules of } CH_4}{1 \text{ mole}}$$

$$= 12 \times 10^{23} \text{ molecules of } CH_4$$

(b) For the second sample of CH_4,

$$? \text{ molecules of } CH_4 = 0.020 \text{ mole} \times \frac{6.0 \times 10^{23} \text{ molecules of } CH_4}{1 \text{ mole}}$$

$$= 1.2 \times 10^{22} \text{ molecules of } CH_4$$

(c) For the H_2O,

$$? \text{ molecules of } H_2O = 4.5 \text{ moles} \times \frac{6.0 \times 10^{23} \text{ molecules of } H_2O}{1 \text{ mole}}$$

$$= 27 \times 10^{23} \text{ molecules of } H_2O$$

(d) For the $Ca(OH)_2$,

$$? \text{ molecules of } Ca(OH)_2$$

$$= 0.50 \text{ mole} \times \frac{6.0 \times 10^{23} \text{ molecules of } Ca(OH)_2}{1 \text{ mole}}$$

$$= 3.0 \times 10^{23} \text{ molecules of } Ca(OH)_2$$

Practice Exercise 3.11   In Practice Exercise 3.10, how many *molecules* are present in each sample?

3.18 Molecular Formulas

Learning Goal 14
Molecular formula of a compound

The **molecular formula** for a compound is *a formula that shows the actual number of atoms of each element that are in one molecule of that compound.* The molecular formula is found from the percentages by weight of the elements *and* the molecular weight of the compound.

Example 3.12

A compound is found to have 40% C, 6.6% H, and 53.4% O by weight. The molecular weight of the compound is 180. Determine its empirical and molecular formulas.

Solution

Find the empirical formula first, by the usual method.

$$? \text{ moles of C atoms} = 40 \cancel{\text{g}} \times \frac{1 \text{ mole of C atoms}}{12 \cancel{\text{g}}}$$

$$= 3.3 \text{ moles of C atoms}$$

$$? \text{ moles of H atoms} = 6.6 \cancel{\text{g}} \times \frac{1 \text{ mole of H atoms}}{1.0 \cancel{\text{g}}}$$

$$= 6.6 \text{ moles of H atoms}$$

$$? \text{ moles of O atoms} = 53.4 \cancel{\text{g}} \times \frac{1 \text{ mole of O atoms}}{16 \cancel{\text{g}}}$$

$$= 3.3 \text{ moles of O atoms}$$

Hence one possible formula is $C_{3.3}H_{6.6}O_{3.3}$, but this does not have whole-number subscripts. Therefore we divide by the smallest number of moles to get

$$C_{3.3/3.3}H_{6.6/3.3}O_{3.3/3.3} \quad \text{or} \quad CH_2O$$

This is the empirical formula. But what is the molecular formula of the compound? It could be any formula wherein the ratio of C:H:O is 1:2:1. Is it also CH_2O, or is it $C_2H_4O_2$, $C_3H_6O_3$, $C_4H_8O_4$, $C_5H_{10}O_5$, $C_6H_{12}O_6$, or what? Let's determine the molecular weight of each proposed formula.

Molecular formula	Molecular weight
CH_2O	30
$C_2H_4O_2$	60
$C_3H_6O_3$	90
$C_4H_8O_4$	120
$C_5H_{10}O_5$	150
$C_6H_{12}O_6$	180

It is clear from our trial-and-error method that the molecular formula is $C_6H_{12}O_6$. But is there an easier way to do this? Yes. We simply ask ourselves, "How many times must we take the formula weight of the empirical formula (30 in this problem) to get the true molecular weight (180 in this problem)?" The answer is obviously *six* times. It takes six times the weight of the empirical formula, CH_2O, to give the true molecular weight, so the true molecular formula must be $C_6H_{12}O_6$.

Practice Exercise 3.12   A compound has a molecular weight of 180 and is composed 60.0% C, 4.48% H, and 35.5% O. Calculate the empirical and molecular formulas for this common compound, which is called aspirin.

3.19 Percentage Composition by Weight

Learning Goal 15
Percentage composition from formula

You have seen how to determine the empirical formula of a compound from its percentage composition. There will also be times when you will have to do the opposite—find the percentage by weight of each element in a compound. To do so, you need to know the empirical or molecular formula for the compound and the atomic weights of its constituent elements.

Example 3.13

Find the percentage by weight of H and O in water.

Solution

In each mole of water (H_2O), there are 2 moles of H, weighing 1 g/mole, and 1 mole of O, weighing 16 g/mole. Therefore

$$\text{Percent H} = \frac{\text{weight of hydrogen in 1 mole of water}}{\text{weight of 1 mole of water}} \times 100$$

$$= \frac{2.0 \text{ g of H}}{18 \text{ g of } H_2O} \times 100 = 11$$

$$\text{Percent O} = \frac{\text{weight of oxygen in 1 mole of water}}{\text{weight of 1 mole of water}} \times 100$$

$$= \frac{16 \text{ g of O}}{18 \text{ g of } H_2O} \times 100 = 89$$

Note that the sum of the percentages has to be 100.

Practice Exercise 3.13   Find the percentage by weight of K and Br in potassium bromide (KBr).

Example 3.14

Find the percentage by weight of each element in the following compounds:
(a) C_8H_{18} (b) NH_3 (c) FeO   (d) Fe_2O_3

Solution
We must find the molecular weight of each compound, and the percent contribution of each element to that weight.
(a) C_8H_{18} contains 96 g of C + 18 g of H = 114 g in each mole. Therefore

$$\text{Percent C} = \frac{96 \text{ g of C}}{114 \text{ g of } C_8H_{18}} \times 100 = 84$$

$$\text{Percent H} = \frac{18 \text{ g of H}}{114 \text{ g of } C_8H_{18}} \times 100 = 16$$

(b) NH_3 contains 14 g of N + 3 g of H = 17 g in each mole. Therefore

$$\text{Percent N} = \frac{14 \text{ g of N}}{17 \text{ g of } NH_3} \times 100 = 82$$

$$\text{Percent H} = \frac{3 \text{ g of H}}{17 \text{ g of } NH_3} \times 100 = 18$$

(c) FeO contains 56 g of Fe + 16 g of O = 72 g in each mole. Therefore

$$\text{Percent Fe} = \frac{56 \text{ g of Fe}}{72 \text{ g of FeO}} \times 100 = 78$$

$$\text{Percent O} = \frac{16 \text{ g of O}}{72 \text{ g of FeO}} \times 100 = 22$$

(d) Fe_2O_3 contains 112 g of Fe + 48 g of O = 160 g in each mole. Therefore

$$\text{Percent Fe} = \frac{112 \text{ g of Fe}}{160 \text{ g of } Fe_2O_3} \times 100 = 70$$

$$\text{Percent O} = \frac{48 \text{ g of O}}{160 \text{ g of } Fe_2O_3} \times 100 = 30$$

Practice Exercise 3.14   Find the percentage by weight of each element in the following compounds: (a) $C_{22}H_{44}$ (b) NaCl   (c) C_6H_6

Example 3.15

How many grams of sulfur are there in 256 g of SO_3?

Solution

Determine the percentage of sulfur in SO_3 and multiply this by the amount of the sample (256 g). SO_3 contains 32 g of S + 48 g of O = 80 g/mole. Therefore

$$\text{Percent S} = \frac{32 \text{ g of S}}{80 \text{ g of SO}_3} \times 100 = 40$$

So 40% of the 256 g of SO_3 is sulfur (S); there are 0.40×256 g = 102.4 g of sulfur (or $1\overline{0}0$ g of sulfur to two significant figures).

Practice Exercise 3.15   How many grams of carbon are there in $1\overline{0}0$ g of CH_4?

Summary

The body of knowledge called science has been developed through the scientific method: observation and classification of data, generalization of observations, and testing of generalizations. Of importance in all the sciences is the idea that the universe is made up of only matter and energy. Matter and energy can be neither created nor destroyed, but they can be changed to other forms. And matter can be transformed into energy, and vice versa. Potential energy is energy that is stored in a body because of its position. Kinetic energy is energy that is due to motion.

Matter may exist in any of three states—solid, liquid, or gas—and may be either heterogeneous (nonuniform) or homogeneous (uniform). Mixtures are combinations of two or more kinds of matter, each retaining its own chemical and physical properties. Mixtures too may be either homogeneous or heterogeneous, and solutions are homogeneous mixtures. Elements are the basic building blocks of matter. The 109 known elements are, in turn, made up of atoms. A compound is a substance that is made up of two or more elements chemically combined in definite proportions by mass or weight. Molecules are the smallest particles that retain the properties of a compound, and atoms are the smallest particles that enter into chemical reactions.

Each element has its own symbol, and every compound has its own chemical formula. Each element has a unique atomic mass or atomic weight. One mole of anything is equal to 6.02×10^{23} things (Avogadro's number of things); 1 mole of atoms of an element has a mass equal to the atomic mass of that element in grams. One mole of molecules of a compound has a mass equal to the molecular mass of that compound in grams.

The empirical and molecular formulas for a compound can be found from the percentages by weight of its elements and its molecular weight (and vice versa). The Law of Definite Proportions states that elements combine in fixed ratios.

Career Sketch

Oceanographer

Oceanographers use the principles of physics, chemistry, mathematics, and engineering to study oceans—their movements, physical properties, and plant and animal life. Oceanographers explore and study the ocean by means of surface ships, aircraft, and various underwater craft, using specialized instruments to measure and record their findings. They use special cameras equipped with strong lights to photograph marine life and the ocean floor, and they use sounding devices to locate, measure, and map things that lie beneath the vast surfaces of the world's oceans. Their

Oceanographers use a sediment trap (the cone-shaped device) to collect samples of sediment at various depths for laboratory analysis.
(Courtesy Woods Hole Oceanographic Institution. Photographer, Nancy Green.)

research not only extends basic scientific knowledge, but it also helps develop methods for forecasting weather, expanding fisheries, mining ocean resources, and improving national defense.

Most oceanographers make observations and conduct experiments at sea, collecting data on tides, currents, and other phenomena. They may study undersea mountain ranges, interactions of the ocean with the atmosphere, or the layers of sediment on and beneath the ocean floor.

Some oceanographers work in laboratories on land where they measure, dissect, and photograph fish, study sea specimens and plankton, and identify, catalog, and analyze different kinds of sea life and minerals. Others plot maps or use computers to test theories about the ocean. To present the results of their studies, they prepare charts, tabulations, and reports and write papers for scientific journals.

Career Sketch

Mortuary Science Technology

Mortuary science technologists work with funeral service directors to embalm the bodies of the deceased and perform burial services. They study mortuary science, human anatomy and physiology, the chemistry of embalming products, embalming procedures, cosmetology, and principles of cremation. They must be able to keep accurate technical records and write technical reports. A knowledge of the pathogens associated with the deceased is also necessary.

More than 40 schools in the United States are accredited by the Committee on Education of the American Board of Funeral Service Education and approved by the State Board of Funeral Directors and Embalmers in the state where the school is located.

Mortuary science technologists must learn and apply their scientific and technical knowledge to prepare bodies for funeral and burial. A high school degree is needed. Mortuary science technology, funeral service sanitation, funeral law, basic chemistry, anatomy and physiology, microbiology, funeral service psychology, and cosmetology are among the college courses one takes in preparing for a career in mortuary science.

Self-Test Exercises

The self-test exercises are arranged in matched pairs. The first exercise of each pair appears in the left column; the exercises whose numbers are in color are answered in the back of the book. Answers to exercises in the right column are not given. Difficult problems are marked with asterisks.

Learning Goal 1: The Scientific Method

1. Explain the difference between a theory and a scientific law.

2. Suppose you are a researcher and you believe you have found a drug that can cure the common cold. How would you use the scientific method to determine whether the drug is effective?

Learning Goal 2: Law of Conservation of Mass and Energy

3. Explain what the Law of Conservation of Mass and Energy means in terms of the workings of our universe.

4. What is the significance of the Law of Conservation of Mass and Energy in terms of your study of chemistry?

Learning Goal 3: Types of Matter

5. Match each word on the left with its definition on the right.

(a) Homogeneous

(b) Heterogeneous

(c) Mixture

(d) Compound

(e) Element

(1) The basic building block of matter _____

(2) The word used to describe matter that is uniform throughout _____

(3) A type of matter in which each part retains its own properties _____

(4) A chemical combination of two or more elements _____

(5) The word used to describe matter that is not uniform throughout _____

6. (a) Distinguish among an element, a compound, and a mixture.
(b) What type of matter is uniform throughout?
(c) What type of matter is not uniform throughout?

Learning Goal 4: Physical and Chemical Properties

7. State whether each of the following processes involves chemical or physical properties.
(a) Paper burning
(b) Glass breaking
(c) An egg being fried
(d) Oil burning
(e) Wood being chopped
(f) Food being digested

8. Determine whether each of the following processes involves chemical or physical properties.
(a) Ice melts.
(b) Sugar dissolves in water.
(c) Milk sours.
(d) Eggs become rotten.
(e) Water boils.
(f) An egg is hard-cooked.

Learning Goal 5: Difference Between Atom and Molecule

9. Describe the difference between an atom and a molecule.

10. (*a*) What is the smallest particle of matter that can enter into chemical combinations?
(*b*) What is the smallest particle of a compound that can enter into chemical combinations?

Learning Goal 6: Law of Definite Composition

11. State the Law of Definite Composition (Proportions).

12. Give an example of the Law of Definite Composition (Proportions).

Learning Goal 7: Atomic and Molecular Weight, Avogadro's Number

13. How many atoms are there in 1 mole of atoms of any element?

14. Twelve items equal one dozen items, but a mole represents a much larger quantity. This quantity is _____ items.

Learning Goal 8: Moles from Grams of an Element

15. A sample of the element osmium (Os) weighs $\overline{4}00$ g. How many moles of osmium atoms is this?

16. How many moles are there in 684 g of $Al_2(SO_4)_3$?

*17. The atomic department store is selling uranium atoms at the reduced price of 100,000,000 atoms for 1 cent. What would be the cost in dollars of 1 milligram of uranium?

18. A sample of $CaSO_4$ weighs 0.1705 g. How many moles of $CaSO_4$ are there in this sample?

19. Determine the number of *moles* of atoms in each of the following samples.
(*a*) 16.2 g of Al   (*b*) 239.5 g of Ti
(*c*) 0.06075 g of Mg   (*d*) 3570 g of U

20. How many moles of HCl are there in $\overline{1}00$ g of this compound?

Learning Goal 9: Weight, in Grams, of a Sample, Given the Moles

21. Dysprosium (Dy), the sixty-sixth element in the periodic table, has an atomic weight of 162.5. What weight of Dy do you need to get 0.5000 mole of Dy atoms?

22. What is the weight in grams of 1.50 moles of Rb atoms?

23. What is the weight in grams of 0.75 mole of rubidium (Rb) atoms?

24. How many grams of SO_2 are contained in 0.020 mole of SO_2?

25. Determine the number of grams in each of the following.
(*a*) 4.600 moles of Ni   (*b*) 0.00300 mole of Br
(*c*) $\overline{2}00$ moles of Ca   (*d*) 0.04000 mole of S

26. How many grams of H_2SO_4 are there in 4.0 moles of this compound?

27. Determine the number of grams in each of the following.
(a) 0.600 mole of H_2O
(b) 20.0 moles of SO_2
(c) 0.00050 mole of CO_2
(d) 0.00600 mole of $Al_2(SO_4)_3$

28. What is the weight in grams of 10.0 moles of H_2O?

Learning Goal 10: Number of Atoms of an Element from Grams of Sample

29. After you've finished an experiment, you find that 69 g of sodium (Na) are left in the reaction vessel. How many atoms of sodium are left?

30. At the end of an experiment you find that there are 13.8 g of sodium left in the reaction vessel. How many atoms of sodium are left?

*31. Diamond is made up of carbon atoms. A two-carat diamond costs $1,800. What is the cost of one carbon atom in the diamond? (Assume that one carat weighs 0.2 g.)

*32. How many atoms are present in a sample of water molecules that weigh $1,8\overline{0}0$ g?

33. The radius of a silver atom is 1.34 angstroms, or 1.34 Å (1 Å = 10^{-8} cm). If the atoms of 0.1 mole of silver atoms were placed in a straight line, touching each other, what distance (in miles) would they cover?

*34. How many atoms of Cl_2 are present in a sample of Cl_2 weighing $35\overline{0}$ g?

Learning Goal 11: Formula from Percentage Composition

35. Sodium, the eleventh element in the periodic table (whose name means "headache" in medieval Latin), is a very dangerous chemical when uncombined. But when 39.6% sodium is combined with 60.4% chlorine, by weight, a common compound is formed. What is the empirical formula of this compound, and what is its common name?

36. When 15.8% Al is combined with 28.2% S and 56.1% O, a compound is formed. Calculate the empirical formula of this compound.

37. Cholesterol, a compound suspected of causing hardening of the arteries, has a molecular weight of 386 and the following percentage composition: 84% C, 11.9% H, and 4.1% O. What is the molecular formula of cholesterol?

38. When 24.2% Ca is combined with 17.1% N and 58.5% O, a common compound is formed. What are its empirical formula and common name?

39. Determine the empirical formula of a compound that has 80.0% C and 20.0% H by weight. The molecular weight of this compound is $3\overline{0}$. What is its molecular formula?

40. When 60.56% C is combined with 11.18% H and 28.26% N, the compound spherophysine, which is used for treating high blood pressure, is formed. What is the empirical formula of this compound?

41. Determine the empirical formula of a compound that has 54.55% C, 9.09% H, and 36.36% O by weight. The molecular weight of this compound is 88. What is its molecular formula?

42. When 64.81% C is combined with 13.60% H and 21.59% O, a compound that is used as a surgical anesthetic is formed. The name of this compound is diethyl ether. What is its empirical formula?

Learning Goal 12: Molecular Weight of a Compound from the Formula

43. Find the atomic weight of each of the following elements. Use the periodic table inside the front cover. (*a*) Rb (*b*) Cr (*c*) U (*d*) Se (*e*) As

44. Find the atomic weights of each of the following elements: (*a*) S (*b*) N (*c*) Li (*d*) Cs Where in the text did you find them?

45. Determine the molecular weight of each of the following compounds: (*a*) Al_2O_3 (*b*) $CuBr_2$ (*c*) H_3PO_4 (*d*) LiCl (*e*) $Ca(OH)_2$ (*f*) $Fe_2(SO_4)_3$ (*g*) $(NH_4)_2S$ (*h*) $CoCl_2$

46. Determine the molecular weight of each of the following compounds: (*a*) H_2O (*b*) H_2SO_4 (*c*) NaCl (*d*) $Ca_3(PO_4)_2$

47. Determine the molecular weight of each of the following compounds. (*a*) SiO_2 (*b*) H_2SO_3 (*c*) $Sr(OH)_2$ (*d*) RbF (*e*) $Cu(NO_3)_2$ (*f*) $CoBr_2$ (*g*) $(NH_4)_3PO_4$ (*h*) $HC_2H_3O_2$

48. Determine the molecular weight of each of the following compounds: (*a*) LiOH (*b*) Na_2CO_3 (*c*) $CoCl_2$ (*d*) NaBr

Learning Goal 13: Molecular Weight, Number of Molecules, or Number of Moles in a Sample

49. Ethyl alcohol, which is found in some popular beverages, has the molecular formula C_2H_6O. You have 9.2 g of ethyl alcohol.
(*a*) How many moles of ethyl alcohol do you have?
(*b*) How many molecules of ethyl alcohol?
(*c*) How many moles of carbon atoms?
(*d*) How many moles of hydrogen atoms?
(*e*) How many moles of oxygen atoms?
(*f*) How many atoms of carbon?
(*g*) How many atoms of hydrogen?
(*h*) How many atoms of oxygen?

50. The formula for methanol, also known as wood alcohol, is CH_3OH. You have 3.2 grams of methanol.
(*a*) How many moles of methanol do you have?
(*b*) How many molecules of methanol?
(*c*) How many moles of carbon atoms?
(*d*) How many moles of hydrogen atoms?
(*e*) How many moles of oxygen atoms?
(*f*) How many atoms of carbon?
(*g*) How many atoms of hydrogen?
(*h*) How many atoms of oxygen?

51. Determine the number of *moles* of molecules in each of the following samples.
(*a*) 7.20 g of H_2O (*b*) 260.7 g of MnO_2
(*c*) 0.070 g of N_2 (*d*) 1980 g of $(NH_4)_2SO_4$

52. Determine the number of moles of molecules in each of the following:
(*a*) 0.34 g of NH_3
(*b*) 1335 g of MnO_2

53. The substance glycine is the simplest amino acid. It has the molecular formula $C_2H_5O_2N$. You have 300 g of glycine.
(*a*) How many moles of glycine do you have?
(*b*) How many molecules of glycine?
(*c*) How many moles of carbon atoms?
(*d*) How many moles of hydrogen atoms?
(*e*) How many moles of oxygen atoms?
(*f*) How many moles of nitrogen atoms?
(*g*) How many grams of carbon?
(*h*) How many grams of hydrogen?
(*i*) How many grams of oxygen?
(*j*) How many grams of nitrogen?
(*k*) What should be true about the sum of the numbers of grams of each element in (*g*), (*h*), (*i*), and (*j*) and the number of grams of glycine we started with?

54. Determine the number of moles of molecules in each of the following:
(*a*) $72\overline{0}$ g of H_2O
(*b*) 13.35 g of MnO_2

55. How many *molecules* of CO_2 are there in 4.50 moles of CO_2?

56. How many *molecules* of SO_2 are there in 6.40 moles of SO_2?

Learning Goal 14: Molecular Formula of a Compound

57. According to Nobel prize winner Linus Pauling, vitamin C, also known as ascorbic acid, is a cure and preventative for colds. The empirical formula of vitamin C is $C_3H_4O_3$ and its molecular weight is 176. What is its molecular formula?

58. A compound is found to have the empirical formula HF and a molecular weight of 40. What is its molecular formula?

59. The empirical formula of a very important compound, called benzene, is CH. Its molecular weight is 78. What is its molecular formula?

60. The compound fumaric acid has the empirical formula CHO. Its molecular weight is 116. What is its molecular formula?

61. The plastic melamine, used to make Melmac® dishes, has the empirical formula CH_2N_2. Its molecular weight is 126. What is its molecular formula?

62. The compound benzidine, used in the manufacture of dyes, has the empirical formula C_6H_6N. Its molecular weight is 184. What is its molecular formula?

63. Dextrose, a type of sugar, has the empirical formula CH_2O. Acetic acid, which is found in vinegar, shares the same empirical formula as dextrose. The molecular weight of dextrose is 180, whereas the molecular weight of acetic acid is 60. Write the molecular formula of each compound.

64. A compound of iron and sulfur with a molecular weight of 208 has an empirical formula of Fe_2S_3. What is its molecular formula?

Learning Goal 15: Percentage Composition from Formula

65. Find the percentages of S, H, and O by weight in sulfuric acid (H_2SO_4).

66. Find the percentage by weight of the elements in each of the following compounds:
(a) C_2H_6 (b) NO_2 (c) CO_2

67. Determine the percentages by weight of the elements in the following compounds: (a) BF_3 (b) UF_6 (c) $C_3H_8O_3$

68. Determine the percentage by weight of oxygen in each of the following compounds: (a) H_2O (b) CO (c) H_2SO_4

*69. Some sodium chloride (NaCl) and some sugar ($C_{12}H_{22}O_{11}$) are mixed together. The mixture is analyzed and found to contain $2\overline{0}\%$ chlorine by weight. The mixture weighs $5\overline{0}$ g. How much sugar is present?

70. How many moles of nitrogen atoms can be obtained from $1,7\overline{0}0$ grams of NH_3?

71. Determine the percentages by weight of the elements in the following compounds.
(a) SO_2 (b) CH_4 (c) $C_{12}H_{22}O_{11}$ (d) $CaCO_3$

72. How many grams of hydrogen can be obtained from $18\overline{0}$ g of water?

73. How many grams of copper can be obtained from $1,6\overline{0}0$ g of CuO?

74. How many grams of oxygen can be obtained from 320 g of SO_2?

75. How many grams of iron can be obtained from 350 g of Fe_2O_3?

76. Determine the percentage by weight of sulfur in H_2S.

77. How many grams of oxygen combine with $2\overline{0}0$ g of sulfur in the compound SO_2? [*Hint:* Do Exercise 71 (a) to obtain the percentages by weight of S and O in SO_2.]

*79. An unknown mixture contains sodium sulfide (Na_2S) and iron(III) oxide (Fe_2O_3). The mixture is analyzed and found to contain 25.0% sulfur. The mixture weighs $5\overline{0}0$ g. How many grams of each compound are present?

78. Determine the percentage by weight of hydrogen in CH_3OH.

80. Determine the percentage by weight of N in HNO_3.

Extra Exercises

81. How would you separate a mixture containing wax and iron filings?

82. If you had Avogadro's number of dollars and spent one dollar every second of every day, how long would it take you to run out of money?

83. What is the simplest formula for the compound that has the following percentage composition: 24.3% C, 4.1% H, and 71.6% Cl?

84. If the molecular weight of the compound in Exercise 83 is determined to be 99, what is its molecular formula?

85. What is the weight of a Fe_3O_4 sample that contains $1\overline{0}0$ g of oxygen?

86. Explain in detail what the chemical formula H_2O means.

87. Make a list of heterogeneous and homogeneous mixtures that you encounter in everyday life. Do the same for elements and compounds.

88. How many carbon atoms are there in a 1.75-carat diamond? [*Hint:* 1.00 carat = $2\overline{0}0$ mg]

89. Two ores of iron are hematite (Fe_2O_3) and magnetite (Fe_3O_4). Calculate the percentage by weight of iron in each ore to determine the richest source of iron.

90. Two ores of copper have the following weight percentage analyses. What is the empirical formula of each ore?

Ore A	Ore B
Cu, 63.3%	Cu, 34.6%
Fe, 11.1%	Fe, 30.4%
S, 25.6%	S, 34.9%

91. The percentages by weight of oxygen and sulfur are the same in the compound SO_2. Calculate the percentage by weight of each.

92. What is the mass in grams of ten atoms of silver?

93. How many moles of phosphorus are there in 4.00 moles of H_3PO_4?

94. In Exercise 93, how many grams of phosphorus would there be?

95. How much is one atomic mass unit in grams?

Cumulative Review / Chapters 1–3

Indicate whether each of the following statements is true or false.

1. Chemistry is the science that deals with matter and the changes it undergoes.

2. Organic chemicals are those chemicals that have been derived from living or once-living organisms.

3. Sugar and salt are examples of inorganic chemicals.

4. The art of *khemia* flourished until the year A.D. 1600.

5. Alchemists were able to transmute lead and iron into gold.

6. The idea that every theory must be proved by experiment was advanced by Robert Boyle.

7. In the year 420 B.C., Greek philosophers were able to prove that atoms existed.

8. Some carbon-containing compounds are classified as inorganic chemicals.

9. The two main branches of chemistry that existed during the 1700s and 1800s still exist today.

10. Nuclear chemistry, biochemistry, and analytical chemistry are three subdivisions into which chemistry can be divided.

Answer the following questions to help sharpen your test-taking skills.

11. Find the area of a rectangular room that measures 12.5 m by 10.2 m. Use the proper number of significant figures in reporting your answer.

12. Calculate the volume of a cardboard box that measures 15.25 cm long, 12.00 cm wide, and 24.85 cm high.

13. Determine the volume of a cylindrical solid whose radius is 14.50 cm and whose height is 25.05 cm.

14. Convert 0.28 kg to (a) grams, (b) decigrams, (c) milligrams.

15. Convert 6.8 m to (a) decimeters, (b) centimeters, (c) millimeters.

16. Convert 125 mm to (a) centimeters, (b) meters, (c) kilometers.

17. Convert 25,595 mL to (a) liters, (b) deciliters.

18. Express 50.0 m in (a) feet, (b) inches.

19. Express 25.55 g in (a) pounds, (b) ounces.

20. Determine the number of cubic centimeters in a cubic inch without using a conversion table.

21. Express 165.00 g in pounds.

22. Express 12.0 gallons in (a) liters, (b) ounces.

23. Determine the density of a cube that has a mass of 500.0 g and measures 12.0 cm on each side.

24. A spherical balloon with a volume of 113.0 mL is filled with an unknown gas weighing 20.0 g. Will the balloon float on air? Can you determine what the gas might be?

25. A small metal sphere weighs 90.0 g. The sphere is placed in a graduated cylinder containing 15.0 mL of water. Once the sphere is submerged, the water in the cylinder measures 30.0 mL. What is the density of the sphere?

26. A small metal sphere with a density of 3.50 g/cm^3 has a mass of 937.83 g. Calculate the radius of the sphere.

27. Solve the following problems using the proper number of significant figures.
(a) 28.64 + 3.2
(b) 125.4 ÷ 13.5
(c) 6.55 × 12.1
(d) 98.4 − 0.12

28. Express each of the following numbers in scientific notation.
(a) 5,000
(b) 0.0005
(c) 602,300,000
(d) 35,$\overline{000}$,000

29. How many significant figures are there in each of the following numbers?
(a) 5,500.0
(b) 0.5123
(c) 12.000
(d) 3,5$\overline{0}$0

30. Convert each of the following temperatures from °F to °C.
(a) 45.0°F
(b) −10.0°F
(c) 450°F
(d) −100.0°F

31. Convert each of the following temperatures from
°C to °F.
(a) 88.0°C
(b) −12.5°C
(c) 65.6°C
(d) $\overline{400}$°C

32. Distinguish between an amorphous solid and a
crystalline solid.

33. Determine whether each of the following processes involves chemical or physical changes.
(a) A match burns.
(b) Glucose dissolves in water.
(c) Bread becomes moldy.
(d) A piece of wood is sawed.

34. Distinguish between heterogeneous and homogeneous matter.

35. A mole represents a given quantity of items.
Name two other words that represent a given quantity of items, and tell how many items each of these
terms denotes.

36. How many moles are there in 3.43 g of
$Al_2(SO_4)_3$?

37. How many moles are there in 4.8 g of C atoms?

38. Determine the number of grams in each of the
following.
(a) 0.750 mole of H_2SO_4
(b) 5.00 moles of H_2O
(c) 0.00700 mole of H_2CO_3
(d) 50.0 moles of $HC_2H_3O_2$

39. After completing an experiment, you find 4.6 g
of unreacted Na in the vessel. How many atoms does
this represent?

40. After a reaction is complete it is determined that
5.328 g of Sn combined with 1.436 g of O. Calculate
the empirical formula of this compound.

41. The empirical formula for the antibiotic nonactin
is $C_{10}H_{16}O_3$. If the molecular weight of nonactin is
736, what is its molecular formula?

42. The empirical formula for fructose, the natural
sugar found in fruit juice, fruits, and honey, is CH_2O.
If the molecular weight of fructose is $18\overline{0}$, what is its
molecular formula?

43. The molecular formula for aspirin is $C_9H_8O_4$. You
have 36.0 g of aspirin.
(a) How many moles of aspirin are there?
(b) How many molecules of aspirin are there?
(c) How many moles of carbon atoms are there?
(d) How many moles of hydrogen atoms are there?
(e) How many moles of oxygen atoms are there?

44. A 6.08-g sample of nitrogen combines with
13.90 g of oxygen to produce a compound whose
molecular weight is 92.0. What is the molecular formula of this compound?

45. Calculate the molecular weight of each of the following compounds.
(a) $Ca_3(PO_4)_2$
(b) CH_4
(c) NO_2
(d) $Ba(C_2H_3O_2)_2$

46. Calculate the percentage composition of each of
the following compounds.
(a) H_2S
(b) NaCl
(c) $Fe(C_2H_3O_2)_3$
(d) K_2SO_4

47. Determine the molecular formula of a compound
that has a molecular weight of 86.0 and is composed
of 83.7% carbon and 16.3% hydrogen.

48. Find the percentage of carbon, hydrogen, and nitrogen in the compound nicotine, given that its empirical formula is C_5H_7N and that its molecular weight
is 162.

49. How many water molecules are there in a
0.720-g sample of H_2O?

50. A patient has been told to take a $1,\overline{000}$-mg supplement of elemental calcium each day. How many
grams of $CaCO_3$ must she take in order to receive
$1,\overline{000}$ mg of elemental calcium?

4 Atomic Theory, Part 1: What's in an Atom?

Learning Goals

After you've studied this chapter, you should be able to:

1. Describe Dalton's atomic theory, Thomson's model of the atom, and Rutherford's model of the atom.
2. Give the charge and mass of the electron, proton, and neutron.
3. Find the number of protons, electrons, and neutrons in an atom of an element.
4. Explain what isotopes are, and give examples.
5. Use the standard isotopic notation for mass number and atomic number.
6. Calculate the average atomic weight of an element when you know the relative abundances of its isotopes.
7. Calculate the percentage abundance of two isotopes when you know the atomic weight of each isotope and the average atomic weight.
8. Use the table of relative abundances of isotopes to calculate the number of atoms or grams of a particular isotope present in a given sample.

John Dalton
(Reprinted with permission from Torchbearers of Chemistry, *by Henry Monmouth Smith,* © *Academic Press, Inc., 1949)*

Introduction

Around 420 B.C., Democritus developed the idea that the atom—indestructible and indivisible—was the smallest particle of matter. Twenty-two hundred years later, John Dalton reinforced this idea in his atomic theory. Yet less than a century after Dalton presented his theory, scientists had adopted a different model, in which the atom was made up of several even smaller, subatomic particles. The development of that model is the subject of this chapter.

4.1 What is a Model?

· · · · · · · · · · · · · · ·
Learning Goal 1
Atomic theory and models

Before we try to discuss the atom, we should recall that until very recently no one had ever seen one. How, then, did we know what an atom was like? We really didn't, exactly. But scientists have put together observations, experimental results, and a good deal of reasoning and have come up with a *model.* This model of the atom is just a representation of what scientists believe the atom is like. Models can have different degrees of accuracy. Take a model of a car, for instance. It could be carved from a solid block of wood or plastic and represent just the shape of the car. Or a more detailed model could show the interior of the car as well. A still more detailed model might have a gasoline-powered motor and move just like a real car. This model would represent the car most accurately. The car model would be a physical model, whereas models of the atom are *intellectual* or *thought* models; however, they too have become more accurate as scientists have gathered more information about atoms.

4.2 Dalton's Atomic Theory

John Dalton (1766–1844) was a British schoolteacher who studied physics and chemistry. On the basis of facts and experimental evidence, he proposed an atomic theory. This theory was an attempt to explain all the different forms of matter. It has turned out to be one of the greatest contributions to chemistry since the time of the ancient Greeks. Dalton theorized that:

1. All elements are composed of tiny, indivisible particles called atoms. This idea was similar to the idea of Democritus. Both men believed there was an ultimate particle—that all matter was composed of tiny indestructible and indivisible spheres.

2. All matter is composed of combinations of these atoms. In Chapter 3 we discussed how the atoms of different elements combine to form the molecules that make up compounds. For example, two atoms of hydrogen and one atom of oxygen combine to form a molecule of water.

3. Atoms of different elements are different. Dalton believed gold was different from silver because somehow the atoms of gold are different from the atoms of silver. For example, he thought they differed in mass.

4. Atoms of the same element have the same size, mass, and form. Dalton believed all gold atoms, for example, are the same as all other gold atoms in every respect.

Dalton's theories laid the foundation for what is believed today. But his basic idea about the indivisibility of the atom was eventually shown to be incorrect, as was his idea that all atoms of the same element have the same size, mass, and form. Revisions of his theory have led to our current model of the atom. Let us review how our present-day model of the atom was developed.

4.3 The Discovery of the Electron

In the mid 1800s scientists began to question whether the atom was really indivisible. They also wanted to know why atoms of different elements had different properties. Some of the answers came with the invention of the Crookes tube, or *cathode-ray tube* (Figure 4.1).

The scientists of the time knew that some substances conduct electric current (that is, they are conductors), whereas other substances do not. And, with enough electrical power, a current can be driven through any substance—solid, liquid, or gas. In the cathode-ray tube, a high-voltage electric current is driven through a nearly empty tube, a near *vacuum*. The tube contains two pieces of metal called *electrodes*. Each electrode is attached by a wire to the source of an electric current. The source has two *terminals*, positive and negative. The electrode attached to the positive electric terminal is called the *anode;* the electrode attached to the negative terminal is called the *cathode*. Crookes showed that when the current was turned on, a beam moved from the cathode to the anode; in other words, the beam moved from the negative to the positive terminal. Because the beam originated at the *negative* terminal, and because it was attracted to the *positive* terminal, the beam had to be negative in nature.

What was this beam? Was it made of particles or waves? Did it come from electricity or from the metal electrodes? Physicists in Crookes's time were not sure about the answers to these questions, but they did make guesses. Whatever this beam was, it traveled in straight lines (they knew

Learning Goal 2
Mass and charges of protons, electrons, and neutrons

Figure 4.1
A Crookes or cathode-ray tube

Figure 4.2

A cathode-ray tube in operation. Compare these pictures with Figure 4.3.
SOURCE: Fundamental Photographs, New York.

that because it cast sharp shadows). For lack of a better name, the German physicist Eugen Goldstein called the beams *cathode rays*, because they came from the cathode.

The German physicists in Crookes's time favored the *wave theory* of cathode rays because the beam traveled in straight lines, like ocean waves. But the English physicists favored the *particle theory*. They said that the beam was composed of tiny particles that moved very quickly—so quickly that they were hardly influenced by gravity. That was why the particles moved in a straight path. (Note how a single experimental observation led to two different theories.)

Crookes proposed a method to solve the dilemma. If the beam was composed of negative particles, a magnet would deflect them (Figure 4.2). But if the beam was a wave, a magnet would cause almost no deflection. Particles would also be more easily deflected by an electric field. In 1897 the English physicist J. J. Thomson used both these techniques—magnetic and electric—to show that the rays were composed of particles (Figure 4.3). Today we call these particles **electrons.** Moreover, we understand that an electric current is actually a flow of electrons.

Were the electrons in the beam coming from the metal electrodes or from the current source? Unless it was the metal that gave off these electrons, the atom could still be the smallest particle of matter. Proof that metals do give off electrons came from the laboratories of Philipp Lenard, a German physicist. In 1902 he showed that ultraviolet light directed onto a metal makes it send out, or emit, electrons. This effect, known as the *photoelectric effect*, indicated that the metal atoms—and the atoms of other elements—contain electrons.

In 1911 a young American physicist named Robert Millikan calculated the mass of the electron: 9.11×10^{-28} grams. (To get an idea of how very small this is, consider the fact that it would take about 16,000,000,000,000,000,000,000,000,000, or 1.6×10^{28}, electrons to equal the weight of a standard half-ounce chocolate bar.)

Figure 4.3

The Thomson experiment: (1) A beam of electrons (dashed line) moves from cathode toward anodes. (2) Some electrons pass between anodes. (3) Electric field causes beam of electrons to bend. This is visible on the screen. (4) But by adding a magnet to counteract deflection caused by electric field, one can make the beam follow a straight path. (5) One can then measure the strengths of the electric and magnetic fields and calculate the charge-to-mass ratio (e/m) of the electron.

4.4 The Proton

The discovery of the electron as part of the atom and as a *negative* particle of electricity was very important, but it raised many questions. Because electrons are part of all atoms, and because atoms in their normal state are electrically *neutral*, scientists reasoned, there must be something else in the atom that balances the negative electron. Otherwise, if matter had a negative charge, you would get a shock every time you touched anything.

This "something else," it seemed, would have to be some other sort of particle that has the same amount of charge as the electron but is positive in nature. Such reasoning led scientists to suppose that there were positive particles, which they called **protons,** in the atom. But then, why could atoms give off only negative particles (electrons) and not positive ones (protons)? The British physicist J. J. Thomson suggested an answer. He thought of the atom as a sphere made up of positive electricity in which electrons were embedded (Figure 4.4). The English called this the *plum-pudding theory,* likening the electrons to the raisins in plum pudding. As Americans, we can find a more familiar analogue in a scoop of chocolate-chip ice cream. The chocolate chips represent the electrons. The ice cream represents a sea of positive electricity (composed of protons). Each element is different from the others because each has a different number of electrons and protons arranged in a different way—like different scoops of ice cream with different numbers of chocolate chips. For example, hydrogen atoms have one electron and one proton. Helium atoms have two electrons and two protons. Thomson seemed to have given scientists an idea they could agree on, but his theory still needed experimental proof.

Lord Rutherford, a former student and colleague of Thomson, attempted to test Thomson's theory. He had been working with positive

Figure 4.4

The Thomson model of the atom

Figure 4.5

According to the Thomson model of the atom, positive particles ought to pass through a metal foil without being deflected.

particles called *alpha* particles (α particles), which are actually helium atoms with their electrons removed. He reasoned that if he shot a stream of these positive particles through a gold foil, the neutral and symmetrical "plum pudding" gold atoms would not greatly affect the path of the alpha particles. Most of the particles would move straight through the foil (Figure 4.5). Rutherford used gold foil in his experiment because it could be hammered into very thin sheets, making the thickness of the foil resemble a single layer of atoms. However, when Rutherford actually performed the experiment, he found that about one alpha particle in 20,000 ricocheted, or bounced back out of the foil! As Rutherford said, "It was almost as incredible as if you fired a fifteen-inch shell at a piece of tissue paper and it came back and hit you."

To explain these peculiar results, Rutherford devised a new model for the atom, called the nuclear model. We subscribe to this model today. Rutherford said that an atom must have a *center of positive charge*, where all the protons must be located. He called this center of positive charge the **nucleus** of the atom. Rutherford felt that the nucleus of the atom must be very small compared to the overall size of the atom. He also said that the nucleus must be where the mass of the atom is concentrated. Rutherford's reasoning was that some alpha particles were tremendously deflected (Figure 4.6) because they were repelled by a high concentration of positive charge with an immovable mass. He also decided that the elec-

Figure 4.6

The actual way that positive particles pass through metal foil, as shown by Rutherford.

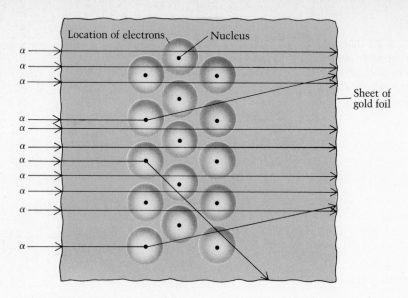

Location of electrons

Nucleus

Sheet of gold foil

trons were located around the nucleus instead of in it and that the mass of the electrons was very small compared to the mass of the protons. His reasoning here was that for so many of the particles to go through the foil undeflected, the electrons had to be located in a relatively large space outside the nucleus but still within the atom.

Later experiments similar to Rutherford's showed that the diameter of the nucleus of an atom is approximately 10^{-13} cm, whereas the diameter of the whole atom is up to 100,000 times as great. Suppose we could expand an atom so that the nucleus was 1 millimeter in diameter (about the size of this dot: ·). Then the nearest electron could be as far as 50 meters (half the length of a football field) away from the nucleus! And between the nucleus and the electrons is only vast empty space.

Rutherford's model of the atom, based as it was on new experimental evidence, made the Thomson model of the atom obsolete.

4.5 The Neutron

Lord Rutherford
(Reprinted with permission from Torchbearers of Chemistry, by Henry Monmouth Smith, © Academic Press, Inc., 1949)

Thus, before 1920, scientists had experimentally confirmed the existence of two basic *subatomic* (a particle that is a constituent of an atom) particles: electrons, which are negatively charged and have a very small mass, and protons, which are positively charged and contain most of the atom's mass. But a problem came up. Scientists had defined a unit of mass called the **atomic mass unit** (abbreviated amu). They knew that the proton had a mass of 1.0072766 amu and that the electron had a much smaller mass of 0.0005486 amu. These numbers all seemed fine at first. The mass

Table 4.1
The Particles in the Atom

Particle	Symbol	Charge	Mass
Proton	p	+1	1.0073 amu
Electron	e	−1	0.0005 amu
Neutron	n	0	1.0087 amu

of hydrogen—which is made up of one proton and one electron—was very close to the sum of the masses of a proton and an electron. (Remember that protons account for just about all of the atom's mass; when we are calculating the mass of an atom we can usually ignore the mass of the electrons.) The problem arose when scientists considered helium. The helium atom has two electrons and two protons, so the scientists thought it should have a mass of about 2 amu. But its mass is 4 amu. And when they looked at other elements, they were even more puzzled. Carbon has six electrons and six protons but a mass of 12 amu. Where was the missing mass?

To account for the extra mass in the atom, scientists proposed another particle, one that had no charge but had a mass equal to the proton's. In 1932, the British physicist James Chadwick (1891–1974) was able to prove the existence of a new subatomic particle called the neutron, which has a mass of 1.008665 amu. If this particle were also present in the atom, it would add to the mass but would not affect the charge (see Table 4.1). Scientists could now picture helium as being composed of two electrons, two protons, and two neutrons, so they could understand helium's mass of 4 amu.

4.6 Atomic Number

Learning Goal 3
Number of protons, electrons, and neutrons

Figure 4.7

In the periodic table used in this text, an element's atomic number appears above its symbol.

The factor that makes one element different from another element is that atoms of different elements have different numbers of electrons and protons. *The number of protons in an atom of an element* is called the **atomic number** of that element and *is symbolized Z.*

The atomic number of an element is also equal to the positive charge on the nucleus and to the number of electrons in the neutral atom of that element. On any periodic table there is a whole number identified with each element. This is the atomic number Z. In the periodic table inside the front cover, the atomic number is given above the symbol for each element. An atom of sodium (Na), which has the atomic number 11, has 11 electrons and 11 protons (Figure 4.7). An atom of magnesium (Mg), which has the atomic number 12, has 12 electrons and 12 protons; and an atom of uranium (U), which has the atomic number 92, has 92 electrons and 92 protons. Note that the elements are listed in the periodic table in the order of their atomic numbers.

4.7 Isotopes

If you look at the periodic table, you will see that the atomic weights of most elements are not whole numbers. For example, sodium (Na) has a weight of 22.990 amu, titanium (Ti) a weight of 47.90 amu, and rubidium (Rb) a weight of 85.468 amu. Why is this so? If each proton and each neutron gives a mass of almost exactly 1 amu to the atom, then the atomic weights of the elements certainly ought to be very close to whole numbers.

Investigation of this question led to the discovery that *each element can exist in more than one form.* When John Dalton said that the atoms of an element must be identical in size, mass, and shape, he did not know that *isotopes,* or different forms of an element, could exist. **Isotopes** are *forms of an element that have the same number of electrons and protons but different numbers of neutrons.*

Although the numbers of negative and positive charges of isotopes are identical, then, isotopes differ in mass. Hydrogen, for example, exists in three isotopic forms (Figure 4.8). The common form of hydrogen (called protium) contains one electron and one proton. Another form of hydrogen (called deuterium) contains one electron, one proton, and one neutron. A third form of hydrogen (called tritium) contains one electron, one proton, and two neutrons.

Note that all isotopes of an element have the same number of protons and thus have the same atomic number. However, it is possible to distinguish among isotopes by means of mass numbers. The **mass number** of an isotope is simply *the sum of the number of protons and the number of neutrons.*

A standard notation has been developed for giving the mass number and atomic number of an isotope. It consists of writing the symbol of the

Learning Goal 4
Definition of an isotope

Learning Goal 5
Standard notation for mass number and atomic number

Protium   Deuterium

Tritium

Figure 4.8
Isotopes of hydrogen

element with its mass number as a *superscript* (above the line) and its atomic number as a *subscript* (below the line). Both numbers are usually written to the left of the symbol for the element.

$$\text{Mass number} \longrightarrow \quad ^{16}_{8}\text{O} \quad ^{12}_{6}\text{C} \quad ^{35}_{17}\text{Cl}$$
$$\text{Atomic number} \longrightarrow$$

With this notation we can immediately tell the composition of the isotope in terms of protons, electrons, and neutrons. The number of protons (or, in the neutral atom, the number of electrons) is simply the atomic number of the element. The number of neutrons is the difference between the mass number and the atomic number. This notation allows us to distinguish among the isotopes of an element.

Example 4.1

Give the number of electrons, protons, and neutrons in each of the following neutral isotopes: (a) $^{238}_{92}\text{U}$ (b) $^{235}_{92}\text{U}$ (c) $^{84}_{36}\text{Kr}$

Solution

The atomic number (subscript) gives the number of electrons in the neutral atom as well as the number of protons. Subtracting the atomic number from the mass number (superscript) gives the number of neutrons.

(a) Because the atomic number is 92, there are 92 electrons and 92 protons. And subtracting the atomic number from the mass number gives the number of neutrons: $238 - 92 = 146$ neutrons. So there are 92 electrons, 92 protons, and 146 neutrons in an atom of $^{238}_{92}\text{U}$.

(b) Again there are 92 electrons and 92 protons. And $235 - 92 = 143$ neutrons. So there are 92 electrons, 92 protons, and 143 neutrons in an atom of $^{235}_{92}\text{U}$. Note that $^{238}_{92}\text{U}$ and $^{235}_{92}\text{U}$ are an *isotope pair*.

(c) Here we have 36 electrons and 36 protons. And $84 - 36 = 48$ neutrons. So there are 36 electrons, 36 protons, and 48 neutrons in an atom of $^{84}_{36}\text{Kr}$.

Practice Exercise 4.1   Give the number of protons, electrons, and neutrons in each of the following isotopes: (a) $^{28}_{14}\text{Si}$ (b) $^{31}_{15}\text{P}$ (c) $^{6}_{3}\text{Li}$

The following notations can be used to distinguish among the three isotopes of hydrogen.

$$^{1}_{1}\text{H} \qquad\qquad ^{2}_{1}\text{H} \qquad\qquad ^{3}_{1}\text{H}$$

Protium   Deuterium   Tritium

Learning Goal 6
Average atomic weight of an element

The deuterium and tritium forms of hydrogen are not so common as the protium form. If a random sample of all hydrogen atoms found on the earth were taken, it would be found to consist of 99.985% protium atoms, 0.015% deuterium atoms, and a vanishingly small amount of tritium atoms. The atomic weight given in the periodic table is the *weighted average of the atomic masses* of all three isotopes. *When calculating the average*

Table 4.2
Naturally Occurring Isotopes of the First 15 Elements

Name	Symbol	Atomic number	Mass number	Percentage natural abundance
Hydrogen-1	$_1^1H$	1	1	99.985
Hydrogen-2	$_1^2H$	1	2	0.015
Hydrogen-3	$_1^3H$	1	3	Negligible
Helium-3	$_2^3He$	2	3	0.00013
Helium-4	$_2^4He$	2	4	99.99987
Lithium-6	$_3^6Li$	3	6	7.42
Lithium-7	$_3^7Li$	3	7	92.58
Beryllium-9	$_4^9Be$	4	9	100
Boron-10	$_5^{10}B$	5	10	19.6
Boron-11	$_5^{11}B$	5	11	80.4
Carbon-12	$_6^{12}C$	6	12	98.89
Carbon-13	$_6^{13}C$	6	13	1.11
Nitrogen-14	$_7^{14}N$	7	14	99.63
Nitrogen-15	$_7^{15}N$	7	15	0.37
Oxygen-16	$_8^{16}O$	8	16	99.759
Oxygen-17	$_8^{17}O$	8	17	0.037
Oxygen-18	$_8^{18}O$	8	18	0.204
Fluorine-19	$_9^{19}F$	9	19	100
Neon-20	$_{10}^{20}Ne$	10	20	90.92
Neon-21	$_{10}^{21}Ne$	10	21	0.257
Neon-22	$_{10}^{22}Ne$	10	22	8.82
Sodium-23	$_{11}^{23}Na$	11	23	100
Magnesium-24	$_{12}^{24}Mg$	12	24	78.70
Magnesium-25	$_{12}^{25}Mg$	12	25	10.13
Magnesium-26	$_{12}^{26}Mg$	12	26	11.17
Aluminum-27	$_{13}^{27}Al$	13	27	100
Silicon-28	$_{14}^{28}Si$	14	28	92.21
Silicon-29	$_{14}^{29}Si$	14	29	4.70
Silicon-30	$_{14}^{30}Si$	14	30	3.09
Phosphorus-31	$_{15}^{31}P$	15	31	100

atomic weight for any element, we must take into account the relative abundance of each isotope of that element. It is for this reason that the atomic weight of hydrogen shown in the periodic table is 1.008 amu. This is nearly the same as the weight of protium, the most common isotope of hydrogen. Table 4.2 gives a breakdown of the isotopes of the first 15 elements. This averaging process also explains why the atomic weights of the elements in the periodic table are not whole numbers.

Learning Goal 7
Percentage abundance of isotopes

Learning Goal 8
Table of relative abundances of isotopes

If we know the percentage abundance of each isotope of an element, we can calculate the average atomic weight of the element. The following example shows how to do this. In the example we use the *mass number* of each isotope to approximate its atomic mass. This reduces the complexity of the calculation but gives good accuracy. (To satisfy yourself that it does, compare the result with the atomic weight given in the periodic table.)

Example 4.2

Using Table 4.2, calculate the atomic weight of boron.

Solution

From Table 4.2, the percentage abundance of ^{10}B is 19.6% and that of ^{11}B is 80.4%. Therefore the atomic weight of

$$B = (10 \text{ amu})(0.196) + (11 \text{ amu})(0.804)$$

$$= 1.96 \text{ amu} + 8.84 \text{ amu} = 10.80 \text{ amu}$$

To review changing percentages to decimal equivalents see Example A–19 in Supplement A.

Practice Exercise 4.2   Using Table 4.2, calculate the atomic weight of lithium.

The concepts of isotopes with different abundances allows us to solve two other kinds of quantitative problems. These are illustrated by the next two examples.

Example 4.3

A hypothetical element Q exists in two isotopic forms, ^{270}Q and ^{280}Q. The atomic weight of Q is 276 amu. What is the percentage abundance of each isotope? (This example is a bit more difficult than the other examples because it requires a good knowledge of algebra and the ability to solve simultaneous equations.)

Solution

Let x equal the fraction of ^{270}Q, and let y equal the fraction of ^{280}Q. Then we have

$$270x + 280y = 276 \quad \text{and} \quad x + y = 1$$

(We get the second equation from the fact that the fractions of all isotopes must add up to 1, or the whole amount, in the same way that all the percentages add up to 100.) Solving these simultaneous equations leads to $x = 0.4$ and $y = 0.6$. In other words, 40% of the element is ^{270}Q and 60% is ^{280}Q.

Practice Exercise 4.3   A hypothetical element Y exists in two isotopic forms, ^{100}Y and ^{110}Y. The atomic weight of Y is 108. What is the percentage abundance of each isotope?

Example 4.4

Using Table 4.2, find how many ^{17}O atoms there are in a sample containing 1,000,000 oxygen atoms.

Solution

From Table 4.2, we find that the natural abundance of ^{17}O is 0.037%. Therefore the number of ^{17}O atoms in one million O atoms equals

$$(0.00037)(1,000,000) = 370 \text{ atoms of } ^{17}O$$

Practice Exercise 4.4   Using Table 4.2, find how many ^{20}Ne atoms there are in a sample containing 1,000,000 neon atoms.

4.8 What Next for the Atom?

We have traced the development of a model for the atom from the time of Dalton, who believed that the atom was an indivisible unit of matter, to the time of Rutherford, who suggested that the atom has a nucleus composed of protons and neutrons and an outer part for the electrons. But many questions were still left unanswered. How were the electrons arranged in an atom? Where were the electrons? How were they moving? How fast were they going? Could they account for the formation of compounds? Why didn't these negative electrons "fall into" the positive nucleus of the atom? After all, opposite charges attract! We will answer these questions in Chapter 5.

Summary

John Dalton's atomic theory was an attempt to explain all the different forms of matter. He theorized (1) that the atom is indivisible; (2) that all atoms of the same element have the same mass, size, and form; (3) that all matter is composed of combinations of atoms; and (4) that atoms of different elements are different. Although the first two of these ideas were later shown to be incorrect, most of Dalton's theory is the basis for what is still believed today. Dalton's model of the atom was modified as various subatomic particles—protons, electrons, and neutrons—were discovered. Thomson's theory of the atom followed his "discovery" of electrons. This British physicist theorized that the atom was a sphere made up of positive electricity in which negatively charged electrons were embedded. Lord Rutherford, a colleague of Thomson, tested Thomson's theory with an experiment in which a stream of positive (alpha) particles were supposed to pass straight through the atoms in a gold foil with only a few particles being slightly deflected. Instead, some of the alpha particles bounced back. To explain this result, Rutherford theorized that an atom has a relatively very small center where all the positively charged protons and most of the atomic mass are located. Electrons, he postulated, are located outside the nucleus, perhaps relatively far from it.

There was one difficulty with the two-particle atomic model. For most atoms the masses of all the protons and electrons did not account for the total mass of the atom. In 1932 James Chadwick was able to prove the existence of a new subatomic particle called the neutron. The neutron has about the same mass as a proton but carries no charge. With neutrons, scientists were able to account for the total mass of atoms.

The number of protons in the atom of an element is called the atomic number of that element. The sum of the number of protons and the number of neutrons is called the mass number of the element. All the atoms of any element have the same atomic number, but they may have different mass numbers. Two or more atoms with the same atomic number but different mass numbers are called isotopes.

. . . .

Career Sketch

Biochemist

Biochemists study the chemical composition and behavior of living things. Most of what we know about reproduction, growth, and heredity we owe to the science of biochemistry. Biochemists also study the effects of food, hormones, and drugs on various organisms. Their work affects areas such as medicine, nutrition, and agriculture. For instance, they in-

Biochemists must be familiar with many research techniques and laboratory instruments, such as the scanning electron microscope (SEM) shown on the left. Biochemists use the SEM to photograph body components. This red blood cell (right) has been magnified 10,000 times.
(Left: Cambridge Stereoscan 200 SEM courtesy Cambridge Instruments, Inc. Right: SEM courtesy John B. Vander Sande, Massachusetts Institute of Technology)

vestigate causes and cures for diseases, identify the nutrients necessary to maintain good health, and develop chemical compounds for the control of pests.

Most biochemists work in research, and some do applied research to solve practical problems. For example, when they find out how an organism forms a hormone, they use that knowledge to synthesize and produce hormones on a mass scale.

Biochemists use a variety of instruments, including electron microscopes and centrifuges, and they sometimes have to devise new instruments and techniques. Their laboratory research involves weighing, filtering, distilling, drying, and culturing (growing microorganisms). Some experiments also require them to design and construct laboratory apparatus or to use radioactive tracers.

Career Sketch

The Technology of Energy Conservation and Use

Technologists employed in this emerging field measure quantities of energy used in heating and cooling or in operating a facility or carrying out an industrial process. They assess efficiency in the use of energy and calculate the amount of energy lost through wasteful processes or the lack of proper insulation. They prescribe remedial steps that can be taken to conserve energy.

Energy conservation and use technologists are trained in techniques to help minimize energy loss. Working as a team led by an engineer, they calculate optimum energy requirements for a given system or process.

These technologists must be able to work with many different types of devices and mechanisms and must understand how computers and these devices interact. They develop, construct, test, install, operate, modify, and maintain the latest equipment in the energy-related fields. A high school diploma is necessary, as is college or technical school training. Physics, chemistry, energy conservation, energy economics, instrumentation, electronics, electromechanical systems, heating, ventilation, and solar power are among the subjects that energy technologists study.

. . . .

Self-Test Exercises

The self-test exercises are arranged in matched pairs. The first exercise of each pair appears in the left column; the exercises whose numbers are in color are answered in the back of the book. Answers to exercises in the right column are not given. Difficult problems are marked with asterisks.

Learning Goal 1: Atomic Theory and Models

1. Explain the idea of a model. Use an example of your own.

2. Give an example of a physical model and an example of an intellectual model.

3. List the points in Dalton's atomic theory. Explain which ones had to be corrected after later experiments were performed.

4. In what ways is Dalton's atomic theory similar to the ideas of Democritus, set forth in 420 B.C.?

5. Contrast the Thomson model and the Rutherford model of the atom.

6. Give a general outline of the construction of a Crookes tube.

*7. Explain how a television set is similar to a Crookes tube. (You may want to check an outside reference source for help in answering this question.)

8. Discuss the experiment that was conducted to test the wave and particle theories of cathode rays.

Learning Goal 2: Mass and Charges of Protons, Electrons, and Neutrons

9. How was the electron discovered?

10. Which of the three subatomic particles has the greatest mass? Which has the least mass?

11. Name the three major subatomic particles. Give their charges and their approximate atomic masses (in amu).

12. Compare the charges of the three subatomic particles and briefly explain how each charge was determined.

Learning Goal 3: Number of Protons, Electrons, and Neutrons

13. Find the number of protons (p), electrons (e), and neutrons (n) in neutral atoms of the following: (a) $^{23}_{11}$Na (b) $^{223}_{87}$Fr (c) $^{238}_{92}$U

14. Determine the number of protons (p), electrons (e), and neutrons (n) in neutral atoms of the following: (a) $^{14}_{7}$N (b) $^{15}_{7}$N

15. Find the number of protons, electrons, and neutrons in neutral atoms of the following: (a) $^{16}_{8}$O (b) $^{17}_{8}$O (c) $^{18}_{8}$O (d) $^{20}_{10}$Ne (e) $^{21}_{10}$Ne (f) $^{22}_{10}$Ne

16. Find the number of protons, electrons, and neutrons in neutral atoms of the following: (a) $^{31}_{15}$P (b) $^{22}_{10}$Ne (c) $^{24}_{12}$Mg

17. From the information that follows, select the element that is not an isotope of the others.
(a) 8p, 8e, 8n (b) 7p, 7e, 8n (c) 8p, 8e, 9n

18. Which of the following describes an atom that is not an isotope of the other two?
(a) 12p, 12e, 14n (b) 12p, 12e, 12n
(c) 13p, 13e, 12n

19. Find the number of protons, electrons, and neutrons in neutral atoms of the following:
(a) $^{57}_{25}$Mn (b) $^{60}_{27}$Co (c) $^{80}_{36}$Kr (d) $^{128}_{52}$Te

20. Determine the number of protons, electrons, and neutrons in neutral atoms of the following hypothetical isotopes: (a) $^{35}_{16}$X (b) $^{300}_{90}$Y

21. The alchemists wanted to turn lead into gold. Suppose that you had the isotope of lead $^{208}_{82}$Pb, and you wanted to turn it into the isotope of gold $^{197}_{79}$Au. How many protons, electrons, and neutrons would you have to remove from the lead atom to turn it into the gold atom?

22. Which of the following describes an atom that is not an isotope of the other two?
(a) 14p, 14e, 15n (b) 15p, 15e, 14n
(c) 15p, 15e, 15n

23. Fill in the blanks in the following table. (Use the periodic table when necessary.)

24. State the number of protons, electrons, and neutrons in $^{42}_{20}$Ca.

Symbol	Protons	Electrons	Neutrons	Mass number	Atomic number
$^{174}_{70}$Yb					
	59			141	
			60		44
			24	45	
		22	28		
$^{?}_{12}$Mg			13		

Learning Goal 4: Definition of an Isotope

25. Define atomic number, mass number, and isotope.

26. Atoms of an element that have the same number of electrons and protons, but different numbers of neutrons, are called _____.

27. Isotopes are atoms with the same number of _____ and _____ but different numbers of _____. (Place the words *protons, electrons,* and *neutrons* in the proper places.)

28. (a) The number of protons in an atom is called the _____ _____ of that element.
(b) The sum of the numbers of protons and neutrons in an isotope is called the _____ of that isotope.

Learning Goal 5: Standard Notation for Mass Number and Atomic Number

29. Write the standard isotopic notation for the following elements.
(a) 48p, 48e, 55n   (b) 55p, 55e, 68n
(c) 87p, 87e, 136n   (d) 17p, 17e, 15n

30. Write the standard isotopic notation for the following elements.
(a) 3p, 3e, 3n   (b) 9p, 9e, 10n
(c) 14p, 14e, 16n

Learning Goal 6: Average Atomic Weight of an Element

31. The element gallium (Ga) exists in two isotopic forms, with the following abundances: 60.16% ^{69}Ga and 39.84% ^{71}Ga. Calculate the atomic weight of gallium.

32. The element carbon (C) exists in two isotopic forms with the following abundances: 98.89% ^{12}C and 1.11% ^{13}C. Calculate the atomic weight of carbon.

33. The element chlorine (Cl) exists naturally in two isotopic forms, with the following abundances: 75.53% ^{35}Cl and 24.47% ^{37}Cl. Calculate the atomic weight of chlorine.

34. The element magnesium (Mg) exists in three isotopic forms with the following abundances: 78.70% ^{24}Mg, 10.13% ^{25}Mg, and 11.17% ^{26}Mg. Calculate the atomic weight of magnesium.

35. The element chromium (Cr) exists naturally in four isotopic forms, with the following abundances: 4.31% ^{50}Cr, 83.76% ^{52}Cr, 9.55% ^{53}Cr, and 2.38% ^{54}Cr. Calculate the atomic weight of chromium.

36. The element copper exists in two isotopic forms with the following abundances: 69.09% ^{63}Cu and 30.91% ^{65}Cu. Calculate the atomic weight of copper.

Learning Goal 7: Percentage Abundance of Isotopes

*37. Bromine exists in two isotopic forms, ^{79}Br and ^{81}Br. The atomic weight of bromine is 79.90 amu. Calculate the percentage abundance of each isotope.

38. Boron (B) exists in two isotopic forms, ^{11}B and ^{10}B. The atomic weight of boron is 10.80. Calculate the percentage abundance of each isotope.

*39. The element antimony occurs naturally in two isotopic forms, ^{121}Sb and ^{123}Sb. The atomic weight of antimony is 121.75. Calculate the percentage abundance of each isotope.

40. Chlorine exists in two isotopic forms, ^{35}Cl and ^{37}Cl. The atomic weight of chlorine is 35.453. Calculate the percentage abundance of each isotope.

Learning Goal 8: Table of Relative Abundances of Isotopes

41. Using Table 4.2, determine how many ^{2}H atoms you would find in 10,000,000 hydrogen atoms.

42. Using Table 4.2, determine how many ^{2}H atoms you would find in 100,000 atoms of hydrogen.

43. Using Table 4.2, determine how many grams of oxygen you would need to get 1 g of ^{17}O.

44. Using Table 4.2, determine how many grams of oxygen you would need to get 1 g of the isotope ^{18}O.

45. Using Table 4.2, determine how many ^{6}Li atoms you would find in 100,000,000 lithium atoms.

46. Using Table 4.2, determine how many ^{7}Li atoms you would find in 1,000 lithium atoms.

47. Using Table 4.2, determine how many ^{26}Mg atoms you would find in 5.00×10^{6} Mg atoms.

48. Using Table 4.2, determine how many ^{14}N atoms you would find in 1.000×10^{14} nitrogen atoms.

49. Using Table 4.2, determine how many ^{13}C atoms you would find in 2.00×10^{4} C atoms.

50. Using Table 4.2, determine how many grams of ^{22}Ne you would find in a sample of neon that has a mass of 10.0 kg.

51. Using Table 4.2, determine how many ^{15}N atoms you would find in 1.0×10^{15} N atoms.

*53. Using Table 4.2, determine how many grams of hydrogen you would need to get $1\bar{0}$ g of ^{2}H.

*55. How many grams of ^{24}Mg would you find in a sample of magnesium that has a mass of 10.0 kg? (Use Table 4.2.)

52. Using Table 4.2, determine how many ^{6}Li atoms you would find in 1,000 atoms of Li.

*54. Using Table 4.2, determine how many ^{14}N atoms you would find in 0.0014 g of N.

*56. Using Table 4.2, determine how many grams of N you would need to get 1.00 g of ^{14}N.

Extra Exercises

57. You have learned that atoms are composed of fundamental particles (electrons, protons, and neutrons), some of which carry an electrical charge. Explain how atoms of an element can be neutral if they consist of charged particles.

58. Consider the following unknown atoms.

$$^{200}_{80}A \qquad ^{208}_{82}B \qquad ^{222}_{86}C \qquad ^{184}_{74}D$$

(a) Which atom has the most electrons in the neutral state?
(b) Which atom has the least neutrons?
(c) What is the mass number of element C?
(d) Which atom has the fewest protons?
(e) Identify element A, using a periodic table.

59. What is the relationship between the atomic number and the number of electrons in a neutral atom?

60. A student produces a beam of *rays* and can't decide whether these *rays* are cathode rays or alpha particles. Explain a simple experiment that the student might perform to determine what these *rays* are.

61. What experiment gave the proof that atoms of all elements contain electrons?

62. Indicate the number of protons, electrons, and neutrons that each of the following atoms contains.
(a) $^{13}_{6}$C   (b) $^{10}_{5}$B   (c) $^{14}_{7}$N

63. What was the problem with Dalton's statement that all the atoms of a particular element are identical in every respect?

64. How many neon-21 atoms are there in one million neon atoms?

65. How many electrons are needed to equal the mass of one proton?

66. Can the atomic number of an element exceed its mass number? Explain.

67. Make a list of some of our modern conveniences that would not function if atoms were indeed indivisible.

68. Referring to Table 4.2, determine all the possible molecular weights for H_2O, using the various isotopes of hydrogen and oxygen.

5 Atomic Theory, Part 2:
Energy Levels and the Bohr Atom

Learning Goals

After you've studied this chapter, you should be able to:

1. Discuss the difference between a line spectrum and a continuous spectrum.
2. Describe the Bohr model of the atom.
3. Explain shells and energy levels.
4. Differentiate between the ground state and an excited state of an atom.
5. Write electron configurations for elements 1 through 20.
6. State what is meant by the octet rule.

Introduction

By the early 1900s, scientists had developed a model of the atom with protons and neutrons in the center, or nucleus, and electrons outside the nucleus. But where, exactly, were the electrons? And why weren't the negative electrons attracted into the nucleus by the positive protons? In 1913 the Danish physicist Niels Bohr proposed a model to answer these and other questions. His model has since been modified in the light of later observations and research, but it still provides an excellent basis for understanding atomic structure. In this chapter we discuss the observations that led Bohr to propose his model, as well as the model itself.

Learning Goal 1
Line spectra and continuous spectra

5.1 Spectra

In the mid-1600s the English scientist, astronomer, and mathematician Sir Isaac Newton allowed a beam of sunlight to pass through a prism. He

Figure 5.1

Sunlight passing through a prism

Robert Bunsen
(Reprinted with permission from Torchbearers of Chemistry, by Henry Monmouth Smith, © Academic Press, Inc., 1949)

found that the prism separated the light into a series of different colors, which we call the visible spectrum (Figure 5.1). His results were published in *Opticks* in 1704. Later experiments showed that a specific amount of energy is associated with each color of light, because light is a form of energy.

In the 1850s the German physicist Robert Kirchhoff and the German chemist Robert Bunsen developed an instrument called the spectroscope. Its main parts were a heat source (the Bunsen burner), a prism, and a telescope (Figure 5.2). Kirchhoff and Bunsen heated samples of various elements. In each case, as an element became hotter and hotter and began to glow, it produced light of its own characteristic color. When the light from the heated element was allowed to pass through the prism, it separated into a series of bright, distinct lines of various colors.

The spectrum produced by sunlight is called a continuous spectrum, because one color merges into the next without any gaps or missing colors. The spectrum that Bunsen and Kirchhoff found is called a line spec-

Figure 5.2

A spectroscope: Element is heated in flame of Bunsen burner. Some light from heated element passes through collimator, which is simply a brass tube with a narrow slit at both ends. The collimator directs a fine beam of light onto the prism. When light passes through the prism, it is broken into its different parts—a spectrum—and observers can see the spectrum through a telescope.

Figure 5.3
(Above) A continuous
spectrum produced by
sunlight. (Below) A line
spectrum produced by
hydrogen.

trum (Figure 5.3), because it is a series of bright lines separated by dark bands. Each of the elements produces its own unique line spectrum.

Scientists find these line spectra very useful for identifying elements in unknown samples. Analysis of line spectra has enabled scientists to identify the elements that compose the sun, the stars, and other extraterrestrial bodies. This technique is also used in chemical laboratories to identify the elements present in unknown samples.

Figure 5.4
The electromagnetic
spectrum

Figure 5.5

Light waves are one form of electromagnetic wave. They consist of an electric field and a magnetic field vibrating at right angles to each other.

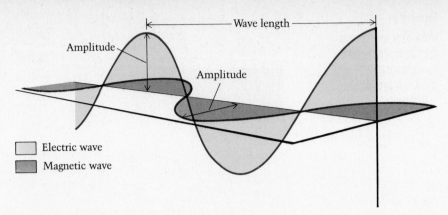

Electric wave

Magnetic wave

The visible spectrum produced by sunlight is only a small part of what is called the electromagnetic spectrum (Figure 5.4). The existence of electromagnetic waves was predicted by the British physicist James Clerk Maxwell in 1864. He assumed that electric and magnetic fields acting together produced radiant energy. He believed that radiant energy took the form of electromagnetic waves and that visible light was only one form of electromagnetic wave (Figure 5.5). Included in the electromagnetic spectrum were both visible waves and invisible waves, each associated with a different amount of energy. For example, x rays are invisible waves with enough energy to pass through skin but not bones. Microwaves are also invisible waves, but they have enough energy to heat food. Electromagnetic waves can travel through the vacuum of space, as well as through air and water.

Maxwell's predictions were verified by the German physicist Heinrich R. Hertz in 1887. He produced electromagnetic waves that were longer than visible light waves by varying an electric charge. This discovery eventually led to the development of radio and television.

The full impact of these experiments with light was not felt until 1913, when it was discovered that the existence of line spectra held the key to the structure of the atom.

5.2 Light as Energy

We have noted that light is a form of energy and that energy can be converted from one form to another. For example, a photoelectric cell converts light energy to electrical energy (in the form of an electric current). Energy may also be absorbed and then re-emitted. For example, the luminous dial on a watch absorbs energy when it is exposed to a strong light. When it is removed from the light source, the dial re-emits this energy in the form of light: It glows.

In the experiments of Kirchhoff and Bunsen, each element was made to absorb energy in the form of heat. When the element got hot enough,

Heat source

Figure 5.6

In the Bohr model of the atom, electrons "jump" from one shell to another when an outside source gives them enough energy to do so.

it emitted this energy as light. But scientists were puzzled about why the elements emitted line spectra rather than continuous spectra. And they wondered what subatomic particle caused the line spectra.

5.3 The Bohr Atom

Early in this century Niels Bohr suggested that electrons are responsible for line spectra. He proposed the idea that electrons travel around the nucleus of the atom in shells, which were described as imaginary spherical surfaces roughly concentric with the nucleus. Bohr also proposed that each shell is associated with a particular energy level, and that shells farther away from the nucleus are associated with higher energy levels. According to this Bohr model, electrons travel in specific shells, just as planets travel in specific orbits around the sun. But electrons, unlike planets, are able to jump from one shell (at one energy level) to another (at a higher energy level) when they absorb enough energy from an outside source to do so (Figure 5.6). They can also fall back to their original shells by emitting this energy.

Look at the hydrogen atom in Figure 5.7. The single electron is moving around the nucleus in the shell closest to the nucleus—the shell associated with the lowest energy. In this situation, the electron is said to be in its *lowest energy level*, and the atom is in its **ground state.** As the electron moves, it neither gains nor loses energy. Its energy of motion exactly counterbalances the attraction of the nucleus. For this reason, it is not pulled into the nucleus.

Now suppose we heat our hydrogen atom. According to Bohr, the electron will jump to a high-energy shell, further from the nucleus, when it has absorbed a certain amount of heat energy. It must absorb the entire amount before it can leave the shell it occupies. It *cannot* absorb some energy, move out toward the next shell, absorb more energy and move closer to that shell, and so on until it reaches the next shell. It must absorb all the energy it needs and then instantaneously jump to the next shell. It may then absorb enough energy at that level to jump to another level (Figure 5.8). Or it may absorb enough energy to jump two or three levels at a time. But in each case it will absorb the energy and jump instantaneously. When an electron is at an energy level above its lowest level, the atom is said to be in an **excited state.** (An electron can also absorb so much energy that it escapes completely from the atom. This process, called *ionization*, is discussed in Chapter 6.)

Eventually, the electron emits some or all of the energy that it absorbed, and it falls back to a lower energy level. The energy is emitted in specific bundles of energy called **quanta** (singular, *quantum*). These quanta are the difference in energy between the higher energy shell where the electron was and the lower energy shell where it lands. If the electron spiraled down without a level-to-level requirement, it would

Figure 5.7
A hydrogen atom in its ground state

Figure 5.8
An excited hydrogen atom

emit a continuous stream of energy. But it moves only from one level to another and emits its energy in specific bursts that may be seen as light energy. It is this light, being emitted by electrons falling back to lower energy levels, that produces the unique line spectrum of hydrogen (or of any other element).

It is hard to visualize electrons moving instantaneously from one energy level to another. Picture a marble rolling down a flight of stairs. The marble can rest on each step (shell), but it can never stop part of the way between steps. And, as the marble falls from step to step, it also loses (emits) energy in bursts.

5.4 Energy Levels

The energy levels in atoms have been given letter names and numbers (Figure 5.9). The farther away a level is from the nucleus, the higher the energy of an electron at that level. The *K* energy level is thus associated with electrons of the lowest energy. And, of the levels shown, the *Q* energy level has electrons of the highest energy. Each energy level can hold only a certain number of electrons at any one time. Table 5.1 gives the theoretical *maximum number* of electrons for each level (or shell).

Figure 5.9
Names of the energy levels in atoms

5.5 Ground-State Electron Configurations of the Elements

The term **electron configuration** refers to the way the electrons fill the various energy levels of the atom. Let's use the Bohr model to look at the configurations of the atoms of some elements in their ground states.

As you saw in Figure 5.7, unexcited hydrogen has its one electron in the *K* level. This seems reasonable; the *K* level is associated with the lowest energy. Both of helium's two electrons are also in the *K* energy level

Table 5.1
Theoretical Maximum Number of Electrons for Each Energy Level

	Energy level	
Letter designation	Number	Maximum number of electrons
K	1	2
L	2	8
M	3	18
N	4	32
O	5	50
P	6	72
Q	7	98

Figure 5.10

Some ground-state electron configurations

(Figure 5.10). Lithium, however, has three electrons. Two of them go in the *K* energy level, but one must go in the *L* energy level, the next-highest level. This is consistent with the fact that the *K* level can hold only two electrons. Atoms with four to ten electrons have two electrons in the *K* level and the remainder in the *L* level. Neon, with ten electrons, fills both the *K* and *L* energy levels (Table 5.2). Sodium, with eleven electrons, has full *K* and *L* levels and one electron in the *M* level.

Table 5.2

Ground-State Electron Configurations of the First 21 Elements

Element	Atomic number	*Electron configuration*			
		K(1)	*L*(2)	*M*(3)	*N*(4)
Hydrogen	1	1			
Helium	2	2			
Lithium	3	2	1		
Beryllium	4	2	2		
Boron	5	2	3		
Carbon	6	2	4		
Nitrogen	7	2	5		
Oxygen	8	2	6		
Fluorine	9	2	7		
Neon	10	2	8		
Sodium	11	2	8	1	
Magnesium	12	2	8	2	
Aluminum	13	2	8	3	
Silicon	14	2	8	4	
Phosphorus	15	2	8	5	
Sulfur	16	2	8	6	
Chlorine	17	2	8	7	
Argon	18	2	8	8	
Potassium	19	2	8	8	1
Calcium	20	2	8	8	2
Scandium	21	2	8	9	2

Table 5.2 shows that—up to element number 18—the electrons fill the energy levels in the order that we might predict. Each level is filled before any electrons are positioned in the next level. But when we write the electron configuration for potassium, $_{19}$K, the pattern changes. We'd expect the ground-state configuration of potassium to be: K, 2 electrons; L, 8 electrons; and M, 9 electrons. But looking at Table 5.2, we see that K contains 2, L contains 8, M contains 8, and N contains 1. Why should this last electron go into the N level when the M level hasn't yet been filled to its capacity? The answer lies in the reasoning behind the *octet rule*.

Learning Goal 5
Electron configurations

5.6 The Octet Rule

Learning Goal 6
Octet rule

You may have noticed that in the atoms we looked at in Table 5.2, there is a tendency for atoms to obtain an energy level with eight electrons. Experiments with helium, neon, argon, krypton, xenon, and radon, which are called the noble gases, show them to be nearly *inert*. That is, they don't react with other elements. With the exception of helium, each of the noble gases has eight electrons in its outermost energy level. (Helium is extremely stable even though it has only two electrons. Its completed K energy level is what makes it so stable.)

It seems that most atoms "attempt" to obtain a stable configuration by achieving eight electrons in the highest energy level. The rule that governs this behavior is called the *octet rule* or the *rule of eight*. Many of the atoms and molecules we will discuss obey the octet rule, though some do not. The electron configuration of calcium, for example, follows the octet rule. Calcium has 20 electrons, distributed as follows: K, 2 electrons; L, 8 electrons; M, 8 electrons; and N, 2 electrons. Even though the M energy level is not completely filled (it could have 18), the last two electrons of calcium go into the N level. However, when we move on to scandium, $_{21}$Sc, other factors come into play, and the M energy level starts to fill to its maximum capacity.

Maybe the octet rule should be called the octet "phenomenon." Like scientific laws, it is something that scientists found out after much study. They discovered that atomic stability is associated with eight electrons in the outer level and simply gave that discovery a name.

5.7 The Importance of Electron Configuration

When we know the electron configurations of elements, we can predict how elements will react with each other.

The number of electrons in the **outermost shell** (the shell farthest away from the nucleus, by position) controls the chemical properties of an element. Elements with similar outer electron configurations behave in very similar ways. For instance, the elements Ne, Ar, Kr, Xe, and Rn are unreactive; in other words, they are almost chemically inert. They all have *eight* electrons in their outermost energy levels. On the other hand, the elements Li, Na, K, Rb, Cs, and Fr all react violently with water; they all have *one* electron in their outermost energy levels.

The electron configurations of the elements also help us predict how some elements will bond to each other. We will discuss that in Chapter 7.

Summary

When a beam of sunlight is passed through a glass prism, the sunlight separates into a rainbow of colors called the visible spectrum. Each color of light is associated with a specific amount of energy. The visible spectrum is a part of the electromagnetic spectrum, and both are continuous spectra—their various parts merge into each other. In the 1850s, Kirchhoff and Bunsen showed that when a sample of an element is heated until it glows, it produces a line spectrum of visible light. Each element produces its own characteristic line spectrum.

In 1913 Niels Bohr proposed a model of the atom to explain these line spectra. The model consisted of shells, or imaginary spherical surfaces, roughly concentric with the nucleus. Each shell was associated with a particular energy level. Bohr postulated that electrons travel around the nucleus of an atom in these shells, just as planets revolve around the sun in their orbits. However, an electron can jump from one shell to another shell, at a higher energy level, when it absorbs enough energy to do so. It can fall back to the original shell by emitting this energy in bursts. It is the separate bursts that show up as line spectra.

The various energy levels of the Bohr atom have been given the letter designations *K* through *Q* or the numbers 1 through 7. (Level *K* or 1 is the lowest-energy level.) Associated with each level is a theoretical maximum electron population. Lower-energy levels are usually filled before electrons are positioned in higher-energy levels. But some elements tend to "complete" a level with only eight electrons when it could accept additional electrons—mainly because a shell with eight electrons is very stable. The electron configurations endow the elements with many of their properties; elements with similar outer electron configurations exhibit similar chemical behavior.

· ·

Career Sketch

Life Scientists Enter the World of Biotechnology

Life scientists look at all aspects of living organisms. They may study a particular animal, a particular plant, or a specific function. They also investigate evolution and the mechanics of heredity. Botanists study plants, zoologists study animals, and microbiologists study microorganisms.

Life scientists must be familiar with many research techniques and complex laboratory equipment, such as this transmission electron microscope (TEM).
(Photograph provided courtesy of Philips Medical Systems, Inc.)

Recently a new and exciting field emerged in the life sciences. It is called genetic engineering. Using advances in biotechnology, researchers are able to transfer hereditary material found within cells to make a whole range of important chemical substances available to us. For example, researchers in California have used genetic engineering to produce a drug that is reported to be a safe and effective means of dissolving blood clots in the coronary arteries of heart attack victims.

Life scientists must be familiar with research techniques and must be able to use a variety of different pieces of laboratory equipment, including the electron microscope and the computer. Twenty-five percent of life scientists teach, and the same number are involved in management and administration. Some work as consultants to business firms or government. Others write for technical publications or test and inspect foods. Some work in technical sales and services. The qualifications for these various positions range from a baccalaureate degree to a Ph.D.

. . . .

Self-Test Exercises

The self-test exercises are arranged in matched pairs. The first exercise of each pair appears in the left column; the exercises whose numbers are in color are answered in the back of the book. Answers to exercises in the right column are not given. Difficult problems are marked with asterisks.

Learning Goal 1: Line Spectra and Continuous Spectra

1. What is a continuous spectrum? a line spectrum?

2. What is a spectrum? What is the difference between the electromagnetic spectrum and the visible spectrum?

3. How could a spectroscope help scientists analyze the sun and other extraterrestrial bodies?

4. Why must an element be heated to glowing before it will exhibit its characteristic spectrum?

5. Can you discover the origin of the word *helium*? [*Hint:* See the *Handbook of Chemistry and Physics.*]

6. Do you think line spectra could be used to identify compounds? Explain.

Learning Goal 2: Bohr Model of the Atom

7. Compare the Rutherford, Thomson, and Bohr models of the atom.

8. Define the terms *ground state* and *excited state*.

9. What is Bohr's most important contribution to our ideas about the structure of the atom?

10. According to the Bohr model of the atom, do electrons "spiral" or "jump" from one energy level to another? What type of spectrum would be produced by each of these types of movements by electrons?

Learning Goal 3: Shells and Energy Levels

11. What subatomic particles produce the line spectrum of an element?

12. Complete this sentence with the word "higher" or "lower." The farther away an energy level is from the nucleus, the _____ the energy of an electron in that level.

13. What is meant by shells and energy levels? How many shells are known to exist in atoms? State their letter and number designations.

14. What is the theoretical maximum number of electrons that can populate each of the first three energy levels?

Learning Goal 4: Difference Between Ground State and Excited State of an Atom

15. What do we mean by the ground state of an atom? the excited state of an atom?

16. Describe the ground-state configuration of a lithium atom. What changes would take place if the lithium atom were in an excited state?

17. If a neutral hydrogen atom has an electron in the L energy level, is the atom in its ground state or an excited state?

18. A particular neutral magnesium atom has two electrons in the K shell, eight electrons in the L shell, and two electrons in the M shell. Would you characterize this atom as in the ground state or an excited state?

Learning Goal 5: Electron Configurations

19. Without using the periodic table, find the electron configurations of the following elements. State which ones are chemically similar. (Don't use Table 5.2 either!) (*a*) $_{11}Na$ (*b*) $_{13}Al$ (*c*) $_8O$ (*d*) $_{16}S$ (*e*) $_5B$ (*f*) $_3Li$

20. Without using the periodic table, find the electron configurations of the following elements, and state which ones are chemically similar. (*a*) $_1H$ (*b*) $_3Li$ (*c*) $_4Be$ (*d*) $_{12}Mg$ (*e*) $_{20}Ca$

21. Determine how many electrons there are in the outermost energy level of a neutral atom of:
(*a*) sodium   (*b*) carbon   (*c*) nitrogen

22. Determine how many electrons there are in the outermost energy level of a neutral atom of:
(*a*) phosphorus   (*b*) fluorine   (*c*) beryllium

23. Name the elements that have the following neutral electron configurations. (You may use the periodic table.)
(*a*) *K*, 2 electrons; *L*, 5 electrons.
(*b*) *K*, 2 electrons; *L*, 8 electrons; and *M*, 2 electrons.
(*c*) *K*, 2 electrons; *L*, 8 electrons; and *M*, 7 electrons.
(*d*) *K*, 2 electrons; *L*, 8 electrons; and *M*, 8 electrons.
(*e*) *K*, 2 electrons; *L*, 8 electrons; *M*, 8 electrons; and *N*, 2 electrons.

24. Using the periodic table, name the elements that have the following ground state electron configurations.
(*a*) *K*, 2 electrons; *L*, 1
(*b*) *K*, 2; *L*, 7
(*c*) *K*, 2; *L*, 8; and *M*, 3

Learning Goal 6: Octet Rule

25. State what is meant by the octet rule.

26. Complete this sentence. The number of electrons in the _____ shell determines the chemical properties of an element.

27. The elements in Group VIIIA of the periodic table (this group is the vertical column of elements farthest to the right in the periodic table) have eight electrons in their outermost energy level. (The exception to this is helium, which has only two electrons in its outermost energy level.) What is *chemically* unique about the Group VIIIA elements.?

28. Separate the following into groups of elements that exhibit similar chemical behavior.
(*a*) Li   (*b*) Ne   (*c*) Ar   (*d*) K   (*e*) Rb   (*f*) Kr
(*g*) Na   (*h*) Cs   (*i*) Rn   (*j*) Xe

Extra Exercises

29. Write electron configurations for all the elements with atomic numbers between 1 and 20 whose behavior is explained by the octet rule (in other words, those that have a stable octet of electrons).

30. Write electron configurations for all elements with atomic numbers between 1 and 20 that are isoelectronic with magnesium. [*Hint:* "Isoelectronic with" in this instance can be taken to mean "has the same number of electrons in the outermost energy level as."]

31. What was the reasoning behind replacement of the Rutherford model of the atom with Bohr's model?

32. What is the distinguishing feature of the electron configuration of the following groups?
(*a*) Noble gases (Group VIIIA)
(*b*) Group IIA elements
(*c*) Group VIIA elements (also called the *halogens*)

33. Explain why the line spectrum of an element is sometimes referred to as its *fingerprint*.

34. Explain briefly how electrons produce a line spectrum.

35. In Chapter 4 you learned that not all atoms of an element need be identical, and that these different atoms are called isotopes. Do isotopes of the same element produce the same line spectrum? Explain.

36. Show by writing electron configurations that the following elements should be in the indicated groups.
(*a*) C (Group IVA)   (*b*) N (Group VA)
(*c*) S (Group VIA)

37. In this chapter we said that a marble rolling down a flight of stairs is similar to an electron jumping from one energy level to the next. Can you think of other analogous phenomena?

38. What can happen to an electron in an atom when it absorbs too much energy while being excited?

39. What would be the atomic number of an element whose energy levels *K, L, M, N, O, P,* and *Q* were filled with the maximum number of electrons?

6 The Periodic Table: Keeping Track of the Elements

Learning Goals

After you've studied this chapter, you should be able to:

1. Discuss the historical basis and significance of the periodic table.
2. Distinguish between a period and a group in the periodic table.
3. State the difference between an A-group (representative) element and a B-group element (transition metal).
4. Predict trends for properties such as atomic radius, ionization potential, and electron affinity.
5. Explain the significance of energy sublevels and their effect on the behavior of an element.
6. Write the electron configurations for various elements, using the filling-order diagram.
7. Define and discuss atomic orbitals.
8. Draw shapes for the *s* and *p* orbitals.
9. Discuss the quantum-mechanical model of the atom.

Introduction

In preceding chapters, we referred to the periodic table for information about the elements, such as atomic number and weight. But the periodic table contains many more facts that can be useful to us. In this chapter we trace the development of the periodic table and see what kinds of additional information it holds. We also discuss a refinement of the Bohr model of the atom—the presently accepted quantum-mechanical model.

6.1 History

Learning Goal 1
Historical basis and
significance of periodic table

From the time of the ancient Greeks up to 1700, only 14 elements were known. Then, in a short span of 10 years, from 1800 to 1810, 14 more elements were discovered. By 1830, 45 elements were known. As new elements were discovered, chemists probably began to feel insecure. All these elements had different properties, and there didn't seem to be any relationship among them. And chemists wondered how many elements actually existed. To answer this question, they had to find some relationship among the known elements.

The first chemist to notice some order among the elements was the German scientist Johann Döbereiner. He published an account of his observations in 1829. It occurred to Döbereiner that bromine had chemical and physical properties somewhere between those of chlorine and iodine and that bromine's atomic weight was almost midway between those of chlorine and iodine (Figure 6.1). Could this be nothing more than coincidence?

Döbereiner searched and found two more groups of similar elements. The first was made up of calcium (Ca), strontium (Sr), and barium (Ba); the second included sulfur (S), selenium (Se), and tellurium (Te) (Figure 6.2). He called these groups *triads* (threes).

In 1864 the English chemist John Newlands tried to arrange all the known elements in order of increasing atomic weight. He came upon the

Element	Color	State	M.P.	B.P.	Atomic weight	Chemical symbol
Chlorine	Green-yellow	Gas	−101°C	−34.6°C	35.5	Cl
Bromine	Brown	Liquid	−7.2°C	58.8°C	79.9	Br
Iodine	Purple	Solid	113.5°C	184.4°C	126.9	I

Figure 6.1
Some similar elements

Figure 6.2
More similar elements

	Color	State	M.P. (°C)	B.P. (°C)	Atomic wt., amu
Ca	Silver-white	Solid	842.8	1487	40.08
Sr	Silver-white to pale yellow	Solid	769	1384	87.62
Ba	Yellow-silvery	Solid	725	1140	137.34
S	Yellow	Solid	113	444	32.06
Se	Blue-gray	Solid	217	685	98.96
Te	Silver-white	Solid	452	1390	127.6

Dmitri Mendeleev (1834–1907). In 1869 he predicted the chemical and physical properties of the element gallium, which he called eka-aluminum.
(Reprinted with permission from Torchbearers of Chemistry, by Henry Monmouth Smith, © Academic Press, Inc., 1949)

idea of arranging them in vertical columns. Because he noticed that the eighth element had chemical and physical properties that were similar to those of the first, he let the eighth element start a new column (Figure 6.3). Newlands called his arrangement the *Law of Octaves*. But there were many places in his arrangement where dissimilar elements were next to each other.

In 1869 the German chemist Julius Lothar Meyer devised an incomplete periodic table consisting of 56 elements. In the same year the Russian chemist Dmitri Mendeleev (pronounced "Menduh-**lay**-eff") arranged the elements in order of increasing atomic weights. In cases where discrepancies arose, *the chemical properties were used to place the elements.* Mendeleev established horizontal rows, or *periods*. Hydrogen by itself made up the first period. Each of the next two periods contained seven elements. The periods after that contained more than seven elements. This is the point at which Mendeleev's table differed from Newlands's table.

Increasing atomic weight							
Li	← Similar →	Na	← Similar →	K
Be	← Similar →	Mg	← Similar →	Ca
B	← Similar →	Al	← Not similar →	Cr
C	← Similar →	Si	← Not similar →	Ti
N	← Similar →	P	← Not similar →	Mn
O	← Similar →	S	← Not similar →	Fe
F	← Similar →	Cl	← Not similar →	Co

Figure 6.3
Newland's Law of Octaves

Figure 6.4

Mendeleev's original periodic table arranged in modern form

Group

Period	I	II	III	IV	V	VI	VII	VIII		
1	1 H							2		
2	3 Li	4 Be	5 B	6 C	7 N	8 O	9 F	10		
3	11 Na	12 Mg	13 Al	14 Si	15 P	16 S	17 Cl	18		
4	19 K	20 Ca	21	22 Ti	23 V	24 Cr	25 Mn	26 Fe	27 Co	28 Ni
4	29 Cu	30 Zn	31	32	33 As	34 Se	35 Br	36		
5	37 Rb	38 Sr	39 Y	40 Zr	41 Nb	42 Mo	43	44 Ru	45 Rh	46 Pd
5	47 Ag	48 Cd	49 In	50 Sn	51 Sb	52 Te	53 I	54		
6	55 Cs	56 Ba	57 La	72	73 Ta	74 W	75	76 Os	77 Ir	78 Pt
6	79 Au	80 Hg	81 Tl	82 Pb	83 Bi	84	85	86		
7	87	88	89	104	105					

58 Ce	59	60	61	62	63	64	65 Tb	66	67	68 Er	69	70	71
90 Th	91	92 U	93	94	95	96	97	98	99	100	101	102	103

Paul Emile Lecoq de Boisbaudran (1838–1912). In 1875 he discovered and named the element gallium. (Reprinted with permission from Torchbearers of Chemistry, by Henry Monmouth Smith, © Academic Press, Inc., 1949)

In arranging his table, Mendeleev occasionally put heavier elements before lighter ones to keep elements with the same chemical properties in the same vertical column, or **group** (Figure 6.4). Sometimes he left open spaces in the table, where he reasoned that unknown elements should go. This was the case with gallium. In 1869, when he made up his table, the element gallium was unknown; however, Mendeleev *predicted* its existence. He based his prediction on the properties of aluminum (which appeared directly above gallium in the table). Mendeleev even went so far as to predict the melting point, boiling point, and atomic weight of the then-unknown gallium, which he called eka-aluminum. Six years later, while analyzing zinc ore, the French chemist Lecoq de Boisbaudran discovered the element gallium. Its properties were almost identical to those Mendeleev had predicted (Figure 6.5).

Figure 6.5
The predicted element called eka-aluminum and the discovered element, gallium, that it turned out to be

	Eka-aluminum	Gallium
Atomic weight →	Predicted:68	Observed: 69.9
Oxide formula →	Predicted: E_2O_3	Observed: Ga_2O_3
Oxide in acid →	Predicted: should dissolve in acid	Observed: dissolves in acid
Salt formation →	Predicted: should form salts as EX_3	Observed: forms salts as GaX_3

The periodic tables proposed by both Mendeleev and Meyer were based on increasing atomic weights. Mendeleev used similarities among chemical properties to account for discrepancies among elements. Certain irregularities in these tables were corrected by the work of Henry G. J. Moseley, a British physicist. Moseley was able to determine the nuclear charge on the atoms of known elements. He concluded that elements should be arranged by increasing atomic number rather than by atomic weight. This research led to the development of the *periodic law,* which states that the chemical properties of the elements are periodic functions of their atomic numbers. Elements with similar chemical properties recur at regular intervals, and they are placed accordingly on the periodic table.

6.2 The Modern Periodic Table

Learning Goal 2
Periods and groups in the periodic table

The periodic table of today is similar to Mendeleev's but has many more elements—those that have been discovered since 1869. Today's table consists of seven horizontal rows called **periods** and a number of vertical columns called **groups** (or **families**) (Table 6.1).

The groups are numbered with roman numerals. All the elements in each group have the same number of electrons in their outermost shells, so they all behave similarly. For example, the Group IA elements react violently when they come into contact with water. And all elements in Group IA have one electron in their outermost shell.

Learning Goal 3
A-group and B-group elements

Some of the groups in the periodic table are labeled with a roman numeral followed by A, others with a roman numeral followed by B. The A groups are called the *representative elements.* The B groups are called the *transition metals.* You can also see that as we move from the top to the bottom of the periodic table (in other words, from period 1 to period 7), the periods get larger—they have more elements in them. In fact, periods 6 and 7 are so large that in order to fit the table on one page, we have to write part of each period *below* the rest of the table (the 15 lanthanides and the 15 actinides).

Table 6.1
Periodic Table of the Elements

Group

Period	IA	IIA											IIIA	IVA	VA	VIA	VIIA	VIIIA
1	1.00794 H 1																	4.002602 He 2
2	6.941 Li 3	9.01218 Be 4			Transition elements								10.811 B 5	12.011 C 6	14.0067 N 7	15.9994 O 8	18.998403 F 9	20.179 Ne 10
3	22.98977 Na 11	24.305 Mg 12	IIIB	IVB	VB	VIB	VIIB	VIIIB			IB	IIB	26.98154 Al 13	28.0855 Si 14	30.97376 P 15	32.066 S 16	35.453 Cl 17	39.948 Ar 18
4	39.0983 K 19	40.078 Ca 20	44.9559 Sc 21	47.88 Ti 22	50.9415 V 23	51.9961 Cr 24	54.9380 Mn 25	55.847 Fe 26	58.9332 Co 27	58.69 Ni 28	63.546 Cu 29	65.38 Zn 30	69.723 Ga 31	72.59 Ge 32	74.9216 As 33	78.96 Se 34	79.904 Br 35	83.80 Kr 36
5	85.4678 Rb 37	87.62 Sr 38	88.9059 Y 39	91.22 Zr 40	92.9064 Nb 41	95.94 Mo 42	(98) Tc 43	101.07 Ru 44	102.9055 Rh 45	106.42 Pd 46	107.8682 Ag 47	112.41 Cd 48	114.82 In 49	118.710 Sn 50	121.75 Sb 51	127.60 Te 52	126.9045 I 53	131.29 Xe 54
6	132.9054 Cs 55	137.33 Ba 56	La–Lu 57–71	178.49 Hf 72	180.9479 Ta 73	183.85 W 74	186.207 Re 75	190.2 Os 76	192.22 Ir 77	195.08 Pt 78	196.9665 Au 79	200.59 Hg 80	204.383 Tl 81	207.2 Pb 82	208.9804 Bi 83	(209) Po 84	(210) At 85	(222) Rn 86
7	(223) Fr 87	(226) Ra 88	Ac–Lr 89–103	(261) Unq 104	(262) Unp 105	(263) Unh 106	(262) Uns 107	(265) Uno 108	(267) Une 109									

Lanthanides

138.9055 La 57	140.12 Ce 58	140.9077 Pr 59	144.24 Nd 60	(145) Pm 61	150.36 Sm 62	151.96 Eu 63	157.25 Gd 64	158.9254 Tb 65	162.50 Dy 66	164.9304 Ho 67	167.26 Er 68	168.9342 Tm 69	173.04 Yb 70	174.967 Lu 71

Actinides

(227) Ac 89	232.0381 Th 90	(231) Pa 91	238.0289 U 92	(237) Np 93	(244) Pu 94	(243) Am 95	(247) Cm 96	(247) Bk 97	(251) Cf 98	(252) Es 99	(257) Fm 100	(258) Md 101	(259) No 102	(260) Lr 103

```
1.00794  ← Atomic weight
   H     ← Symbol
   1     ← Atomic number
```

Metal   Metalloid   Nonmetal

6.3 Periodic Trends

Learning Goal 4
Periodic trends

When we examine the periodic table, we can see certain trends in the properties of the elements. For example, the elements in Group IA (at the left) are all **metals.** As we move across to the right, we continue to encounter metals through more than half the width of the table. Then we reach elements that are not metals but that have some properties like those of metals and some properties like those of nonmetals. These are the **metalloids,** which were mentioned earlier. As we move still further to the right, we finally reach the **nonmetals.** (See Section 3.9 for a review of these terms.)

Other trends that show up in the periodic table have to do with electron configuration, atomic radius, the tendency to lose or to gain electrons, and some physical properties. In the remainder of this chapter, we discuss these trends as well as the properties themselves and their structural basis.

6.4 Periodicity and Electron Configuration

All the elements within each A group have the same number of electrons in their outermost shell. Moreover, for the A-group elements, *the group number tells us how many outer electrons there are.* For example, all the elements in Group VIA have six electrons in their outermost energy level. All the elements in Group IA have one electron in their outermost energy level.

For the B-group elements, there is not necessarily any connection between group number and the number of electrons in the outermost energy level. For example, Group IVB elements *do not* have four electrons in their outermost energy level.

6.5 Energy Sublevels

Learning Goal 5
Energy sublevels and
chemical properties

Electron configuration is *the* factor that determines how an atom behaves chemically, so let us look deeper into the make-up of an energy level. If we do this, we find that each energy level can be divided into *sublevels*. (Scientists have developed models for describing these energy sublevels. We will cover them in the final section of this chapter.)

To visualize the idea of energy sublevels, think of a tall office building with the unusual architectural feature depicted in Figure 6.6: The "build-

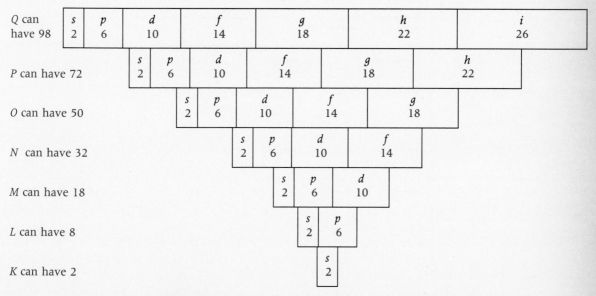

Figure 6.6
Energy levels and sublevels, showing maximum numbers of electrons in each

ing" is upside down. Each floor of the building represents an energy level (K, L, M, and so on). Offices on each floor represent the energy sublevels. Each floor is a different size, so each has a different number of offices. In the same way, each energy level has a different number of sublevels. The sublevels are represented by the letters s, p, d, f, g, h, and i. The K energy level (the ground floor) has only one sublevel, the s sublevel. The L energy level has two, the s and p sublevels. And so on.

Each office on a floor of our "energy-level building" is a different size, and can hold only a certain number of desks. That is, each sublevel is a different size and can hold only a certain number of electrons. The s sublevel can hold a maximum of two electrons and the p sublevel a maximum of six. Figure 6.6 shows the maximum number of electrons that each sublevel can hold. In Chapter 5 we discussed the maximum number of electrons for each energy level: K can hold 2; L, 8; M, 18; N, 32; O, 50; P, 72; and Q, 98. Now we see that the total number of electrons that a

Figure 6.7

How energy levels overlap

main energy level can hold is simply the sum of the numbers of electrons that all of its *sublevels* can hold.

Keeping Figure 6.6 in mind, we can now write the electron configurations of the elements as we did in Chapter 5, only this time we can be more exact, because we can use *sub*levels. To make it easier to write these configurations, we will represent the *K* energy level by the number 1, the *L* energy level by the number 2, the *M* level by the number 3, and so on. Then, for example, a 2*p* electron is an electron in the *p* sublevel of level 2 (or *L*).

Learning Goal 6
Electron configurations

In Chapter 5 we wrote electron configurations by filling the lowest energy levels first. Up to element number 18, this procedure worked well. But for elements above number 18, that method didn't produce the right configuration. We discussed the octet rule, but even that couldn't get us past element 20. We attributed the problem to "other factors." What we didn't take into account was the *overlapping of energy levels*, shown in Figure 6.7. For example, the 4*s* sublevel has a lower energy than the 3*d* sublevel. Therefore the 4*s* sublevel fills with electrons before the 3*d* sublevel does—even though the 4*s* sublevel is farther away from the nucleus than the 3*d* sublevel. And the 5*s* sublevel fills with electrons before the 4*d* and 4*f* sublevels, for the same reason.

Now you can write the electron configuration for an element with any atomic number—by simply filling the lowest energy *sublevel* first. The filling-order diagram in Figure 6.8 will help you do this. All you have to do is follow the diagonal arrows, starting at the upper left. That is, the 1*s* sublevel is filled first, then the 2*s* sublevel, followed by the 2*p*, 3*s*, 3*p*, 4*s*, 3*d*, and 4*p* sublevels, and so on. (Unfortunately, even using Figure 6.8, you will not be able to predict the electron configurations for all the elements. There are a few exceptions.)

Figure 6.8
The order in which electrons fill energy sublevels

Here are two examples of the notation we shall use to write electron configuration:

Electrons in sublevel

$_1$H: $\mathbf{1s^1}$ $_2$He: $\mathbf{1s^2}$

Main energy level   Sublevel

Note that the superscript to the right of the sublevel letter tells us the number of electrons in that sublevel. Here are some more examples:

$$_3\text{Li: } 1s^2\, 2s^1$$

$$_4\text{Be: } 1s^2\, 2s^2$$

$$_5\text{B: } 1s^2\, 2s^2\, 2p^1$$

$$_6\text{C: } 1s^2\, 2s^2\, 2p^2$$

$$_{18}\text{Ar: } 1s^2\, 2s^2\, 2p^6\, 3s^2\, 3p^6$$

$$_{19}\text{K: } 1s^2\, 2s^2\, 2p^6\, 3s^2\, 3p^6\, 4s^1$$

$$_{20}\text{Ca: } 1s^2\, 2s^2\, 2p^6\, 3s^2\, 3p^6\, 4s^2$$

$$_{21}\text{Sc: } 1s^2\, 2s^2\, 2p^6\, 3s^2\, 3p^6\, 4s^2\, 3d^1$$

Because the $4s$ sublevel is lower in energy than the $3d$ sublevel, the last electron in potassium (K) is in the $4s$ sublevel and not the $3d$ sublevel (check Figure 6.8 again). The same is true of calcium (Ca). But beginning with scandium (Sc), the $3d$ sublevel starts to fill. This is because the $4s$ sublevel is now completely filled, and the $3d$ sublevel is now the lowest unfilled energy sublevel. Sublevels continue to fill in this manner. Table 6.2 gives the resulting electron configurations for all the elements.

6.6 Similarities Among Elements in a Group and Period

The configuration of the outermost energy level determines the chemical behavior of an element. Thus all elements in the same group should behave (react) similarly, because they all have the same outer-shell electron configuration. But what about elements in the same period? In general, the elements in any period have different outer-shell electron configurations, so these elements have different chemical properties.

For example, in period 3, Al (with two $3s$ electrons and one $3p$ electron) is a metal, whereas Cl (with two $3s$ electrons and five $3p$ electrons) is a nonmetal. However, this is not so for the transition metals in periods 4 through 7. Consider the elements Mn, Fe, Co, and Ni in period 4. Their electron configurations differ because they have different numbers of electrons in their $3d$ sublevels. However, level 3 is *not* their outermost energy level. The outermost level for these transition metals is the $4s$ sublevel. We would therefore expect these elements to exhibit somewhat

Table 6.2
Electron Configuration of the Elements

Atomic number	Element	1 s	2 s	2 p	3 s	3 p	3 d	4 s	4 p	4 d	4 f	5 s	5 p	5 d	5 f	6 s	6 p	6 d	6 f	7 s
1	H	1																		
2	He	2																		
3	Li	2	1																	
4	Be	2	2																	
5	B	2	2	1																
6	C	2	2	2																
7	N	2	2	3																
8	O	2	2	4																
9	F	2	2	5																
10	Ne	2	2	6																
11	Na	2	2	6	1															
12	Mg	2	2	6	2															
13	Al	2	2	6	2	1														
14	Si	2	2	6	2	2														
15	P	2	2	6	2	3														
16	S	2	2	6	2	4														
17	Cl	2	2	6	2	5														
18	Ar	2	2	6	2	6														
19	K	2	2	6	2	6		1												
20	Ca	2	2	6	2	6		2												
21	Sc	2	2	6	2	6	1	2												
22	Ti	2	2	6	2	6	2	2												
23	V	2	2	6	2	6	3	2												
24	Cr	2	2	6	2	6	5	1												
25	Mn	2	2	6	2	6	5	2												
26	Fe	2	2	6	2	6	6	2												
27	Co	2	2	6	2	6	7	2												
28	Ni	2	2	6	2	6	8	2												
29	Cu	2	2	6	2	6	10	1												
30	Zn	2	2	6	2	6	10	2												
31	Ga	2	2	6	2	6	10	2	1											
32	Ge	2	2	6	2	6	10	2	2											
33	As	2	2	6	2	6	10	2	3											
34	Se	2	2	6	2	6	10	2	4											
35	Br	2	2	6	2	6	10	2	5											
36	Kr	2	2	6	2	6	10	2	6											

(cont.)

Table 6.2
Electron Configuration of the Elements (Continued)

Atomic number	Element	1	2		3			4				5				6				7
		s	*s*	*p*	*s*	*p*	*d*	*s*	*p*	*d*	*f*	*s*	*p*	*d*	*f*	*s*	*p*	*d*	*f*	*s*
37	Rb	2	2	6	2	6	10	2	6			1								
38	Sr	2	2	6	2	6	10	2	6			2								
39	Y	2	2	6	2	6	10	2	6	1		2								
40	Zr	2	2	6	2	6	10	2	6	2		2								
41	Nb	2	2	6	2	6	10	2	6	4		1								
42	Mo	2	2	6	2	6	10	2	6	5		1								
43	Tc	2	2	6	2	6	10	2	6	6		1								
44	Ru	2	2	6	2	6	10	2	6	7		1								
45	Rh	2	2	6	2	6	10	2	6	8		1								
46	Pd	2	2	6	2	6	10	2	6	10										
47	Ag	2	2	6	2	6	10	2	6	10		1								
48	Cd	2	2	6	2	6	10	2	6	10		2								
49	In	2	2	6	2	6	10	2	6	10		2	1							
50	Sn	2	2	6	2	6	10	2	6	10		2	2							
51	Sb	2	2	6	2	6	10	2	6	10		2	3							
52	Te	2	2	6	2	6	10	2	6	10		2	4							
53	I	2	2	6	2	6	10	2	6	10		2	5							
54	Xe	2	2	6	2	6	10	2	6	10		2	6							
55	Cs	2	2	6	2	6	10	2	6	10		2	6			1				
56	Ba	2	2	6	2	6	10	2	6	10		2	6			2				
57	La	2	2	6	2	6	10	2	6	10		2	6	1		2				
58	Ce	2	2	6	2	6	10	2	6	10	2	2	6			2				
59	Pr	2	2	6	2	6	10	2	6	10	3	2	6			2				
60	Nd	2	2	6	2	6	10	2	6	10	4	2	6			2				
61	Pm	2	2	6	2	6	10	2	6	10	5	2	6			2				
62	Sm	2	2	6	2	6	10	2	6	10	6	2	6			2				
63	Eu	2	2	6	2	6	10	2	6	10	7	2	6			2				
64	Gd	2	2	6	2	6	10	2	6	10	7	2	6	1		2				
65	Tb	2	2	6	2	6	10	2	6	10	9	2	6			2				
66	Dy	2	2	6	2	6	10	2	6	10	10	2	6			2				
67	Ho	2	2	6	2	6	10	2	6	10	11	2	6			2				
68	Er	2	2	6	2	6	10	2	6	10	12	2	6			2				
69	Tm	2	2	6	2	6	10	2	6	10	13	2	6			2				
70	Yb	2	2	6	2	6	10	2	6	10	14	2	6			2				
71	Lu	2	2	6	2	6	10	2	6	10	14	2	6	1		2				
72	Hf	2	2	6	2	6	10	2	6	10	14	2	6	2		2				

(cont.)

Table 6.2
Electron Configuration of the Elements (Continued)

Atomic number	Element	1	2		3			4				5				6				7
		s	s	p	s	p	d	s	p	d	f	s	p	d	f	s	p	d	f	s
73	Ta	2	2	6	2	6	10	2	6	10	14	2	6	3		2				
74	W	2	2	6	2	6	10	2	6	10	14	2	6	4		2				
75	Re	2	2	6	2	6	10	2	6	10	14	2	6	5		2				
76	Os	2	2	6	2	6	10	2	6	10	14	2	6	6		2				
77	Ir	2	2	6	2	6	10	2	6	10	14	2	6	7		2				
78	Pt	2	2	6	2	6	10	2	6	10	14	2	6	9		1				
79	Au	2	2	6	2	6	10	2	6	10	14	2	6	10		1				
80	Hg	2	2	6	2	6	10	2	6	10	14	2	6	10		2				
81	Tl	2	2	6	2	6	10	2	6	10	14	2	6	10		2	1			
82	Pb	2	2	6	2	6	10	2	6	10	14	2	6	10		2	2			
83	Bi	2	2	6	2	6	10	2	6	10	14	2	6	10		2	3			
84	Po	2	2	6	2	6	10	2	6	10	14	2	6	10		2	4			
85	At	2	2	6	2	6	10	2	6	10	14	2	6	10		2	5			
86	Rn	2	2	6	2	6	10	2	6	10	14	2	6	10		2	6			
87	Fr	2	2	6	2	6	10	2	6	10	14	2	6	10		2	6			1
88	Ra	2	2	6	2	6	10	2	6	10	14	2	6	10		2	6			2
89	Ac	2	2	6	2	6	10	2	6	10	14	2	6	10		2	6	1		2
90	Th	2	2	6	2	6	10	2	6	10	14	2	6	10		2	6	2		2
91	Pa	2	2	6	2	6	10	2	6	10	14	2	6	10	2	2	6	1		2
92	U	2	2	6	2	6	10	2	6	10	14	2	6	10	3	2	6	1		2
93	Np	2	2	6	2	6	10	2	6	10	14	2	6	10	4	2	6	1		2
94	Pu	2	2	6	2	6	10	2	6	10	14	2	6	10	6	2	6			2
95	Am	2	2	6	2	6	10	2	6	10	14	2	6	10	7	2	6			2
96	Cm	2	2	6	2	6	10	2	6	10	14	2	6	10	7	2	6	1		2
97	Bk	2	2	6	2	6	10	2	6	10	14	2	6	10	8	2	6	1		2
98	Cf	2	2	6	2	6	10	2	6	10	14	2	6	10	10	2	6			2
99	Es	2	2	6	2	6	10	2	6	10	14	2	6	10	11	2	6			2
100	Fm	2	2	6	2	6	10	2	6	10	14	2	6	10	12	2	6			2
101	Md	2	2	6	2	6	10	2	6	10	14	2	6	10	13	2	6			2
102	No	2	2	6	2	6	10	2	6	10	14	2	6	10	14	2	6			2
103	Lr	2	2	6	2	6	10	2	6	19	14	2	6	10	14	2	6	1		2
104	Unq	2	2	6	2	6	10	2	6	10	14	2	6	10	14	2	6	2		2
105	Unp	2	2	6	2	6	10	2	6	10	14	2	6	10	14	2	6	3		2
106	Unh	2	2	6	2	6	10	2	6	10	14	2	6	10	14	2	6	4		2
107	Uns	2	2	6	2	6	10	2	6	10	14	2	6	10	14	2	6	5		2
108	Uno	2	2	6	2	6	10	2	6	10	14	2	6	10	14	2	6	6		2
109	Une	2	2	6	2	6	10	2	6	10	14	2	6	10	14	2	6	7		2

similar behavior, and they do. Moreover, the same is true of the transition elements in periods 5 through 7.

To summarize:

1. Elements in any one group behave in similar ways, because they have *identical electron configurations in the outermost shell.*

2. Elements in any one period generally do not behave similarly.

3. However, because of the manner in which electrons fill their shells (filling some inner shells after outer shells), the transition metals—that is, the B groups—show somewhat similar chemical behavior.

6.7 Atomic Radius

The periodic table is helpful in predicting the relative sizes of atoms within any period or group. We shall consider the **atomic radius** (radius of an atom) to be the distance from the center of the nucleus to the outermost electron. Atomic radii are measured with a unit called the *angstrom*, named after a Swedish physicist. One angstrom (1 Å) is equal to 10^{-10} m, or one hundred-millionth of a centimeter.

In any group in the periodic table, each element has one more energy level than the element above it (Figure 6.9). Therefore, as we move down through a particular group, the atomic radii of the elements should increase.

But what happens to the atomic radii of the elements as we move from left to right across a period? You might assume that, because the number of electrons in the atom is increasing, the size of the atom should also increase. This is not the case. Remember that, as we move across a period, the number of protons increases along with the number of electrons. Therefore there is a greater pull on the electrons of the inner energy levels, which usually causes a shrinking of the atomic radius (Figure 6.10).

Using the elements in period 2, we can explain what happens. In lithium (atomic number 3), three protons attract two K electrons and one L electron. In beryllium (atomic number 4), four protons attract two K

Figure 6.9

Relative sizes of some group IIA elements

Beryllium, Be   Magnesium, Mg   Calcium, Ca

Figure 6.10

The atomic radii of the elements decrease as we move from left to right across a period. The increased nuclear charge results in a greater pull on the electrons of the inner energy levels, so the atom shrinks.

Lithium, Li
(Atomic radius
1.23 Å)

Beryllium, Be
(Atomic radius
0.89 Å)

Boron, B
(Atomic radius
0.82 Å)

Carbon, C
(Atomic radius
0.77 Å)

electrons and two *L* electrons. In boron, five protons attract two *K* electrons and three *L* electrons. This increased positive charge acting on the *K* electrons makes the *K* energy level shrink. The *L* energy level also shrinks, and this results in a decreased atomic radius (Figure 6.11).

Also keep in mind that for the transition metals, atomic radii change very little because electrons are filling the *d* and *f* subshells. Table 6.3 shows the atomic radii of all the elements.

6.8 Ionization Potential

When an electron is pulled completely away from a neutral atom, what remains behind is a positively charged particle called an *ion*. A **positive ion** (also called a *cation*), is *an atom (or group of atoms) that has lost one or more electrons*. A **negative ion** (also called an *anion*) is *an atom (or group of atoms) that has acquired one or more electrons*.

The process of removing an electron from a neutral atom is called *ionization*. Here is what happens when a neutral atom of potassium becomes ionized:

$$K \longrightarrow K^{1+} + 1e^{1-}$$

The equation shows the neutral potassium atom (on the left) losing one electron to become a potassium ion (on the right). The superscript 1+ on

123 Å
Lithium, Li

0.89 Å
Beryllium, Be

0.82 Å
Boron, B

0.77 Å
Carbon, C

0.75 Å
Nitrogen, N

0.73 Å
Oxygen, O

0.72 Å
Fluorine, F

0.71 Å
Neon, Ne

Figure 6.11

The decrease of atomic radii within a period

Table 6.3
Periodic Table of Atomic Radii (in Angstroms)

Group IA	IIA	IIIB	IVB	VB	VIB	VIIB	VIII			IB	IIB	IIIA	IVA	VA	VIA	VIIA	VIIIA
Period 1 0.32 H 1																	0.31 He 2
2 1.23 Li 3	0.89 Be 4					Transition elements						0.82 B 5	0.77 C 6	0.75 N 7	0.73 O 8	0.72 F 9	0.71 Ne 10
3 1.54 Na 11	1.36 Mg 12						VIII			IB	IIB	1.18 Al 13	1.11 Si 14	1.06 P 15	1.02 S 16	0.99 Cl 17	0.98 Ar 18
4 2.03 K 19	1.74 Ca 20	1.44 Sc 21	1.32 Ti 22	1.22 V 23	1.18 Cr 24	1.17 Mn 25	1.17 Fe 26	1.16 Co 27	1.15 Ni 28	1.17 Cu 29	1.25 Zn 30	1.26 Ga 31	1.22 Ge 32	1.20 As 33	1.17 Se 34	1.14 Br 35	1.12 Kr 36
5 2.16 Rb 37	1.91 Sr 38	1.62 Y 39	1.45 Zr 40	1.34 Nb 41	1.30 Mo 42	1.27 Tc 43	1.25 Ru 44	1.25 Rh 45	1.28 Pd 46	1.34 Ag 47	1.48 Cd 48	1.44 In 49	1.40 Sn 50	1.40 Sb 51	1.36 Te 52	1.33 I 53	1.31 Xe 54
6 2.35 Cs 55	1.98 Ba 56	La– Lu 57–71	1.44 Hf 72	1.34 Ta 73	1.30 W 74	1.28 Re 75	1.26 Os 76	1.27 Ir 77	1.30 Pt 78	1.34 Au 79	1.49 Hg 80	1.48 Tl 81	1.47 Pb 82	1.46 Bi 83	1.46 Po 84	1.45 At 85	Rn 86
7 Fr 87	2.20 Ra 88	Ac– Lr 89–103	Unq 104	Unp 105	Unh 106	Uns 107	Uno 108	Une 109									

1.69 La 57	1.65 Ce 58	1.64 Pr 59	1.64 Nd 60	1.63 Pm 61	1.62 Sm 62	1.85 Eu 63	1.62 Gd 64	1.61 Tb 65	1.60 Dy 66	1.58 Ho 67	1.58 Er 68	1.58 Tm 69	1.70 Yb 70	1.56 Lu 71
2.0 Ac 89	1.65 Th 90	Pa 91	1.42 U 92	Np 93	Pu 94	Am 95	Cm 96	Bk 97	Cf 98	Es 99	Fm 100	Md 101	No 102	Lr 103

0.77 ←—Atomic radius
C
6 ←—Atomic number

the potassium shows that it has become a *positively charged particle*. That is, it has lost an electron ($1e^{1-}$), which is negatively charged. Thus the potassium is left with one more positive charge (proton) than negative charge (electron). It is no longer electrically neutral.

The **ionization potential** of an element is *the energy needed to pull an electron away from an isolated ground-state atom of that element.* The farther away from the nucleus an electron is, the less it is attracted by the positive charge of the nucleus, and the less energy is needed to pull it away from the atom. For this reason, the electron is always pulled away from the outermost energy level. In any group of the periodic table, the ionization potential decreases as we move down the group (Figure 6.12).

Figure 6.12
The decrease of ionization
potential within a group

Hydrogen, H
314 kcal/mole

Lithium, Li
124 kcal/mole

Sodium, Na
119 kcal/mole

Table 6.4
Periodic Table of Ionization Potentials (in Kilocalories Per Mole)

Period	Group IA	IIA	IIIB	IVB	VB	VIB	VIIB	VIII			IB	IIB	IIIA	IVA	VA	VIA	VIIA	VIIIA
1	314 H 1																	566 He 2
2	124 Li 3	215 Be 4					Transition elements						191 B 5	260̄ C 6	335 N 7	312 O 8	402 F 9	498 Ne 10
3	119 Na 11	176 Mg 12											138 Al 13	188 Si 14	254 P 15	239 S 16	300̄ Cl 17	363 Ar 18
4	100̄ K 19	141 Ca 20	151 Sc 21	158 Ti 22	155 V 23	156 Cr 24	171 Mn 25	182 Fe 26	181 Co 27	176 Ni 28	178 Cu 29	216 Zn 30	138 Ga 31	187 Ge 32	242 As 33	225 Se 34	273 Br 35	323 Kr 36
5	96 Rb 37	131 Sr 38	152 Y 39	160̄ Zr 40	156 Nb 41	166 Mo 42	172 Tc 43	173 Ru 44	178 Rh 45	192 Pd 46	174 Ag 47	207 Cd 48	133 In 49	169 Sn 50	199 Sb 51	208 Te 52	241 I 53	280̄ Xe 54
6	90̄ Cs 55	120̄ Ba 56	Lu 57–71	127 Hf 72	140 Ta 73	184 W 74	181 Re 75	201 Os 76	212 Ir 77	208 Pt 78	212 Au 79	241 Hg 80	141 Tl 81	171 Pb 82	184 Bi 83	196 Po 84	At 85	248 Rn 86
7	Fr 87	122 Ra 88	Ac– Lr 89–103	Unq 104	Unp 105	Unh 106	Uns 107	Uno 108	Une 109									

129 La 57	159 Ce 58	133 Pr 59	145 Nd 60	Pm 61	129 Sm 62	131 Eu 63	142 Gd 64	155 Tb 65	157 Dy 66	Ho 67	Er 68	Tm 69	143 Yb 70	115 Lu 71
162 Ac 89	Th 90	Pa 91	92 U 92	Np 93	Pu 94	Am 95	Cm 96	Bk 97	Cf 98	Es 99	Fm 100	Md 101	No 102	Lr 103

260̄ ←— Ionization potential
C
6 ←— Atomic number

Ionization energy is usually measured in units called calories. A *calorie* is the amount of heat needed to raise one gram of water one degree Celsius. A more lengthy discussion of the calorie appears in Chapter 10.

Within each period, ionization potential increases from left to right (Table 6.4). This results from the decrease in atomic radius (and thus the stronger attraction between nucleus and outer electrons) as we move from left to right in the periodic table. The ionization potentials of the transition metals in each period do not vary much, because their atomic radii do not vary much (see Section 6.7).

6.9 Electron Affinity

An atom's ability to acquire an additional electron is called its **electron affinity** and is another important factor that determines its chemical properties. The process in which a neutral atom acquires an electron can be represented in this way:

$$Cl + e^{1-} \longrightarrow Cl^{1-}$$

The equation shows that when a neutral chlorine atom picks up an electron, it becomes a negatively charged chloride ion. When the chlorine atom picks up an electron to form a chloride ion, it gives up something in return: *energy*. The amount of energy the chlorine atom releases is a measure of its electron affinity. And electron affinity depends on the attraction between an electron and the nucleus of an atom.

As we move down a group, the electron affinity of the elements generally decreases. If we look at some Group VIIA elements, we can see a decrease in electron affinity as we move from chlorine to iodine (Table 6.5). This is because, in iodine, the additional electron is added to the fifth energy level, whereas in chlorine the additional electron is added to the third energy level. (Remember that the closer the electron is to the nucleus, the more it is attracted to the nucleus, and electron affinity depends on how strongly an electron is attracted to the nucleus.) Fluorine's unexpectedly low electron affinity is caused by the special nature of this ele-

Table 6.5
Electron Affinities of Group VIIA Elements

Element	Electron affinity in electron volts	Outer electron configuration
F	3.6	$2s^2\ 2p^5$
Cl	3.75	$3s^2\ 3p^5$
Br	3.53	$4s^2\ 4p^5$
I	3.2	$5s^2\ 5p^5$

Table 6.6
Summary of Trends in the Periodic Table

Trend of	Top to bottom in a group	Left to right in a period
Atomic radius	increases	decreases
Ionization potential	decreases	increases
Electron affinity	decreases	increases

ment. (Some scientists feel that the electron affinity value for fluorine is reasonable but that the electron affinity value for chlorine is too high.)

If we examine the trend of electron affinity within a period, we find that it *increases* as we go from left to right. This is because the atom gets smaller as we go from left to right in a period, and the attraction between the nucleus and the electron becomes greater.

Table 6.6 summarizes what we have discussed about trends in the periodic table.

6.10 Orbitals—The Quantum-Mechanical View of the Atom

Learning Goal 7
Atomic orbitals

In the Bohr model of the atom, we look at electrons as though they moved around the nucleus within discretely defined spherical shells or energy levels. For most of our work in this course, this is the model that best suits our needs. However, a more modern model of the atom, the **quantum-mechanical model,** has been developed since Bohr's day. In this model of the atom, we think of an electron *not* as being in a carefully defined orbit, but instead as occupying a volume of space. This volume of space is called an **orbital.** The orbital represents a region of space in which the electron can be found with a 95% probability. (We'll have more to say about probability shortly.)

Learning Goal 8
Shapes of atomic orbitals

Each type of orbital has a specific shape and can hold a maximum of two electrons (Figure 6.13). For example, all *s* orbitals are spherical in shape. However, the different levels of *s* orbitals occupy different regions of space around the nucleus of the atom. Figure 6.14 shows how the 1*s*, 2*s*, and 3*s* orbitals are related to each other.

The *p* orbitals are dumbbell-shaped, as Figure 6.13 shows. There are three *p* orbitals in each major energy level except *K*. They are called p_x, p_y, and p_z, and they correspond to the three axes in space. Each of these *p* orbitals can hold a maximum of two electrons, so the three *p* orbitals combined can hold a maximum of six electrons. (This total corresponds nicely to the maximum number of electrons that a *p* sublevel can hold.) The *p* orbitals that are associated with the different energy levels occupy different regions of space around the nucleus of the atom. Figure 6.15 shows

Figure 6.13

The shapes of the s, p_x, p_y, and p_z orbitals. The shape of the orbital defines the region of space in which an electron may be found with a probability of 95%. Darker areas are regions of higher probability.

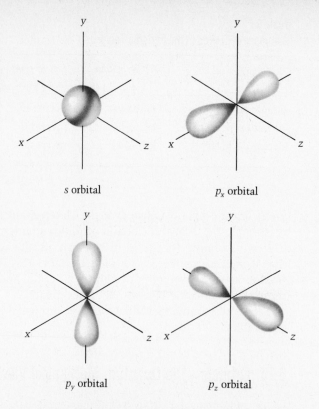

s orbital p_x orbital

p_y orbital p_z orbital

the relationship between the $2p_x$ and $3p_x$ orbitals, and Figure 6.16 shows the relationship among the 1s, 2s, and $2p_x$ orbitals.

Although we could extend this discussion to the d and f orbitals, we won't do so in this text. What is important here is that you have a fundamental understanding of the quantum-mechanical model of the atom.

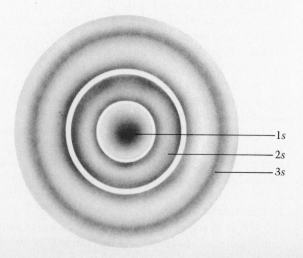

1s

2s

3s

Figure 6.14

The relationship of the 1s, 2s, and 3s orbitals to each other

Figure 6.15
The spatial relationship of the $2p_x$ and $3p_x$ orbitals to each other

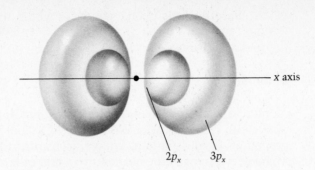

6.11 Orbital Probability—An Explanation

· · · · · · · · · · · · · ·
Learning Goal 9
Quantum-mechanical model of the atom

To clarify the idea of an electron being found in a specific orbital with a probability of 95%, consider the following analogy. George, a bank teller, works at the First National Bank between the hours of 10:00 A.M. and 3:00 P.M. The probability (or chance) of finding George anywhere in the bank between 10:00 A.M. and 3:00 P.M. on a given day is extremely high, let's say 95%. (It's not 100%, because there's always a chance that George may be sick on a given day, or that he may take the day off.)

Now suppose we make a map of the inside of the bank. Every ten minutes during the course of the day, we mark George's position on the map. For most of the day, we would find him at window 3, where he works. This is where most of our marks would be on the map. However, there would also be marks on the map that would show George in other parts of the bank. Perhaps he took a coffee break or left his window to get some information from the bank manager.

This map would be very much like the orbital pictures shown in Figures 6.13 to 6.16. Where there are many marks or darker shading, we have a better chance of finding both George and the electron—for example, at window 3 or at the edges of the orbitals in Figure 6.15. Where there are fewer marks or lighter shading, the probability of finding George or an electron is low. And the probability of finding an electron any place

Figure 6.16
The relationship among the $1s$, $2s$, and $2p_x$ orbitals

within the orbital (or George any place within the bank) is 95%. Just as George can take a day off, an electron can move out of its orbital and then—just as easily—move back in.

Summary

As more and more elements were discovered during the eighteenth and nineteenth centuries, scientists attempted to find relationships among the properties of the various known elements. Finally, in 1869, Dmitri Mendeleev arranged the elements in a grid or table form, with horizontal rows called periods and vertical columns called groups. Elements with similar chemical properties were positioned in the same group. The table ordered the known elements and their properties so well that Mendeleev was able to use it to predict the properties of yet undiscovered elements. The periodic table of today is similar to Mendeleev's. It consists of seven periods and sixteen groups. For elements in the A groups (representative elements), the group number is the same as the number of electrons in the outermost shell. For elements in the B groups (transition metals), there is not a simple connection between group number and number of electrons in the outermost energy level. However, because the transition metals fill inner shells after outer shells, they show quite similar chemical behavior.

Each energy level (K, L, M, etc., or 1, 2, 3, etc.) in an atom consists of one or more sublevels. The sublevels are represented by the letters s, p, d, f, g, h, and i. Each sublevel can hold a certain maximum number of electrons, and the total number of electrons that a main energy level can hold is simply the sum of these maximums for all its sublevels. The sublevels of some energy levels overlap, so that, for example, the $4s$ sublevel has a lower energy than the $3d$ sublevel. This accounts for the positioning of some electrons in energy level 4 before level 3 is filled. However, with very few exceptions, the lowest-energy sublevels are always filled first.

The atomic radius of an atom is the distance from the center of the nucleus to the outermost electron. As we move from the top to the bottom of any group in the periodic table, atomic radius generally increases. And as we move from left to right across a period, atomic radius generally decreases. The ionization potential for an element is the energy needed to pull an electron away from an isolated atom of that element. As we move from the top to the bottom of a group, the ionization potential decreases. And as we move from left to right across a period, the ionization potential generally increases. Electron affinity is a measure of an atom's ability to acquire additional electrons. This measure is related to the energy released when an additional electron is added to a neutral atom. As we move down a group in the periodic table, the electron affinity of the elements generally decreases. And as we move from left to right across a period, the electron affinity increases.

In the quantum-mechanical model of the atom, electrons are thought of as occupying a certain volume of space about the nucleus of the atom.

The region of space in which an electron can be found with 95% probability is called an orbital. Each orbital can hold a maximum of two electrons. All orbitals of a certain type (for example, all *s* orbitals) have the same shape, but orbitals associated with different energy levels (for example, 1*s* and 2*s* orbitals) differ in size and sometimes in the direction in which they travel about the nucleus.

Career Sketch

Food Scientist

Food scientists investigate the chemical, physical, and biological nature of food and apply this knowledge to the cooking, preserving, packaging, distributing, and storing of food. About three-fifths of all scientists in food processing work in research and development. Others work in quality-control laboratories or in processing areas of food-producing plants. Some teach or do basic research in colleges and universities.

Food scientists in basic research study the structure and composition of food and the changes it undergoes in storage and cooking. For example, they may develop new sources of proteins, study the effects of processing on microorganisms, or search for factors that affect the flavor, texture, or appearance of foods. Food scientists who work in applied research and

Food scientists carry out research in well-equipped chemical laboratories.
(Kraft Corporate Marketing)

development create new foods and develop new processing methods. They also seek to improve existing foods by making them more nutritious and enhancing their flavor, color, and texture.

. . . .

Self-Test Exercises

Learning Goal 1: Historical Basis and Significance of Periodic Table

1. Locate at least two instances in the periodic table (Table 6.1) in which a heavier element appears before a lighter one.

2. What was Mendeleev's reason for placing some heavier elements before lighter ones in his table? Why didn't this upset the "logic" of his table?

3. Using the periodic table (Table 6.1), determine the total number of: (*a*) metallic elements, (*b*) metalloids, (*c*) nonmetallic elements.

4. Why aren't elements 57 to 71 and 89 to 103 shown with the other elements in the body of Table 6.1?

5. Discuss the contributions to the development of the periodic table made by:
(*a*) Döbereiner, (*b*) Newlands, (*c*) Mendeleev.

6. How did Mendeleev's table differ from Newlands's table of the elements?

Learning Goal 2: Periods and Groups in the Periodic Table

7. In the periodic table, a horizontal row is called a _____ . A vertical column is called a _____ .

8. Which elements would behave similarly: those within any one group or those within any one period? Why?

9. What similarity is there among the outermost energy levels of elements within the same group?

10. "Elements in the B groups (the transition metals) show somewhat similar chemical behavior." Explain why this statement is true.

Learning Goal 3: A-Group and B-Group Elements

11. What does the group number tell us about the A-group elements?

12. Where are the A-group elements located on the periodic table? Where are the B-group elements located?

13. The *representative* elements are also known as the _____ (A/B) group elements. The *transition metals* are also known as the _____ (A/B) group elements.

14. "Elements within any one group have the same number of electrons in their outermost energy levels." Is this statement true for the A-group or the B-group elements?

Learning Goal 4: Periodic Trends

15. List the following elements in order of increasing ionization potential: Cl, Ca, Sr, Rb, and S. Use only the periodic table (Table 6.1); don't look up the ionization potentials.

16. What is an angstrom, and how big is it?

17. List the following elements in order of increasing atomic radius: In, P, I, Sb, and As. Once again, use only the periodic table.

19. Define *electron affinity.*

21. After you've removed one electron from a neutral atom, would you need more or less energy to remove a second electron? Why?

23. Where in the periodic table do we find the following?
(*a*) Elements with the highest electron affinity
(*b*) Elements with the lowest ionization potential
(*c*) The most unreactive elements

25. Complete the following statements:
(*a*) The ionization potential _____ (increases/decreases) as one moves down a *group* of elements.
(*b*) The ionization potential _____ (increases/decreases) as one moves from left to right across a *period* of elements.
(*c*) The atomic radius _____ (increases/decreases) as one moves from left to right across a *period* of elements.

27. List the following elements in order of *increasing* atomic radius: Cl, Ca, Sr, Rb, and S. Use only the periodic table (Table 6.1); don't look up the atomic radii.

29. List the following elements in order of *increasing* ionization potential: P, I, Sb, and As. Once again, use only the periodic table.

31. A positive ion is an atom (or group of atoms) that has: (*a*) gained a proton, (*b*) lost a proton, (*c*) gained an electron, (*d*) lost an electron.

33. A Cl^{1-} ion is a Cl atom that has: (*a*) gained a proton, (*b*) lost a proton, (*c*) gained an electron, (*d*) lost an electron.

35. An Mg^{2+} ion is an Mg atom that has: (*a*) gained two protons, (*b*) lost two protons, (*c*) gained two electrons, (*d*) lost two electrons.

18. Does atomic radius increase or decrease as we move from the top to the bottom of a group in the periodic table? Why?

20. As we move from left to right across a period in the periodic table, does atomic radius increase or decrease? Why?

22. List the following elements in order of increasing atomic radius: In, Tl, H, F, S. (Use the periodic table.)

24. What name is given to the energy needed to pull an electron away from an isolated atom?

26. Explain why it is more difficult to remove an electron from an atom that is located at the top of a group in the periodic table than to remove an electron from an atom that is located at the bottom of a group.

28. The elements sodium and chlorine are both in period 3 of the periodic table. Sodium has an ionization potential of 119 kcal/mole, and chlorine has an ionization potential of $30\overline{0}$ kcal/mole. Why is the ionization potential of chlorine so much greater?

30. List the following elements in order of increasing ionization potential: Br, Ge, Se, Cs, F.

32. The quantity that depends on the attraction between an electron and the nucleus of an atom is called _____ .

34. Electron affinity _____ (increases/decreases) as one moves down a *group* of elements.

36. Electron affinity _____ (increases/decreases) as one moves from left to right across a *period* of elements.

Learning Goal 5: Energy Sublevels and Chemical Properties

37. Elements 57 to 71 are called the *lanthanide* series, and elements 89 to 103 are called the *actinide* series. Why do the elements in these series have similar chemical properties?

38. What is unique about the filling order of electrons in the 3*d* and 4*s* sublevels of elements 21 to 30? What would you expect about the behavior of these elements?

39. Element 111 will behave like what elements in the periodic table?

40. Element 112 should behave like what other element in the periodic table?

41. Write the expected electron configuration for element 118. What is the group number of this element? How many electrons should be in its outermost energy level?

42. Write the expected electron configuration for element 117. What is the group number of this element? How many electrons should be in its outermost energy level?

Learning Goal 6: Electron Configurations

43. Using Figure 6.8, write the electron configurations of the following elements: (a) $_6C$ (b) $_{22}Ti$ (c) $_{36}Kr$ (d) $_{53}I$

44. Using Figure 6.8, write the electron configurations of the following elements: (a) $_5B$ (b) $_{16}O$

45. Name the neutral elements with the following electron configurations. (Use Table 6.1.)
(a) $1s^2\ 2s^2\ 2p^6\ 3s^2\ 3p^6\ 3d^{10}\ 4s^2\ 4p^6\ 4d^{10}\ 5s^2\ 5p^6\ 6s^2$
(b) $1s^2\ 2s^2\ 2p^6\ 3s^2\ 3p^6\ 4s^2$
(c) $1s^2\ 2s^2\ 2p^6\ 3s^2\ 3p^6\ 3d^{10}\ 4s^2\ 4p^6\ 5s^2$
(d) $1s^2\ 2s^2$

46. Name the neutral elements with the following electron configurations. (Use Table 6.1.)
(a) $1s^2$
(b) $1s^2\ 2s^2\ 2p^3$
(c) $1s^2\ 2s^2\ 2p^6\ 3s^2\ 3p^6\ 3d^{10}\ 4s^1$
(d) $1s^2\ 2s^2\ 2p^6\ 3s^2\ 3p^6\ 3d^{10}\ 4s^2\ 4p^4$

47. A certain neutral element has 2 electrons in the K level, 8 electrons in the L level, 18 electrons in the M level, and 3 electrons in the N level. List the following information for this element: (a) atomic number, (b) total number of s electrons, (c) total number of p electrons, (d) total number of d electrons, (e) number of protons.

48. A certain neutral element has 2 electrons in the K level, 8 electrons in the L level, 18 electrons in the M level, and 2 electrons in the N level. List the following information for this element: (a) atomic number, (b) total number of s electrons, (c) total number of p electrons, (d) total number of d electrons, (e) number of protons.

49. By writing electron configurations, determine which of the following elements belong to the same groups: elements with atomic number 6, 19, 17, 14, 3, and 35.

50. By writing electron configurations, determine which of the following elements belong to the same groups: elements with atomic number 5, 18, 16, 36, 10, and 13.

51. An easy way to find the maximum number of electrons for an energy level is to use the formula

Maximum number of electrons $= 2n^2$

where n is the number of the energy level (for example, $K = 1$, $L = 2$, and so on). Use this formula to calculate the maximum number of electrons that the Q energy level can hold.

52. An easy way to find the maximum electron number for an energy level is to use the formula

Maximum number of electrons $= 2n^2$

where n is the number of the energy level (for example, $K = 1$, $L = 2$, and so on). Use this formula to calculate the maximum number of electrons that the N energy level can hold.

53. Using Figure 6.8, write the electron configurations for each of the following elements: (a) $_{31}Ga$ (b) $_{51}Sb$ (c) $_{82}Pb$ (d) $_{88}Ra$

54. Using Figure 6.8, write the electron configurations for the following elements: (a) $_3Li$ (b) $_{12}Mg$ (c) $_{56}Ba$

55. Name the neutral elements with the following electron configurations. (Use Table 6.1.)
(a) $1s^2\ 2s^2\ 2p^6\ 3s^2$
(b) $1s^2\ 2s^2\ 2p^6\ 3s^2\ 3p^6\ 3d^{10}\ 4s^2\ 4p^6\ 4d^{10}\ 4f^{14}$
(c) $1s^2\ 2s^2\ 2p^6\ 3s^2\ 3p^6\ 3d^{10}\ 4s^2\ 4p^6\ 4d^{10}\ 4f^{14}\ 5s^2$ $5p^6\ 6s^2$
(d) $1s^2\ 2s^2\ 2p^6\ 3s^2\ 3p^6\ 3d^{10}\ 4s^2\ 4p^6\ 4d^{10}\ 4f^{14}\ 5s^2$ $5p^6\ 5d^{10}\ 6s^1$

56. Name the neutral elements with the following electron configurations. (Use Table 6.1.)
(a) $1s^2\ 2s^2\ 2p^6\ 3s^1$
(b) $1s^2\ 2s^2\ 2p^6\ 3s^2\ 3p^6\ 3d^{10}\ 4s^2\ 4p^6\ 4d^{10}\ 4f^{12}$
(c) $1s^2\ 2s^2\ 2p^6\ 3s^2\ 3p^6\ 3d^{10}\ 4s^2\ 4p^6\ 4d^{10}\ 4f^{14}\ 5s^2$ $5p^6\ 6s^1$
(d) $1s^2\ 2s^2\ 2p^6\ 3s^2\ 3p^6\ 3d^{10}\ 4s^2\ 4p^6\ 4d^{10}\ 4f^{14}\ 5s^2$ $5p^6\ 5d^{10}$

57. A given neutral element has 2 electrons in the *K* level, 8 electrons in the *L* level, 18 electrons in the *M* level, 22 electrons in the *N* level, 8 electrons in the *O* level, and 2 electrons in the *P* level. List the following information for this element: (*a*) atomic number, (*b*) total number of *s* electrons, (*c*) total number of *p* electrons, (*d*) total number of *d* electrons, (*e*) total number of *f* electrons, (*f*) number of protons.

58. A given neutral element has 2 electrons in the *K* level, 8 electrons in the *L* level, 18 electrons in the *M* level, 32 electrons in the *N* level, 9 electrons in the *O* level, and 2 electrons in the *P* level. List the following information for this element: (*a*) atomic number, (*b*) total number of *s* electrons, (*c*) total number of *p* electrons, (*d*) total number of *d* electrons, (*e*) total number of *f* electrons, (*f*) number of protons.

59. *True or False.* We know that the number of electrons in the outermost energy level of the Group IIIB elements is 3 because the group number reflects the number of electrons in the outermost energy level. [Give the reason(s) for your choice.]

60. *True or False.* We know that the number of electrons in the outermost energy level of the Group IIIA elements is 3 because the group number reflects the number of electrons in the outermost energy level. [Give the reason(s) for your choice.]

61. By writing electron configurations, determine which of the following elements belong to the same groups: elements with atomic number 15, 38, 50, 56, 14, and 33.

62. By writing electron configurations, determine which of the following elements belong to the same groups: elements with atomic number 16, 36, 39, 8, and 34.

Learning Goal 7: Atomic Orbitals

63. Define *atomic orbital*.

64. The volume of space occupied by an electron is called a(n) _____ .

65. How many electrons are permitted in an orbital?

66. *True or False.* The maximum number of electrons that an orbital can hold is three.

67. According to the Bohr model of the atom, the *d* sublevel can hold a maximum of ten electrons. How many *d* atomic orbitals do you think there are?

68. According to the Bohr model of the atom, the *p* sublevel can hold a maximum of six electrons. How many *p* atomic orbitals do you think there are?

Learning Goal 8: Shapes of Atomic Orbitals

69. Draw the shape of (*a*) an *s* orbital, (*b*) a p_x orbital, (*c*) a p_y orbital, (*d*) a p_z orbital.

70. *True or False.* Regardless of what energy level they are associated with, all *p* orbitals have the same dumbbell shape.

Learning Goal 9: Quantum-Mechanical Model of the Atom

***71.** Using an analogy similar to the one in the text, explain what is meant by 95% probability, and describe how it relates to the quantum-mechanical model of the atom.

72. What is meant by the phrase "region of highest probability of finding an electron"?

Extra Exercises

73. Explain why chlorine is a smaller atom than sodium, even though it contains more subatomic particles.

74. How is the octet rule related to the periodic table?

75. What relationship is there among (a) all elements in a particular period? (b) all elements in a particular group?

76. Draw a simple diagram showing the relative size, shape, and position of all orbitals *through* 3p.

77. What is the difference between an *orbit* and an *orbital?*

78. What would you expect the atomic number to be for the noble gas that follows radon?

79. Based on periodic trends, which group of atoms is most likely to form (a) 1+ ions? (b) 1− ions?

80. Explain the difference between a chlorine atom and a chloride ion.

81. Which member of each of the following pairs should have the larger radius? Why?
(a) K, Br (b) K, Na (c) K, K^{1+}

82. Which has the higher ionization potential, calcium or magnesium? Why?

83. How many elements are in the (a) first period? (b) second period? (c) third period? (d) fourth period? (e) fifth period? (f) sixth period?

84. Which element of each pair has the smaller ionization potential? (a) K, Ca (b) O, Se

85. Which element of each pair has the larger atomic radius? (a) K, Ca (b) O, Se

86. What would be the group number for element 114?

87. Why are the *atomic weights* of elements 94 through 109 listed in parentheses?

88. If you had the responsibility of naming the next element to be discovered, what would you name it? (Please keep your favorite textbook authors in mind when making your decision!)

Cumulative Review / Chapters 4–6

1. How are protons, neutrons, and electrons distributed in the atom?

2. How did Eugen Goldstein choose the term *cathode rays?*

3. What experiment did Sir William Crookes perform to determine whether cathode rays were composed of particles or waves?

4. What is the photoelectric effect?

5. What reasoning led physicist J. J. Thomson to develop the "plum pudding" theory of the atom?

6. When Rutherford tested Thomson's theory of the atom with his own experiments, what results led him to devise a new model of the atom?

7. Why did Rutherford's model of the atom render Thomson's atomic model obsolete?

8. What type of particles add to the mass of the atom without affecting its charge?

9. John Dalton said that atoms of an element must be identical in size, mass, and shape. How does the existence of isotopes disprove this theory?

10. Give the number of protons that are found in each of the following isotopes: (a) $^{59}_{27}Co$ (b) $^{90}_{40}Zr$ (c) $^{74}_{34}Se$

11. Using Table 4.2, calculate the atomic weight of neon.

12. Using Table 4.2, determine how many ^{15}N atoms there are in a sample containing 100,000 nitrogen atoms.

13. Write the standard isotopic notation for the following elements. (Use the periodic table to get the symbols of the elements.) (a) 17p, 17e, 18n (b) 14p, 14e, 14n

14. The hypothetical element Z exists in two isotopic forms, ^{50}Z and ^{52}Z. The atomic weight of Z is 50.5. What is the percentage abundance of each isotope?

15. State the number of protons, electrons, and neutrons in $^{73}_{32}Ge$.

16. How many electrons are there in a neutral $^{60}_{26}Fe$ atom?

17. Of the three isotopes of oxygen, which is the most abundant: oxygen-16, oxygen-17, or oxygen-18?

18. Using Table 4.2, determine how many ^7Li atoms there are in a sample containing 2.0×10^3 lithium atoms.

19. Using Table 4.2, determine how many grams of ^7Li you would find in a sample of lithium that has a mass of 100.0 kg.

20. Using Table 4.2, determine how many grams of Li you would need to get 1.00 gram of ^7Li.

21. Do electrons, protons, or neutrons produce the line spectrum of an element?

22. As an electron travels to an energy level farther from the nucleus, the energy of that electron _____ _____. (a) increases, (b) decreases, (c) remains the same

23. How many electrons are in the outermost energy level of an oxygen atom?

24. Without using the periodic table, determine the electron configurations of the following elements: (a) $_1H$ (b) $_3Li$ (c) $_{11}Na$ (d) $_{19}K$

25. A neutral atom of magnesium has _____ electrons in its outermost energy level. (You may consult the periodic table.)

26. Name the element that has the following ground-state configuration. You may consult the periodic table.

    K, 2 electrons; L, 8 electrons; M, 6 electrons

27. The chemical properties of an element are determined by the number of electrons in the _____. (a) energy level closest to the nucleus, (b) outermost energy level, (c) second energy level, (d) none of these

28. Why is phosphorus in group VA?

29. Write electron configurations for all elements with atomic numbers between 1 and 20 that are isoelectronic (have the same number of electrons in the outermost energy level) with $_3Li$.

30. What is the atomic number of an element whose energy levels K, L, M, and N are filled to capacity?

31. When a rainbow is produced naturally, what acts as the prism?

32. How are line spectra used in chemical laboratories today?

33. In a detective story it was discovered that the victim died of arsenic poisoning. This information was made available by analysis of samples of the victim's hair. How could spectral analysis have made this discovery possible?

34. How is radiant energy produced?

35. Waves on a lake travel through water, and sound waves travel through air. Through what medium do electromagnetic waves travel?

36. What types of waves are included in the electromagnetic spectrum? (a) visible, (b) invisible, (c) both of these, (d) neither of these

37. How did Heinrich R. Hertz verify the theory of electromagnetic radiation proposed by James Clerk Maxwell?

38. Give an example of light energy being absorbed and re-emitted.

39. Give an example of heat energy being absorbed and re-emitted as light energy.

40. Bundles of energy absorbed or emitted by electrons are called _____ .

41. State the periodic law and explain how it is the basis of the modern-day periodic table.

42. What was the contribution of Henry G. J. Moseley in terms of the periodic table?

43. The element with atomic number 107 is an artificially produced radioactive element. Both Soviet and West German scientists claim to have discovered it, but neither claim has been officially accepted. To produce an isotope of the element with atomic number 107, a sample of bismuth-209 was bombarded with chromium-54. Write the electron configuration of the element with atomic number 107. What is its group number?

44. In 1974 a team of U.S. scientists working under Albert Ghiorso at the Lawrence Berkeley Laboratory in Berkeley, California, bombarded the element californium-249 with a beam of oxygen-18 to produce an isotope of the element with atomic number 106. Write the electron configuration for the element with atomic number 106. What is its group number?

45. Our modern periodic table contains cases where elements are not in proper order according to atomic weight. Where do these cases occur?

46. *True or False.* Ionization potential increases as one moves down a group of elements.

47. *True or False.* Atomic radius decreases as one moves from left to right across a period of elements.

48. *True or False.* A Ca^{2+} ion is a calcium atom that has gained two protons.

49. The element with atomic number 108 should behave like what other elements in the periodic table?

50. Name the neutral element with the electron configuration $1s^2 2s^2 2p^6 3s^1$.

51. *True or False.* The elements on the periodic table show properties that are related to their atomic numbers.

52. *True or False.* The element with the electron configuration $1s^2 2s^2 2p^6$ belongs in group VIA.

53. Lithium and fluorine are both in period 2 of the periodic table. Which element has the higher ionization potential? Why?

54. Write the electron configuration for $_{14}Si$.

55. *True or False.* The volume of space occupied by an electron is called an orbital.

56. On the basis of periodic trends, determine which group of atoms is most likely to form: (a) 2+ ions (b) 2− ions

57. Which member of each of the following pairs should have the larger radius?
(a) Li, Na (b) Ca, Ca^{2+}

58. Which element has the lower ionization potential, Mg or Cl?

59. Which element has the larger atomic radius, Li or Cs?

60. Which element has the larger atomic radius, Li or F?

7 Chemical Bonding: How Atoms Combine

Learning Goals

After you've studied this chapter, you should be able to:

1. Write Lewis (electron-dot) structures for the A-group elements.
2. Define and give an example of a covalent bond and a coordinate covalent bond.
3. Name the diatomic elements from memory.
4. Write electron-dot structures for various covalent compounds.
5. Define and give an example of an ionic bond.
6. Use the concept of electronegativity to find the ionic percentage and covalent percentage of a chemical bond.
7. Write the chemical formula for some compounds when you are given their chemical names.
8. Find the oxidation numbers of less familiar elements when you are given chemical formulas of compounds containing these elements.
9. Write the chemical name of some compounds when you are given the formula for the compounds.
10. Distinguish between a polar and a nonpolar bond.
11. Distinguish between a polar and a nonpolar molecule.

Introduction

Matter is held together by forces acting between atoms. When chemical reactions occur, atoms can combine to form compounds. In doing so, they

Figure 7.1

Electron-dot diagrams of some A-group elements

IA	IIA	IIIA	IVA	VA	VIA	VIIA
H						
Li	Be ·	B ·	·C·	·N·	·O:	·F:

join together in such a way as to attain more stable configurations that are at lower levels of chemical potential energy.

Atoms have the ability to gain, lose, or share electrons. This enables them to form compounds in two different ways. When electrons are gained and lost, the compounds formed are held together by forces that bind oppositely charged ions. When electrons are shared, the compounds formed are held together by forces that bind atoms together to form molecules.

Whether the binding forces are due to electron loss and gain or to sharing, when atoms combine they are said to be joined together by a **chemical bond.** They have attained a more stable arrangement—usually eight electrons in their outermost shell. In this chapter we examine the types of bonds that tie atoms together in molecules and unite elements in compounds.

·F:

·Br:

Figure 7.2

The electron-dot diagrams of all Group VIIA elements are the same.

7.1 Lewis (Electron-Dot) Structures

. . . .

Learning Goal 1
Lewis structures

Because it is the electrons in the outermost energy levels that form the bonds, we will be concerned only with these electrons in the elements we study. A very useful notation for showing the outermost electrons of an atom (and the bonds they form) was devised in 1916 by G. N. Lewis, an American physical chemist. In this *electron-dot* notation, the outer electrons are shown as dots on the sides of the symbol for the element. (Figure 7.1 provides several examples.) Each dot corresponds to an outer-shell electron. The dots are placed one to a side, beginning at the top and moving clockwise. A second dot is placed on a side only after there is one dot on each of the four sides of the symbol. If you move from left to right in Figure 7.1, you will see how each additional dot (as necessary) is placed on the diagram.

·N·

·As·

Figure 7.3

The electron-dot diagrams of all Group VA elements are the same.

Recall now that all the elements in the same A group have the same number of electrons in their outermost energy level. This means that all the elements in any A group have the same electron-dot diagram. Figure 7.2 shows the diagrams for fluorine and bromine, which are both in Group VIIA. Both have seven dots. And Figure 7.3 shows the diagrams for nitrogen and arsenic; both are in Group VA, so both have five dots.

7.2 The Covalent Bond

. . . .

Learning Goal 2
Covalent bonds

It has long been known that the Group VIIIA elements (noble gases) tend to be very unreactive and stable. Because of their tendency to keep to themselves, these elements are called the "noble gases." Researchers be-

Figure 7.4
Bonding between fluorine atoms

$$:\overset{..}{\underset{..}{F}}\cdot \;+\; \cdot\overset{..}{\underset{..}{F}}: \;\longrightarrow\; :\overset{..}{\underset{..}{F}}:\overset{..}{\underset{..}{F}}:$$

lieve that this stability is connected with the fact that each of these Group VIIIA elements has eight electrons in its outermost energy level. (The element helium is an exception, because its atom has only two electrons, but these two electrons fill the outermost energy level of the helium atom.) The atoms of other elements can achieve the same type of stability by obtaining eight electrons in their outermost energy levels. One way they can do this is to *share* electrons with other atoms, thus forming a *covalent bond.* A **covalent bond** is *a chemical bond in which two or more atoms share electrons.*

To see how electrons are shared, imagine a sample of fluorine atoms in a closed container. A fluorine atom has seven electrons in its outermost energy level. It needs eight electrons in that energy level to complete its octet. One way to obtain another electron is to share an outer electron belonging to another fluorine atom (Figure 7.4). (Examining Figure 7.4, think of the arrow as indicating that the items on the left combine to form what is on the right. We shall discuss such chemical equations in more detail in Chapter 8.) When a fluorine atom shares one of its electrons with another fluorine atom, a fluorine molecule (F_2) is formed. But now each atom has a stable octet. Evidence of this stability is the fact that fluorine is always found as F_2 molecules.

The type of covalent bond formed between the two fluorine atoms is called a **single covalent bond,** because each atom has shared a *single* electron with the other to form the bond. The single bond is the most common type of covalent bond.

$$:\overset{..}{\underset{..}{F}}\cdot \;+\; \cdot\overset{..}{\underset{..}{F}}: \;\longrightarrow\; :\overset{..}{\underset{..}{F}}:\overset{..}{\underset{..}{F}}:$$

$$:\overset{..}{\underset{..}{Cl}}\cdot \;+\; \cdot\overset{..}{\underset{..}{Cl}}: \;\longrightarrow\; :\overset{..}{\underset{..}{Cl}}:\overset{..}{\underset{..}{Cl}}:$$

$$:\overset{..}{\underset{..}{Br}}\cdot \;+\; \cdot\overset{..}{\underset{..}{Br}}: \;\longrightarrow\; :\overset{..}{\underset{..}{Br}}:\overset{..}{\underset{..}{Br}}:$$

Figure 7.5
Bonding diagrams of Group VIIA diatomic elements

The Diatomic Elements

Learning Goal 3
Diatomic elements

Diatomic elements are elements that are found naturally in molecules with two atoms each. Fluorine is thus a *diatomic element.* In fact, all the Group VIIA elements are diatomic. They all form single covalent bonds (Figure 7.5). Hydrogen, oxygen, and nitrogen also exist in diatomic forms (Figure 7.6), because they share each other's electrons.

An easy way to remember the diatomic elements is with the memory aid HONClBrIF (pronounced "honkelbrif"). Remember that these atoms exist as diatomic molecules because doing so yields a stable octet of electrons, or, in the case of hydrogen, a stable duet (two electrons, like helium).

$$H\cdot \; + \; \cdot H \longrightarrow H:H$$
Hydrogen (single bond)

$$:\ddot{O}\cdot \; + \; \cdot \ddot{O}: \longrightarrow :\ddot{O}::\ddot{O}:$$
Oxygen (double bond)

$$\cdot \ddot{N}\cdot \; + \; \cdot \ddot{N}\cdot \longrightarrow \ddot{N}::\ddot{N}$$
Nitrogen (triple bond)

Figure 7.6
Bonding diagrams for other diatomic elements. Hydrogen needs only two electrons to complete its outer energy level. In O_2, two electrons from each atom contribute to the bond. Three electrons from each nitrogen atom contribute to the bond.

. . . .
Learning Goal 4
Electron-dot structures of covalent compounds

Other Molecules Having Single Covalent Bonds

Now let's see how some elements obtain eight electrons (or, in the case of hydrogen, two electrons) in their outermost energy levels by forming compounds.

Water, which has the formula H_2O, bonds in this way:

$$\dot{H} + \dot{H} + \cdot \ddot{O}: \longrightarrow \begin{array}{c} H:\ddot{O}: \\ H \end{array}$$

The two hydrogens share their single electrons with the oxygen. So the oxygen obtains eight electrons in its outermost energy level, and each hydrogen now has its two electrons.

In ammonia (NH_3), methane (CH_4), and hydrogen chloride (HCl), the bonding looks like this:

$$3\dot{H} + \cdot \ddot{N}\cdot \longrightarrow \begin{array}{c} H:\ddot{N}:H \\ H \end{array}$$

$$4\dot{H} + \cdot \dot{C}\cdot \longrightarrow \begin{array}{c} H \\ H:\ddot{C}:H \\ \ddot{H} \end{array}$$

$$\dot{H} + \cdot \ddot{Cl}: \longrightarrow H:\ddot{Cl}:$$

The N, C, and Cl atoms now have eight electrons in their outermost energy level. The H atoms complete their outer shells with two electrons. The molecules are more stable than the atoms from which they are formed.

The Use of Dashes

Instead of using electron-dot notation to show the bonding in compounds, we sometimes use dashes. The dash (—) has the same meaning as a double dot (:), but it simplifies the diagram. The dash, then, represents a single bond, or a shared pair of electrons. The following examples will show you how the dash is used in combination with dot notation.

$$:\ddot{Cl}:\ddot{Cl}: \qquad or \qquad :\ddot{Cl}-\ddot{Cl}:$$

$$\begin{array}{c} H:\ddot{O}: \\ H \end{array} \qquad or \qquad \begin{array}{c} H-\ddot{O}: \\ | \\ H \end{array}$$

$$\begin{array}{c} H:\ddot{N}:H \\ H \end{array} \qquad or \qquad \begin{array}{c} H-\ddot{N}-H \\ | \\ H \end{array}$$

$$\begin{array}{c} H \\ H:\ddot{C}:H \\ \ddot{H} \end{array} \qquad or \qquad \begin{array}{c} H \\ | \\ H-C-H \\ | \\ H \end{array}$$

$$H:\ddot{Cl}: \qquad or \qquad H-\ddot{Cl}:$$

This notation is especially useful when you want to represent the structures of compounds that have double and triple bonds.

The Double Bond

A **double covalent bond** is *one in which two pairs of electrons are shared between two atoms.* For example, in oxygen (O_2) the bonding is

$$\cdot \ddot{O} \colon + \cdot \ddot{O} \colon \longrightarrow \colon \ddot{O} \colon \ddot{O} \colon \quad \text{or} \quad \colon \ddot{O} {=} \ddot{O} \colon$$

By sharing two electrons, each oxygen atom obtains eight electrons in its outermost energy level. (Actually, this bonding structure for O_2 is simplified and not entirely correct, but the simplified structure communicates the general idea.)

The bonding in the compound ethylene (C_2H_4), which is used to make polyethylene plastic, can be written as

$$
\begin{array}{cc}
\text{H} & \text{H} \\
\cdot x & \cdot x \\
\text{C} \colon & \colon \text{C} \\
\cdot x & \cdot x \\
\text{H} & \text{H}
\end{array}
\qquad \text{or} \qquad
\begin{array}{cc}
\text{H} & \text{H} \\
| & | \\
\text{C} & {=} & \text{C} \\
| & | \\
\text{H} & \text{H}
\end{array}
$$

(In this diagram an *x* is used to represent the electron in each hydrogen.) Here the hydrogen atoms are satisfied with two electrons, and each of the carbon atoms has the eight electrons it needs.

The Triple Bond

A **triple covalent bond** is *one in which each atom shares three of its electrons with the other.* For example, in nitrogen (N_2) the bonding can be diagrammed in this way:

$$\cdot \ddot{N} \cdot + \cdot \ddot{N} \cdot \longrightarrow \ddot{N} \colon \colon \ddot{N} \quad \text{or} \quad \ddot{N} {\equiv} \ddot{N}$$

With the triple bond, each nitrogen atom obtains eight electrons in its outermost energy level. The original N atom has only five electrons, but when two N atoms get together by triple bonding, each has a share in eight electrons.

In the compound acetylene (C_2H_2), the bonding is

$$\text{H} \overset{x}{\cdot} \text{C} \colon \colon \text{C} \overset{x}{\cdot} \text{H} \qquad \text{or} \qquad \text{H} {-} \text{C} {\equiv} \text{C} {-} \text{H}$$

Example 7.1

Write the electron-dot structures for each of the following covalent compounds.
(a) CO_2 (Each oxygen is bonded to the carbon, but the oxygens aren't bonded to each other.)
(b) CH_3I (The iodine and all the hydrogens are bonded to the carbon.)

(c) $COCl_2$ (This is phosgene. The oxygen and both chlorines are bonded to the carbon.)
(d) $AsCl_3$ (Each chlorine is bonded to the arsenic.)
(e) H_2S (Both hydrogens are bonded to the sulfur.)
(f) CH_2O (This is formaldehyde, which we will talk more about in our discussion of aldehydes in Chapter 19. The oxygen and both hydrogens are bonded to the carbon.)
(g) C_2Cl_2 (One chlorine is bonded to each carbon.)

Solution

(a) :Ö=C=Ö:

(b) H—C—Ï: (with H above and H below C)

(c) :Cl—C=Ö (with :Cl: below C)

(d) :Cl—As—Cl: (with :Cl: below As)

(e) H—S̈—H

(f) H—C=Ö: (with H below C)

(g) :Cl—C≡C—Cl:

Practice Exercise 7.1   Write the electron-dot structures for the following covalent compounds: (*a*) NH_3 (*b*) HCl (*c*) SO_2

7.3 The Coordinate Covalent Bond

In the examples of covalent bonds that we have discussed so far, both atoms furnished electrons to form the bond. This can be thought of as a "Dutch-treat" bond, in which each atom pays its share of the bill. Another type of covalent bond exists. It is called a **coordinate covalent bond,** and in it, only *one atom donates the electrons to form the bond.*

A coordinate covalent bond can be thought of as a "you-treat" bond: One atom pays the entire bill. An example of this type of bond is found in the compound phosphoric acid (H_3PO_4). A coordinate covalent bond forms between the phosphorus atom and the single oxygen atom, to which it gives two electrons.

Coordinate covalent bond

H:Ö×P×Ö:H

Ordinary covalent bond

Ordinary (not coordinate) covalent bonds form between the phosphorus atom and the other three oxygen atoms (those with hydrogens attached to them). Once formed, the coordinate covalent bond is indistinguishable from an ordinary covalent bond. However, in drawing bonding diagrams of compounds that have coordinate covalent bonds, we sometimes use an arrow to designate the coordinate covalent bond. For example, H_3PO_4 may be diagrammed as follows:

$$
\begin{array}{c}
O \\
\uparrow \\
H-O-P-O-H \\
| \\
O \\
| \\
H
\end{array}
$$

(The arrow does not mean that the bond is in any way different.)

Example 7.2

Draw bonding diagrams for the following covalent compounds that have coordinate covalent bonds included in their structures.
(a) SO_3 (Each oxygen is bonded to the sulfur.)
(b) H_2SO_4 (Each oxygen is bonded to the sulfur, and each of the hydrogens is bonded to a different oxygen.)

Solution

(a)

$$
\begin{array}{c}
\ddot{O}: \\
\| \\
:\ddot{O} \leftarrow S \rightarrow \ddot{O}:
\end{array}
$$

There are two coordinate covalent bonds in this compound.

(b)

$$
\begin{array}{c}
:\ddot{O}: \\
\uparrow \\
H-\ddot{O}-S-\ddot{O}-H \\
\downarrow \\
:\ddot{O}:
\end{array}
$$

There are two coordinate covalent bonds in this compound.

Practice Exercise 7.2   Draw a bonding diagram for H_2SO_3, sulfurous acid, which has a coordinate covalent bond included in its structure. [*Hint:* Each oxygen is bonded to the sulfur and each of the hydrogens is bonded to a different oxygen.]

7.4 Ionic Bonding

One atom can actually *give* electrons to another atom so that they both obtain an octet of electrons in their outermost energy level. This process is called **ionic bonding.** An example of ionic bonding is the formation of

Figure 7.7
Ionic bonding in sodium fluoride

sodium fluoride (Figure 7.7). A bond formed by sharing electrons (a covalent bond) would help the fluorine *but not the sodium*. If a sodium atom shared a pair of electrons, it would have only two electrons in its outermost energy level:

$$\text{Na} : \ddot{\text{F}} :$$

This does not satisfy the octet rule.

Learning Goal 5
Ionic bonds

However, note what happens if the sodium *gives* or *transfers* its outermost electron to the fluorine. First, the fluorine obtains eight electrons in its outermost energy level. And, in the sodium atom, the *L* energy level (which contains eight electrons) now becomes the outermost energy level; so the sodium, too, has a completed octet.

But what holds the atoms together? Remember that the fluorine now has an extra electron, which used to belong to the sodium. This means that the fluorine now has a total of ten electrons and still has only nine protons. Therefore the fluorine atom *has a net electric charge* of 1− (because it has one extra electron). The sodium atom has given up an electron, so it now has eleven protons and only ten electrons. Therefore the sodium atom *has a net electric charge* of 1+ (because it has one less electron). The two atoms have become ions. An **ion** is *an atom or group of atoms that has gained or lost electrons and therefore has a net negative or positive charge. A* **monatomic ion** *is an ion composed of a single atom.* The positively charged sodium atom is called a *sodium ion*, written Na^{1+}. The negatively charged fluorine atom is called a fluor*ide* ion, written F^{1-}. Opposite charges attract, so the Na^{1+} and F^{1-} ions are attracted to each other. This attraction makes them form the neutral salt sodium fluoride (NaF). The positive–negative attraction is what holds these ions together. An **ionic bond** is *a chemical bond in which one atom transfers one or more electrons to another atom and the resulting ions are held together by the attraction of opposite charges.*

Here's another example: Calcium and chlorine combine by forming an ionic bond to make calcium chloride ($CaCl_2$), but it takes two chlorines to do the job (Figure 7.8). In this compound the calcium atom gives up its

$$\overset{.}{C}a \cdot \; + \; \cdot \overset{..}{\underset{..}{C}l} \colon \; + \; \cdot \overset{..}{\underset{..}{C}l} \colon \; \longrightarrow \; Ca^{2+} \; + \; \colon \overset{..}{\underset{..}{C}l} \colon^{1-} \; + \; \colon \overset{..}{\underset{..}{C}l} \colon^{1-}$$

Figure 7.8
Ionic bonding in calcium chloride

two outermost electrons, transferring one electron to each chlorine atom. The calcium atom becomes the positively charged Ca^{2+} ion, because it still has 20 protons and now only 18 electrons. Each chlorine atom now has an extra electron, for a total of 18, but it still has only 17 protons. As a result, each chlorine atom becomes a negatively charged chloride ion. The calcium ion and the chloride ions are attracted to each other, forming $CaCl_2$.

Figure 7.9
Covalent bonding in boron trifluoride

7.5 Exceptions to the Octet Rule

There are exceptions to every rule, and not all elements obey the octet rule. For example, in the compound boron trifluoride (BF_3), the boron atom has only six electrons in its outermost energy level (Figure 7.9).

Arsenic pentachloride is another exception to the rule. (This name means that one arsenic is bonded to five chlorines; *penta* is from the Greek word meaning "five.") In this compound, arsenic has *ten* outer electrons (Figure 7.10). In fact, many of the transition metals fail to follow the octet rule. Still, in the vast majority of cases, the octet rule will give you the correct compound.

Figure 7.10
Covalent bonding in arsenic pentachloride

7.6 Covalent or Ionic Bonds: The Concept of Electronegativity

At this point you may be asking, "How can I tell whether two elements bond ionically or covalently?" To answer this question, we have to introduce the concept of electronegativity. **Electronegativity** is *the attraction that an atom has for the electrons it is sharing with another atom.* The relative electronegativities for atoms of the different elements are listed in Table 7.1. (Note that no electronegativity values are listed for the Group VIIIA elements, because they tend not to form chemical compounds.)

The electronegativity scale was devised by Linus Pauling, winner of the 1954 Nobel prize in chemistry and the 1962 Nobel peace prize. The scale is based on fluorine having an assigned value of 4.0. Fluorine is the most electronegative element. It tends to attract electrons more strongly than any other element does.

Learning Goal 6
Ionic and covalent bonds

The metal elements generally have low electronegativities, whereas the nonmetals have high electronegativities. This means that the atoms of metals tend to lose electrons more readily than the atoms of nonmetals. In fact, the atoms of nonmetals have a strong tendency to pick up electrons. By looking at Table 7.1, you can see the trend for electronegativity, which generally increases as one moves from left to right across a given period and decreases as one moves from the top to the bottom of a group.

Table 7.1
Periodic Table of Electronegativities

Period	IA	IIA	IIIB	IVB	VB	VIB	VIIB	VIII			IB	IIB	IIIA	IVA	VA	VIA	VIIA	VIIIA
1	2.1 H 1																	He 2
2	1.0 Li 3	1.5 Be 4											2.0 B 5	2.5 C 6	3.0 N 7	3.5 O 8	4.0 F 9	Ne 10
3	0.9 Na 11	1.2 Mg 12											1.5 Al 13	1.8 Si 14	2.1 P 15	2.5 S 16	3.0 Cl 17	Ar 18
4	0.8 K 19	1.0 Ca 20	1.3 Sc 21	1.5 Ti 22	1.6 V 23	1.6 Cr 24	1.5 Mn 25	1.8 Fe 26	1.8 Co 27	1.8 Ni 28	1.9 Cu 29	1.6 Zn 30	1.6 Ga 31	1.8 Ge 32	2.0 As 33	2.4 Se 34	2.8 Br 35	Kr 36
5	0.8 Rb 37	1.0 Sr 38	1.2 Y 39	1.4 Zr 40	1.6 Nb 41	1.8 Mo 42	1.9 Tc 43	2.2 Ru 44	2.2 Rh 45	2.2 Pd 46	1.9 Ag 47	1.7 Cd 48	1.7 In 49	1.8 Sn 50	1.9 Sb 51	2.1 Te 52	2.5 I 53	Xe 54
6	0.7 Cs 55	0.9 Ba 56	La– Lu 57–71	1.3 Hf 72	1.5 Ta 73	1.7 W 74	1.9 Re 75	2.2 Os 76	2.2 Ir 77	2.2 Pt 78	2.4 Au 79	1.9 Hg 80	1.8 Tl 81	1.8 Pb 82	1.9 Bi 83	2.0 Po 84	2.2 At 85	Rn 86
7	0.7 Fr 87	0.9 Ra 88	Ac– Lr 89–103	Unq 104	Unp 105	Unh 106	Uns 107	Uno 108	Une 109									

Transition elements

1.1 La 57	1.1 Ce 58	1.1 Pr 59	1.1 Nd 60	1.1 Pm 61	1.1 Sm 62	1.1 Eu 63	1.1 Gd 64	1.1 Tb 65	1.1 Dy 66	1.1 Ho 67	1.1 Er 68	1.1 Tm 69	1.1 Yb 70	1.2 Lu 71
1.1 Ac 89	1.3 Th 90	1.5 Pa 91	1.7 U 92	1.3 Np 93	1.3 Pu 94	1.3 Am 95	1.3 Cm 96	1.3 Bk 97	1.3 Cf 98	1.3 Es 99	1.3 Fm 100	1.3 Md 101	1.3 No 102	Lr 103

2.5
C
6

←— Electronegativity
←— Atomic number

    To decide whether a bond is ionic or covalent, we find the difference
between the electronegativities of the elements forming the bond.
Chemists generally agree that if the difference between the electronegativities of the two elements is 2.0 or greater, the bond between the two
elements is primarily ionic. If the difference is less than 2.0, the bond is
primarily covalent. (Note that 2.0 is simply a convenient number to use
as the cutoff point between ionic and covalent bonding. There is nothing
special about this number. In fact, there is a more sophisticated way of
looking at ionic and covalent bonds, which we will discuss shortly.)

Example 7.3

Determine whether the bonds in the following compounds are ionic or covalent (use Table 7.1): (*a*) H_2S (*b*) KCl (*c*) MgO (*d*) H_2

Solution

(a) There are two H—S bonds in hydrogen sulfide (H_2S). We want to determine whether these are ionic or covalent. From Table 7.1, we find that the electronegativity of H is 2.1, and that of S is 2.5. The difference in electronegativities is

$$2.5 - 2.1 = 0.4$$

Because the difference is only 0.4, the bond between hydrogen and sulfur is covalent. (If the difference is less than 2.0, the bond is considered covalent.)

(b) From Table 7.1, we find that the electronegativity of K is 0.8, and that of Cl is 3.0. The difference in electronegativities is

$$3.0 - 0.8 = 2.2$$

Because the difference is 2.2, the bond between potassium and chlorine is ionic.

(c) From Table 7.1, we find that the electronegativity of Mg is 1.2, and that of O is 3.5. The difference in electronegativities is

$$3.5 - 1.2 = 2.3$$

Because the difference is 2.3, the bond between magnesium and oxygen is ionic.

(d) From Table 7.1 we find that the electronegativity of H is 2.1. The difference between the electronegativities of the two H atoms is

$$2.1 - 2.1 = 0.0$$

The bond between the two H atoms is very definitely a covalent bond. (It's as covalent as it can get!)

Practice Exercise 7.3   Determine whether the bonds in each of the following compounds are ionic or covalent (use Table 7.1): (*a*) H_2O (*b*) KF (*c*) MgS (*d*) Cl_2

7.7 Ionic Percentage and Covalent Percentage of a Bond

So far we have thought of compounds as either ionically bonded or covalently bonded. We decided which *type* of bond by calculating the difference in electronegativities of the elements forming the bond. A difference of less than 2.0 between the electronegativities of the two elements meant that the bond was covalent, whereas a difference of 2.0 or more meant that the bond was ionic. But this is an oversimplified approach.

Table 7.2
The Relationship Between Electronegativity Difference and the Ionic Percentage and
Covalent Percentage of a Chemical Bond

Difference in electronegativity	Ionic percentage	Covalent percentage
0.0	0.0	100
0.1	0.5	99.5
0.2	1.0	99.0
0.3	2.0	98.0
0.4	4.0	96.0
0.5	6.0	94.0
0.6	9.0	91.0
0.7	12.0	88.0
0.8	15.0	85.0
0.9	19.0	81.0
1.0	22.0	78.0
1.1	26.0	74.0
1.2	30.0	70.0
1.3	34.0	66.0
1.4	39.0	61.0
1.5	43.0	57.0
1.6	47.0	53.0
1.7	51.0	49.0
1.8	55.0	45.0
1.9	59.0	41.0
2.0	63.0	37.0
2.1	67.0	33.0
2.2	70.0	30.0
2.3	74.0	26.0
2.4	76.0	24.0
2.5	79.0	21.0
2.6	82.0	18.0
2.7	84.0	16.0
2.8	86.0	14.0
2.9	88.0	12.0
3.0	89.0	11.0
3.1	91.0	9.0
3.2	92.0	8.0

Chemists find it better to express chemical bonds as a certain percentage ionic and a certain percentage covalent. Table 7.2 relates difference in electronegativities to the ionic and covalent percentages of any bond. With this table, we can state that the hydrogen–oxygen bond in water (with an electronegativity difference of 1.4) is 39% ionic and 61% covalent. (The percentages listed in Table 7.2 come directly from Pauling's work.)

Example 7.4

Determine the ionic and covalent percentages for the bonds in each of the following compounds: (a) NH_3 (b) SO_2 (c) HCl (d) NaF

Solution

(a) Look up the electronegativities of nitrogen and hydrogen, determine the difference, and use Table 7.2 to find the ionic and covalent percentages of the bond. The electronegativity of N is 3.0, and that of H is 2.1. The difference between these electronegativities is

$$3.0 - 2.1 = 0.9$$

This corresponds to a bond that is 19% ionic and 81% covalent.
(b) The electronegativity of S is 2.5 and that of O is 3.5. The difference between these electronegativities is

$$3.5 - 2.5 = 1.0$$

This corresponds to a bond that is 22% ionic and 78% covalent.
(c) The electronegativity of H is 2.1 and that of Cl is 3.0. The difference between these electronegativities is

$$3.0 - 2.1 = 0.9$$

This corresponds to a bond that is 19% ionic and 81% covalent.
(d) The electronegativity of Na is 0.9 and that of F is 4.0. The difference between these electronegativities is

$$4.0 - 0.9 = 3.1$$

This corresponds to a bond that is 91% ionic and 9% covalent.

Practice Exercise 7.4   Determine the ionic and covalent percentages for the bonds in each of the following compounds:
(a) K_2O (b) $CaCl_2$ (c) Mg_3N_2 (d) HF

Example 7.5

List the following compounds in order of decreasing covalence of their bonds: H_2S, H_2O, CO_2, and N_2.

Solution

Using Tables 7.1 and 7.2, determine the percentage covalence of each compound. In the compound H_2S, the H—S bond has an electronegativity difference of 0.4, which means it is 96% covalent.

In the compound H_2O, the O—H bond has an electronegativity difference of 1.4, which means it is 61% covalent.

In the compound CO_2, the C=O bond has an electronegativity difference of 1.0, which means it is 78% covalent.

In the compound N_2, the N≡N bond has an electronegativity difference of 0.0, which means it is 100% covalent.

Therefore the order of covalence is

N_2	H_2S	CO_2	H_2O
Most covalent			Least covalent

Practice Exercise 7.5   List the following compounds in order of decreasing covalence of their bonds: CO_2, $AsCl_3$, CS_2, H_2S.

7.8 Polyatomic Ions

Over the years, in studying the composition of many compounds, chemists have found that certain groups of covalently bonded atoms appear over and over again. Because these groups are electrically charged, they are not molecules but ions. More specifically, they are *polyatomic ions* (ions that are combinations of many atoms); the prefix *poly* is from the Greek word for "many." A **polyatomic ion** is a *charged group of covalently bonded atoms.*

Examples of polyatomic ions are: nitrate, $(NO_3)^{1-}$; sulfite, $(SO_3)^{2-}$; sulfate, $(SO_4)^{2-}$; carbonate, $(CO_3)^{2-}$; and phosphate, $(PO_4)^{3-}$. Polyatomic ions are common in minerals, plants, animals, and human beings. *They never exist in an uncombined state* but are always part of a chemical compound. For example, one can't isolate nitrate ions and put them in a bottle, but there are hundreds of chemical compounds made up, in part, of nitrate ions. Some examples of nitrate compounds are potassium nitrate, KNO_3 (used in chemical fertilizers); calcium nitrate, $Ca(NO_3)_2$ (used in chemical fertilizers and also in matches); and silver nitrate, $AgNO_3$ (used in photography, mirror manufacturing, hair dyeing, and silver plating and as an external medicine).

7.9 Oxidation Numbers

Chemists who have studied how elements combine to form compounds have discovered certain trends. It seems that elements tend to form ions with specific charges or they tend to form only a certain number of covalent bonds. To describe this phenomenon, chemists have devised a system to indicate how elements combine to form compounds. This system involves the assignment of oxidation numbers to the substances involved in forming the compound. An **oxidation number** is *the positive or negative whole number that expresses the combining capacity of an element in a particular compound.* Oxidation numbers are not real, but are "tools" created by chemists for bookkeeping purposes.

Learning Goal 7
Writing chemical formulas

The oxidation number can be positive or negative, depending on whether the element tends to attract electrons strongly or give them up. In carbon tetrachloride (Figure 7.11), the carbon has an oxidation number of 4+ and the chlorine has an oxidation number of 1−. Elements with high electronegativity values usually have negative oxidation numbers, and elements with low electronegativity values usually have positive oxidation numbers.

Table 7.3 lists the oxidation numbers of some important ions. Note that for many monatomic ions, the oxidation number is the same as the charge on the ion. In other words, because the sodium atom tends to give up one electron when forming a chemical compound, it has a charge of 1+ as a sodium ion. Therefore its oxidation number is also 1+. Elements in Group IA, which tend to become ions with 1+ charges, all have oxidation numbers of 1+. Elements in Group IIA, which tend to become ions with 2+ charges, all have oxidation numbers of 2+. Elements in Group IIIA, which tend to become ions with 3+ charges, all have oxidation numbers of 3+.

Group IVA elements do not usually form ions. However, they still have oxidation numbers, because they do form chemical compounds. The most common oxidation numbers of the Group IVA elements are 4+ and 4−.

The Group VA elements are a bit more difficult to explain. When they form simple ions, they usually do so by gaining three electrons. Therefore they have a charge of 3−. In this case, their oxidation number is also 3−. However, in other compounds in which they bond covalently, their oxidation numbers can be anywhere from 5+ to 5−.

Elements in Group VIA, which tend to become ions with 2− charges, all have oxidation numbers of 2−. (Remember, Group VIA elements tend to gain two electrons to complete their octet.) Elements in Group VIIA, which tend to become ions with 1− charges, all have oxidation numbers of 1−.

Once you know the oxidation numbers of the various ions, you can predict the chemical formulas of any compounds they form. This is because all chemical compounds must be electrically neutral; in other words, they must have no net (overall) charge.

The examples that follow on pages 156–160 will help you learn how to write the chemical formulas of compounds when you know the compound names.

$$Cl$$
$$|$$
$$Cl—C—Cl$$
$$|$$
$$Cl$$

Figure 7.11
Covalent bonding in carbon tetrachloride

Table 7.3
Charges of Ions Frequently Used in Chemistry

1+	2+	3+
Hydrogen H^{1+}	Calcium Ca^{2+}	Iron(III) Fe^{3+}
Lithium Li^{1+}	Magnesium Mg^{2+}	Aluminum Al^{3+}
Sodium Na^{1+}	Barium Ba^{2+}	
Potassium K^{1+}	Zinc Zn^{2+}	
Mercury(I)* Hg_2^{2+}	Mercury(II) Hg^{2+}	
Copper(I) Cu^{1+}	Tin(II) Sn^{2+}	
Silver Ag^{1+}	Iron(II) Fe^{2+}	
Ammonium $(NH_4)^{1+}$	Lead(II) Pb^{2+}	
	Copper(II) Cu^{2+}	

1−	2−	3−
Fluoride F^{1-}	Oxide O^{2-}	Nitride N^{3-}
Chloride Cl^{1-}	Sulfide S^{2-}	Phosphide P^{3-}
Hydroxide $(OH)^{1-}$	Sulfite $(SO_3)^{2-}$	Phosphate $(PO_4)^{3-}$
Nitrite $(NO_2)^{1-}$	Sulfate $(SO_4)^{2-}$	Arsenate $(AsO_4)^{3-}$
Nitrate $(NO_3)^{1-}$	Carbonate $(CO_3)^{2-}$	Borate $(BO_3)^{3-}$
Chlorate $(ClO_3)^{1-}$	Chromate $(CrO_4)^{2-}$	
Permanganate $(MnO_4)^{1-}$	Dichromate $(Cr_2O_7)^{2-}$	
Acetate $(C_2H_3O_2)^{1-}$	Oxalate $(C_2O_4)^{2-}$	
Hydrogen carbonate $(HCO_3)^{1-}$ (bicarbonate)		
Hydrogen sulfate $(HSO_4)^{1-}$ (bisulfate)		
Cyanide $(CN)^{1-}$		

*Note that the mercury(I) ion is a diatomic ion. In other words, you never find an Hg^{1+} ion alone, but always as Hg^{1+}—Hg^{1+}. The two Hg^{1+} ions are bonded to each other.

Example 7.6

Write the chemical formula for sodium chloride.

Solution

Because the sodium ion has a charge of 1+ (Na^{1+}) and the chloride ion has a charge of 1− (Cl^{1-}), the formula for sodium chloride is

$$Na_1^{1+}Cl_1^{1-}$$

The subscripts show that we need one ion of each element to obtain a neutral compound. The subscript 1 is usually not written. Likewise, the charges $1+$ and $1-$ are left out, giving us the chemical formula: $NaCl$.

Practice Exercise 7.6   Write the chemical formula for potassium fluoride.

Let's look more deeply into the formation of chemical compounds. We assume that any compound has a positive part and a negative part. The oxidation number tells us *how* positive or negative a specific ion is. Our job is to choose the right subscripts to balance the charges, so that the compound is electrically neutral. Here are suggestions to help you choose the right subscripts.

1. Write the symbol of each element of the compound, along with its oxidation number. For example, in the compound zinc chloride, write

$$Zn^{2+} Cl^{1-}$$

2. You want the positive side of the compound to balance the negative side. You can bring this about by crisscrossing the numbers.

$$Zn^{2+}Cl^{1-} \qquad \text{or} \qquad Zn_1^{2+} Cl_2^{1-} \qquad \text{or} \qquad ZnCl_2$$

3. Note that the subscript numbers are written without charge (that is, without a plus or minus sign). For example, the formula for aluminum oxide is

$$Al^{3+}O^{2-} \qquad \text{or} \qquad Al_2^{3+}O_3^{2-} \qquad \text{or} \qquad Al_2O_3$$

4. Also note that subscript numbers should be in least-common-denominator form. For example, the formula for barium oxide is

$$Ba^{2+}O^{2-} \qquad \text{or} \qquad Ba_2^{2+}O_2^{2-} \qquad \text{or} \qquad Ba_1^{2+}O_1^{2-} \qquad \text{or} \qquad BaO$$

And the formula for zinc sulfate is

$$Zn^{2+}(SO_4)^{2-} \qquad \text{or} \qquad Zn_2^{2+}(SO_4)_2^{2-}$$
$$\text{or} \qquad Zn_1^{2+}(SO_4)_1^{2-} \qquad \text{or} \qquad ZnSO_4$$

5. Be sure to use parentheses around the polyatomic ions when the subscript outside the parentheses is greater than 1. For example, in sodium nitrate,

$$Na^{1+}(NO_3)^{1-} \qquad \text{or} \qquad Na_1^{1+}(NO_3)_1^{1-} \qquad \text{or} \qquad NaNO_3$$

no parentheses are needed around the nitrate group. However, in calcium nitrate,

$$Ca^{2+}(NO_3)^{1-} \quad \text{or} \quad Ca_1{}^{2+}(NO_3)_2{}^{1-} \quad \text{or} \quad Ca(NO_3)_2$$

parentheses *are* needed around the nitrate group to indicate that the entire group is taken twice.

Example 7.7

Write the formula for barium chloride.

Solution

Table 7.3 lists the charges of the barium and chloride ions, and we begin by writing $Ba^{2+}Cl^{1-}$. Then we choose the right subscripts to balance the charges:

$$Ba_1{}^{2+}Cl_2{}^{1-} \quad \text{or} \quad BaCl_2$$

Practice Exercise 7.7   Write the formula for magnesium oxide.

Example 7.8

Write the formula for aluminum sulfide.

Solution

First we write $Al^{3+}S^{2-}$. Then we choose the right subscripts to balance the charges:

$$Al_2{}^{3+}S_3{}^{2-} \quad \text{or} \quad Al_2S_3$$

Practice Exercise 7.8   Write the formula for strontium sulfide.

You can use the same procedure when you're dealing with polyatomic ions. If two or more of a polyatomic ion are needed to balance the charges, remember to put parentheses around the entire group.

Example 7.9

Write the chemical formula for sodium carbonate.

Solution

First we write $Na^{1+}(CO_3)^{2-}$. Then we choose the right subscripts to balance the charges:

$$Na_2{}^{1+}(CO_3)_1{}^{2-} \quad \text{or} \quad Na_2CO_3$$

Practice Exercise 7.9   Write the formula for potassium dichromate.

Example 7.10

Write the chemical formula for aluminum sulfate.

Solution

First we write $Al^{3+}(SO_4)^{2-}$. Then we choose the right subscripts to balance the charges:

$$Al_2{}^{3+}(SO_4)_3{}^{2-} \qquad \text{or} \qquad Al_2(SO_4)_3$$

Practice Exercise 7.10   Write the formula for aluminum carbonate.

Table 7.3 shows that some atoms can form more than one kind of ion. This is especially true of the transition metals (the ones in the middle of the periodic table) because of their unique electron configurations. An element such as copper can combine with other elements in two different ways. It can combine as Cu^{1+}, called copper(I), or as Cu^{2+} called copper(II).

Example 7.11

Write the chemical formula for copper chloride.

Solution

We can have either copper(I) chloride or copper(II) chloride. The roman numeral tells us the oxidation number of the copper compound that is formed.

If we want copper(I) chloride, we find the formula in the following way. We start with $Cu^{1+}Cl^{1-}$. Then we choose the right subscripts to balance the charges:

$$Cu_1{}^{1+}Cl_1{}^{1-} \qquad \text{or} \qquad CuCl$$

If we want copper(II) chloride, we find the formula in the following way. We start with $Cu^{2+}Cl^{1-}$. Then, again, we choose the right subscripts to balance the charges:

$$Cu_1{}^{2+}Cl_2{}^{1-} \qquad \text{or} \qquad CuCl_2$$

Each of these copper chlorides is a distinctly different compound, with its own properties.

Practice Exercise 7.11   Write the formula for copper(I) bromide.

Example 7.12

Write the chemical formula for iron(III) oxide.

Solution

The (III) tells us that the iron we want is the Fe^{3+} ion, so we begin with $Fe^{3+}O^{2-}$. Then we choose the right subscripts to balance the charges:

$$Fe_2^{3+}O_3^{2-} \quad \text{or} \quad Fe_2O_3$$

Practice Exercise 7.12   Write the formula for iron(II) oxide.

Example 7.13

Write the chemical formula for copper(I) phosphate.

Solution

The (I) tells us that the copper we want is the Cu^{1+} ion, so we first write $Cu^{1+}(PO_4)^{3-}$. Then we choose the right subscripts to balance the charges:

$$Cu_3^{1+}(PO_4)_1^{3-} \quad \text{or} \quad Cu_3PO_4$$

Practice Exercise 7.13   Write the formula for copper(I) nitride.

. . . .
Learning Goal 8
Finding oxidation numbers

Whenever you work in chemistry, you will need to write chemical formulas. And this task is much easier and faster if you have the information in Table 7.3 right at your fingertips. So spend some time mulling over this table and getting acquainted with the oxidation numbers. Actually, the charges or oxidation numbers of most A-group elements can be predicted straight from the periodic table.

For Group IA: 1+
For Group IIA: 2+
For Group IIIA: 3+
For Group IVA: 4±
For Group VA: usually 3± or 5±
For Group VIA: usually 2−
For Group VIIA: usually 1−

However, it is a good idea to memorize the oxidation numbers of the polyatomic ions and transition metals in Table 7.3.

Finding Oxidation Numbers

Table 7.3 lists oxidation numbers only for some of the more common ions. However, once you know these, you can use them to find the oxidation numbers of less common ions from the compounds they form. You need only remember that every compound must be neutral. That is, the sum of the oxidation numbers of all the atoms in the compound must be zero.

Example 7.14

What is the oxidation number of cobalt in $CoCl_3$?

Solution

We know that each chloride ion has a charge of $1-$. The molecule contains three chloride ions, for a total charge of $3-$. For the compound to be neutral, the single cobalt ion must have a charge of $3+$: $Co_1^{3+}Cl_3^{1-}$.

Practice Exercise 7.14   What is the oxidation number of copper in $CuCl_2$?

Example 7.15

Find the oxidation number of each underlined element:
(a) $\underline{Mn}O_2$ (b) $K\underline{Mn}O_4$ (c) $\underline{Cs}_2SO_3$ (d) $\underline{Ga}_2O_3$
We follow the same procedure used in the previous example.

Solution

(a) We know that the oxidation number of oxygen is $2-$: $Mn_1^?O_2^{2-}$
    For the compound to be electrically neutral, the manganese must have a charge of $4+$: $Mn_1^{4+}O_2^{2-}$
(b) We know that the oxidation number of potassium is $1+$, and that of oxygen is $2-$: $K_1^{1+}Mn_1^?O_4^{2-}$
    For the compound to be electrically neutral, the manganese must have a charge of $7+$: $K_1^{1+}Mn_1^{7+}O_4^{2-}$
(c) We know that the oxidation number of the sulfite group is $2-$: $Cs_2^?(SO_3)_1^{2-}$
    For the compound to be electrically neutral, the cesium must have a charge of $1+$: $Cs_2^{1+}(SO_3)_1^{2-}$
(d) We know that the oxidation number of oxygen is $2-$: $Ga_2^?O_3^{2-}$
    For the compound to be electrically neutral, the gallium must have a charge of $3+$: $Ga_2^{3+}O_3^{2-}$

Practice Exercise 7.15   Find the oxidation number of each underlined element: (a) $\underline{Sn}F_4$ (b) $\underline{W}Cl_5$

7.10 Chemical Nomenclature: The Names of Chemical Compounds

Naming chemical compounds is a vast, but systematic, endeavor in chemistry. Chemists from all over the world get together to agree on rules for naming compounds. This is done at meetings of the IUPAC (the International Union of Pure and Applied Chemistry). Some of the more important rules follow.

Rule 1 ▶ The positive ion is always written first. For example, we write NaCl, not ClNa. (If the compound contains a metal and a nonmetal, the name of the metal comes first.)

Rule 2 ▶ For ionic compounds, the name of the compound consists of the names of both *ions*. (Once the elements join together to form compounds, they are not the same elements any more, but ions.) For example, NaCl is sodium chlor*ide*, not sodium chlor*ine*. Names of compounds made up of only two elements always end in *-ide*.

Rule 3 ▶ Compounds made up of elements with variable oxidation numbers must include the roman numeral as part of the compound name. For example, $Fe(NO_3)_3$ is called iron(III) nitrate, not just iron nitrate.

Rule 4 ▶ For compounds containing oxide ions or composed of two nonmetal elements, a Greek prefix is sometimes used to indicate number. For example, CO is carbon *mono*xide, CO_2 is carbon *di*oxide, and PCl_3 is phosphorus *tri*chloride. (The Greek prefixes are given in Table 7.4.)

Rule 5 ▶ With some compounds you have a choice; you can use either the Greek prefix or the roman numeral. For example, PCl_5 can be called either phosphorus(V) chloride or phosphorus pentachloride.

A more advanced discussion of chemical nomenclature appears in Supplement D.

Table 7.4
Greek Prefixes and Their Meanings

mono	= 1
di	= 2
tri	= 3
tetra	= 4
penta	= 5
hexa	= 6
hepta	= 7
octa	= 8
nona	= 9
deca	= 10

Example 7.16

Name the following compounds: (*a*) $CuCl_2$ (*b*) $HgSO_4$
(*c*) $Cu(NO_3)_2$ (*d*) $Fe_2(SO_3)_3$ (*e*) Ag_2CrO_4 (*f*) $Ba_3(PO_4)_2$
(*g*) $MnCl_2$ (*h*) MnO_2 (*i*) Cu_2S (*j*) SnO_2 (*k*) Cs_2CO_3 (*l*) UBr_4

Solution

(a) Copper(II) chloride   **(b)** Mercury(II) sulfate
(c) Copper(II) nitrate   **(d)** Iron(III) sulfite
(e) Silver chromate   **(f)** Barium phosphate
(g) Manganese(II) chloride   **(h)** Manganese(IV) oxide or
                                       manganese dioxide
(i) Copper(I) sulfide   **(j)** Tin(IV) oxide or tin dioxide
(k) Cesium carbonate   **(l)** Uranium(IV) bromide or ura-
                                       nium tetrabromide

Practice Exercise 7.16   Name the following compounds: (*a*) $FeBr_3$
(*b*) Zn_3N_2 (*c*) $Mg(NO_2)_2$ (*d*) $NaHSO_4$

Example 7.17

Write the chemical formula for each of the following compounds:
(a) Aluminum bromide   **(b)** Iron(III) sulfide
(c) Lead(II) nitrate   **(d)** Calcium acetate
(e) Selenium dioxide   **(f)** Arsenic(V) sulfite
(g) Ammonium nitrate   **(h)** Arsenic(III) chloride
(i) Phosphorus pentabromide **(j)** Sulfur trioxide

Solution

(*a*) $AlBr_3$ (*b*) Fe_2S_3 (*c*) $Pb(NO_3)_2$ (*d*) $Ca(C_2H_3O_2)_2$ (*e*) SeO_2
(*f*) $As_2(SO_3)_5$ (*g*) NH_4NO_3 (*h*) $AsCl_3$ (*i*) PBr_5 (*j*) SO_3

Practice Exercise 7.17   Write the chemical formula for each of the following compounds: (*a*) silver sulfate   (*b*) zinc nitride
(*c*) magnesium chloride   (*d*) sodium chlorate

7.11 Shapes and Polarities of Molecules

Up to now, we have used only two-dimensional diagrams to show
how atoms are bonded together in molecules. But our world is *three*-
dimensional, and atoms and molecules, like other objects, have three dimensions. In fact, the three-dimensional shapes of some molecules lead
to additional properties of covalently bonded compounds.

Figure 7.12

Three-dimensional models of some diatomic elements

H:H

:Ö::Ö: :N⋮⋮N:

H:Cl:
   ..

H_2O

—105°—

H   109.5°

H   C   H

H

CH_4

Figure 7.13

Three-dimensional models of hydrogen chloride (HCl), water (H_2O), and methane (CH_4) molecules

δ^+ δ^-

HBr

Figure 7.14

A three-dimensional drawing of HBr. The δ^+ indicates that the hydrogen side of the molecule has a partial positive charge. The δ^- shows that the bromine side of the molecule has a partial negative charge. (δ is the lower-case Greek letter delta.)

Three-Dimensional Characteristics of Molecules

If we could enlarge a single molecule of any diatomic element (for instance, H_2, O_2, or N_2), it would seem to be linear (in a line) (Figure 7.12). This would also be true for covalently bonded compounds consisting of two *un*like atoms, such as HCl (Figure 7.13).

But if we could enlarge a single molecule of water, we would see a bent molecule (Figure 7.13). Experiments have shown that the angle between the hydrogen atoms is about 105 degrees.

If we looked at a molecule of methane (CH_4), we would see a pyramid-shaped symmetrical molecule (Figure 7.13). The angle between neighboring hydrogen atoms in methane has been shown to be 109.5 degrees.

Molecules of other compounds exhibit many other shapes. The important thing about these shapes is that they can affect bonding and the properties of the molecules.

The Not-So-Covalent Bond

Recall that when two elements form a covalent bond, they share electrons. But sometimes they don't share the electrons equally. For example, in the compound hydrogen bromide (HBr), the bromine has a greater attraction for the electrons than the hydrogen does. In other words, bromine has a greater electronegativity than hydrogen. Because of the bromine's greater attraction for electrons, the two electrons being shared spend more time with the bromine than with the hydrogen. This unequal sharing makes the hydrogen seem to have a partial positive charge, and the bromine a partial negative charge (Figure 7.14). This kind of covalent bond, in which there is unequal sharing of electrons, is called a **polar covalent bond.** This is what we meant when we said a bond was a certain percent ionic.

Most covalent bonds are polar. However, bonds formed by the diatomic elements (H—H, O=O, and so on) are nonpolar. This is because there is no difference in electronegativity between the two elements forming the bond. In a **nonpolar covalent bond,** the electrons are shared equally by the atoms.

Example 7.18

Determine whether each of the following covalent bonds is polar or nonpolar. [*Hint:* Use Table 7.1 to determine the electronegativity of each element.] (*a*) N—O bond (*b*) S—O bond (*c*) Cl—Cl bond (*d*) P—O bond (*e*) C—S bond

Solution

We look up the electronegativity values of each element in Table 7.1. If a difference in electronegativity exists between the two elements forming the bond, then the bond is polar. If no difference in electronegativity exists, then the bond is nonpolar.

(a) N——O
   3.0   3.5
  0.5 difference

An electronegativity difference exists, so the bond is *polar.*

(b) S——O
   2.5   3.5
  1.0 difference

An electronegativity difference exists, so the bond is *polar.*

(c) Cl——Cl
   3.0   3.0
  0.0 difference

No electronegativity difference exists, so the bond is *nonpolar.*

(d) P——O
   2.1   3.5
  1.4 difference

An electronegativity difference exists, so the bond is *polar.*

(e) C——S
   2.5   2.5
  0.0 difference

No electronegativity difference exists, so the bond is *nonpolar.*

Practice Exercise 7.18   Determine whether each of the following covalent bonds is polar or nonpolar. [*Hint:* Use Table 7.1.] (*a*) F—F bond (*b*) C—O bond (*c*) B—O bond

Learning Goal 10
Polar and nonpolar bonds

Polar and Nonpolar Molecules

A **polar molecule** (or **dipole**) is a molecule that is positive at one point and negative at another point. For example, hydrogen bromide (Figure 7.14) contains a polar covalent bond. As a result, the hydrogen side of the molecule is positive and the bromine side is negative. Thus HBr is a polar molecule.

And then there is water, a compound with two polar bonds. A molecule of water is also polar, because there is a center of positive charge between the two hydrogens and a center of negative charge on the oxygen (Figure 7.15).

Learning Goal 11
Polar and nonpolar molecules

Figure 7.15

Three-dimensional models of water and carbon tetrachloride molecules. The water molecule is polar; the carbon tetrachloride nonpolar.

H_2O

A molecule of carbon tetrachloride (CCl_4) has four carbon–chlorine bonds. Each bond is a *polar* covalent bond. But the molecule itself is *nonpolar* (Figure 7.15). Here's why: The carbon tetrachloride molecule is shaped like a perfectly symmetrical pyramid. The positive charge is on the carbon atom, but so is the center of the negative charges of the four Cl atoms. The two charges coincide and cancel each other, so the molecule is not polar.

Example 7.19

Determine whether the bonds in each of the following molecules are polar or nonpolar. Then determine whether the molecule itself is polar or nonpolar.

(a) F_2 (linear in shape)

(b) $CHCl_3$, chloroform (shaped like a pyramid)

(c) *Cis*-dichloroethene (the *cis* means that the two chlorines are on the same side of the double bond).

$$\begin{array}{ccc} H & & H \\ & \diagdown \quad \diagup & \\ & C=C & \\ & \diagup \quad \diagdown & \\ Cl & & Cl \end{array}$$

(d) *Trans*-dichloroethene (the *trans* means that the two chlorines are on opposite sides of the double bond).

$$\begin{array}{ccc} H & & Cl \\ & \diagdown \quad \diagup & \\ & C=C & \\ & \diagup \quad \diagdown & \\ Cl & & H \end{array}$$

Solution

(a) Fluorine is a diatomic element; therefore the F—F bond is nonpolar and the F_2 molecule is nonpolar.

(b) The bonding in chloroform is similar to the bonding in carbon tetrachloride. Each bond in chloroform is a polar covalent bond. But the molecule of chloroform, unlike that of carbon tetrachloride, is polar. This is because the center of positive charge and the center of negative charge do not coincide. The molecule is not symmetrical:

$$\begin{array}{ccc} \delta+H & & Cl \\ & \diagdown \quad \diagup & \\ & C & \\ & \diagup \quad \diagdown & \\ Cl & & \delta- Cl \end{array}$$

(c) In *cis*-dichloroethene, each C—H and C—Cl bond is polar. The C=C bond is nonpolar. The molecule is polar, because it has a center of positive charge between the two hydrogens and a center of negative charge between the two chlorines. Because the centers of positive charge and negative charge do not coincide, the molecule is polar.

(d) In *trans*-dichloroethene, the C—H and C—Cl bonds are polar and the C=C bond is nonpolar. The molecule is also nonpolar, because it is shaped in such a way that the center of positive charge coincides with the center of negative charge. The charges coincide as shown in the double bond between the two carbon atoms.

Practice Exercise 7.19 Determine whether the bonds in each of the following molecules are polar or nonpolar. Then determine whether the molecule itself is polar or nonpolar.

Molecule	Shape
BH_3	
Cl_2O	
CO_2	O=C=O

Summary

In a Lewis (electron-dot) diagram of an element, the symbol for the element is surrounded by dots. The dots correspond to the electrons in the outermost energy level of an atom of the element. Electron-dot diagrams are used to show how atoms combine to form molecules.

It has long been known that Group VIIIA elements tend to be very stable. Scientists believe this stability is due to the fact that each element in this group (except helium) has eight electrons in its outermost energy level. The atoms of other elements can achieve this same sort of stability by obtaining eight electrons in their outermost energy level. To do so, two atoms may share electrons in a covalent bond. A single covalent bond is formed when each atoms donates one electron to the bond, a double covalent bond is formed when each atom donates two electrons, and a triple covalent bond is formed when each atom donates three electrons.

In a coordinate covalent bond, only one atom donates the two electrons required to form the bond. However, both atoms obtain an octet of electrons in their outermost shell.

Elements that are found in molecules with two atoms each (such as H_2) are called diatomic elements; these elements are hydrogen, oxygen, nitrogen, and the Group VIIA elements. The bonds between their atoms are covalent bonds.

An ion is an atom or group of atoms that has gained or lost electrons and therefore has a negative or positive electric charge. In ionic bonding, atoms of one element donate electrons to the atoms of a second element. Both become ions, but ions of opposite charges. These opposite charges attract, and this attraction holds the atoms together in a molecule.

A polyatomic ion is a charged group of covalently bonded atoms. The oxidation numbers of monatomic ions express their charges and are used in writing the formulas for molecules.

Electronegativity is the attraction that an atom has for the electrons it is sharing with another atom. If the difference between the electronegativities of two elements is 2.0 or greater, the bond between the two elements is said to be ionic. If the difference is less than 2.0, the bond is covalent. The concept of electronegativity is also used in determining the ionic and covalent percentages of bonds.

Because of their three-dimensional character, some molecules exhibit unequal sharing of electrons, resulting in polar covalent bonds. Such molecules may have a positive side and a negative side; if so, they are called polar molecules or dipoles.

. . . .

Career Sketch

Astronomers

Astronomers explore the fundamental nature of the universe. They seek to know how it originated and how our solar system evolved. Astronomers—sometimes called *astrophysicists*—use the principles of physics and mathematics to study the behavior of matter and energy in distant galaxies. Some use the information they gain to support (or disprove) theories about the nature of matter and energy. An example is Einstein's theory of relativity.

To observe the universe, astronomers use large telescopes, radiotelescopes, and other instruments that can detect electromagnetic radiation from distant sources. Today's astronomers spend little time actually looking at stars through telescopes, because photographic and electronic light-detecting equipment is more effective with dim or distant stars and galaxies. By using spectroscopes to analyze light from stars, astronomers can determine their chemical composition. Astronomers also use radiotele-

With the use of sophisticated equipment, astronomers continually monitor our universe.
(Mount Wilson and Palomar Observatories)

scopes and other electronic means to observe radio waves, x rays, and cosmic rays. They use electronic computers to analyze data and to solve complex mathematical equations that they develop to represent various theories. They also use computers for processing astronomical data so that they can calculate the orbits of asteroids or comets, guide spacecraft, and work out tables for navigational handbooks.

Self-Test Exercises

Learning Goal 1: Lewis Structures

1. Write electron-dot structures for the following elements: (*a*) Sb (*b*) I (*c*) Ba (*d*) Rb (*e*) Kr

2. Write the electron-dot structure for each of the following elements: (*a*) F (*b*) Ca (*c*) K (*d*) C (*e*) B

3. Write the electron-dot structure for: (*a*) a chlorine atom, (*b*) a chloride ion.

4. Write the electron-dot structure for: (*a*) a bromine atom, (*b*) a fluoride ion.

5. Write the electron-dot structure for: (*a*) a beryllium atom, (*b*) a beryllium ion.

6. Write the electron-dot structure for: (*a*) a barium atom, (*b*) a barium ion.

7. Write electron-dot structures for the following elements: (*a*) S (*b*) Al (*c*) Mg (*d*) Ge

8. Write the electron-dot structure for each of the following elements: (*a*) Si (*b*) Te (*c*) At

9. Write electron-dot structures for the following ions: (*a*) magnesium ion (*b*) sodium ion (*c*) aluminum ion

10. Write the electron-dot structure for each of the following ions: (*a*) calcium ion (*b*) cesium ion (*c*) boron ion

11. Write electron-dot structures for the following ions: (*a*) bromide ion (*b*) selenide ion (*c*) phosphide ion

12. Write the electron-dot structure for each of the following ions: (*a*) sulfide ion (*b*) telluride ion (*c*) nitride ion

Learning Goal 2: Covalent Bonds

13. Define the term *covalent bond*.

14. Define the term *coordinate covalent bond*.

15. When each atom donates *two* electrons to form a covalent bond, the bond formed is a _____ (single, double, quadruple) covalent bond.

16. When each atom donates *three* electrons to form a covalent bond, the bond formed is a _____ (single, triple, sextuple) covalent bond.

Learning Goal 3: Diatomic Elements

17. Name the diatomic elements from memory.

18. Which of the diatomic elements contain single covalent bonds? Which contain double bonds? Which contain triple bonds?

Learning Goal 4: Electron-Dot Structures of Covalent Compounds

19. Write the electron-dot diagram for each of the following compounds.
(*a*) $CHBr_3$ (*b*) C_2H_2 (*c*) PH_3 (*d*) HCN

20. Write the electron-dot diagram for each of the following covalent compounds: (*a*) HCl (*b*) CBr_4 (*c*) SO_3 (*d*) $SbCl_3$

21. Write the electron-dot diagram for each of the following covalent compounds.
(*a*) CF_4 (*b*) C_2H_4 (*c*) AsH_3 (*d*) H_2S

22. Write the electron-dot diagram for each of the following covalent compounds: (*a*) CS_2 (*b*) CO_2 (*c*) Cl_2O (*d*) C_3H_8

23. Not all compounds follow the octet rule. Show the bonding diagram for the covalent compound PCl_5. How many electrons surround the P atom in this compound?

24. Not all compounds follow the octet rule. Show the bonding diagram for the compound BF_3. How many electrons surround the boron atom in this compound?

25. Draw the bonding diagram for each of the following covalent compounds, which have coordinate covalent bonds: (*a*) HNO_3 (*b*) H_2SO_3

26. Draw the bonding diagram for the compound chloric acid, $HClO_3$, which has a coordinate covalent bond. [*Hint:* The chlorine is bonded to three oxygen atoms, and the hydrogen is bonded to one of the oxygen atoms.]

Learning Goal 5: Ionic Bonds

27. Define the term *ionic bond*.

28. How is an ionic bond different from a covalent bond?

Learning Goal 6: Ionic and Covalent Bonds

29. Determine whether each of the following compounds is bonded ionically or covalently.
(*a*) KCl (*b*) NH₃ (*c*) CaF₂ (*d*) HBr (*e*) OF₂

30. Determine whether each of the following compounds is bonded ionically or covalently:
(*a*) MgO (*b*) BaCl₂ (*c*) BaO (*d*) CaF₂ (*e*) SO₂

31. Find the ionic and covalent percentages for the bond in each of the following compounds.
(*a*) CH₄ (*b*) N₂ (*c*) FeO (*d*) Al₂O₃

32. Find the ionic and covalent percentages for the bonds in each of the following compounds:
(*a*) HCl (*b*) NH₃ (*c*) KI (*d*) CS₂

33. List the following compounds in order of *decreasing* covalence: HI, HCl, HF and HBr.

34. List the following compounds in order of *decreasing* covalence: Br₂, SO₂, NH₃, HCl.

35. Find the ionic and covalent percentages for the bonds in each of the following compounds.
(*a*) CF₄ (*b*) AsH₃ (*c*) CsCl (*d*) CuO

36. Find the ionic and covalent percentages for the bonds in each of the following compounds:
(*a*) CO₂ (*b*) F₂ (*c*) MgCl₂ (*d*) BaF₂

37. List the following compounds in order of *increasing* covalence: H₂Se, H₂O, H₂Te, H₂S.

38. List the following compounds in order of *increasing* covalence: HF, NH₃, BH₃, CH₄, H₂O.

39. Which of the following compounds is the *most* ionic? LiCl NaCl KCl RbCl CsCl

40. Which of the following compounds is the *most* ionic: HF, HCl, HBr, or HI?

41. Determine whether the bonds in each of the following compounds are ionic or covalent (use Table 7.1). (*a*) SO₃ (*b*) BiCl₃ (*c*) LiF (*d*) NO

42. Determine whether the bonds in each of the following compounds are ionic or covalent (use Table 7.1). (*a*) LiH (*b*) OF₂ (*c*) CaI₂ (*d*) MgH₂

Learning Goal 7: Writing Chemical Formulas

43. Write the chemical formula for each of the following compounds:
(*a*) copper(I) sulfite
(*b*) mercury(II) phosphate
(*c*) tin(II) arsenate
(*d*) iron(III) acetate
(*e*) aluminum oxide
(*f*) ammonium carbonate
(*g*) uranium hexafluoride
(*h*) sulfur dioxide
(*i*) zinc nitrite
(*j*) lithium hydroxide

44. Write the chemical formula for each of the following compounds:
(*a*) iron(III) carbonate
(*b*) sulfur trioxide
(*c*) aluminum chloride
(*d*) lithium phosphate

45. Write the chemical formula for each of the following compounds:
(*a*) aluminum sulfite
(*b*) copper(I) arsenate
(*c*) tin(II) acetate
(*d*) ammonium sulfate
(*e*) osmium tetraoxide
(*f*) manganese dioxide
(*g*) antimony pentachloride
(*h*) sulfur trioxide
(*i*) iron(III) sulfate
(*j*) copper(II) nitrite

46. Write the chemical formula for each of the following compounds:
(*a*) sodium nitrate
(*b*) calcium hydroxide
(*c*) copper(II) sulfate
(*d*) calcium oxide
(*e*) calcium carbonate
(*f*) dinitrogen monoxide
(*g*) ammonium chloride
(*h*) potassium carbonate
(*i*) lead(II) oxide
(*j*) iron disulfide

47. Write the number that corresponds to each of the following prefixes:
(*a*) octa   (*b*) tri   (*c*) tetra   (*d*) nona   (*e*) penta
(*f*) hepta

48. Match each chemical formula on the left with the proper common and compound name on the right.
(1) $MgSO_4 \cdot 7H_2O$
(2) N_2O
(3) $CaSO_4 \cdot \frac{1}{2}H_2O$
(4) CaO

(*a*) lime (calcium oxide)
(*b*) epsom salt (magnesium sulfate heptahydrate)
(*c*) laughing gas (dinitrogen monoxide)
(*d*) plaster of Paris (calcium sulfate hemihydrate)

Learning Goal 8: Finding Oxidation Numbers

49. Find the oxidation number of the underlined element or polyatomic ion.
(*a*) Na$\underline{Cl}O_4$ (*b*) $\underline{Pb}_3O_4$ (*c*) $\underline{Ge}S_2$ (*d*) $\underline{V}OCl_3$
(*e*) Ca$(\underline{HCO_3})_2$

50. Find the oxidation number of the underlined element: (*a*) H$\underline{Cl}$O   (*b*) H$\underline{Cl}$O$_2$ (*c*) H$\underline{Cl}$O$_3$
(*d*) H$\underline{Cl}$O$_4$

51. Find the oxidation number of the underlined element or polyatomic ion.
(*a*) $\underline{V}_2O_5$ (*b*) $\underline{In}_2O$ (*c*) $\underline{N}_2O$ (*d*) $\underline{Pb}P_5$
(*e*) $\underline{Mg}_3(BO_3)_2$ (*f*) Al$(\underline{ClO_3})_3$

52. Find the oxidation number of the underlined element or polyatomic ion:
(*a*) $H_3\underline{B}O_3$ (*b*) H$\underline{Br}$O$_3$ (*c*) H$\underline{I}$O$_3$ (*d*) $\underline{Cr}PO_3$
(*e*) Al$(\underline{HCO_3})_3$ (*f*) Li$\underline{Cl}$O

Learning Goal 9: Writing Chemical Names

53. Write the names of the following compounds:
(*a*) Ag_2CO_3
(*b*) OsO_4
(*c*) $Hg_3(PO_4)_2$
(*d*) $Fe_2(SO_4)_3$
(*e*) $NaNO_3$
(*f*) SiO_2
(*g*) $CoCl_2$
(*h*) Cu_3N
(*i*) $CuCrO_4$
(*j*) $Zn(OH)_2$

54. Write the names of the following compounds:
(*a*) PbO
(*b*) $AgCl$
(*c*) $CaCO_3$
(*d*) $NaNO_2$
(*e*) $LiCl$
(*f*) CuS
(*g*) Al_2O_3
(*h*) $NaOH$
(*i*) $Mg(OH)_2$
(*j*) Na_2SO_4

55. Name the following compounds:
(*a*) Li_2SO_4
(*b*) $Cu_3(PO_4)_2$
(*c*) $CoBr_2$
(*d*) $AuBr_3$
(*e*) FeN
(*f*) $FeSO_3$
(*g*) $Hg(ClO_3)_2$
(*h*) $Fe(OH)_3$
(*i*) CO
(*j*) Cu_2CrO_4

56. Name the following compounds:
(*a*) FeS
(*b*) $CaSO_4$
(*c*) NaH
(*d*) CaC_2
(*e*) LiI
(*f*) PbS
(*g*) KCN
(*h*) NH_4I
(*i*) H_2S
(*j*) $AlPO_4$

Learning Goals 10 and 11: Polar and Nonpolar Bonds and Molecules

57. What are polar molecules? What are nonpolar molecules?

58. Determine whether the bond in each of the following molecules is polar or nonpolar:
(*a*) H_2 (*b*) NO (*c*) HCl

59. Determine whether the *bonds* in each of the following molecules are polar or nonpolar. Then determine whether the *molecule* itself is polar or nonpolar.

Molecule	Shape	Molecule	Shape
(a) PBr_3	Br、 Br—P—Br	(b) Cl_2O	O Cl Cl
(c) CH_3Cl	H、 H C H Cl	(d) O_2	O=O

60. Determine whether the *bonds* in each of the following molecules are polar or nonpolar. Then determine whether the *molecule* is polar or nonpolar.

Molecule	Shape
(a) BeF_2	F—Be—F
(b) BrF_3	F、 F Br F
(c) CCl_4	Cl Cl—C—Cl Cl
(d) Cl_2	Cl—Cl

61. Determine whether the *bonds* in each of the following molecules are polar or nonpolar. Then determine whether the *molecule* itself is polar or nonpolar.

Molecule	Shape	Molecule	Shape
(a) C_2H_2	H—C≡C—H	(b) NH_3	H、 N H H
(c) CH_4	H H—C—H H	(d) N_2	N≡N

62. Determine whether the *bonds* in each of the following molecules are polar or nonpolar. Then determine whether the *molecule* is polar or nonpolar.

Molecule	Shape
(a) H_2	H—H
(b) CO_2	O=C=O
(c) C_6H_6	H C H—C C—H H—C C—H C H
(d) CH_2O	H—C=O H

Extra Exercises

63. Predict the oxidation number that each of the following elements is *most likely* to have: (*a*) atomic number 53, (*b*) atomic number 15, (*c*) atomic number 38, (*d*) atomic number 34, (*e*) atomic number 87, (*f*) atomic number 13.

64. Write the chemical formulas for the following combinations:
(*a*) Element A with 1 electron in the outermost level and element B with 6 electrons in the outermost level
(*b*) Element C with 7 electrons in the outermost level and element D with 2 electrons in the outermost level

65. Write the chemical formulas for the following combinations:
(*a*) Element A, whose electron configuration is $1s^2 2s^2 2p^4$, and element B, whose electron configuration is $1s^2 2s^2 2p^6 3s^1$
(*b*) Element A, whose electron configuration is $1s^2 2s^2 2p^6 3s^2 3p^1$, and element B, whose electron configuration is $1s^2 2s^2 2p^4$

66. Give the oxidation numbers of elements with the atomic numbers 53, 54, and 55, and explain why they have those particular oxidation numbers.

67. "In both ionic and covalent bonds, *electron deficiencies are satisfied.*" Explain what is meant by this statement.

68. Draw the electron-dot diagram for each of the following: (a) $(PH_4)^{1+}$ (b) CO_2 (c) OF_2

69. Explain why the formula for sodium chloride, NaCl, involves one sodium ion and one chloride ion, but the formula for calcium chloride, $CaCl_2$, has two chloride ions with every one calcium ion.

70. Determine the order of increasing bond polarity among the following groups: $O{=}O$ $N{-}O$ $C{-}O$

71. Which of the following bonds is the most covalent? $C{-}H$ $C{-}O$ $N{-}O$ $C{-}S$

72. Draw electron-dot diagrams for each of the following: (a) $(NH_4)^{1+}$ (b) ClBr

73. Predict the compound that is *most likely* to be formed from each of the following pairs of elements:
(a) Mg and O   (b) Al and O   (c) Cs and S

74. Write the formula for the compound formed from: (a) Al^{3+} and N^{3-} (b) V^{5+} and O^{2-}

75. Name each of the compounds in Exercise 74.

76. Write the formulas for the following compounds:
(a) tin(II) ion plus sulfide ion
(b) copper(I) ion plus phosphate ion
(c) iron(III) ion plus nitrite ion
(d) magnesium ion plus acetate ion

8

The Chemical Equation:
Recipe for a Reaction

Learning Goals

After you've studied this chapter, you should be able to:

1. Write formula equations from word equations, and vice versa.
2. Balance formula equations.
3. Recognize and give examples of some general types of chemical reactions (combination, decomposition, single-replacement, and double-replacement).
4. Predict the products for the various types of reactions.
5. Define the terms *acid*, *base*, and *salt*, and recognize each from its formula.
6. Use the activity series and solubility table to predict reactions.
7. Define the terms *oxidation, reduction, oxidizing agent,* and *reducing agent*.
8. Determine which substance has been oxidized and which substance has been reduced in a redox reaction.
9. Determine which substance is the reducing agent and which substance is the oxidizing agent in a redox reaction.

Introduction

Combine four cups of sifted flour and one cake of yeast soaked in one and one-third cups of 35°C water. Add two tablespoons of olive oil and one teaspoon of salt. Roll out the dough, add tomato sauce and mozzarella cheese, and bake. When these ingredients are baked, has a chemical reac-

Sulfur oxides are formed in the atmosphere when pollutants containing the element sulfur react with oxygen in the air. When these sulfur oxides react with moisture in the air, acids are formed. The final result is acid rain.
(Steve Delaney, EPA)

tion occurred? Yes. What's more, this recipe is a set of directions that tells you how to mix certain *reactants* to yield a delectable *product*, a pizza.

In a sense, we can think of a chemical reaction as a set of directions. When we follow these directions and mix the proper chemicals, we get the products that we want. In this chapter, you will study some important classes of chemical reactions. You will also learn how to write chemical equations and how to interpret the information they provide.

8.1 Word Equations

Learning Goal 1
Formula and word equations

One way to represent a chemical reaction is with words. For example, we can say

Sulfur trioxide + water $\longrightarrow$ sulfuric acid

In a chemical equation such as this one, the reacting substances or **reactants** are shown on the left-hand side. The **products** are shown on the right-hand side. The plus sign means *and*, and the arrow means *yields* or *gives*. Our word equation, then, means that "sulfur trioxide and water react to yield sulfuric acid." (By the way, this equation represents part of the process by which "acid rain" develops from one by-product of fuel combustion.)

Our word equation tells us what happens when sulfur trioxide combines with water. But it tells us nothing about the chemical formulas of the reactants and products. Nor does it tell us anything about *how* the reactants combine to form the products. That's why chemists prefer to use formula notation to write chemical reactions. And, like other people, chemists prefer to use the simpler shorthand notation whenever possible.

8.2 The Formula Equation

Let us substitute the formulas for the reactants and product in our word equation:

$$\text{Sulfur trioxide} + \text{water} \longrightarrow \text{sulfuric acid}$$

$$SO_3 \quad + H_2O \longrightarrow \quad H_2SO_4$$

Now let's do some "atomic bookkeeping" to see whether we have complied with the Law of Conservation of Mass. On the left-hand side of this equation are one sulfur atom, four oxygen atoms, and two hydrogen atoms. And there are exactly the same numbers of these atoms on the right-hand side. So we have not "created" or "destroyed" atoms in writing the formula equation (see Figure 8.1). The equation is *balanced*. A chemical equation is **balanced** when *the total number of atoms of each element on the left-hand side is exactly equal to the total number of atoms of each element on the right-hand side.*

Let's now look at the chemical reaction in which hydrogen and oxygen form water. Because hydrogen and oxygen are both diatomic elements, we must write the equation for this chemical reaction as follows:

$$H_2 + O_2 \xrightarrow[\text{spark}]{\text{Electric}} H_2O$$

(Many reactions occur only under special conditions. We specify these conditions by writing them above or around the arrow.) Now for the

Learning Goal 2
Balancing equations

Figure 8.1
Models show how sulfur trioxide and water produce sulfuric acid.

$$SO_3 \quad + \quad H_2O \longrightarrow H_2SO_4$$

bookkeeping: There are two hydrogens on the left side and two hydrogens on the right side of the equation. However, there are two oxygens on the left side but only *one* oxygen on the right side. This equation is *not balanced,* because the number of atoms of each element on the left *does not* equal the number of atoms of each element on the right.

We can't let the equation stand this way. Every chemical equation must be balanced, because chemical reactions satisfy the Law of Conservation of Mass. As it stands, our reaction falsely implies that an atom of oxygen was lost or destroyed.

8.3 Balancing a Chemical Equation

To balance a chemical equation, we determine how many molecules of each substance are needed to satisfy the Law of Conservation of Mass. We indicate this by placing *coefficients,* or small whole numbers, before the formulas in the equation. These coefficients are usually obtained by trial and error. For example, in our equation for the formation of water, we have

$$H_2 + O_2 \xrightarrow[\text{spark}]{\text{Electric}} H_2O \qquad \text{(not balanced)}$$

Because this unbalanced equation lacks one oxygen atom on the right, we place the coefficient 2 before the formula for water on the right:

$$H_2 + O_2 \xrightarrow[\text{spark}]{\text{Electric}} 2H_2O \qquad \text{(not balanced)}$$

Now we have two oxygen atoms on the left and two on the right. But the coefficient 2 has unbalanced the hydrogen atoms. There are now four on the right, but only two on the left. To remedy this, we use two molecules of H_2:

$$2H_2 + O_2 \xrightarrow[\text{spark}]{\text{Electric}} 2H_2O \qquad \text{(balanced)}$$

Now the equation is balanced: There are four hydrogen atoms and two oxygen atoms on the left and the same numbers on the right. The balanced equation states that two hydrogen molecules and one oxygen molecule react to form two water molecules (Figure 8.2).

There are two rules to keep in mind when you balance a chemical equation:

Rule 1 ▶ The equation must be accurate. Correct formulas for all reactants and products must be shown.

Rule 2 ▶ The Law of Conservation of Mass must be observed. The same number of atoms of each element must appear on both sides of the balanced equation.

Figure 8.2

The formation of water from
hydrogen and oxygen

$$2H_2 \qquad + \qquad O_2 \qquad \longrightarrow \qquad 2H_2O$$

Example 8.1

Write a balanced chemical equation for the reaction

$$\text{Mercury(II) oxide} \xrightarrow{\text{Heat}} \text{mercury metal} + \text{oxygen gas}$$

Solution

We begin by writing the formulas for the reactants and products:

$$HgO \xrightarrow{\text{Heat}} Hg + O_2$$

There is only one oxygen atom on the left, but there are two oxygens
on the right. We balance the oxygens by placing the coefficient 2 in
front of the HgO:

$$2HgO \xrightarrow{\text{Heat}} Hg + O_2$$

Now the oxygens are balanced, but there are two mercury atoms on
the left and only one on the right. We balance the mercury atoms by
placing a 2 in front of the Hg:

$$2HgO \xrightarrow{\text{Heat}} 2Hg + O_2 \qquad \text{(balanced)}$$

The balanced chemical equation shows two mercury atoms and two
oxygen atoms on each side of the equation.

Practice Exercise 8.1   Write a balanced equation for the following reaction.

$$\text{Copper(II) oxide} \longrightarrow \text{copper metal} + \text{oxygen gas}$$

Here are two important points to remember.

1. You *cannot* balance an equation using *subscripts*, because this would
change the formulas of the compounds. For example,

$$HgO \xrightarrow{\text{Heat}} Hg + O_2$$

cannot be balanced by doing this:

$$HgO_2 \xrightarrow{\text{Heat}} Hg + O_2$$

because HgO_2 is not the formula for mercury(II) oxide.

2. You *cannot* insert a coefficient between elements in a chemical formula. For example,

$$HgO \xrightarrow{\text{Heat}} Hg + O_2$$

cannot be balanced by doing this:

$$Hg2O \xrightarrow{\text{Heat}} Hg + O_2$$

because such an expression has no meaning.

After you have balanced a number of equations, you will find it easy to choose the proper coefficients. Here are some suggestions that may help:

1. Balance metal and nonmetal atoms first.

2. Balance hydrogen and oxygen atoms last.

3. Balance complex ions that contain more than one atom *as a group* if they appear unchanged on both sides of the equation.

Example 8.2

Write a balanced chemical equation for the reaction

Aluminum + sulfuric acid $\longrightarrow$ aluminum sulfate + hydrogen gas

Solution

We first write the correct formula for each compound:

$$Al + H_2SO_4 \longrightarrow Al_2(SO_4)_3 + H_2$$

There is one aluminum atom on the left-hand side but two aluminum atoms on the right-hand side. We balance the aluminums by placing the coefficient 2 in front of the Al on the left-hand side:

$$2Al + H_2SO_4 \longrightarrow Al_2(SO_4)_3 + H_2$$

There is one SO_4 group on the left-hand side and three SO_4 groups on the right-hand side. We balance the SO_4 groups by placing a 3 in front of the H_2SO_4:

$$2Al + 3H_2SO_4 \longrightarrow Al_2(SO_4)_3 + H_2$$

Now there are six hydrogens on the left-hand side and two hydrogens on the right-hand side. We balance the hydrogens by placing a 3 in front of the H_2:

$$2Al + 3H_2SO_4 \longrightarrow Al_2(SO_4)_3 + 3H_2 \quad \text{(balanced)}$$

There are now two aluminum atoms, three sulfate groups, and six hydrogen atoms on each side of the equation. And so we have a balanced chemical equation.

Practice Exercise 8.2   Write a balanced equation for the following reaction.

$$\text{Zinc} + \text{sulfuric acid} \longrightarrow \text{zinc sulfate} + \text{hydrogen gas}$$

Example 8.3

Write a balanced equation for each of the following reactions.

(a) Hydrogen gas + nitrogen gas $\xrightarrow[\text{Pressure}]{\text{Heat}}$ ammonia(NH_3)

(b) Sodium oxide + water $\longrightarrow$ sodium hydroxide

(c) Calcium bromide + chlorine $\longrightarrow$ calcium chloride + bromine

(d) Sodium hydroxide + phosphoric acid (H_3PO_4) $\longrightarrow$

sodium phosphate + water

(e) Iron(III) oxide + carbon monoxide $\xrightarrow{\text{Heat}}$ iron + carbon dioxide

Solution

We use the same procedure as in the previous examples: Write the formula for each substance, and then balance the equation.

(a) $H_2 + N_2 \xrightarrow[\text{Pressure}]{\text{Heat}} NH_3$ (unbalanced)

$3H_2 + N_2 \xrightarrow[\text{Pressure}]{\text{Heat}} 2NH_3$ (balanced)

(b) $Na_2O + H_2O \longrightarrow NaOH$ (unbalanced)
$Na_2O + H_2O \longrightarrow 2NaOH$ (balanced)

(c) $CaBr_2 + Cl_2 \longrightarrow CaCl_2 + Br_2$ (balanced)

(d) $NaOH + H_3PO_4 \longrightarrow Na_3PO_4 + H_2O$ (unbalanced)
$3NaOH + H_3PO_4 \longrightarrow Na_3PO_4 + 3H_2O$ (balanced)

(e) $Fe_2O_3 + CO \xrightarrow{\text{Heat}} Fe + CO_2$ (unbalanced)

$Fe_2O_3 + 3CO \xrightarrow{\text{Heat}} 2Fe + 3CO_2$ (balanced)

Practice Exercise 8.3   Write a balanced equation for each of the following word equations.

(a) Aluminum + chlorine gas $\longrightarrow$ aluminum chloride

(b) Sodium + water $\longrightarrow$ sodium hydroxide + hydrogen gas

(c) Potassium nitrate $\longrightarrow$ potassium nitrite + oxygen gas

(d) Nitric acid + barium hydroxide $\longrightarrow$ barium nitrate + water

8.4 Types of Chemical Reactions

The millions of known chemical reactions may be classified in many ways. We shall use a classification scheme that is helpful in predicting the products of many chemical processes. In this scheme, there are four basic classes:

1. Combination reactions

2. Decomposition reactions

3. Single-replacement reactions

4. Double-replacement (or double-exchange) reactions

All of these chemical reactions may also be classified as redox or nonredox reactions. (*Redox* is short for "oxidation–reduction.") We have more to say about these reactions later in the chapter.

Combination Reactions

Combination reactions are *reactions in which two or more substances combine to form a more complex substance* (Figure 8.3). The general formula is

$$A + B \longrightarrow AB$$

In some combination reactions, two *elements* join to form a compound. For example,

$$4Fe + 3O_2 \longrightarrow 2Fe_2O_3$$

Iron   Oxygen   Iron(III)
       gas   oxide

Iron(III) oxide is iron rust. When a product made of iron contacts moist air, the iron rusts, forming a crust of iron(III) oxide. The crust flakes off, a new surface of iron becomes visible, and it too begins to rust. The iron can keep rusting until the entire piece of metal is eaten away.

Another example of a combination reaction that joins two elements is

$$C + 2S \longrightarrow CS_2$$

Carbon   Sulfur   Carbon
                       disulfide

Carbon disulfide is an important chemical used in manufacturing rayon and in electronic vacuum tubes. However, it is a highly poisonous substance and should be used only in well-ventilated areas.

In some combination reactions, two *compounds* join to form a more complex compound. For example,

$$SO_3 + H_2O \longrightarrow H_2SO_4$$

Sulfur   Water   Sulfuric
trioxide   acid

$$Na_2O + H_2O \longrightarrow 2NaOH$$

Sodium   Water   Sodium
oxide   hydroxide

Figure 8.3
A combination reaction:
iron + sulfur ⟶ iron(II)
sulfide

Iron   Sulfur   Iron(II) sulfide

Learning Goal 4
Predicting products of
chemical reactions

Example 8.4

Complete and balance the following equations for combination reactions.

(a) $H_2 + Cl_2 \longrightarrow$?
(b) $K + Br_2 \longrightarrow$?
(c) $Ba + O_2 \longrightarrow$?

Solution

We are told that these are combination reactions, so we know that the products are formed by combination of the elements involved. We first write the correct formula for the product. Then we balance the equation.

(a) $H_2 + Cl_2 \longrightarrow HCl$ (unbalanced)
    $H_2 + Cl_2 \longrightarrow 2HCl$ (balanced)
(b) $K + Br_2 \longrightarrow KBr$ (unbalanced)
    $2K + Br_2 \longrightarrow 2KBr$ (balanced)
(c) $Ba + O_2 \longrightarrow BaO$ (unbalanced)
    $2Ba + O_2 \longrightarrow 2BaO$ (balanced)

Practice Exercise 8.4   Complete and balance the following combination reactions.

(a) $K + O_2 \longrightarrow$?
(b) $Ca + O_2 \longrightarrow$?
(c) $H_2 + Br_2 \longrightarrow$?

Decomposition Reactions

Decomposition reactions are *reactions in which a complex substance is broken down into simpler substances.* They are thus the reverse of combination reactions. The general formula for a decomposition reaction is

$$AB \longrightarrow A + B$$

Some examples of common decomposition reactions follow.

1. The decomposition of a metal oxide (Figure 8.4):

$$2HgO \xrightarrow{\text{Heat}} 2Hg + O_2$$

Mercury(II)   Mercury   Oxygen
oxide   gas

2. The decomposition of a nonmetal oxide:

$$2H_2O \xrightarrow{\text{Electricity}} 2H_2 + O_2$$

3. The decomposition of a metal chlorate:

$$2KClO_3 \xrightarrow{\text{Heat}} 2KCl + 3O_2$$

Potassium   Potassium
chlorate   chloride

Figure 8.4

The decomposition of mercury(II) oxide molecules

$2HgO \longrightarrow 2Hg + O_2$

Mercury(II) oxide   Mercury metal   Oxygen gas

4. The decomposition of a metal carbonate:

$$MgCO_3 \xrightarrow{Heat} MgO + CO_2$$

Magnesium   Magnesium
carbonate   oxide

5. The decomposition of a metal hydroxide:

$$Ca(OH)_2 \xrightarrow{Heat} CaO + H_2O$$

Calcium   Calcium
hydroxide   oxide

Decomposition reactions are a very important class of reactions. They are the ones that produce pure samples of various elements. For example, very pure hydrogen and oxygen can be obtained from the decomposition of water. Pure copper can be produced by the decomposition of copper(II) oxide.

Example 8.5

Complete and balance the following equations for decomposition reactions. [*Hint:* See the examples listed above for the kinds of compounds undergoing decomposition.]

(a) $Mg(OH)_2 \xrightarrow{Heat} ? + ?$

(b) $CaCO_3 \xrightarrow{Heat} ? + ?$

(c) $NaClO_3 \xrightarrow{Heat} ? + ?$
(d) $HCl \longrightarrow ? + ?$

Solution
We first write the correct formulas for the products and then balance the equation.

(a) $Mg(OH)_2 \xrightarrow{Heat} MgO + H_2O$ (balanced)

(b) $CaCO_3 \xrightarrow{Heat} CaO + CO_2$ (balanced)

(c) $NaClO_3 \xrightarrow{\text{Heat}} NaCl + O_2$ (unbalanced)

$2NaClO_3 \xrightarrow{\text{Heat}} 2NaCl + 3O_2$ (balanced)

(d) $HCl \longrightarrow H_2 + Cl_2$ (unbalanced)

$2HCl \longrightarrow H_2 + Cl_2$ (balanced)

Practice Exercise 8.5   Complete and balance the following equations for decomposition reactions.

(a) $Sr(OH)_2 \longrightarrow ? + ?$

(b) $SrCO_3 \longrightarrow ? + ?$

(c) $KClO_3 \longrightarrow ? + ?$

(d) $KCl \longrightarrow ? + ?$

Single-Replacement Reactions

A **single-replacement reaction** is *one in which an uncombined element replaces another element that is in a compound* (Figure 8.5). The general formula for this type of reaction is

$$A + BC \longrightarrow AC + B$$

Examples of single-replacement reactions follow.

1. The replacement of one metal by another metal that is more active (the activity of metals is discussed in Section 8.5):

$$Mg + Cu(NO_3)_2 \longrightarrow Mg(NO_3)_2 + Cu$$

2. The replacement of hydrogen in water by a Group IA metal:

$$2Na + 2H_2O \longrightarrow 2NaOH + H_2$$

3. The replacement of hydrogen in an acid by an active metal:

$$Zn + 2HCl \longrightarrow ZnCl_2 + H_2$$

Figure 8.5

A single-replacement reaction: zinc + copper(II) sulfate $\longrightarrow$ copper + zinc sulfate

Zinc   Copper(II) sulfate   Copper   Zinc sulfate

Example 8.6

Complete and balance the following equations for single-replacement reactions.

(a) $K + H_2O \longrightarrow ? + ?$

(b) $Mg + H_2SO_4 \longrightarrow ? + ?$

(c) $Zn + NiCl_2 \longrightarrow ? + ?$

Solution
We first write the correct formulas for the products and then balance the equation.
(a) $K + H_2O \longrightarrow KOH + H_2$ (unbalanced)
    $2K + 2H_2O \longrightarrow 2KOH + H_2$ (balanced)
(b) $Mg + H_2SO_4 \longrightarrow MgSO_4 + H_2$ (balanced)
(c) $Zn + NiCl_2 \longrightarrow ZnCl_2 + Ni$ (balanced)

Practice Exercise 8.6   Complete and balance the following equations for single-replacement reactions.
(a) $Li + H_2O \longrightarrow ? + ?$
(b) $Al + H_2SO_4 \longrightarrow ? + ?$
(c) $Mg + Al(NO_3)_3 \longrightarrow ? + ?$

Single-replacement reactions are also a very important class of reactions. Chemists often use single-replacement reactions to obtain various metals from compounds containing these metals. For example, you can obtain pure silver metal from the compound silver nitrate by causing a solution of silver nitrate to react with copper metal.

$$Cu + 2AgNO_3 \longrightarrow Cu(NO_3)_2 + 2Ag$$

Double-Replacement Reactions

A **double-replacement** (or *double-exchange*) **reaction** is *one in which two compounds exchange ions with each other* (Figure 8.6). The general formula is

$$A^+B^- + C^+D^- \longrightarrow A^+D^- + C^+B^-$$

Double-replacement reactions usually take place in water. To show that they do, we use the notation (*aq*), which stands for *aqueous,* or in water.
    There are two important kinds of double-replacement reactions. One kind takes place between two salts. A **salt** is a compound composed of a positive and a negative ion. The positive ion may be a monatomic ion of a metal (for example, Na^{1+} or Ca^{2+}) or a polyatomic ion such as $(NH_4)^{1+}$.

. . . .
Learning Goal 5
Acids, bases, and salts

Figure 8.6

A double-replacement reaction: potassium chloride + silver nitrate $\longrightarrow$ potassium nitrate + silver chloride

Potassium chloride   Silver nitrate   Potassium nitrate   Silver chloride

The negative ion may be a monatomic ion of a nonmetal (for example, Cl^{1-}) or a polyatomic ion such as $(SO_4)^{2-}$ or $(NO_3)^{1-}$.

A double-replacement reaction between two salts can occur if one of the *products* separates from the solution—in other words, if it is insoluble in the solution (or if a gas is formed). When a product is insoluble, it separates out of the solution in solid form (called a **precipitate**). To show this, we use the notation (*s*), meaning *solid*. Table B.1 in Supplement B tells you whether a particular product is soluble in water.

When a salt dissolves in water it **dissociates,** or breaks up into its ions. So, for example, when we dissolve silver nitrate ($AgNO_3$) in water, there are Ag^+ and $(NO_3)^-$ ions in the solution. There are no $AgNO_3$ units as such in the water.

Now let's see what happens when we mix an aqueous solution of silver nitrate and an aqueous solution of potassium chloride. Both salts are soluble in water, so both solutions are clear. But when we mix the two solutions and allow them to react, we get a solid precipitate. This precipitate is silver chloride, which is *not* soluble in water. The reaction is

$$KCl(aq) + AgNO_3(aq) \longrightarrow KNO_3(aq) + AgCl(s)$$

The other product, potassium nitrate, remains in solution [as K^+ and $(NO_3)^-$ ions] because it *is* soluble in water.

The second important kind of double-replacement reaction takes place between an acid and a base. The simplest **acids** are *compounds that dissociate in water into hydrogen ions and negatively charged ions of a nonmetal (or negatively charged polyatomic ions)*. Acids release hydrogen ions in water. The simplest **bases** are *composed of metallic ions (or positively charged polyatomic ions) combined with hydroxide ions*. Bases also release hydroxide ions in water.

In an acid–base reaction, the acid and base react to form a salt and water. This class of reaction is also called a **neutralization reaction.** Some examples of neutralization reactions are

$$\underset{\substack{\text{Hydrochloric}\\\text{acid}}}{HCl} + \underset{\substack{\text{Sodium}\\\text{hydroxide}}}{NaOH} \longrightarrow NaCl + H_2O$$

$$\underset{\substack{\text{Sulfuric}\\\text{acid}}}{H_2SO_4} + 2NaOH \longrightarrow Na_2SO_4 + 2H_2O$$

$$\underset{\substack{\text{Phosphoric}\\\text{acid}}}{H_3PO_4} + \underset{\substack{\text{Lithium}\\\text{hydroxide}}}{3LiOH} \longrightarrow Li_3PO_4 + 3H_2O$$

We make much use of neutralization reactions in our daily lives. For example, if you have an acid stomach (caused by excess hydrochloric acid), you take a product such as milk of magnesia to neutralize the excess acidity. Milk of magnesia contains a base—magnesium hydroxide, $Mg(OH)_2$.

Example 8.7

Complete and balance the following equations for double-replacement reactions.

(a) $BaCl_2(aq) + Na_2SO_4(aq) \longrightarrow ? + ?$

(b) $H_3PO_4 + Ca(OH)_2 \longrightarrow ? + ?$

Solution

As usual, we first write the correct formula for each product and then balance the equation.

(a) $BaCl_2(aq) + Na_2SO_4(aq) \longrightarrow$
$$NaCl(aq) + BaSO_4(s) \qquad \text{(unbalanced)}$$
$BaCl_2(aq) + Na_2SO_4(aq) \longrightarrow$
$$2NaCl(aq) + BaSO_4(s) \qquad \text{(balanced)}$$

(b) $H_3PO_4 + Ca(OH)_2 \longrightarrow Ca_3(PO_4)_2 + H_2O \qquad \text{(unbalanced)}$
$2H_3PO_4 + 3Ca(OH)_2 \longrightarrow Ca_3(PO_4)_2 + 6H_2O \qquad \text{(balanced)}$

Practice Exercise 8.7   Complete and balance the following equations for double-replacement reactions.

(a) $BiCl_3 + H_2S \longrightarrow ? + ?$

(b) $Fe(OH)_3 + H_2SO_4 \longrightarrow ? + ?$

8.5 The Activity Series

Learning Goal 6
Using activity series and solubility table

The **activity series** (Table 8.1) is *a list of elements in decreasing order of their reactivity, or ability to react chemically.* The most active elements (those that react most readily) appear at the top of the table, and reactivity decreases as we move down the list. Under ordinary conditions, each element in the table can replace any element that appears below it by taking part in a single-replacement reaction. For example, aluminum is higher in Table 8.1 than zinc, so it replaces zinc in the compound zinc nitrate:

$$2Al + 3Zn(NO_3)_2 \longrightarrow 2Al(NO_3)_3 + 3Zn$$

The activity series is very helpful in determining whether a particular reaction—especially a single-replacement reaction—can take place under ordinary conditions.

Table 8.1
The Activity Series

Metals	Nonmetals
lithium	fluorine
potassium	chlorine
calcium	bromine
sodium	iodine
magnesium	
aluminum	
zinc	
chromium	
iron	
nickel	
tin	
lead	
HYDROGEN*	
copper	
mercury	
silver	
platinum	
gold	

*Hydrogen is in capital letters because the activities of the other metals were calculated relative to the activity of hydrogen.

Example 8.8

Predict whether each of the following reactions can occur.

(a) $Cu + 2HCl \xrightarrow{\text{Heat}} CuCl_2 + H_2$

(b) $H_2 + CuO \xrightarrow{\text{Heat}} H_2O + Cu$

Solution

(a) Table 8.1 shows that copper is *below* hydrogen in the activity series. Therefore copper cannot replace hydrogen in the compound HCl, and

$$Cu + HCl \longrightarrow \text{no reaction}$$

(b) Hydrogen is above copper in the activity series, so it can replace copper in the compound CuO. This reaction can occur:

$$H_2 + CuO \longrightarrow H_2O + Cu$$

Practice Exercise 8.8   Predict whether each of the following reactions can occur.
(a) $Cu + H_2SO_4 \longrightarrow CuSO_4 + H_2$
(b) $H_2 + CuCl_2 \longrightarrow 2HCl + Cu$

8.6 Oxidation–Reduction (Redox) Reactions

We've just finished classifying reactions according to one scheme—as combination, decomposition, single-replacement, or double-replacement reactions. Now we will introduce another scheme for classifying reactions—as either oxidation–reduction (redox, for short) reactions or nonredox reactions. Let's begin with some definitions.

Oxidation and Reduction

Learning Goal 7
Oxidation, reduction;
oxidizing and reducing agents

In the past, chemists defined oxidation as the combining of a substance with oxygen. (That's how the process got its name.) Now, however, the concept of oxidation is expanded to include reactions that do not involve oxygen at all. **Oxidation** is *the loss of electrons by a substance undergoing a chemical reaction. (Alternatively, it is an increase in the oxidation number of a substance.)* (If you don't recall the meaning of *oxidation number*, refer to Section 7.9, and note that the oxidation number of an element in the uncombined state is zero.) As an example, look at the following equation, in which the oxidation number of the iron is shown over its symbol:

$$\overset{0}{4Fe} + 3O_2 \longrightarrow \overset{3+}{2Fe_2O_3}$$

The oxidation number of the iron changes from 0 to +3. By our definition, this means that the iron is oxidized. But what happens to the oxygen? We say that the oxygen has been *reduced*, which leads us to our definition of reduction. **Reduction** is *the gain of electrons by a substance undergoing a chemical reaction. (Alternatively, it is a decrease in the oxidation number of a substance.)*

In the reaction between iron and oxygen, the oxidation number of the oxygen changes from 0 to −2, because each oxygen atom gains two electrons from the iron:

$$4Fe + \overset{0}{3O_2} \longrightarrow \overset{2-}{2Fe_2O_3}$$

Note that, in this reaction, one substance (iron) is oxidized, while the other substance (oxygen) is reduced. This is not a coincidence: The processes of oxidation and reduction always occur together. In other words, oxidation cannot take place in a reaction unless there is also reduction. And reduction can't take place in a reaction unless there is also oxidation. A reaction in which oxidation and reduction take place is called an **oxidation–reduction** (or **redox**) reaction. All other reactions are **nonredox** reactions.

Example 8.9

In each of the following reactions, determine which substance is oxidized and which substance is reduced.

(a) $2K + Br_2 \longrightarrow 2KBr$

(b) $S + O_2 \longrightarrow SO_2$

(c) $Mg + H_2SO_4 \longrightarrow MgSO_4 + H_2$

(d) $2H_2O \longrightarrow 2H_2 + O_2$

Learning Goal 8
Determining oxidized and
reduced substances

Solution

In each case, we use the ideas of Section 7.9 (and the fact that the oxidation number of an element in its uncombined state is zero) to write in the oxidation numbers for the elements.

(a)
$$\overset{0}{2K} + \overset{0}{Br_2} \longrightarrow \overset{1+\;1-}{2\,K\,Br}$$

The oxidation number of the potassium changes from 0 to +1. This increase means that the potassium is oxidized. The oxidation number of the bromine changes from 0 to −1. This decrease means that the bromine is reduced.

(b)
$$\overset{0}{S} + \overset{0}{O_2} \longrightarrow \overset{4+\;2-}{S\,O_2}$$

The oxidation number of the sulfur changes from 0 to +4, so the sulfur is oxidized. The oxidation number of the oxygen changes from 0 to −2, so the oxygen is reduced.

(c)
$$\overset{0}{Mg} + \overset{1+\;\;2-}{H_2(SO_4)} \longrightarrow \overset{2+\;\;2-}{Mg(SO_4)} + \overset{0}{H_2}$$

The oxidation number of the magnesium changes from 0 to +2, so the magnesium is oxidized. The oxidation number of the hydrogen changes from +1 to 0. This decrease means that the hydrogen is reduced. The oxidation number of the sulfate group does not change, so it is neither oxidized nor reduced.

(d)

$$\overset{\text{1+ \quad 2-}}{2H_2\,O} \longrightarrow \overset{\text{0}}{2H_2} + \overset{\text{0}}{O_2}$$

The oxidation number of the oxygen changes from -2 to 0, so the oxygen is oxidized. The oxidation number of the hydrogen changes from $+1$ to 0, so the hydrogen is reduced.

Practice Exercise 8.9   In each reaction, determine which substance is oxidized and which substance is reduced.
(a) $Cu + 2AgNO_3 \longrightarrow 2Ag + Cu(NO_3)_2$
(b) $2Na + Cl_2 \longrightarrow 2NaCl$
(c) $Cl_2 + 2NaBr \longrightarrow 2NaCl + Br_2$
(d) $Zn + S \longrightarrow ZnS$

Oxidizing and Reducing Agents

An **oxidizing agent** is *a substance that causes something else to be oxidized.* A **reducing agent** is *a substance that causes something else to be reduced.* As things work out, the oxidizing agent is the substance being reduced, and the reducing agent is the substance being oxidized. Let's take another look at the reaction in which iron and oxygen form iron(III) oxide:

$$4Fe + 3O_2 \longrightarrow 2Fe_2O_3$$

The iron is the substance being oxidized, and the oxygen is the substance being reduced. Therefore the iron is the *reducing agent* and the oxygen is the *oxidizing agent.*

$$\overset{\text{0}}{4Fe} \quad + \quad \overset{\text{0}}{3O_2} \quad \longrightarrow \quad \overset{\text{3+ \quad 2-}}{2Fe_2O_3}$$

Substance oxidized (reducing agent) Substance reduced (oxidizing agent)

Example 8.10

For each reaction in Example 8.9, state which substance is the oxidizing agent and which substance is the reducing agent.

Solution

Just remember that the substance being oxidized is the reducing agent and the substance being reduced is the oxidizing agent.

Learning Goal 9
Determining oxidizing and
reducing agents

Actually, in this case, the whole compound, H_2SO_4, is considered the oxidizing agent.

The hydrogen is reduced and the oxygen is oxidized. The compound water acts as both oxidizing agent and reducing agent.

Practice Exercise 8.10   For each reaction in Practice Exercise 8.9, state which substance is the oxidizing agent and which substance is the reducing agent.

Summary

A chemical equation is a representation of a chemical reaction. A word equation indicates (in words) which reactants combine (or decompose) and which products are formed as a result of the reaction. In a formula equation, the words are replaced with chemical formulas. In addition, the formula equation must be balanced, so that the number of atoms of each element on the left side of the equation equals the number of atoms of that element on the right side. We balance an equation by placing the proper coefficients before elements or compounds in the equation. Coefficients are generally chosen by trial and error.

Chemical reactions may be classified as combination reactions, decomposition reactions, single-replacement reactions, or double-replacement reactions. Combination reactions are those in which two or more substances combine to form a more complex substance. Decomposition reactions (the reverse of combination reactions) involve the breakdown of a complex substance into simpler substances. A single-replacement reaction is one in which an uncombined element replaces another element in a compound. A double-replacement reaction is one in which two compounds exchange ions with each other.

To predict whether or not a reaction will occur under ordinary conditions, it is important to know the relative activities of the elements involved in the reaction. The activity series is a list of elements in decreasing order of activity or reactivity. It can be used to predict whether or not a single-replacement reaction will occur. That is, each element in the list can replace any element below it in a single-replacement reaction.

Another way of classifying chemical reactions is by dividing them into oxidation–reduction (redox) reactions and non-oxidation–reduction (nonredox) reactions. Oxidation is the loss of electrons by a substance undergoing a chemical reaction (in other words, an increase in the oxidation number of the substance). Reduction is the gain of electrons by a substance undergoing a chemical reaction (in other words, a decrease in the oxidation number of the substance). Oxidation and reduction always occur together. An oxidizing agent is a substance that causes something else to be oxidized, and a reducing agent is a substance that causes something else to be reduced. A redox reaction is a reaction in which oxidation and reduction take place. In such a reaction, the oxidizing agent is always reduced, and the reducing agent is usually oxidized.

Career Sketch

Drafters and Design Technicians

Converting an idea in a designer's mind into precise specifications from which a product can be made is the work of drafters and design technicians. They assist engineers and architects in the design and drafting of electrical circuits, machines, structures, weldments, or architectural plans. They also prepare engineering plans, layouts, and detailed drawings.

Over 300,000 drafters are employed in the United States in government and private industry. These people must be able to "think in three dimensions" and apply mathematics and physical science to such diverse endeavors as construction projects, laboratory design, and the designing of scientific instruments.

Drafters prepare detailed drawings showing exact dimensions and specifications required for building. Here an architectural project is being prepared with the use of a computer.
(CADAM, Inc.)

Some are employed in the field of photographic drawing, wherein photographs replace conventional drawings in the areas of revision work and model designing. In addition, the new field of computer graphics employs those who use the computer to make design sketches, plan layouts, and specify details.

Good hand–eye coordination is required for this type of work, and precise coordination is needed to make the drawings accurate. Understanding of algebra, trigonometry, mechanics, engineering materials, engineering drawing, and descriptive geometry is needed for this field. A knowledge of how to use engineering handbooks, tables, and calculators to help solve technical problems is also required.

Self-Test Exercises

Learning Goals 1 and 2: Formula and Word Equations, Balancing Equations

1. Write a balanced chemical equation for each of the following reactions.

(*a*) Lithium + water $\longrightarrow$
           lithium hydroxide + hydrogen gas

(*b*) Nitric acid (HNO_3) + calcium hydroxide $\longrightarrow$
               calcium nitrate + water

(*c*) Magnesium + zinc nitrate $\longrightarrow$
           magnesium nitrate + zinc

(*d*) Potassium oxide + water $\longrightarrow$
           potassium hydroxide

(*e*) Zinc sulfide + oxygen gas $\overset{\text{Heat}}{\longrightarrow}$
           zinc oxide + sulfur dioxide

(*f*) Aluminum + iron(III)oxide $\longrightarrow$
           aluminum oxide + iron

2. Write a balanced chemical equation for each of the following reactions.

(*a*) Aluminum hydroxide + sulfuric acid $\longrightarrow$
           aluminum sulfate + water

(*b*) Calcium + oxygen $\longrightarrow$ calcium oxide

(*c*) Barium chlorate $\overset{\text{Heat}}{\longrightarrow}$
           barium chloride + oxygen gas

(*d*) Iron(III) hydroxide + sulfuric acid $\longrightarrow$
           iron(III) sulfate + water

(*e*) Lithium + oxygen gas $\longrightarrow$ lithium oxide

(*f*) Phosphorus + iodine $\longrightarrow$ phosphorus triiodide

3. Write a balanced chemical equation for each of the following reactions.

(*a*) Iron + oxygen gas $\longrightarrow$ iron(III) oxide

(*b*) Sulfuric acid + aluminum hydroxide $\longrightarrow$
           aluminum sulfate + water

(*c*) Silver nitrate + barium chloride $\longrightarrow$
           silver chloride + barium nitrate

(*d*) Copper(I) sulfide + oxygen gas $\longrightarrow$
           copper + sulfur dioxide

4. Write a balanced chemical equation for each of the following reactions.

(*a*) Hydrogen gas + oxygen gas $\longrightarrow$ water

(*b*) Potassium hydroxide + acetic acid $\longrightarrow$
           potassium acetate + water

(*c*) Nitrogen gas + hydrogen gas $\longrightarrow$ ammonia

(*d*) Potassium chlorate $\longrightarrow$
           potassium chloride + oxygen gas

Learning Goal 3: Recognizing Types of Chemical Reactions

5. Identify each of the following reactions as a combination, decomposition, single-replacement, or double-replacement reaction.

(*a*) $2HgO \longrightarrow 2Hg + O_2$

(*b*) $H_2SO_4 + Ca(OH)_2 \longrightarrow CaSO_4 + 2H_2O$

(*c*) $2Al + 3ZnCl_2 \longrightarrow 2AlCl_3 + 3Zn$

(*d*) $2Na + Cl_2 \longrightarrow 2NaCl$

(*e*) $H_2 + Cl_2 \longrightarrow 2HCl$

(*f*) $AgNO_3 + HCl \longrightarrow HNO_3 + AgCl$

6. Identify each of the following reactions as a combination, decomposition, single-replacement, or double-replacement reaction.

(*a*) $H_2 + Br_2 \longrightarrow 2HBr$

(*b*) $2PbO_2 \overset{\text{Heat}}{\longrightarrow} 2PbO + O_2$

(*c*) $2KI + Br_2 \longrightarrow 2KBr + I_2$

(*d*) $2Fe(OH)_3 + 3H_2SO_4 \longrightarrow Fe_2(SO_4)_3 + 6H_2O$

(*e*) $2Al + 3Cl_2 \longrightarrow 2AlCl_3$

(*f*) $2K_3PO_4 + 3BaCl_2 \longrightarrow 6KCl + Ba_3(PO_4)_2$

7. Identify the type (combination, decomposition, single-replacement, or double-replacement) of each reaction in Exercise 13.

8. Identify the type (combination, decomposition, single-replacement, or double-replacement) of each reaction in Exercise 14.

9. Identify each of the reactions in Exercise 3 as a combination, decomposition, single-replacement, or double-replacement reaction.

10. Identify each of the reactions in Exercise 2 as a combination, decomposition, single-replacement, or double-replacement reaction.

Learning Goal 4: Predicting Products of Chemical Reactions

11. Complete and balance each of the following equations. Use the activity series and solubility table when necessary.

Combination reactions

(a) $H_2 + Br_2 \longrightarrow$?
(b) $BaO + H_2O \longrightarrow$?
(c) $Na + Cl_2 \longrightarrow$?
(d) Nitrogen gas + oxygen gas $\longrightarrow$
$\qquad\qquad\qquad\qquad$ nitrogen(IV) oxide

Decomposition reactions

(e) $CaCO_3 \xrightarrow{\text{Heat}}$? + ?
(f) $KOH \longrightarrow$? + ?
(g) $Hg(ClO_3)_2 \xrightarrow{\text{Heat}}$? + ?
(h) $PbCl_2 \longrightarrow$? + ?

Single-replacement reactions

(i) $Li + H_2O \longrightarrow$? + ?
(j) Zinc + sulfuric acid $\longrightarrow$? + ?
(k) $Ni + Al(NO_3)_3 \longrightarrow$? + ?
(l) $Al + Hg(C_2H_3O_2)_2 \longrightarrow$? + ?

Double-replacement reactions

(m) Sulfuric acid + ammonium hydroxide $\longrightarrow$
$\qquad\qquad\qquad\qquad\qquad$? + ?
(n) Silver nitrate + barium chloride $\longrightarrow$? + ?
(o) H_2SO_3 (sulfurous acid) + $Al(OH)_3 \longrightarrow$? + ?
(p) $NaNO_3(aq) + KCl(aq) \longrightarrow$? + ?

13. Complete and balance each of the following equations. Use the activity series and solubility table when necessary.

(a) $N_2O \xrightarrow{\text{Heat}}$? + ?
(b) $H_2CO_3 \xrightarrow{\text{Heat}}$? + ?
(c) Sodium nitrate $\xrightarrow{\text{Heat}}$ sodium nitrite + oxygen gas
(d) $H_2 + F_2 \longrightarrow$?
(e) $N_2 + H_2 \xrightarrow[\text{High pressure}]{\text{Heat}}$?
(f) $Ca + HCl \longrightarrow$? + ?
(g) $Cu + NiCl_2 \longrightarrow$? + ?
(h) $AgNO_3(aq) + K_3AsO_4(aq) \longrightarrow$? + ?

15. Complete and balance each of the following equations.

12. Complete and balance each of the following equations. Use the activity series and solubility table when necessary.

Combination reactions

(a) $Cu + S \longrightarrow$?
(b) $CaO + H_2O \longrightarrow$?
(c) $Na + F_2 \longrightarrow$?
(d) Phosphorus + iodine $\longrightarrow$?

Decomposition reactions

(e) $MgCO_3 \xrightarrow{\text{Heat}}$? + ?
(f) $LiOH \longrightarrow$? + ?
(g) $Mg(ClO_3)_2 \longrightarrow$? + ?
(h) $Mg_3N_2 \longrightarrow$? + ?

Single-replacement reactions

(i) $Al + H_3PO_4 \longrightarrow$? + ?
(j) $Zn + Pb(NO_3)_2 \longrightarrow$? + ?
(k) $Cl_2 + KBr \longrightarrow$? + ?
(l) $Br_2 + NaCl \longrightarrow$? + ?

Double-replacement reactions

(m) $AgNO_3 + H_2S \longrightarrow$? + ?
(n) Barium chloride + ammonium carbonate $\longrightarrow$
$\qquad\qquad\qquad\qquad\qquad$? + ?
(o) $CaCO_3 + H_3PO_4 \longrightarrow$? + ?
(p) Silver nitrate + magnesium chloride $\longrightarrow$? + ?

14. Complete and balance each of the following equations. Use the activity series and solubility table when necessary.
(a) $Na_2O \longrightarrow$
(b) $H_2 + I_2 \longrightarrow$
(c) Phosphorus + oxygen gas $\longrightarrow$
$\qquad\qquad\qquad\qquad$ diphosphorus trioxide
(d) $K_3PO_4 + BaCl_2 \longrightarrow$? + ?
(e) $KClO_3 \longrightarrow$? + ?
(f) $Ca + O_2 \longrightarrow$?
(g) $Cl_2 + NH_4I \longrightarrow$? + ?
(h) $Ag_2O \xrightarrow{\text{Heat}}$? + ?

16. Complete and balance each of the following equations.

Combination reactions

(a) $K + Cl_2 \longrightarrow$?
(b) $K_2O + H_2O \longrightarrow$?
(c) $MgO + H_2O \longrightarrow$?
(d) $CaO + CO_2 \longrightarrow$?

Decomposition reactions

(e) $NH_3 \longrightarrow$? + ?
(f) $SrCO_3 \longrightarrow$? + ?
(g) $NaClO_3 \longrightarrow$? + ?
(h) $Mg(OH)_2 \longrightarrow$? + ?

Combination reactions

(a) $K + Br_2 \longrightarrow$?
(b) $Li_2O + H_2O \longrightarrow$?
(c) $BaO + H_2O \longrightarrow$?
(d) $MgO + CO_2 \longrightarrow$?

Decomposition reactions

(e) $MgCO_3 \longrightarrow$? + ?
(f) Copper(I) sulfide $\longrightarrow$? + ?
(g) Chromium(III) carbonate $\longrightarrow$? + ?
(h) $Al(ClO_3)_3 \overset{\text{Heat}}{\longrightarrow}$? + ?

Learning Goal 5: Acids, Bases, and Salts

17. The reaction of an acid and a base produces a salt and water. Find the acid and base needed to produce each of the following salts:
(a) $NaCl$, (b) K_2SO_4, (c) $Al(NO_3)_3$.

18. The reaction of an acid and a base produces a salt and water. Find the acid and base needed to produce each of the following salts:
(a) KCl, (b) $MgSO_4$, (c) $Fe_2(SO_4)_3$.

Learning Goal 6: Using Activity Series and Solubility Table

19. Complete and balance each of the following equations. Check the activity series and solubility table to be sure that the reaction can take place.

Single-replacement reactions

(a) $Zn + HNO_2 \longrightarrow$? + ?
(b) $Ag + NiCl_2 \longrightarrow$? + ?
(c) $Zn + AgNO_3 \longrightarrow$? + ?
(d) $Cs + H_2O \longrightarrow$? + ?

Double-replacement reactions

(e) $HCl + Al(OH)_3 \longrightarrow$? + ?
(f) $KNO_3 + ZnCl_2 \longrightarrow$? + ?
(g) $Al(NO_3)_3 + NaOH \longrightarrow$? + ?
(h) $K_2CrO_4 + Pb(NO_3)_2 \longrightarrow$? + ?

20. Complete and balance each of the following equations. Check the activity series and solubility table to be sure that the reaction can take place.

Single-replacement reactions

(a) $Cu + AgCl \longrightarrow$? + ?
(b) $Cl_2 + KF \longrightarrow$? + ?
(c) $Zn + HNO_3 \longrightarrow$? + ?
(d) $Cu + HCl \longrightarrow$? + ?

Double-replacement reactions

(e) $NH_4NO_3 + H_3PO_4 \longrightarrow$? + ?
(f) $ZnCl_2 + KOH \longrightarrow$? + ?
(g) $Ni_3(PO_4)_2 + HCl \longrightarrow$? + ?
(h) $AgNO_3 + KCl \longrightarrow$? + ?

Learning Goal 7: Oxidation, Reduction; Oxidizing and Reducing Agents

21. Define the following terms, and give an example of each: (a) oxidation, (b) reduction, (c) oxidizing agent, (d) reducing agent.

22. Complete each of the following statements.
(a) The loss of electrons is called _____ .
(b) A gain of electrons is called _____ .
(c) A substance that causes something else to be oxidized is called a(n) _____ .
(d) A substance that causes something else to be reduced is called a(n) _____ .

Learning Goal 8: Determining Oxidized and Reduced Substances

23. Balance each of the following reactions, and determine which substance is oxidized and which is reduced.
(a) $K + Cl_2 \longrightarrow KCl$
(b) $NH_3 \longrightarrow N_2 + H_2$
(c) $CuO + H_2 \longrightarrow Cu + H_2O$
(d) $Sn + Cl_2 \longrightarrow SnCl_4$

24. Balance each of the following reactions, and determine which substance is oxidized and which is reduced.
(a) $Na + F_2 \longrightarrow NaF$
(b) $Al + C \longrightarrow Al_4C_3$
(c) $Cu + S \longrightarrow CuS$
(d) $Ag_2O \longrightarrow Ag + O_2$

Learning Goal 9: Determining Oxidizing and Reducing Agents

25. For each of the reactions in Exercise 23, determine the oxidizing agent and the reducing agent.

26. For each of the reactions in Exercise 24, determine the oxidizing agent and the reducing agent.

27. For each single-replacement reaction in Exercise 19, determine the oxidizing agent and the reducing agent.

28. For each single-replacement reaction in Exercise 20, determine the oxidizing agent and the reducing agent.

Extra Exercises

29. For the reaction
$$5Zn + V_2O_5 \longrightarrow 5ZnO + 2V$$
list (a) the oxidizing agent, (b) the reducing agent, (c) the element oxidized, and (d) the element reduced.

30. Magnesium carbonate and calcium carbonate, when heated, both decompose to form carbon dioxide and the metal oxides. Write these two reactions.

31. Determine whether the following reactions are balanced. If they are not, correct them.
(a) $N_2 + 3H_2 \longrightarrow NH_3$
(b) $Fe + H_2O \longrightarrow Fe_3O_4 + H_2$
(c) $PCl_3 + Cl_2 \longrightarrow PCl_5$

32. A student is asked to balance the following:
Calcium hydroxide + hydrochloric acid $\longrightarrow$
                              calcium chloride + water
The student works the problem and gets the following solution:
$$CaOH + HCl \longrightarrow CaCl + H_2O$$
This attempt is marked incorrect. Can you explain why the answer is wrong?

33. Write a balanced equation for each of the following. If no reaction occurs, write *no reaction*.
(a) Carbon burning in oxygen gas to form carbon monoxide

(b) Hydrogen gas and chlorine gas reacting to form hydrogen chloride gas
(c) Metallic sodium dropped into water to create sodium hydroxide and hydrogen gas
(d) Silver metal and hydrochloric acid reacting to produce _____

34. Consider an experiment in which metal A is immersed in a water solution of ions of metal B. For each of the following pairs A and B, determine whether or not a reaction occurs. If a reaction does occur, write the balanced equation, assuming that the B ions are part of nitrate salts.

	A	B
(a)	Cu	Zn
(b)	Mg	Zn
(c)	Cu	Ni
(d)	Zn	Cu

35. Explain why an oxidizing agent is reduced.

36. Write balanced equations for the following word equations and state what type of reaction each is.
(a) Calcium hydroxide + hydrochloric acid $\longrightarrow$
(b) Zinc metal + sulfuric acid $\longrightarrow$
(c) Barium metal + oxygen gas $\longrightarrow$
(d) Cesium oxide + water $\longrightarrow$

37. Balance the equation

$NH_3(g) + O_2(g) \longrightarrow N_2O_4(g) + H_2O(g)$

38. For the redox reaction

$(Cr_2O_7)^{2-} + 14H^{1+} + 6Fe^{2+} \longrightarrow$
$$2Cr^{3+} + 7H_2O + 6Fe^{3+}$$

(a) Determine which substance is oxidized.
(b) Determine which substance is reduced.
(c) Determine the oxidizing agent.
(d) Determine the reducing agent.

39. Balance the following equation, and write the chemical name of each product and reactant in the balanced equation.

$$Na + P_4 \longrightarrow Na_3P$$

40. Write a balanced equation for a reaction that produces magnesium chloride. [*Hint:* In other words, magnesium chloride has to be at least one of the products.]

41. Write a balanced equation for an acid–base reaction that produces lithium chloride as one of the products.

42. Are single-replacement reactions also redox reactions? Explain with examples.

43. Write a balanced equation for a reaction that produces carbon dioxide as one of the products.

44. A reaction produces zinc phosphate. What were the reactants? Write a balanced equation for the reaction.

9 Stoichiometry: The Quantities in Reactions

Learning Goals

After you've studied this chapter, you should be able to:

1. Calculate the quantities of reactants needed or products yielded in a chemical reaction.
2. Determine which starting material is the limiting reactant in a chemical reaction.

Introduction

In Chapter 8 we wrote balanced equations showing how many atoms or molecules of each substance take part in a reaction. For example, we considered the reaction

$$N_2 + 3H_2 \xrightarrow[\text{Pressure}]{\text{Heat}} 2NH_3$$

In this reaction, one molecule of nitrogen combines with three molecules of hydrogen to form two molecules of ammonia (Figure 9.1). Industry

Figure 9.1
The Haber process

$$3H_2 \qquad + \qquad N_2 \longrightarrow 2NH_3$$

uses this reaction to make ammonia for fertilizers, explosives, and plastics, among many other products. The industrial process based on this reaction is called the *Haber process* for the synthesis of ammonia. (To *synthesize* means to combine the parts into the whole.)

Suppose we also wanted to produce ammonia in large quantities, for commercial use. To prevent waste and thereby minimize costs, we would like to combine just enough of each reactant. But how much is "just enough of each reactant"? And how much ammonia would we obtain? Or suppose we wanted to produce just 100 kilograms of ammonia: How much nitrogen and hydrogen would we need?

To answer such questions, we need to "translate" balanced chemical equations into quantities of reactants and products. In this chapter you will learn how to do so by using stoichiometry, which is pronounced "stoy-key-**om**-eh-tree."

9.1 The Mole Method

Stoichiometry is defined as *the calculation of the quantities of substances involved in chemical reactions.* An important idea in stoichiometry is that a balanced equation tells us how many *moles* of the various substances take part in the reaction. For example, the balanced equation for the Haber process,

$$N_2 + 3H_2 \xrightarrow[\text{Pressure}]{\text{Heat}} 2NH_3$$

tells us that 1 *mole* of nitrogen and 3 *moles* of hydrogen react to form 2 *moles* of ammonia. (If you've forgotten what a mole is, check back to Chapter 3.)

The reasoning behind this idea is as follows: The balanced chemical equation tells us the number of molecules or atoms of each substance that are involved in a single instance of the reaction. It also tells us in what *ratio* these molecules or atoms are involved. In the Haber process, the molecules are involved in the ratio one N_2 to three H_2 to two NH_3, or $1N_2 : 3H_2 : 2NH_3$. Using this ratio, we can also say that 1 *Avogadro's number* of nitrogen molecules and 3 *Avogadro's numbers* of hydrogen molecules produce 2 *Avogadro's numbers* of ammonia molecules. Because 1 Avogadro's number of molecules is 1 mole, we can say that 1 mole of nitrogen and 3 moles of hydrogen react to produce 2 moles of ammonia.

This same sort of reasoning holds true for every balanced equation. *The coefficients in a balanced chemical equation indicate the number of moles of each substance that take part in the equation.*

Once we know the number of moles of each substance in a reaction, we can find out how many grams (or kilograms or pounds) of each substance are involved. For example, we know that the atomic weight of nitrogen is 14 and that the atomic weight of hydrogen is 1. Then we can say that since 1 mole of N_2 weighs 28 grams, 1 mole of H_2 weighs 2 grams, and 1 mole of NH_3 weighs 17 grams, 28 grams of nitrogen react with 6 grams of hydrogen to produce 34 grams of ammonia:

$$N_2 + 3H_2 \xrightarrow[\text{Pressure}]{\text{Heat}} 2NH_3$$

1 mole	3 moles	2 moles
28 grams	6 grams	34 grams

Note that we have satisfied the Law of Conservation of Mass. The mass of the product is equal to the sum of the masses of the reactants.

In the remainder of this chapter, we shall apply this basic idea to several types of stoichiometric problems.

9.2 Quantities of Reactants and Products

Learning Goal 1
Simple stoichiometry

The best way to learn how to solve stoichiometric problems is to work through a number of them. As we do this, remember that the balanced equation for a reaction contains the basic information about the relative quantities of substances involved. Also note that, if we know the quantity of even a single reactant or product, we can obtain the quantities of all other reactants or products.

Let us assume that we want to synthesize 136 grams of ammonia. How many grams of nitrogen and hydrogen do we need? The first thing we do is set up the balanced chemical equation.

$$N_2 + 3H_2 \longrightarrow 2NH_3$$

Next we record what is given and what must be found.

$$N_2 + 3H_2 \longrightarrow 2NH_3$$

$$?\,g \quad ?\,g \qquad 136\,g$$

The next step is to convert grams of ammonia into moles (because the chemical equation relates substances by moles, not grams).

$$? \text{ moles of } NH_3 = (136 \text{ g})\left(\frac{1 \text{ mole}}{17 \text{ g}}\right)$$

$$= 8.00 \text{ moles of } NH_3$$

$$N_2 + 3H_2 \longrightarrow 2NH_3$$

$$?\,g \quad ?\,g \qquad 136\,g \text{ (which is 8.00 moles)}$$

A balance is used to weigh precise quantities of materials in the laboratory.
(Courtesy of Mettler Instrument Corporation)

Now we use the coefficients in the balanced equation to calculate the number of moles of hydrogen and nitrogen necessary to form 8 moles of ammonia. We do it in three steps. First, we determine the number of moles of nitrogen, as shown at the top of the next page.

$$? \text{ moles of } N_2 = (8.00 \text{ moles NH}_3)\left(\frac{1 \text{ mole } N_2}{2 \text{ moles NH}_3}\right)$$

$$= 4.00 \text{ moles of } N_2$$

We obtained the factor

$$\left(\frac{1 \text{ mole } N_2}{2 \text{ moles NH}_3}\right)$$

from the coefficients of the balanced equation. This factor tells us that 1 mole of nitrogen is necessary to produce 2 moles of ammonia. Now we are ready for the second step.

$$? \text{ moles of } H_2 = (8.00 \text{ moles NH}_3)\left(\frac{3 \text{ moles } H_2}{2 \text{ moles NH}_3}\right)$$

$$= 12.0 \text{ moles of } H_2$$

We also obtained the factor

$$\left(\frac{3 \text{ moles } H_2}{2 \text{ moles NH}_3}\right)$$

from the balanced equation, which tells us that 3 moles of hydrogen are necessary to produce 2 moles of ammonia. We now have

$$N_2 \quad + \quad 3H_2 \quad \longrightarrow \quad 2NH_3$$

$$? \text{ g} \qquad\qquad ? \text{ g} \qquad\qquad 136 \text{ g}$$

$$4.00 \text{ moles} \quad 12.0 \text{ moles} \qquad 8.00 \text{ moles}$$

Another way to obtain the same information is to remember that the moles of reactants and products must be in the ratio of

$$N_2 : H_2 : NH_3$$

$$1 : 3 : 2$$

We could ask ourselves the question,

$$1 : 3 : 2 \quad \text{ is the same as } \quad ? : ? : 8$$

To answer this, we would note that

$$1 : 3 : 2$$

$$\downarrow \times 4$$

$$? : ? : 8$$

or

$$1 : 3 : 2 \quad \text{ is the same as } \quad 4 : 12 : 8$$

Our final step is to change *moles* of nitrogen and hydrogen into *grams*.

$$? \text{ grams of } N_2 = (4.00 \text{ moles})\left(\frac{28.0 \text{ g}}{1 \text{ mole}}\right) = 112 \text{ g of } N_2$$

$$? \text{ grams of } H_2 = (12.0 \text{ moles}) \left(\frac{2.00 \text{ g}}{1 \text{ mole}} \right) = 24.0 \text{ g of } H_2$$

The problem is solved.

$$N_2 \quad + \quad 3H_2 \quad \longrightarrow \quad 2NH_3$$

112 g	24.0 g	136 g
4.00 moles	12.0 moles	8.00 moles

We can check our mathematics by seeing whether this reaction obeys the Law of Conservation of Mass. The grams of nitrogen and hydrogen we start out with should equal the grams of ammonia produced.

Mass of reactants	*Mass of products*
Grams of N_2 = 112 g	Grams of NH_3 = 136 g
Grams of H_2 = 24 g	
Total mass = 136 g	Total mass = 136 g

To practice our stoichiometry, let's solve some additional problems.

Many scientists have suggested using propane gas (C_3H_8) in motor vehicles. Because propane burns cleaner than today's gasolines, they feel that the use of propane would drastically reduce the amount of automobile pollution. When propane reacts with sufficient oxygen, the resulting products are carbon dioxide and water, in the form of water vapor. Our problem is this: How many grams of oxygen are needed for combustion of 22 grams of propane? How many grams of carbon dioxide and how many grams of water would be produced? (Remember that C_3H_8 is the formula for propane.)

Step 1

Write a balanced equation, listing what is given and what we want to find.

$$C_3H_8 + 5O_2 \longrightarrow 3CO_2 + 4H_2O$$

22 g	? g	? g	? g

Step 2

Change the grams of C_3H_8 into moles.

$$? \text{ moles of } C_3H_8 = (22 \text{ g}) \left(\frac{1 \text{ mole}}{44 \text{ g}} \right) = 0.50 \text{ mole of } C_3H_8$$

$$C_3H_8 \quad + 5O_2 \longrightarrow 3CO_2 + 4H_2O$$

22 g	? g	? g	? g
0.50 mole			

Step 3

Use the coefficients of the equation to determine the moles of oxygen needed, as well as the moles of carbon dioxide and water produced.

$$? \text{ moles of } O_2 \text{ needed} = (0.50 \text{ mole } C_3H_8)\left(\frac{5 \text{ moles } O_2}{1 \text{ mole } C_3H_8}\right)$$

$$= 2.5 \text{ moles of } O_2 \text{ needed}$$

$$? \text{ moles of } CO_2 \text{ produced} = (0.50 \text{ mole } C_3H_8)\left(\frac{3 \text{ moles } CO_2}{1 \text{ mole } C_3H_8}\right)$$

$$= 1.5 \text{ moles of } CO_2 \text{ produced}$$

$$? \text{ moles of } H_2O \text{ produced} = (0.50 \text{ mole } C_3H_8)\left(\frac{4 \text{ moles } H_2O}{1 \text{ mole } C_3H_8}\right)$$

$$= 2.0 \text{ moles of } H_2O \text{ produced}$$

$$C_3H_8 \quad + \quad 5O_2 \quad \longrightarrow \quad 3CO_2 \quad + \quad 4H_2O$$

22 g	? g	? g	? g
0.50 mole	2.5 moles	1.5 moles	2.0 moles

Note that you can obtain the same result by the other method.

$$1 \; : \; 5 \; : \; 3 \; : \; 4$$

$$0.50 : \; ? \; : \; ? \; : \; ?$$

or

$$1 \; : \; 5 \; : \; 3 \; : \; 4$$

$$0.50 : 2.5 : 1.5 : 2.0$$

Step 4

Change moles of O_2, CO_2, and H_2O into grams.

$$? \text{ grams of } O_2 = (2.5 \text{ moles})\left(\frac{32 \text{ g}}{1 \text{ mole}}\right) = 8\overline{0} \text{ g of } O_2$$

$$? \text{ grams of } CO_2 = (1.5 \text{ moles})\left(\frac{44 \text{ g}}{1 \text{ mole}}\right) = 66 \text{ g of } CO_2$$

$$? \text{ grams of } H_2O = (2.0 \text{ moles})\left(\frac{18 \text{ g}}{1 \text{ mole}}\right) = 36 \text{ g of } H_2O$$

$$C_3H_8 \quad + \quad 5O_2 \quad \longrightarrow \quad 3CO_2 \quad + \quad 4H_2O$$

22 g	8$\overline{0}$ g	66 g	36 g
0.50 mole	2.5 moles	1.5 moles	2.0 moles

Check your results.

Mass of reactants		Mass of products	
Grams of C_3H_8 =	22 g	Grams of CO_2 =	66 g
Grams of O_2 =	$8\overline{0}$ g	Grams of H_2O =	36 g
Total mass	= 102 g	Total mass	= 102 g

Example 9.1

The *thermite* reaction is a chemical reaction that is important to railroads, because railroad workers use it to weld railroad tracks. In this reaction, aluminum metal and iron(III) oxide yield aluminum oxide and iron metal. Once begun, the reaction releases a tremendous amount of heat—so much that the molten iron produced fuses the railroad tracks (Figure 9.2).

How many grams of aluminum and iron(III) oxide are needed to produce $28\overline{0}$ grams of iron metal? How many grams of aluminum oxide will be produced?

Figure 9.2
The thermite reaction in the lab

Solution
Follow the procedure we used before. Do it in steps. [*Note:* Fe_2O_3 stands for iron(III) oxide and Al_2O_3 stands for aluminum oxide.]

Step 1

$$2Al + Fe_2O_3 \longrightarrow Al_2O_3 + 2Fe$$
$$? \text{ g} \quad\quad ? \text{ g} \quad\quad\quad ? \text{ g} \quad\quad 28\overline{0} \text{ g}$$

Step 2

$$? \text{ moles of Fe} = (28\overline{0} \text{ g})\left(\frac{1 \text{ mole}}{56.0 \text{ g}}\right) = 5.00 \text{ moles of Fe}$$

$$2Al + Fe_2O_3 \longrightarrow Al_2O_3 + 2Fe$$

$$? \text{ g} \qquad ? \text{ g} \qquad\qquad ? \text{ g} \qquad 28\overline{0} \text{ g}$$

$$5.00 \text{ moles}$$

Step 3

$$? \text{ moles of Al} = (5.00 \text{ moles Fe})\left(\frac{2 \text{ moles Al}}{2 \text{ moles Fe}}\right)$$

$$= 5.00 \text{ moles of Al}$$

$$? \text{ moles of Fe}_2O_3 = (5.00 \text{ moles Fe})\left(\frac{1 \text{ mole Fe}_2O_3}{2 \text{ moles Fe}}\right)$$

$$= 2.50 \text{ moles of Fe}_2O_3$$

$$? \text{ moles of Al}_2O_3 = (5.00 \text{ moles Fe})\left(\frac{1 \text{ mole Al}_2O_3}{2 \text{ moles Fe}}\right)$$

$$= 2.50 \text{ moles of Al}_2O_3$$

$$2Al \quad + \quad Fe_2O_3 \quad \longrightarrow \quad Al_2O_3 \quad + \quad 2Fe$$

$$? \text{ g} \qquad\qquad ? \text{ g} \qquad\qquad\qquad ? \text{ g} \qquad\qquad 28\overline{0} \text{ g}$$

$$5.00 \text{ moles} \quad 2.50 \text{ moles} \qquad\qquad 2.50 \text{ moles} \quad 5.00 \text{ moles}$$

Step 4

$$? \text{ grams of Al} = (5.00 \text{ moles})\left(\frac{27.0 \text{ g}}{1 \text{ mole}}\right) = 135 \text{ g of Al}$$

$$? \text{ grams of Fe}_2O_3 = (2.50 \text{ moles})\left(\frac{16\overline{0} \text{ g}}{1 \text{ mole}}\right) = 4\overline{00} \text{ g of Fe}_2O_3$$

$$? \text{ grams of Al}_2O_3 = (2.50 \text{ moles})\left(\frac{102 \text{ g}}{1 \text{ mole}}\right) = 255 \text{ g of Al}_2O_3$$

$$2Al \quad + \quad Fe_2O_3 \quad \longrightarrow \quad Al_2O_3 \quad + \quad 2Fe$$

$$135 \text{ g} \qquad 4\overline{00} \text{ g} \qquad\qquad 255 \text{ g} \qquad 28\overline{0} \text{ g}$$

$$5.00 \text{ moles} \quad 2.50 \text{ moles} \qquad\qquad 2.50 \text{ moles} \quad 5.00 \text{ moles}$$

Step 5

Mass of reactants	*Mass of products*
Grams of Al = 135 g	Grams of Al_2O_3 = 255 g
Grams of Fe_2O_3 = 4$\overline{00}$ g	Grams of Fe = 28$\overline{0}$ g
Total mass = 535 g	Total mass = 535 g

Let's do another problem, so we can go through the procedure once more.

Practice Exercise 9.1 Magnesium oxide is a compound that occurs in nature as the mineral periclase. It is composed of the elements magnesium and oxygen. In industry, magnesium oxide is used in the manufacture of refractory crucibles, fire bricks, and casein glue. It is also used as a reflector in optical instruments. It combines with water to form magnesium hydroxide, which acts as an antacid in low concentrations. How many grams of magnesium and oxygen gas are needed to produce 4.032 grams of magnesium oxide?

Example 9.2

Polyvinyl chloride (PVC for short) is one of the best-known and most widely used plastics in the world today. PVC is used to make food wrappings, luggage, plastic garbage bags, phonograph records, and automobile upholstery. It was originally thought that PVC was an extremely safe compound. But studies now indicate that there may be some hidden hazards. The problem is not with the PVC itself, but with the compound vinyl chloride (C_2H_3Cl), from which the PVC is made. Vinyl chloride is a colorless gas, and studies show that it may cause liver damage and some forms of cancer in people who are in constant contact with it. Another problem with PVC and vinyl chloride is that, when the compounds are burned, an extremely poisonous gas (hydrogen chloride, HCl) can be produced.

Our problem is this: Assume that the following reaction occurs when we burn vinyl chloride (C_2H_3Cl).

$$2C_2H_3Cl + 5O_2 \longrightarrow 4CO_2 + 2H_2O + 2HCl$$

How many grams of hydrogen chloride (HCl) will be produced if 6.25 g of vinyl chloride are burned? How many grams of O_2 are needed, and how many grams of CO_2 and water are produced?

Follow the procedure we used before. Do it in steps.

Solution

Step 1

$$2C_2H_3Cl + 5O_2 \longrightarrow 4CO_2 + 2H_2O + 2HCl$$
$$6.25\ g \quad\ ?\ g \qquad\qquad ?\ g \quad\ ?\ g \quad\ ?\ g$$

Step 2

$$? \text{ moles of } C_2H_3Cl = (6.25\ \cancel{g})\left(\frac{1\ \text{mole}}{62.5\ \cancel{g}}\right) = 0.100 \text{ mole of } C_2H_3Cl$$

$$2C_2H_3Cl + 5O_2 \longrightarrow 4CO_2 + 2H_2O + 2HCl$$
$$6.25\ g \quad\ ?\ g \qquad\qquad ?\ g \quad\ ?\ g \quad\ ?\ g$$
$$0.100 \text{ mole}$$

Step 3

$$? \text{ moles of } O_2 = (0.100 \text{ mole } C_2H_3Cl)\left(\frac{5 \text{ moles } O_2}{2 \text{ moles } C_2H_3Cl}\right)$$

$$= 0.250 \text{ mole of } O_2$$

$$? \text{ moles of } CO_2 = (0.100 \text{ mole } C_2H_3Cl)\left(\frac{4 \text{ moles } CO_2}{2 \text{ moles } C_2H_3Cl}\right)$$

$$= 0.200 \text{ mole of } CO_2$$

$$? \text{ moles of } H_2O = (0.100 \text{ mole } C_2H_3Cl)\left(\frac{2 \text{ moles } H_2O}{2 \text{ moles } C_2H_3Cl}\right)$$

$$= 0.100 \text{ mole of } H_2O$$

$$? \text{ moles of } HCl = (0.100 \text{ mole } C_2H_3Cl)\left(\frac{2 \text{ moles } HCl}{2 \text{ moles } C_2H_3Cl}\right)$$

$$= 0.100 \text{ mole of } HCl$$

$2C_2H_3Cl$	+	$5O_2$	$\longrightarrow$	$4CO_2$	+	$2H_2O$	+	$2HCl$
6.25 g		? g		? g		? g		? g
0.100 mole		0.250 mole		0.200 mole		0.100 mole		0.100 mole

Step 4

$$? \text{ grams of } O_2 = (0.250 \text{ mole})\left(\frac{32.0 \text{ g}}{1 \text{ mole}}\right) = 8.00 \text{ g of } O_2$$

$$? \text{ grams of } CO_2 = (0.200 \text{ mole})\left(\frac{44.0 \text{ g}}{1 \text{ mole}}\right) = 8.80 \text{ g of } CO_2$$

$$? \text{ grams of } H_2O = (0.100 \text{ mole})\left(\frac{18.0 \text{ g}}{1 \text{ mole}}\right) = 1.80 \text{ g of } H_2O$$

$$? \text{ grams of } HCl = (0.100 \text{ mole})\left(\frac{36.5 \text{ g}}{1 \text{ mole}}\right) = 3.65 \text{ g of } HCl$$

Step 5

Mass of reactants		*Mass of products*	
Grams of C_2H_3Cl =	6.25 g	Grams of CO_2 =	8.80 g
Grams of O_2 =	8.00 g	Grams of H_2O =	1.80 g
		Grams of HCl =	3.65 g
Total mass =	14.25 g	Total mass =	14.25 g

Practice Exercise 9.2   Aluminum nitrate can be produced in a single-replacement reaction. Aluminum replaces zinc in the compound zinc nitrate, because aluminum is higher than zinc in the activity series. The reaction is as follows:

$$2Al + 3Zn(NO_3)_2 \longrightarrow 2Al(NO_3)_3 + 3Zn$$

Aluminum nitrate is used in tanning leather, in antiperspirant preparations, and as a corrosion inhibitor. How many grams of aluminum nitrate are produced when 9.469 g of zinc nitrate react with aluminum? How many grams of aluminum are needed and how many grams of zinc are produced?

9.3 The Limiting-Reactant Problem

Learning Goal 2
Limiting-reactant problems

Sometimes a chemist who wants to synthesize a compound finds that the quantities of the starting materials are limited. In order to find out how much product can be formed, the chemist must first determine which one of the starting materials will be completely used up. The other will be in *excess*, so not all of it will be used up to form the product. In other words, the chemist must determine which one of the starting materials is the *limiting reactant*.

To clarify this, let's imagine an automobile manufacturer, the Ajax Automobile Company, which makes a model of a car that is called the Ajax. In order to make just *one* of these automobiles, the assembly line needs (among other things) the following parts: *one* body and *four* wheels. We can represent this statement as follows:

$$1 \text{ body} + 4 \text{ wheels} \longrightarrow 1 \text{ automobile}$$

On a particular day, the assembly line has 50 bodies and 160 wheels. How many automobiles can be produced? If you use the number of bodies as the basis for your answer, you might think Ajax could produce 50 automobiles. However, if you use the number of wheels as the basis for your answer, you'll find that Ajax can produce only 40 automobiles. (Remember, it takes 4 wheels to produce an automobile—not counting spare tires. Therefore $160 \div 4 = 40$.) So the company has 10 extra bodies, but not enough wheels to use them. As a result, the most automobiles it can produce is 40. The "limiting reactant" in this case is the wheels, and the bodies are the "reactant" that is in excess.

Let's go through this entire procedure again, as if it were a chemical problem.

Step 1

We follow our usual procedure

$$1 \text{ body} + 4 \text{ wheels} \longrightarrow 1 \text{ automobile}$$
$$50 \text{ bodies} \quad 160 \text{ wheels} \quad\quad ? \text{ automobiles}$$

Step 2

In a chemical problem of this type, we would have to change grams into moles at this point. However, in our car problem, we already know the relationship of bodies to wheels. Therefore all we have to do is determine which substance gets used up first (bodies or wheels).

(a) Determine how many bodies would be needed to use up 160 wheels.

$$? \text{ bodies needed} = (160 \text{ wheels})\left(\frac{1 \text{ body}}{4 \text{ wheels}}\right) = 40 \text{ bodies}$$

(b) Determine how many wheels would be needed to use up 50 bodies.

$$? \text{ wheels needed} = (50 \text{ bodies})\left(\frac{4 \text{ wheels}}{1 \text{ body}}\right) = 200 \text{ wheels}$$

But the Ajax Company doesn't have 200 wheels—only 160. Therefore the wheels limit the number of automobiles it can produce.

Step 3

We may use the number of wheels to calculate the number of automobiles Ajax can produce and the number of bodies it will use. [Note that we've already calculated the number of bodies that the company will use in Step 2(a).]

$$? \text{ automobiles} = (160 \text{ wheels})\left(\frac{1 \text{ automobile}}{4 \text{ wheels}}\right) = 40 \text{ automobiles}$$

Now let's do a chemical problem. Assume that a chemist has 203 grams of $Mg(OH)_2$ (magnesium hydroxide) and 164 grams of HCl (hydrogen chloride). The chemist wants to make these substances react to form $MgCl_2$ (magnesium chloride), which can be used for fireproofing wood, or in a disinfectant. How much $MgCl_2$ can the chemist produce, and how much water is formed?

Step 1

We follow our usual procedure

$$Mg(OH)_2 + 2HCl \longrightarrow MgCl_2 + 2H_2O$$

$$203 \text{ g} \qquad 164 \text{ g} \qquad\qquad ? \text{ g} \qquad ? \text{ g}$$

Step 2

Determine the number of moles of each starting substance.

$$? \text{ moles of } Mg(OH)_2 = (203 \text{ g})\left(\frac{1 \text{ mole}}{58.0 \text{ g}}\right) = 3.50 \text{ moles of } Mg(OH)_2$$

$$? \text{ moles of HCl} = (164 \text{ g})\left(\frac{1 \text{ mole}}{36.5 \text{ g}}\right) = 4.50 \text{ moles of HCl}$$

Step 3

To find which starting substance is used up first, use the following procedure.

(a) Write the balanced equation, and place the number of moles of each substance below its formula.

$$Mg(OH)_2 \; + \quad 2HCl \quad \longrightarrow \quad MgCl_2 + 2H_2O$$

203 g   164 g   ? g   ? g

3.50 moles   4.50 moles

(b) Determine how much HCl would be needed to use up all the $Mg(OH)_2$.

$$? \text{ moles of HCl needed} = (3.50 \; \cancel{\text{moles Mg(OH)}_2}) \left(\frac{2 \text{ moles HCl}}{1 \; \cancel{\text{mole Mg(OH)}_2}} \right)$$

$$= 7.00 \text{ moles of HCl needed}$$

But we have only 4.50 moles of HCl. Therefore:

(c) Our limiting reactant must be HCl, and the $Mg(OH)_2$ must be the reactant in excess. We can check this by calculating the amount of $Mg(OH)_2$ needed to react with all the HCl.

$$? \text{ moles of Mg(OH)}_2 \text{ needed} = (4.50 \; \cancel{\text{moles HCl}}) \left(\frac{1 \text{ mole Mg(OH)}_2}{2 \; \cancel{\text{moles HCl}}} \right)$$

$$= 2.25 \text{ moles of Mg(OH)}_2 \text{ needed (or 131 g)}$$

We have 3.50 moles of $Mg(OH)_2$, so we can expect to have $3.50 - 2.25 = 1.25$ moles of $Mg(OH)_2$ left over, or in excess.

Step 4

Using the number of moles of HCl that we have, calculate the moles of $MgCl_2$ and H_2O that will be formed.

$$? \text{ moles of MgCl}_2 = (4.50 \; \cancel{\text{moles HCl}}) \left(\frac{1 \text{ mole MgCl}_2}{2 \; \cancel{\text{moles HCl}}} \right)$$

$$= 2.25 \text{ moles of MgCl}_2$$

$$? \text{ moles of H}_2O = (4.50 \; \cancel{\text{moles HCl}}) \left(\frac{2 \text{ moles H}_2O}{2 \; \cancel{\text{moles HCl}}} \right)$$

$$= 4.50 \text{ moles of H}_2O$$

Step 5

Calculate the number of grams of $MgCl_2$ and H_2O formed.

$$? \text{ grams of MgCl}_2 = (2.25 \; \cancel{\text{moles}}) \left(\frac{95.0 \text{ g}}{1 \; \cancel{\text{mole}}} \right) = 214 \text{ g of MgCl}_2$$

$$? \text{ grams of H}_2O = (4.50 \; \cancel{\text{moles}}) \left(\frac{18.0 \text{ g}}{1 \; \cancel{\text{mole}}} \right) = 81.0 \text{ g of H}_2O$$

$$Mg(OH)_2 \; + \quad 2HCl \quad \longrightarrow \quad MgCl_2 \quad + \quad 2H_2O$$

131 g   164 g   214 g   81.0 g

2.25 moles   4.50 moles   2.25 moles   4.50 moles

Step 6

Check your mathematics by seeing whether the reaction obeys the Law of Conservation of Mass.

Mass of reactants	Mass of products
Grams of $Mg(OH)_2$ = 131 g	Grams of $MgCl_2$ = 214 g
Grams of HCl used = 164 g	Grams of H_2O = 81 g
Total mass used = 295 g	Total mass produced = 295 g

As we expected, we had 1.25 moles (72.5 g) of extra $Mg(OH)_2$. This material remained unchanged during the course of the reaction.

Example 9.3

Chloroform ($CHCl_3$), a quick-acting, powerful anesthetic used in spy stories to incapacitate the victim, can decompose when it reacts with oxygen. The products formed are HCl and deadly phosgene gas ($COCl_2$), a substance so harmful when inhaled that it was used as a poison gas against enemy troops in World War I.

How many grams of $COCl_2$ and HCl can be formed from 35.4 grams of $CHCl_3$ and 6.40 grams of O_2?

Solution

Follow the procedure we used before.

Step 1

$$2CHCl_3 + O_2 \longrightarrow 2COCl_2 + 2HCl$$
$$35.4 \text{ g} \quad 6.40 \text{ g} \qquad ? \text{ g} \qquad ? \text{ g}$$

Step 2

$$? \text{ moles of } CHCl_3 = (35.4 \text{ g})\left(\frac{1 \text{ mole}}{118 \text{ g}}\right) = 0.300 \text{ mole of } CHCl_3$$

$$? \text{ moles of } O_2 = (6.40 \text{ g})\left(\frac{1 \text{ mole}}{32.0 \text{ g}}\right) = 0.200 \text{ mole of } O_2$$

Step 3

Determine which of the substances is the limiting reactant.

$$2CHCl_3 + O_2 \longrightarrow 2COCl_2 + 2HCl$$
$$35.4 \text{ g} \qquad 6.40 \text{ g} \qquad\quad ? \text{ g} \qquad ? \text{ g}$$
$$0.300 \text{ mole} \quad 0.200 \text{ mole}$$

$$\begin{array}{l} ? \text{ moles of} \\ CHCl_3 \text{ needed} \end{array} = (0.200 \text{ mole } O_2)\left(\frac{2 \text{ moles } CHCl_3}{1 \text{ mole } O_2}\right)$$

$$= 0.400 \text{ mole of CHCl}_3 \text{ needed}$$

We would need 0.400 mole of $CHCl_3$ to use all the oxygen. However, we do not have 0.400 mole of $CHCl_3$. Therefore the $CHCl_3$ must be the limiting reactant, and the O_2 must be the reactant in excess. Let's check this.

$$? \text{ moles of } O_2 \text{ needed} = (0.300 \text{ mole CHCl}_3)\left(\frac{1 \text{ mole } O_2}{2 \text{ moles CHCl}_3}\right)$$

$$= 0.150 \text{ mole of } O_2 \text{ needed (or 4.80 g)}$$

We have more than enough oxygen present to use all the chloroform, because we have 0.200 mole of O_2 and we need only 0.150 mole of it.

Step 4

Using the moles of $CHCl_3$ that are present, calculate the moles of $COCl_2$ and HCl that will be formed.

$$? \text{ moles of } COCl_2 = (0.300 \text{ mole CHCl}_3)\left(\frac{2 \text{ moles } COCl_2}{2 \text{ moles CHCl}_3}\right)$$

$$= 0.300 \text{ mole of } COCl_2$$

$$? \text{ moles of } HCl = (0.300 \text{ mole CHCl}_3)\left(\frac{2 \text{ moles } HCl}{2 \text{ moles CHCl}_3}\right)$$

$$= 0.300 \text{ mole of } HCl$$

Step 5

$$? \text{ grams of } COCl_2 = (0.300 \text{ mole})\left(\frac{98.0 \text{ g}}{1 \text{ mole}}\right) = 29.4 \text{ g of } COCl_2$$

$$? \text{ grams of } HCl = (0.300 \text{ mole})\left(\frac{36.0 \text{ g}}{1 \text{ mole}}\right) = 10.8 \text{ g of } HCl$$

$$2CHCl_3 \quad + \quad O_2 \quad \longrightarrow \quad 2COCl_2 \quad + \quad 2HCl$$

2CHCl₃	O₂	2COCl₂	2HCl
35.4 g	4.80 g	29.4 g	10.8 g
0.300 mole	0.150 mole	0.300 mole	0.300 mole

Step 6

Check your mathematics by seeing whether the reaction obeys the Law of Conservation of Mass.

Mass of reactants		*Mass of products*	
Grams of CHCl₃ used = 35.4 g		Grams of COCl₂	= 29.4 g
Grams of O₂ used = 4.8 g		Grams of HCl	= 10.8 g
Total mass used = 40.2 g		Total mass produced = 40.2 g	

Note that in this reaction we had an excess of 0.050 mole (1.6 g) of O_2. This excess oxygen remained unchanged during the course of the reaction.

Practice Exercise 9.3   Potassium acetate, $KC_2H_3O_2$, is a salt with a molecular weight of 98.14. It can be produced by the reaction of potassium hydroxide with acetic acid. Potassium acetate has been used in veterinary medicine, in combating cardiac arrhythmias, and as an expectorant. It acts as a diuretic in animals. How many grams of potassium acetate can be produced from 28.0 g of potassium hydroxide and 120.0 g of acetic acid?

Summary

Stoichiometry is the calculation of the quantities of elements or compounds involved in chemical reactions. Knowledge of stoichiometry enables chemists to determine the quantities of reactants needed and the quantities of products that are formed. It also permits chemists to produce industrial and commercial chemicals with little or no waste.

The coefficients in a balanced equation indicate the ratio in which molecules (and thus moles) of substances take part in a chemical reaction. The mole method for computing the quantities of reactants and products makes use of this fact. First we find the number of moles of each substance involved in the reaction from this ratio. Then, from this, we find the number of grams of each substance. The same method may be used to determine which of several reactants is the limiting reactant. It may also be used to determine the quantity of product that can be formed with the available amount of the limiting reactant.

- -

Career Sketch

Dentists

It is hardly necessary to describe the duties of dentists. Not only do they fill cavities in our teeth, but dentists also examine teeth and tissues of the mouth to diagnose diseases or abnormalities. They take x rays, straighten teeth, and treat gum diseases. They extract teeth and substitute artificial ones—dentures designed for the individual patient. In addition, they may clean teeth. Some dentists (dental surgeons) also perform corrective surgery of the gums and supporting bones. The professional training of dentists places great stress on biology and chemistry.

*Dentists examine teeth and
tissues of the mouth to
diagnose diseases or
abnormalities.*
(Courtesy Park Dental Group. Photo
S. Sherman)

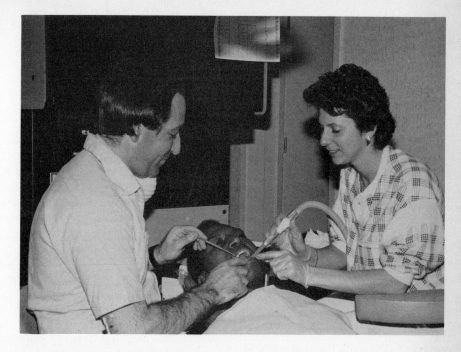

Dentists spend most of their time with patients, but they may devote some time to laboratory work, such as making dentures and inlays. Most dentists, however (particularly those in large cities) send their laboratory work to commercial firms. Some dentists employ dental hygienists to clean patients' teeth and instruct patients in self-care. (See Career Sketch on dental hygienists, page 245.) They may employ others who book appointments, file records, send out bills, help with "chairside" duties, and provide therapeutic services under the supervision of the dentist.

. . . .

Self-Test Exercises

Learning Goal 1: Simple Stoichiometry

1. In some parts of the world, when a person is convicted of a capital crime and is sentenced to death, the execution is carried out with poison gas. The following chemical reaction is used to produce the poison gas:

$$H_2SO_4 + KCN \longrightarrow K_2SO_4 + HCN$$

Hydrogen
cyanide

2. Carbon disulfide can be produced by the reaction

$$C + S \longrightarrow CS_2$$

CS_2 is a very flammable liquid that is also poisonous. It is used in the manufacture of rayon, carbon tetrachloride, and soil disinfectants. Poisoning usually occurs from inhalation, but can also occur due to ingestion and absorption through the skin.

(*a*) Balance the equation.

(a) Balance the equation.
(b) How many grams of sulfuric acid are needed to react with 13.0 g of potassium cyanide?
(c) How many grams of potassium sulfate and hydrogen cyanide are produced?

3. Nitrous oxide (N_2O), commonly called laughing gas, can be prepared from the decomposition of ammonium nitrate:

$$NH_4NO_3 \xrightarrow{\text{Heat}} N_2O + H_2O$$

(a) Balance the equation.
(b) How many grams of ammonium nitrate are needed to prepare 2.2 g of nitrous oxide?
(c) How many grams of water are produced?

5. We can prepare the chemical explosive trinitrotoluene (TNT) by the reaction of toluene (C_7H_8) with nitric acid:

$$C_7H_8 + HNO_3 \longrightarrow \underset{\text{TNT}}{C_7H_5N_3O_6} + H_2O$$

(a) Balance the equation.
(b) How many grams of toluene and nitric acid are necessary to manufacture 1,000 g of TNT?
(c) How many grams of water will be produced in the reaction?

7. The compound ammonium sulfate can be used as a local analgesic (pain reliever). This compound can be prepared by a reaction of ammonium chloride and sulfuric acid.
(a) Write a balanced equation for this reaction.
(b) How many grams of sulfuric acid are needed to react with 15.9 g of ammonium chloride?
(c) How many grams of ammonium sulfate and hydrogen chloride are produced?

*9. Suppose you are a chemist who has been asked to analyze a sample containing potassium chlorate ($KClO_3$) and sodium chloride. You are asked to determine the percentage of potassium chlorate in the sample. You can do this by heating a weighed amount of the sample. The potassium chlorate will decompose to potassium chloride and oxygen gas. You can reweigh the sample after heating, to tell how much oxygen was lost. Then you can backtrack to find the number of grams of potassium chlorate that decomposed to yield this weight of oxygen. From this information, you can then calculate the percentage of potassium chlorate in the sample.

(b) How many grams of carbon are needed to react with 0.160 g of sulfur?
(c) How many grams of carbon disulfide are produced?

4. Lithium phosphate is a white crystalline powder that is soluble in dilute acids. It can be prepared by the reaction

$$H_3PO_4 + LiOH \longrightarrow Li_3PO_4 + H_2O$$

(a) Balance the equation.
(b) How many grams of lithium phosphate can be produced from 168.0 g of lithium hydroxide?
(c) How many grams of phosphoric acid are needed, and how many grams of water are produced?

6. Calcium hydroxide, also known as slaked lime, is used in mortar, plaster, cement, and other building and paving materials. When it reacts with phosphoric acid, it produces calcium phosphate and water.
(a) Write a balanced chemical equation for this reaction.
(b) How many grams of calcium phosphate can be produced from 148 g of calcium hydroxide?
(c) How many grams of phosphoric acid are needed, and how many grams of water are produced?

8. Barium chloride is used in manufacturing pigments and colors, in weighting and dyeing textile fabrics, and in tanning and finishing leather. When it reacts with potassium phosphate, the products formed are potassium chloride and barium phosphate.
(a) Write a balanced equation for this reaction.
(b) How many grams of potassium chloride can be produced from 0.208 g of barium chloride?
(c) How many grams of potassium phosphate are needed, and how many grams of barium phosphate are produced?

*10. A sample containing $NaClO_3$ and KCl must be analyzed to determine the percentage of $NaClO_3$. The sample is weighed and then heated, and the $NaClO_3$ decomposes to NaCl and O_2. The sample is weighed after heating to determine the amount of O_2 lost during decomposition. The number of grams of $NaClO_3$ that decomposed to yield the weight of oxygen is calculated. The percentage of $NaClO_3$ in the sample is then calculated. Perform this analysis using the following information:

Grams of sample before heating, 10.00 g
Grams of sample after heating, 8.00 g

Assume that the weights of your sample are as follows, and find the percentage of potassium chlorate in the mixture of potassium chlorate and sodium chloride.

Grams of sample before heating = 5.00 g

Grams of sample after heating = 4.00 g

11. Hydrogen peroxide (H_2O_2) is a compound with many uses. In dilute solutions (3% H_2O_2 in water), it can be used as an antiseptic for cuts. In a solution of 90% concentration, it can be used as a fuel for rockets. H_2O_2 is prepared in the following way:

$$BaO_2 + H_3PO_4 \longrightarrow H_2O_2 + Ba_3(PO_4)_2$$

Barium
peroxide

(*a*) Balance the equation.
(*b*) How many grams of hydrogen peroxide can be made from 338 g of barium peroxide?
(*c*) How much phosphoric acid is needed to react with 338 g of barium peroxide?

*13. Suppose you have a solution containing 588 g of sulfuric acid, and you put $2\overline{0}0$ g of magnesium into it. A reaction occurs. After the reaction is complete, 56 g of the original magnesium remain. How many grams of hydrogen were produced?

*15. How many grams of *air* are necessary for a complete combustion of $9\overline{0}$ g of ethane (C_2H_6), producing carbon dioxide and water vapor? Assume that the air contains 23% oxygen by weight.

17. Butane, C_4H_{10}, is an organic compound used as a fuel in some types of cigarette lighters. When butane is burned with oxygen, the following reaction takes place.

$$C_4H_{10} + O_2 \longrightarrow CO_2 + H_2O \text{ (unbalanced)}$$

(*a*) Balance the equation.
(*b*) How many grams of oxygen are needed to burn 23.2 g of butane?
(*c*) How many grams of carbon dioxide and water are produced?

19. Ozone, O_3, is a pale-blue gas that can be formed in the atmosphere from O_2. The reaction is

$$3O_2 \longrightarrow 2O_3$$

How many grams of ozone can be formed from 64.0 g of O_2?

21. Aluminum sulfate has been used to tan leather, to treat sewage, and as an antiperspirant. The compound can be formed by the acid–base reaction

12. Lead(II) iodide is a chemical used in bronzing, printing, and photography, as well as in making gold pencils. It is composed of lead and iodine.
(*a*) Write a balanced equation for the formation of lead(II) iodide from its elements.
(*b*) How many grams of lead and iodine are needed to produce 1.38 g of lead(II) iodide?

*14. Suppose you have a solution containing 109.5 g of HCl, and you put 100.0 g of magnesium into it. A reaction occurs. After the reaction is complete, 28.00 g of the original magnesium remain. How many grams of hydrogen were produced?

*16. How many grams of *air* are necessary for the complete combustion of 64 g of methane (CH_4), producing carbon dioxide and water vapor? Assume that the air contains 23% oxygen by weight.

18. Ethane, C_2H_6, is an organic compound that, together with 90% propane and 5% butane, composes bottled gas. When ethane is burned with oxygen, the following reaction takes place:

$$C_2H_6 + O_2 \longrightarrow CO_2 + H_2O$$

(*a*) Balance the equation.
(*b*) How many grams of oxygen are needed to burn $18\overline{0}$ g of ethane?
(*c*) How many grams of carbon dioxide and water are produced?

20. Ozone (O_3) is formed from oxygen gas in the reaction

$$3O_2 \longrightarrow 2O_3$$

How many grams of oxygen gas are needed to form 0.048 g of ozone?

22. Zinc sulfide is used as a pigment for paints, oilcloths, linoleum, and leather. It is composed of the elements zinc and sulfur.

$Al(OH)_3 + H_2SO_4 \longrightarrow$
$$Al_2(SO_4)_3 + H_2O \quad \text{(unbalanced)}$$

(a) Balance the equation.

(b) How many grams of aluminum hydroxide and sulfuric acid are needed to produce $50\overline{0}$ g of aluminum sulfate?

(c) How many grams of water are produced?

23. In a blast furnace, iron(III) oxide is converted to iron metal. The overall reaction for this process is

$$Fe_2O_3 + CO \longrightarrow Fe + CO_2 \text{ (unbalanced)}$$

(a) Balance the equation.

(b) How many grams of iron(III) oxide are needed to produce 454.0 g of iron metal (about one pound)?

(c) How many grams of CO are needed, and how many grams of CO_2 are produced?

25. Nitric acid is prepared commercially by the *Ostwald* process. The first part of this process involves the reaction of ammonia with oxygen to produce nitric oxide, using platinum as a catalyst:

$$NH_3 + O_2 \xrightarrow{\text{Pt}} NO + H_2O \text{ (unbalanced)}$$

(a) Balance the equation.

(b) How many grams of oxygen are needed to react with 425 g of ammonia?

(c) How many grams of nitric oxide and water are produced?

*27. Suppose that you have a solution of $50\overline{0}$ g of dissolved $Cu(NO_3)_2$ and you add $30\overline{0}$ g of Zn metal to it. A reaction occurs. After the reaction is complete, 127 g of Zn remain. How many grams of $Zn(NO_3)_2$ are produced?

(a) Write a balanced equation for the formation of zinc sulfide from its elements.

(b) How many grams of zinc and sulfur are needed to produce 2.44 g of zinc sulfide?

24. Silver nitrate, a topical anti-infective agent, reacts with copper to form copper(II) nitrate and silver metal.

(a) Write a balanced equation for this reaction.

(b) How many grams of copper are needed to react with 1.70 g of silver nitrate?

(c) How many grams of silver metal and copper(II) nitrate are formed from the reaction?

26. Glucose, $C_6H_{12}O_6$, reacts with oxygen to produce carbon dioxide and water.

(a) Write a balanced equation for this reaction.

(b) How many grams of glucose are needed to react with 3.20 g of oxygen?

(c) How many grams of carbon dioxide and water are produced?

28. Hydrogen gas reacts with iodine to produce hydrogen iodide. Hydrogen iodide is a strong irritant that is used in the manufacture of hydroiodic acid.

(a) Write a balanced equation for the reaction between hydrogen and iodine.

(b) How many grams of hydrogen gas and iodine are needed to produce $32\overline{0}$ g of hydrogen iodide?

Learning Goal 2: Limiting-Reactant Problems

29. The compound acetylene (C_2H_2) is used as a fuel for welding (for example, in the oxyacetylene torch). In years gone by, acetylene was used as a surgical anesthetic. The compound results from the reaction

$$CaC_2 + H_2O \longrightarrow Ca(OH)_2 + C_2H_2 \text{ (unbalanced)}$$
Calcium   Acetylene
carbide

(a) Balance this equation.

(b) If you have 128 g of calcium carbide and 144 g of water, how many grams of calcium hydroxide and acetylene can you produce?

30. Chloroform, $CHCl_3$, reacts with oxygen to form HCl and deadly phosgene gas, $COCl_2$, according to the equation

$$2CHCl_3 + O_2 \longrightarrow 2COCl_2 + 2HCl$$

How many grams of $COCl_2$ and HCl can be formed from 236 g of $CHCl_3$ and 64.0 g of O_2?

31. Methyl alcohol (CH_3OH), sometimes called wood alcohol, is an extremely poisonous substance when taken internally. Death can result from drinking just 30 mL of it; smaller amounts can produce nausea, convulsions, blindness, and respiratory failure. Methyl alcohol is usually obtained by distilling wood. But it can be synthesized in the laboratory by a reaction of carbon monoxide and hydrogen gas (under high temperature and pressure).

$$CO + H_2 \xrightarrow[\text{3,000 lb/in}^2 \text{ pressure}]{\text{300 to 400°C}} CH_3OH \text{ (unbalanced)}$$

(a) Balance the equation.
(b) How many grams of methyl alcohol can be produced from 200 g of carbon monoxide and 40 g of hydrogen?

33. Water can be produced from its elements by the reaction

$$H_2 + O_2 \xrightarrow{\text{Electric spark}} H_2O \text{ (unbalanced)}$$

(a) Balance the equation.
(b) How many grams of water can be produced from 10 g of H_2 and 64 g of O_2?

35. When calcium phosphide is placed in water, it forms phosphine gas (PH_3). In air, PH_3 usually bursts into flames. In fact, this substance (PH_3) may be responsible for people observing faint flickers of light in marshes. The PH_3 could be produced by the reduction of naturally occurring phosphorus compounds. The reaction for the formation of PH_3 from calcium phosphide is

$$Ca_3P_2 + H_2O \longrightarrow PH_3 + Ca(OH)_2 \text{ (unbalanced)}$$

(a) Balance the equation.
(b) How many grams of PH_3 can be produced from 515 g of calcium phosphide and 216 g of water?
(c) How many grams of calcium hydroxide are produced?

37. Elementary phosphorus can be produced by the reaction

$$Ca_3(PO_4)_2 + 3SiO_2 + 5C \longrightarrow 3CaSiO_3 + 5CO + P_2$$

Determine the number of moles of each product formed from 8.0 moles of $Ca_3(PO_4)_2$, 20 moles of SiO_2, and 45 moles of C.

32. Nitrogen gas reacts with hydrogen gas to form ammonia gas (NH_3). How many grams of ammonia gas can be formed from 0.014 g of N_2 and 0.020 g of H_2?

34. Propane, C_3H_8, reacts with oxygen gas to form carbon dioxide and water according to the equation

$$C_3H_8 + 5O_2 \longrightarrow 3CO_2 + 4H_2O$$

How many grams of carbon dioxide and water can be formed from 11.0 g of propane and 32 g of oxygen gas?

36. Aluminum reacts with iron(III) oxide to produce iron and aluminum oxide.
(a) Write a balanced equation for this reaction.
(b) How many grams of aluminum oxide and iron can be produced from 135 g of Al and 800 g of iron(III) oxide?

38. Barium metal reacts with oxygen gas to form barium oxide, a chemical used for drying gases and solvents.
(a) Write a balanced equation for this reaction.
(b) How many grams of barium oxide can be produced from 100 g of barium and 100 g of oxygen gas?

39. An inexpensive way of preparing pure hydrogen gas involves the following reaction between iron and steam:

$$Fe + H_2O \longrightarrow Fe_3O_4 + H_2 \text{ (unbalanced)}$$

(a) Balance the equation.

(b) How many grams of hydrogen gas can be prepared from 225 g of Fe and 225 g of H_2O?

(c) How many grams of Fe_3O_4 are produced?

40. Sulfur dioxide is a chemical used in preserving fruits and vegetables and as a disinfectant in breweries and food factories. It is produced from the reaction of sulfur and oxygen gas.

(a) Write a balanced equation for this reaction.

(b) How many grams of sulfur dioxide can be prepared from 100.0 g of sulfur and 200.0 g of oxygen gas?

Extra Exercises

41. (a) Write the balanced equation for the reaction in which chlorine, Cl_2, reacts with potassium metal, K, to produce potassium chloride. How many moles of chlorine are needed to react with 10.0 g of potassium?

(b) How many grams of potassium chloride are produced?

42. Consider the reaction

$$Cu(NO_3)_2 \longrightarrow CuO + NO_2 + O_2 \text{ (unbalanced)}$$

(a) Balance the reaction.

(b) What weight of O_2 is produced if $1\overline{0}0$ g of copper(II) nitrate react?

(c) How much copper(II) nitrate must react if 54.0 g of NO_2 are formed?

43. Assume that $1\overline{0}0$ g of carbon and $15\overline{0}$ g of oxygen are available for the reaction $C + O_2 \longrightarrow CO_2$.

(a) What is the limiting reactant?

(b) How much CO_2 is formed?

44. A reaction is carried out with 8.70 g of hydrogen and 64.0 g of oxygen to produce water.

(a) Write a balanced equation for this reaction.

(b) What amount of each substance will be on hand when the reaction is complete?

45. Calculate the number of grams of oxygen gas produced by the decomposition of 1.00 kg of potassium chlorate. [*Hint:* $KClO_3 \longrightarrow KCl + O_2$ (*unbalanced*)]

46. Write the equation for the reaction between zinc metal and sulfuric acid. What mass of zinc is required to produce $6\overline{0}0$ g of hydrogen gas?

47. Two experiments are performed. In the first, a 50.0-g plate of silver is immersed in a solution of copper(II) chloride. In the second experiment, a 50.0-g plate of lead is immersed in the same solution.

(a) Determine in which experiment a reaction occurs, and write a balanced equation for this reaction.

(b) Assuming that the metal is the limiting reactant, calculate the masses of all products formed in the reaction that occurs.

48. Aluminum reacts with sulfuric acid to produce aluminum sulfate and hydrogen gas. How many grams of hydrogen are produced from the reaction of 60.0 g of aluminum with excess sulfuric acid?

49. Hydrochloric acid reacts with sodium hydroxide to produce sodium chloride and water. Suppose that 36.5 g of HCl react with 80.0 g of NaOH. How many grams of sodium chloride are produced? Also, how many grams of the excess reactant are left at the end of the reaction?

50. Carbon reacts with oxygen under certain conditions to produce carbon monoxide. Suppose that 48.0 g of carbon react with 64.0 g of oxygen gas to produce *only* carbon monoxide. How many grams of carbon monoxide are produced?

Cumulative Review / Chapters 7–9

1. Write the electron-dot structure for (a) a calcium atom, (b) a calcium ion.

2. Write the electron-dot structure for (a) a chlorine atom, (b) a chloride ion.

3. Distinguish between an ionic bond, a covalent bond, and a coordinate covalent bond.

4. Write the electron-dot diagram for C_3H_8.

5. List the following compounds in order of decreasing covalence: HCl, $BaCl_2$, CaF_2, $CaCl_2$.

6. Which of the following compounds is most ionic, LiF, NaF, RbF, or CsF?

7. Write the chemical formula for each of the following compounds.
(a) copper(I) sulfite
(b) osmium tetraoxide
(c) ammonium carbonate
(d) dinitrogen monoxide

8. Find the oxidation number of each underlined element.
(a) $\underline{W}Cl_5$
(b) $\underline{Sn}F_4$
(c) $\underline{V}_2O_5$
(d) $\underline{In}I_3$

9. Write the name of each of the following compounds.
(a) NH_4I
(b) CaC_2
(c) Cu_2CrO_4

10. Determine the formula of a hypothetical compound that would be formed between elements X and Y if X has three electrons in the outer shell and Y has six electrons in the outer shell.

11. True or False: The electron-dot structure for carbon dioxide is

$$:\overset{\times}{\underset{\times}{O}}\overset{\times}{\underset{\times}{C}}\overset{\times}{\underset{\times}{O}}:$$

12. True or False: An atom having the electron configuration $1s^2\,2s^2\,2p^6\,3s^2$ would form an ion with a 2+ charge.

13. True or False: A Br^{1-} ion, an atom of Kr, and a Rb^{1+} ion have the same electron structure.

14. True or False: The bonds in a molecule of

H
 \
  N   are nonpolar.
 /
H   H

15. True or False: A molecule of $CHCl_3$ is polar.

16. Write the chemical formula for the following combination:
$$1s^2\,2s^2\,2p^4 \quad \text{and} \quad 1s^2\,2s^2\,2p^6\,3s^2$$

17. Which of the following bonds is most covalent, C—H, C—Cl, or C—O?

18. Write the formula for the compound formed from Na^{1+} and SO_4^{2-}.

19. Predict the oxidation number that the element with atomic number 16 is most likely to have.

20. Draw the electron-dot diagram for
(a) NO_3^{1-}
(b) NH_4^{1+}

21. Write a balanced chemical equation for the reaction in which calcium bromide + sulfuric acid forms hydrogen bromide + calcium sulfate.

22. When magnesium metal reacts with oxygen, magnesium oxide is formed. Write a balanced equation for this reaction.

23. A reagent bottle found in a chemistry laboratory was coated with a thin film that had a cloudy appearance. Chemical analysis of the film showed it to be ammonium chloride formed by reaction of ammonia gas and hydrogen chloride gas. Write the balanced equation for this reaction.

24. Identify each of the following reactions as a combination, decomposition, single-replacement, or double-replacement reaction.
(a) $C_{12}H_{22}O_{11} \longrightarrow 12C + 11H_2O$
(b) $H_2 + Br_2 \longrightarrow 2HBr$
(c) $2KBr + Cl_2 \longrightarrow 2KCl + Br_2$
(d) $2HCl + Ca(OH)_2 \longrightarrow CaCl_2 + 2H_2O$

25. Complete and balance each of the following equations. Use the activity series and solubility table when necessary.
(a) $H_2 + Cl_2 \longrightarrow ?$
(b) $NaCl(aq) + Pb(NO_3)_2(aq) \longrightarrow ? + ?$
(c) $Al + H_2SO_4 \longrightarrow ? + ?$
(d) $SrCO_3 \overset{heat}{\longrightarrow} ? + ?$

26. Complete and balance the following equation.
$$CaCO_3 \longrightarrow ? + ?$$

27. Balance the following equations.
(a) $TiCl_4 + H_2O \longrightarrow TiO_2 + HCl$
(b) $P_4O_{10} + H_2O \longrightarrow H_3PO_4$

28. Complete and balance the following equations. Check the activity series and solubility table to be sure that the reaction can take place.
(a) $Fe(OH)_3 + H_3PO_4 \longrightarrow ? + ?$
(b) $Pb(OH)_2 + HNO_3 \longrightarrow ? + ?$

29. Acid rain is formed in the atmosphere by the reaction of sulfur trioxide (an air pollutant) and water, forming sulfuric acid. Complete and balance the reaction describing this process.

30. In the engine of an automobile, fuel that contains a chemical called octane (C_8H_{18}) reacts with oxygen to form carbon dioxide and water. Write a balanced equation for this reaction.

31. Isopropyl alcohol (C_3H_7OH—rubbing alcohol) reacts with oxygen to produce carbon dioxide and water. Write a balanced equation for this reaction.

32. Hard water containing $CaSO_4$ can be softened by adding washing soda (Na_2CO_3). Write a balanced chemical equation for this reaction. Which chemical precipitates out, causing the water to soften?

33. When sugar ($C_{12}H_{22}O_{11}$) is heated, it decomposes to form carbon and water. Write a balanced chemical equation for this reaction.

34. True or False: In the following reaction, the total number of moles of product plus reactant is 9. [Balance the equation before answering.]
$$Al + H_2SO_4 \longrightarrow H_2 + Al_2(SO_4)_3$$

35. True or False: In the following reaction, the total number of moles of product plus reactant is 4. [Balance the equation before answering.]
$$NaNO_3 \longrightarrow NaNO_2 + O_2$$

36. True or False: In the following reaction, metallic zinc is oxidized and hydrogen ions are reduced.
$$Zn + H_2SO_4 \longrightarrow ZnSO_4 + H_2$$

37. True or False: In the following reaction, sodium is oxidized and chloride is reduced.
$$Na + Cl_2 \longrightarrow NaCl$$

38. True or False: In the following reaction, tin is the reducing agent and HNO_3 is the oxidizing agent.
$$Sn + HNO_3 \longrightarrow SnO_2 + NO_2 + H_2O$$

39. True or False: The sum of the coefficients for the following oxidation–reduction reaction is 38.
$$K_2Cr_2O_7 + FeCl_2 + HCl \longrightarrow$$
$$CrCl_3 + KCl + FeCl_3 + H_2O \quad \text{(unbalanced)}$$

40. Balance the following redox reaction:
$$C + H_2SO_4 \longrightarrow CO_2 + SO_2 + H_2O$$

41. When copper(II) oxide reacts with hydrogen, copper and water are produced. How many moles of copper oxide are needed to completely react with 5 moles of hydrogen?

42. Consider the following chemical equation:
$$2Na + 2H_2O \longrightarrow 2NaOH + H_2$$
How many moles of sodium hydroxide are produced when 0.46 mole of sodium reacts with 0.20 mole of water?

43. Calculate the number of grams of propane needed to produce 6.0 moles of carbon dioxide by the following reaction.
$$C_3H_8 + 5O_2 \longrightarrow 3CO_2 + 4H_2O$$

44. How many grams of hydrochloric acid are needed to produce 340 g of hydrogen sulfide by the following reaction?
$$FeS + HCl \longrightarrow FeCl_2 + H_2S$$

45. A 30.0-g sample of ethane (C_2H_6) is allowed to react with 16.0 g of oxygen to produce carbon dioxide and water. How many grams of carbon dioxide will this reaction produce?

46. How many grams of phosphoric acid are needed to react with 1,360 g of lithium hydroxide in the following reaction?
$$H_3PO_4 + 3LiOH \longrightarrow Li_3PO_4 + 3H_2O$$

47. Calcium hydroxide reacts with phosphoric acid to produce calcium phosphate and water. How many grams of calcium phosphate can be produced from 49.0 grams of phosphoric acid? How many grams of calcium hydroxide are needed?

48. When Epsom salt (magnesium sulfate heptahydrate—$MgSO_4 \cdot 7H_2O$) is heated, 1 mole of magnesium sulfate and 7 moles of water are produced. How many moles of magnesium sulfate are produced when 15.0 moles of Epsom salt are heated? How many grams is this?

49. When hydrogen peroxide (H_2O_2) decomposes, water and oxygen are produced. How many grams of hydrogen peroxide are needed to produce 63.0 grams of water?

50. When sodium chloride and lead(II) nitrate react, lead(II) chloride solid and sodium nitrate result. When 66.4 g of lead(II) nitrate react with 5.85 g of sodium chloride, how many grams of lead(II) chloride are formed?

51. Carbon dioxide gas reacts with lithium hydroxide to form lithium carbonate and water. How many grams of lithium hydroxide are needed to produce 148 g of lithium carbonate?

52. Magnesium and oxygen react to form magnesium oxide. How many grams of magnesium are required to react with 48.0 g of oxygen?

53. Barium plus water form barium hydroxide and hydrogen gas upon reaction. When 137.3 g of barium are mixed with 72.00 g of water, how much barium hydroxide and how much hydrogen are produced? Which reactant is in excess and by how much?

10 Heats of Reaction: Chemistry and Energy

Learning Goals

After you have studied this chapter, you should be able to:

1. Distinguish between exothermic reactions and endothermic reactions.
2. Define the terms *heat of reaction, heat of formation,* and *enthalpy.*
3. Define the terms *calorie* and *specific heat.*
4. Calculate the amount of heat gained or lost by a substance when its temperature changes.
5. Write a thermochemical equation.
6. Explain the significance of positive and negative values of ΔH.
7. Calculate the heat of formation or heat of reaction when you are given a balanced equation and the appropriate information.
8. Calculate heats of reaction using Hess's Law.

Introduction

Turn on a *light,* take a ride in your *car,* cook a roast in your *oven,* or just turn on your portable *radio.* All these devices use energy, and this energy usually comes from chemical reactions. The entire world operates on energy. In fact, the use of energy from chemical reactions is the basis of life itself. In this chapter we discuss the changes in energy that occur in chemical reactions, and we examine some of their practical uses.

Solar energy is our most powerful energy source. Solar panels on the roof of this experimental house provide the occupants with most of the heat and hot water they need. (Courtesy Shell Oil Company)

10.1 Some Take and Some Give

Learning Goal 1
Exothermic and endothermic reactions

Associated with every substance is a quantity of energy (a sort of chemical potential energy). When a substance reacts to form a new substance, it can either lose energy to the surroundings or gain energy from the surroundings. This is because, in chemical reactions, old bonds between atoms are broken and new bonds are formed. Breaking a bond usually requires additional energy, whereas forming a bond usually releases energy. Depending on the number of bonds broken and the number formed, and on the strengths of the bonds, energy is either released to the surroundings or absorbed from them. This energy is usually in the form of heat.

Reactions that release energy to the surroundings are called **exothermic reactions.** For example, the explosion of a stick of dynamite is a chemical reaction that releases a tremendous quantity of heat energy to the environment. *Reactions that absorb energy from the surroundings* are called **endothermic reactions.** For example, plants live and grow by a process called photosynthesis. In this process plants use carbon dioxide, they use water, and they absorb *energy from the sun* (sunlight) to make glucose and oxygen.

10.2 Heat and the Chemical Reaction

· · · · · · · · · · · · · ·
Learning Goal 2
Definitions of energy terms

The energy that is associated with a particular substance is called its **enthalpy.** (The symbol for enthalpy is H.) One mole of any substance at a given temperature and pressure has a specific enthalpy, which may be thought of as its heat content. We cannot measure the heat content (enthalpy) of a substance itself. However, we can measure the *change* in heat content that takes place during a chemical reaction. That is, the heat that is absorbed or released during the reaction can be measured. *The change in heat content that results from a chemical reaction* is defined as the **heat of reaction.**

The heat of reaction is usually measured at 25°C and 1 atmosphere (atm) pressure. (The temperature 25°C was probably chosen because it is close to room temperature. The pressure 1 atmosphere—normal atmospheric pressure at sea level—is the pressure in most laboratories. This temperature and pressure are called *standard-state conditions*.) The reactants are kept at a fixed temperature before the reaction, and the products are brought back to the same temperature after the reaction. This allows us to measure only the heat given up or absorbed during the reaction, and not heat involved in temperature or pressure changes. We measure this heat in units of calories. One **calorie** is *the quantity of heat needed to raise the temperature of 1 gram of water by 1 Celsius degree.* Large heat changes are measured in kilocalories (1,000 cal = 1 kcal). (It is the kilocalorie, also known among nutritionists as the large calorie or Calorie, that dieters count when they are trying to get in shape.)

· · · · · · · · · · · · · ·
Learning Goal 3
Definition of calorie and specific heat

With their adoption of the SI metric system (the revised metric system) a few years ago, scientists recognized the *joule* as the basic unit of energy, replacing the calorie. However, the calorie is still used by many scientists to express the quantity of energy something contains, and it may be some time before it disappears from use. In case you find it necessary to convert from calories to joules or vice versa, remember that 1 calorie equals 4.18 joules.

$$1 \text{ calorie} = 4.18 \text{ joules}$$

Example 10.1

One liter of water ($1,\overline{0}00$ g) is heated from $2\overline{0}$°C to $8\overline{0}$°C. How much energy does the water absorb?

Solution

In terms of our definition of the calorie, we can say that every gram of water that we heat by 1°C absorbs 1 calorie of heat energy. We can state this relationship as follows:

Heat, in calories = (mass of water, in grams)
                    × (temperature change, in °C)

or, in abbreviated form, as

$$Q_{cal} = (m)(\Delta t) \qquad \text{(for water only)}$$

The symbol Δ (the Greek capital letter delta) means "change in." Using the abbreviated form, we have

$$Q_{cal} = (m)(\Delta t) \qquad \text{(for water only)}$$
$$= (1,\overline{0}00 \text{ g})(8\overline{0}°C - 2\overline{0}°C) = (1,\overline{0}00 \text{ g})(6\overline{0}°C)$$
$$= 6\overline{0},000 \text{ cal} \qquad \text{or} \qquad 6\overline{0} \text{ kcal}$$

Practice Exercise 10.1   You heat 2,$\overline{0}00$ mL of water (2,$\overline{0}00$ g) from 5$\overline{0}$°C to 7$\overline{0}$°C. How much energy does the water absorb?

10.3 Specific Heat

The formula used in Example 10.1,

$$Q_{cal} = (m)(\Delta t)$$

works only for water. However, we can use this formula to perform calculations on other substances once we understand the concept of specific heat. The **specific heat** of a substance is *the number of calories needed to raise the temperature of 1 gram of the substance by 1 Celsius degree.* The specific heat of water is taken to be 1.00 calorie per gram per degree Celsius, or 1.00 cal/(g)(°C). Table 10.1 gives the specific heats of some other substances.

The general formula for calculating the heat (in calories) gained or lost by any substance is then

· · · · · · · · · · · · · · ·

Learning Goal 4
Calculating calories

$$Q_{cal} = (c)(m)(\Delta t)$$

Table 10.1
Specific Heat of Some Substances at 25°C

Substance	Specific heat (c) in cal/(g)(°C)
Aluminum	0.215
Copper	0.0920
Ethyl alcohol	0.581
Gold	0.0308
Iron	0.107
Nickel	0.106
Zinc	0.0922
Water (liquid)	1.000

where c is the symbol for specific heat. Let's work some examples using this formula.

Example 10.2

How much heat is absorbed when $5\overline{0}0$ g of water are heated from 25.0°C to 75.0°C?

Solution

We write the general formula and substitute known values. Remember that the specific heat c for water is 1.00 cal/(g)(°C).

$$Q_{cal} = (c)(m)(\Delta t) = \left(1.00\ \frac{\text{cal}}{(\cancel{\text{g}})(\cancel{°\text{C}})}\right)(5\overline{0}0\ \cancel{\text{g}})(50.0°\cancel{\text{C}})$$

$$= 25{,}\overline{0}00\ \text{cal} \quad \text{or} \quad 25.0\ \text{kcal}$$

Practice Exercise 10.2   How much heat is absorbed when 250.0 g of water are heated from 10.0°C to 60.0°C?

Example 10.3

How much heat is needed to raise the temperature of a $2\overline{0}0$-g sample of aluminum from 20.0°C to 50.0°C?

Solution

Again we write the general formula and then substitute known values. Table 10.1 tells us that c for aluminum is 0.215 cal/(g)(°C). And the change in temperature is 50.0°C − 20.0°C, or 30.0°C. Then,

$$Q_{cal} = (c)(m)(\Delta t) = \left(0.215\ \frac{\text{cal}}{(\cancel{\text{g}})(\cancel{°\text{C}})}\right)(2\overline{0}0\ \cancel{\text{g}})(30.0°\cancel{\text{C}})$$

$$= 1{,}290\ \text{cal} \quad \text{or} \quad 1.29\ \text{kcal}$$

Practice Exercise 10.3   Find the amount of heat needed to raise the temperature of a 500.0-g sample of aluminum from 25.0°C to 75.0°C.

Example 10.4

If we had heated iron in place of aluminum in Example 10.3, how much heat would we have needed?

Solution

We solve this problem exactly as we did Example 10.3. However, the specific heat of iron is 0.107 cal/(g)(°C).

$$Q_{cal} = (c)(m)(\Delta t) = \left(0.107 \frac{cal}{(\text{g})(°\text{C})}\right)(2\overline{0}0 \text{ g})(30.0°\text{C})$$

$$= 642 \text{ cal} \quad \text{or} \quad 0.642 \text{ kcal}$$

(Iron requires about half as much heat as aluminum requires.)

Practice Exercise 10.4   Repeat Practice Exercise 10.3, substituting iron for aluminum.

10.4 The Calorimeter

An instrument called a *calorimeter* can be used to measure changes in the heat content of substances when they undergo chemical reactions. For example, the *bomb calorimeter* shown in Figure 10.1 is used to measure heat-content changes in substances reacting with oxygen.

Figure 10.1

A bomb calorimeter to determine the heat of combustion of C(*s*)

Suppose we want to measure the heat of reaction of glucose as it reacts with oxygen. We obtain a sample of glucose, say 1.80 g. We place it inside a special heat-transferring container that fits into the calorimeter. Then we position the container in the calorimeter along with an ignition device that will cause a spark, which will allow the sugar and oxygen to react. The calorimeter is filled with a known amount of water, say 1,000 g. We measure the temperature of the water and find it to be 25.0°C. We press the ignition switch, and a large quantity of heat is liberated by the reaction between glucose and oxygen.

$$C_6H_{12}O_6(s) + 6O_2(g) \longrightarrow 6CO_2(g) + 6H_2O(l) + \text{Energy}$$

The heat passes through the heat-transferring container to the water, the temperature of which rises to 31.7°C. Now we can calculate how much heat is required to raise the temperature of 1,000 g of water from 25.0°C to 31.7°C. This is exactly the amount of heat that was liberated by the glucose–oxygen reaction. We proceed as follows:

$$Q_{cal} \text{ (absorbed by water)} = (c_{water})\,(m_{water})\,(\Delta t_{water})$$

$$= \left(1.0\,\frac{\text{cal}}{(\text{g})(°C)}\right)(1{,}000 \text{ g})(31.7°C - 25.0°C)$$

$$= \left(1.0\,\frac{\text{cal}}{(\cancel{\text{g}})(°\cancel{C})}\right)(1{,}000\;\cancel{\text{g}})(6.7°\cancel{C})$$

$$= 6{,}700 \text{ cal} \quad \text{or} \quad 6.7 \text{ kcal}$$

Thus 6.7 kcal of heat energy must have been available to raise the temperature of the water. But this energy could have come only from the glucose–oxygen reaction. We may therefore say that the reaction of 1.8 g of glucose with oxygen released 6.7 kcal of energy.

It is customary to express heats of reaction in units of kilocalories per mole. To do so for the glucose–oxygen reaction, we first note that the molecular weight of glucose is 180 g/mole. Then,

$$?\,\frac{\text{kcal}}{\text{mole}} = \left(\frac{6.7 \text{ kcal}}{1.80\;\cancel{\text{g}}}\right)\left(\frac{18\overline{0}\;\cancel{\text{g}}}{1 \text{ mole}}\right) = 670\,\frac{\text{kcal}}{\text{mole}}$$

Now, in the equation for the glucose–oxygen reaction, we can indicate the exact heat of reaction:

$$C_6H_{12}O_6(s) + 6O_2(g) \longrightarrow 6CO_2(g) + 6H_2O(l) + 670 \text{ kcal}$$

An equation in this form is called a thermochemical equation. A **thermochemical equation** is *a chemical equation that is balanced, that indicates the physical state of each substance (solid s, liquid l, or gas g), and that shows the amount of heat liberated or absorbed during the reaction.* The foregoing equation states that 1 mole of glucose in solid form reacts with 6 moles of gaseous oxygen to form 6 moles of carbon dioxide gas and 6 moles of liquid water, releasing 670 kcal for each mole of glucose that reacts. Note that this also means that 670 kcal are released for every 6 moles of oxygen that react with the glucose, or for every 6 moles of CO_2 or every 6 moles of water that are formed.

10.5 Writing Thermochemical Reactions

Learning Goal 5
Writing thermochemical equations

Writing a thermochemical reaction is simply a matter of obtaining the necessary information and placing it in the proper form. Here are some examples that illustrate the procedure and a few of the "fine points."

Example 10.5

When 1 mole of carbon (graphite) reacts with oxygen at 25°C and 1 atm pressure, carbon dioxide is formed. In addition, 94.1 kcal of heat are released per mole of carbon dioxide formed. Write the thermochemical equation for this reaction.

Solution

$$C(\text{graphite}) + O_2(g) \longrightarrow CO_2(g) + 94.1 \text{ kcal}$$

The reactants (C and O_2) have a total heat content that is 94.1 kcal greater than the product (CO_2). The heat is given up to the surroundings when the product is formed.

Practice Exercise 10.5   When ethane, $C_2H_6(g)$, reacts with oxygen gas, the products formed are carbon dioxide gas and liquid water. When this reaction is carried out at 25°C and 1 atm pressure, 186 kcal of heat are released per mole of carbon dioxide formed. Write the thermochemical equation for this reaction.

Heats of reaction are usually expressed in kilocalories per mole of product formed. However, this is not always the case, so each equation must be examined carefully. Example 10.6 will help explain this point.

Example 10.6

When hydrogen gas reacts with oxygen gas, water is produced and heat is given off. Experiments have shown that 136.6 kcal of heat are produced when *two* moles of water are formed. Write the thermochemical equation for this reaction.

234

Chapter 10 Heats of Reaction: Chemistry and Energy

Solution

$$2H_2(g) + O_2(g) \longrightarrow 2H_2O(l) + 136.6 \text{ kcal}$$

This equation tells us that 136.6 kcal are released to the environment when 2 moles of water are formed. However, you may sometimes see the equation written in the following way:

$$H_2(g) + \tfrac{1}{2}O_2(g) \longrightarrow H_2O(l) + 68.3 \text{ kcal}$$

This equation tells us that 68.3 kcal are released to the environment when 1 mole of water is formed. Because the quantity of heat produced is proportional to the amount of water formed, both equations are correct.

Practice Exercise 10.6 When 2.00 moles of hexane, $C_6H_{14}(l)$, burn in sufficient oxygen gas, the products are carbon dioxide gas and water. In addition, 1,980 kcal of heat are given up to the environment. Write a balanced thermochemical equation for this reaction.

Once you have found the heat of reaction when 1 mole of a substance is formed, you can use this information to find the heat produced or absorbed when *any* amount of that substance is formed.

Example 10.7

How much heat is produced when 144 g of $H_2O(l)$ are formed from the reaction between hydrogen and oxygen?

Solution

Because the formation of 1 mole of water produces 68.3 kcal, we must find out how many moles of water there are in 144 g.

$$? \text{ moles of } H_2O = (144 \text{ g})\left(\frac{1 \text{ mole}}{18.0 \text{ g}}\right) = 8.00 \text{ moles of } H_2O$$

$$? \text{ kcal of heat produced} = (8.00 \text{ moles})\left(\frac{68.3 \text{ kcal}}{1 \text{ mole}}\right)$$

$$= 546 \text{ kcal of heat produced}$$

Practice Exercise 10.7 How much heat is produced when 54.0 g of $H_2O(l)$ are formed from the reaction between hydrogen gas and oxygen gas?

Up to now we have neglected endothermic reactions (reactions that absorb heat from the surroundings). Here's how to write a thermochemical equation for this type of reaction.

Example 10.8

When hydrogen iodide is formed from hydrogen and iodine, heat is *absorbed* from the surroundings. Experiments have shown that 6.2 kcal of heat are absorbed per mole of hydrogen iodide produced. Write the thermochemical equation for the reaction.

Solution

There are two ways to write this type of equation.
(a) Write it with the heat of reaction on the left-hand side of the equation. This shows that the reactants (H_2 and I_2) require heat to yield the product (HI).

$$6.2 \text{ kcal} + \tfrac{1}{2}H_2(g) + \tfrac{1}{2}I_2(g) \longrightarrow HI(g)$$

(b) Write it with the heat of reaction on the right-hand side of the equation, as a negative number. Think of this as indicating that the product must *lose* heat for the reaction to be completed.

$$\tfrac{1}{2}H_2(g) + \tfrac{1}{2}I_2(g) \longrightarrow HI(g) - 6.2 \text{ kcal}$$

Chemists usually prefer writing heats of reaction on the right-hand side of the equation.

Practice Exercise 10.8   When $PCl_5(g)$ breaks down to form $PCl_3(g)$ and $Cl_2(g)$, heat is absorbed from the surroundings. Experiments have shown that 22.2 kcal of heat are absorbed per mole of $PCl_5(g)$ broken down. Write the thermochemical equation for this reaction.

10.6 Another Way of Writing Heats of Reaction

Learning Goal 6
Significance of ΔH

The symbol for heat content (enthalpy) is H, so we can represent a *change* in heat content with the symbol ΔH. (Again we are using the Greek letter Δ to mean "change in.") Because a heat of reaction is a change in heat content during a reaction, we represent it with ΔH.

Chemists define ΔH as the sum of the heat contents of the products less the sum of the heat contents of the reactants:

$$\Delta H = H(\text{products}) - H(\text{reactants})$$

Figure 10.2
Reaction diagrams

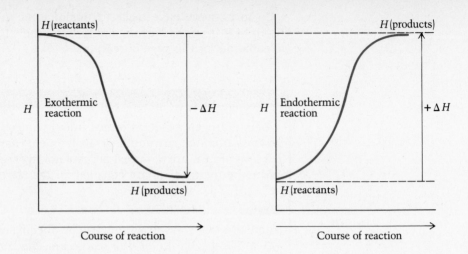

With this notation, ΔH is negative for an exothermic reaction and positive for an endothermic reaction (Figure 10.2).

In the formation of 1 mole of $CO_2(g)$, 94.1 kcal of heat are released to the environment. This means that the products have a lower heat content than the reactants, and $\Delta H = -94.1$ kcal. This may be indicated by writing

$$C(\text{graphite}) + O_2(g) \longrightarrow CO_2(g) \qquad \Delta H = -94.1 \text{ kcal}$$

In the formation of 1 mole of $HI(g)$ under standard conditions, heat is absorbed, so $\Delta H = +6.2$ kcal. This may be shown as

$$\tfrac{1}{2}H_2(g) + \tfrac{1}{2}I_2(s) \longrightarrow HI(g) \qquad \Delta H = +6.2 \text{ kcal}$$

10.7 Heats of Formation

Up to this point we have used the term **heat of reaction** for the change in heat content (or enthalpy) during the course of a reaction. However, the examples we used all involved the formation of a compound from its elements:

$$\tfrac{1}{2}H_2(g) + \tfrac{1}{2}I_2(g) \longrightarrow HI(g) - 6.2 \text{ kcal}$$

$$H_2(g) + \tfrac{1}{2}O_2(g) \longrightarrow H_2O(l) + 68.3 \text{ kcal}$$

$$C(\text{graphite}) + O_2(g) \longrightarrow CO_2(g) + 94.1 \text{ kcal}$$

When a compound is formed from its elements in their usual states at 1 atm pressure, the heat of reaction is called the **heat of formation** of the compound and is given the symbol ΔH_f^0. Heats of formation have been determined experimentally for many compounds, and Table 10.2

Table 10.2
Heats of Formation at 25°C and 1 atm Pressure

Substance	ΔH_f^0 (kcal/mole)	Substance	ΔH_f^0 (kcal/mole)
$BaCl_2(s)$	−205.56	$H_2O(g)$	− 57.8
$BaO_2(s)$	−150.5	$H_2O(l)$	− 68.3
$CH_4(g)$	− 17.9	$H_2O_2(l)$	− 44.8
$C_2H_2(g)$	+ 54.2	$H_2S(g)$	− 4.8
$C_2H_6(g)$	− 20.2	$H_2SO_4(aq)$	−217
$C_8H_{18}(l)$	− 59.97	$H_3PO_4(aq)$	−308.2
$C_6H_{12}O_6(s)$	−304.4	$SO_2(g)$	− 70.96
$CCl_4(l)$	− 33.4	$SO_3(g)$	− 94.45
$CO(g)$	− 26.4	$NH_3(g)$	− 11.04
$CO_2(g)$	− 94.1	$NO(g)$	+ 21.5
$CaO(s)$	−151.9	$NO_2(g)$	+ 8.1
$CaCO_3(s)$	−288.5	$NaCl(aq)$	− 97.3
$HF(g)$	− 64.2	$NaOH(aq)$	−112
$HCl(g)$	− 22.1	$Fe_2O_3(s)$	−197
$HBr(g)$	− 8.7	$KOH(aq)$	−115
$HI(g)$	+ 6.2	$K_2O(s)$	− 86.4
$HCl(aq)$	− 39.6		

lists some of them. These heats of formation can be used to determine ΔH for complex reactions.

10.8 Determining Heats of Reaction

You will find that Table 10.2 makes it easy to determine heats of reaction. However, you must keep the following rules in mind.

Rule 1 ▶ The heat of reaction is equal to the heat of formation of the products less the heat of formation of the reactants:

$$\Delta H = \Delta H_f^0(\text{products}) - \Delta H_f^0(\text{reactants})$$

Rule 2 ▶ The heat of formation of an element at standard-state conditions is zero:

$$\Delta H_f^0(\text{element}) = 0$$

Also keep in mind that the values in Table 10.2 are *per mole of compound*. This is important when more or less than 1 mole is involved in the balanced equation.

Example 10.9

Determine ΔH for the reaction

$$CaO(s) + CO_2(g) \longrightarrow CaCO_3(s)$$

Solution

We obtain the ΔH_f^0 value for each compound from Table 10.2 and then use Rule 1 above.

Substance	ΔH_f^0(kcal/mole)
$CaO(s)$	-151.9
$CO_2(g)$	-94.1
$CaCO_3(s)$	-288.5

. . . .

Learning Goal 7
Heats of formation and heats of reaction

$$\Delta H = \Delta H_f^0(\text{products}) - \Delta H_f^0(\text{reactants})$$

$$= (\Delta H_{CaCO_3}^0) - (\Delta H_{CaO}^0 + \Delta H_{CO_2}^0)$$

$$= (-288.5 \text{ kcal}) - (-151.9 \text{ kcal} - 94.1 \text{ kcal})$$

$$= (-288.5 \text{ kcal}) - (-246.0 \text{ kcal}) = -288.5 \text{ kcal} + 246.0 \text{ kcal}$$

$$= -42.5 \text{ kcal}$$

This reaction is exothermic. It produces 42.5 kcal of heat per mole of $CaCO_3$ formed.

Practice Exercise 10.9   Use Table 10.2 to determine ΔH for the reaction

$$K_2O(s) + H_2O(l) \longrightarrow 2KOH(aq)$$

Example 10.10

Determine ΔH for the reaction

$$2C_2H_6(g) + 7O_2(g) \longrightarrow 4CO_2(g) + 6H_2O(l)$$
$$\text{Ethane}$$

Solution

Again we obtain ΔH_f^0 values from Table 10.2:

Substance	ΔH_f^0(kcal/mole)
$C_2H_6(g)$	-20.2
$O_2(g)$	0.0
$CO_2(g)$	-94.1
$H_2O(l)$	-68.3

Now, because 6 moles of water and 4 moles of CO_2 are formed, we have to multiply ΔH_f^0 for water by 6 and ΔH_f^0 for CO_2 by 4. And because 2 moles of ethane and 7 moles of oxygen are used, we have to multiply ΔH_f^0 for ethane by 2 and ΔH_f^0 for oxygen by 7.

$$\Delta H = \Delta H_f^0(\text{products}) - \Delta H_f^0(\text{reactants})$$

$$= [6(\Delta H_{H_2O}^0) + 4(\Delta H_{CO_2}^0)] - [2(\Delta H_{C_2H_6}^0) + 7(\Delta H_{O_2}^0)]$$

$$= [6(-68.3 \text{ kcal}) + 4(-94.1 \text{ kcal})]$$
$$- [2(-20.2 \text{ kcal}) + 7(0.0 \text{ kcal})]$$

$$= -786.2 \text{ kcal} + 40.4 \text{ kcal} = -745.8 \text{ kcal}$$

The reaction is exothermic. The combustion of *two* moles of ethane releases 745.8 kcal of heat to the environment.

Practice Exercise 10.10 Use Table 10.2 to determine ΔH for the reaction

$$2CO(g) + O_2(g) \longrightarrow 2CO_2(g)$$

10.9 Practical Applications of Heat from Chemical Reactions

At the beginning of this chapter, we mentioned some uses of energy in our everyday lives. In most cases this energy is obtained from fuels, substances that are rich in chemical potential energy—for example, oil, gas, and coal. We usually obtain heat energy from fuels by making them react in the presence of oxygen; in everyday language, we burn them. Then we use the heat energy they produce to do work.

Example 10.11

Modern gas ranges use methane (CH_4) for fuel. Calculate ΔH for the combustion of 1 mole of methane:

$$CH_4(g) + 2O_2(g) \longrightarrow CO_2(g) + 2H_2O(l)$$

Solution
Obtain ΔH_f^0 for each compound from Table 10.2, and then use Rule 1:

Substance	ΔH_f^0(kcal/mole)
$CH_4(g)$	−17.9
$O_2(g)$	0.0
$CO_2(g)$	−94.1
$H_2O(l)$	−68.3

$$\Delta H = \Delta H_f^0(\text{products}) - \Delta H_f^0(\text{reactants})$$

$$= [2(\Delta H^0_{H_2O}) + 1(\Delta H^0_{CO_2})] - [1(\Delta H^0_{CH_4}) + 2(\Delta H^0_{O_2})]$$

$$= [2(-68.3 \text{ kcal}) + 1(-94.1 \text{ kcal})]$$
$$- [1(-17.9 \text{ kcal}) + 2(0.0 \text{ kcal})]$$

$$= -230.7 \text{ kcal} + 17.9 \text{ kcal} = -212.8 \text{ kcal}$$

This reaction is exothermic. When 1 mole of methane burns, 212.8 kcal of heat are released to the environment.

Practice Exercise 10.11   The first step in the synthesis of nitric acid is the reaction of ammonia gas and oxygen gas to form nitric oxide gas and steam. Calculate ΔH for this reaction:

$$4NH_3(g) + 5O_2(g) \longrightarrow 4NO(g) + 6H_2O(g)$$

Example 10.12

The combustion of methane releases considerable energy to do work. Let's compare this with the energy released by the combustion of 1 mole of octane, C_8H_{18} (used in automobile fuels):

$$C_8H_{18}(l) + 12\tfrac{1}{2}O_2(g) \longrightarrow 8CO_2(g) + 9H_2O(l)$$

Solution

As usual, obtain ΔH^0_f for each compound from Table 10.2, and then use Rule 1.

Substance	ΔH^0_f(kcal/mole)
$C_8H_{18}(l)$	−59.97
$O_2(g)$	0.0
$CO_2(g)$	−94.1
$H_2O(l)$	−68.3

$$\Delta H = \Delta H^0_f(\text{products}) - \Delta H^0_f(\text{reactants})$$

$$= [8(\Delta H^0_{CO_2}) + 9(\Delta H^0_{H_2O})] - [1(\Delta H^0_{C_8H_{18}}) + 12\tfrac{1}{2}(\Delta H^0_{O_2})]$$

$$= [8(-94.1 \text{ kcal}) + 9(-68.3 \text{ kcal})]$$
$$- [1(-59.97 \text{ kcal}) + 12\tfrac{1}{2}(0.0 \text{ kcal})]$$

$$= -1,367.5 \text{ kcal} + 59.97 \text{ kcal} = -1,307.5 \text{ kcal}$$

The reaction is exothermic. When 1 mole of octane burns, 1,307.5 kcal of heat are released to the environment. This is more than six times the energy produced by the combustion of 1 mole of methane. But tread softly: Because 1 mole of octane weighs seven times as much as 1 mole of methane, we actually get a little more heat from 1 g of *methane* than we do from 1 g of octane.

Practice Exercise 10.12   Iron ore (Fe_2O_3) is reduced in the reaction

$$Fe_2O_3(s) + 3CO(g) \longrightarrow 3CO_2(g) + 2Fe(s)$$

Using Table 10.2, determine ΔH for this reaction.

Example 10.13

We all know that sugar is a source of quick energy. In our bodies sucrose (sugar) is broken down into carbon dioxide and water by the very complicated process of cellular respiration. Sucrose itself is made up of two simpler sugars, glucose and fructose. Let's determine how much energy our bodies obtain from the oxidation of 1 mole of the simpler sugar, glucose ($C_6H_{12}O_6$), in the reaction

$$C_6H_{12}O_6(s) + 6O_2(g) \longrightarrow 6CO_2(g) + 6H_2O(l)$$

Solution
Again we use Table 10.2 and Rule 1.

Substance	ΔH_f^0 (kcal/mole)
$C_6H_{12}O_6(s)$	−304.4
$O_2(g)$	0.0
$CO_2(g)$	− 94.1
$H_2O(l)$	− 68.3

$$\Delta H = \Delta H_f^0(\text{products}) - \Delta H_f^0(\text{reactants})$$

$$= [6(\Delta H_{CO_2}^0) + 6(\Delta H_{H_2O}^0)] - [1(\Delta H_{C_6H_{12}O_6}^0) + 6(\Delta H_{O_2}^0)]$$

$$= [6(-94.1 \text{ kcal}) + 6(-68.3 \text{ kcal})]$$
$$- [1(-304.4 \text{ kcal}) + 6(0.0 \text{ kcal})]$$

$$= -974.4 \text{ kcal} + 304.4 \text{ kcal} = -67\overline{0} \text{ kcal}$$

The reaction is exothermic. Our bodies obtain $67\overline{0}$ kcal of heat from the oxidation of 1 mole of glucose.

Practice Exercise 10.13   Acetylene gas, which is used for welding, reacts with hydrogen to produce ethane gas. Calculate ΔH for this reaction:

$$C_2H_2(g) + 2H_2(g) \longrightarrow C_2H_6(g)$$

10.10 Hess's Law of Heat Summation

Learning Goal 8
Using Hess's Law

Sometimes it isn't possible to find the heat of reaction or the heat of formation of a compound directly. But an indirect way has been developed, using *Hess's Law of Heat Summation*. It involves the summing of heats of known reactions—in other words, treating chemical equations as algebraic equations. An example will illustrate how this is done.

Example 10.14

Carbon monoxide has been found to have a heat of formation of -26.4 kcal per mole. This means that

$$C(\text{graphite}) + \tfrac{1}{2}O_2(g) \longrightarrow CO(g) + 26.4 \text{ kcal}$$

However, it is almost impossible to find the heat of formation of CO by performing this reaction, because some CO_2 also forms. But we can find it by using the following two equations:

$$C(\text{graphite}) + O_2(g) \longrightarrow CO_2(g) + 94.05 \text{ kcal} \qquad (1)$$

$$CO(g) + \tfrac{1}{2}O_2(g) \longrightarrow CO_2(g) + 67.64 \text{ kcal} \qquad (2)$$

Solution

We simply rewrite Equation 1 as given, reverse Equation 2, *and then add "algebraically."*

$$C(\text{graphite}) \qquad + \cancel{O_2(g)}^{+\tfrac{1}{2}O_2(g)} \longrightarrow \cancel{CO_2(g)} \qquad + 94.05 \text{ kcal} \qquad (1)$$

$$\cancel{CO_2(g)} \qquad \longrightarrow CO(g) + \tfrac{1}{2}\cancel{O_2(g)} - 67.64 \text{ kcal} \qquad (2)$$

$$C(\text{graphite}) \qquad + \tfrac{1}{2}O_2(g) \longrightarrow CO(g) \qquad + 26.41 \text{ kcal} \qquad (3)$$

Equation 3 was obtained by adding Equations 1 and 2 and canceling similar terms on opposite sides of the equations. Equation 3 shows the heat of formation (ΔH_f^0) of carbon monoxide. Note that, when we reversed Equation 2, we changed the sign of its heat of reaction.

Practice Exercise 10.14   Consider the following equations:

$$S(s) + O_2(g) \longrightarrow SO_2(g) + 70.96 \text{ kcal} \qquad (1)$$

$$SO_2(g) + \tfrac{1}{2}O_2(g) \longrightarrow SO_3(g) + 23.49 \text{ kcal} \qquad (2)$$

$$SO_3(g) + H_2O(l) \longrightarrow H_2SO_4(l) + 31.14 \text{ kcal} \qquad (3)$$

$$H_2(g) + \tfrac{1}{2}O_2(l) \longrightarrow H_2O(l) + 68.32 \text{ kcal} \qquad (4)$$

Calculate ΔH_f^0 for $H_2SO_4(l)$ using Hess's Law. [*Hint:* The equation for the formation of $H_2SO_4(l)$ from its elements is $H_2(g) + S(s) + 2O_2(g) \longrightarrow H_2SO_4(l)$.]

The principle on which this method is based was discovered by the Swiss-Russian chemist Germain Henri Hess, in 1840. Hess noted that the heat of a chemical reaction must be the same, regardless of the steps by which the reaction occurs.

Example 10.15

Given the following equations, calculate ΔH_f^0 for $SO_2(g)$ as formed from $S(s)$ and $O_2(g)$.

$$SO_2(g) + \tfrac{1}{2}O_2(g) \longrightarrow SO_3(g) + 23.49 \text{ kcal} \qquad (1)$$

$$S(s) + 1\tfrac{1}{2}O_2(g) \longrightarrow SO_3(g) + 94.45 \text{ kcal} \qquad (2)$$

Solution

We reverse Equation 1, rewrite both equations, and add.

$$SO_3(g) \longrightarrow SO_2(g) + \tfrac{1}{2}O_2(g) - 23.49 \text{ kcal} \qquad (1)$$

$$\underline{S(s) + 1\tfrac{1}{2}O_2(g) \longrightarrow SO_3(g) + 94.45 \text{ kcal}} \qquad (2)$$

$$S(s) + O_2(g) \longrightarrow SO_2(g) + 70.96 \text{ kcal} \qquad (3)$$

Equation 3 was obtained by adding Equations 1 and 2 and canceling similar terms on opposite sides of the equations. Note that, when we reversed Equation 1, we changed the sign of its heat of reaction.

Practice Exercise 10.15   Consider the following equations:

$$C(s) + O_2(g) \longrightarrow CO_2(g) + 94.1 \text{ kcal} \qquad (1)$$

$$H_2(g) + \tfrac{1}{2}O_2(g) \longrightarrow H_2O(l) + 68.3 \text{ kcal} \qquad (2)$$

$$C_6H_{12}O_6(s) + 6O_2(g) \longrightarrow 6CO_2(g) + 6H_2O(l) + 670 \text{ kcal} \quad (3)$$

Calculate ΔH_f^0 for $C_6H_{12}O_6(s)$, using Hess's Law. [*Hint:* The equation for the formation of $C_6H_{12}O_6(s)$ from its elements is $6C(s) + 6H_2(g) + 3O_2(g) \longrightarrow C_6H_{12}O_6(s)$.]

Example 10.16

The compound nitrogen(II) oxide, NO, is an air pollutant produced by the burning of fuels. Nitrogen(II) oxide reacts with oxygen in the atmosphere to produce deadly nitrogen(IV) oxide, NO_2, a yellow-brown gas. Given the following equations and using Hess's Law, determine the heat of formation of nitrogen(II) oxide.

$$\tfrac{1}{2}N_2(g) + O_2(g) \longrightarrow NO_2(g) - 8.1 \text{ kcal} \qquad (1)$$

$$NO(g) + \tfrac{1}{2}O_2(g) \longrightarrow NO_2(g) + 13.5 \text{ kcal} \qquad (2)$$

Solution

We reverse Equation 2, rewrite both equations, and add.

$$\frac{1}{2}N_2(g) + O_2(g) \longrightarrow NO_2(g) - 8.1 \text{ kcal} \qquad (1)$$

$$NO_2(g) \longrightarrow NO(g) + \frac{1}{2}O_2(g) - 13.5 \text{ kcal} \qquad (2)$$

$$\frac{1}{2}N_2(g) + \frac{1}{2}O_2(g) \longrightarrow NO(g) - 21.6 \text{ kcal} \qquad (3)$$

Equation 3 was obtained by adding Equations 1 and 2 and canceling similar terms on opposite sides of the equations. Note that, when we reversed Equation 2, we changed the sign of its heat of reaction.

Practice Exercise 10.16   Consider the following equations:

$$C(s) + O_2(g) \longrightarrow CO_2(g) + 94.1 \text{ kcal} \qquad (1)$$

$$H_2(g) + \frac{1}{2}O_2(g) \longrightarrow H_2O(l) + 68.3 \text{ kcal} \qquad (2)$$

$$CH_4(g) + 2O_2(g) \longrightarrow CO_2(g) + 2H_2O(l) + 212.8 \text{ kcal} \qquad (3)$$

Calculate ΔH_f^0 for $CH_4(g)$, using Hess's Law.

Summary

In a chemical reaction, heat is either absorbed from the environment or released to the environment as old bonds are broken and new bonds are formed. A reaction in which energy is absorbed from the surroundings is called endothermic, and a reaction in which energy is released is called exothermic. Associated with every substance is a heat content called its enthalpy. Although we cannot measure enthalpy directly, we can measure the amount of heat released or absorbed during a reaction. This change in heat content that occurs during a chemical reaction is called the heat of reaction. Heats of reaction are usually measured at standard-state conditions: 25°C and 1 atm pressure. Heat (including heat of reaction) is measured in calories. One calorie is the quantity of heat needed to raise the temperature of 1 gram of water by 1 Celsius degree. The specific heat of a substance is defined as the number of calories needed to raise the temperature of 1 gram of the substance by 1 Celsius degree.

A chemical equation that indicates the heat of reaction and the physical state of each reactant and product is called a thermochemical equation.

Heat of reaction ΔH may be defined as the sum of the heat contents of the products less the sum of the heat contents of the reactants. For an exothermic reaction this difference is negative; for an endothermic reaction it is positive. When a compound is formed from its elements, the heat of reaction is called the heat of formation ΔH_f^0 for the compound. The heat of formation for an element (at standard-state conditions) is zero. Hess's Law of Heat Summation provides an algebraic method for finding the heat of formation of a compound indirectly.

Career Sketch

Dental hygienists

Dental hygienists help the public achieve and maintain good oral health. As members of the dental health team, dental hygienists perform preventive and therapeutic dental services under the supervision of a dentist. Responsibilities of the hygienist vary, depending on the laws of the state in which the hygienist works. However, hygienists' duties usually include removing deposits and stains from patient's teeth, instructing patients in how to care for their teeth and giving them dietetic and nutritional counseling, and applying medicine to prevent tooth decay. Dental hygienists take patients' medical and dental histories, take dental x-ray pictures and develop the film, make impressions of teeth for the dentist to study, and prepare other diagnostic aids for the dentist. In some states dental hygienists also help the dentist carry out pain-controlling and restorative procedures.

Dental hygienists who work in school systems examine children's teeth, help the dentist determine the dental treatment needed, and report findings to parents. They also scale and polish teeth and instruct the children in proper mouth care. They often present programs on oral health to students.

A few dental hygienists assist in research projects. Those with advanced training may teach in schools of dental hygiene.

As members of the dental health team, dental hygienists perform preventive and dental therapeutic services under the supervision of a dentist.
(Courtesy Park Dental Group. Photo S. Sherman)

Self-Test Exercises

Learning Goals 1 and 2: Exothermic and Endothermic Reactions, Definitions of Energy Terms

1. Define the following terms: (*a*) exothermic reaction, (*b*) endothermic reaction, (*c*) heat of formation, (*d*) heat of reaction.

2. The symbol H stands for the term _____ .

3. How does the sum of the heat contents of the products compare with that of the reactants in (*a*) an endothermic reaction, (*b*) an exothermic reaction?

4. The change in heat content during a chemical reaction is called the _____ .

Learning Goal 3: Definition of Calorie and Specific Heat

5. Define the terms *calorie* and *specific heat.*

6. What are standard-state conditions, and why is it necessary to adhere to them when calculating changes in heat?

Learning Goal 4: Calculating Calories

7. How much heat is needed to raise the temperature of 250 g of water from $10.0°C$ to $90.0°C$?

8. How much heat is needed to raise the temperature of 500.0 g of water from $0°C$ to $75.0°C$?

9. Calculate the amount of heat needed to raise the temperature of 50.0 g of copper from $25.0°C$ to $100.0°C$.

10. Calculate the amount of heat needed to raise the temperature of 100.0 g of copper from $45.0°C$ to $80.0°C$.

*11. How much heat is required to raise the temperature of 100.0 mL of ethyl alcohol from $15.0°C$ to $60.0°C$? The density of ethyl alcohol is 0.800 g/mL.

*12. How much heat is required to raise the temperature of 50.0 cm^3 of ethyl alcohol from $25.0°C$ to $55.0°C$? The density of ethyl alcohol is 0.800 g/cm^3.

13. Suppose that 308 cal of heat are added to a sample of gold that weighs $1,000$ g. What is the final temperature of the gold, given that its initial temperature is $20.0°C$?

14. Suppose that 5.00 kcal of heat are added to a sample of gold that weighs 2.00 kg. What is the final temperature of the gold if its initial temperature is $10.0°C$?

*15. A 0.146-g sample of H_2 gas is placed in a calorimeter, where it is allowed to react with oxygen gas. The heat generated by the reaction is transferred to the calorimeter water. The water weighs 500 g. The temperature of the water increases from $20.0°C$ to $30.0°C$. From this information, calculate the heat of reaction in kilocalories per mole.

16. How much heat is needed to raise the temperature of 1.000 kg of water from $32.0°C$ to $99.9°C$?

Learning Goal 5: Writing Thermochemical Equations

17. Using Table 10.2, write thermochemical equations for the formation of the following compounds from their elements:
(*a*) $CO(g)$,   (*b*) $HCl(g)$,   (*c*) $H_2O_2(l)$.

18. Using Table 10.2, write thermochemical equations for the formation of the following compounds from their elements:
(*a*) $HF(g)$,   (*b*) $HCl(g)$,   (*c*) $HBr(g)$.

Learning Goal 6: Significance of ΔH

19. Explain the significance of positive and negative values of ΔH. Use a chemical equation to explain your answer.

20. (*a*) What is meant by the term *enthalpy*? (*b*) What is the difference between the symbols H and ΔH?

Learning Goal 7: Heats of Formation and Heats of Reaction

21. Calculate the heat released when 102 g of hydrogen sulfide (H_2S) are produced from its elements. *Note:* ΔH_f^0 for H_2S is -4.8 kcal/mole. [*Hint:* Calculate the number of moles of H_2S.]

22. Calculate the heat released when 15.3 g of barium oxide (BaO) are produced from its elements. (ΔH_f^0 for BaO is -133.4 kcal/mole.)

23. Rocket engines often use liquid hydrogen and liquid oxygen for fuel. These reactants are heated, mixed, and united to produce steam.

$$H_2(g) + \tfrac{1}{2}O_2(g) \longrightarrow H_2O(g) + 57.8 \text{ kcal}$$

Calculate the amount of heat released when 1 kg of steam is produced.

24. Hexane is a chemical that is sometimes used in thermometers instead of mercury. Hexane burns in oxygen to produce carbon dioxide and water according to the equation

$$2C_6H_{14}(l) + 19O_2(g) \longrightarrow$$
$$12CO_2(g) + 14H_2O(l) + 1{,}980 \text{ kcal}$$

How many grams of hexane are needed to produce $19{,}80\overline{0}$ kcal of heat?

25. (*a*) Using Table 10.2, calculate ΔH for the reaction

$$2C_2H_2(g) + 5O_2(g) \longrightarrow 4CO_2(g) + 2H_2O(l)$$

(*b*) Calculate how much heat is produced when $5{,}20\overline{0}$ g of C_2H_2 are burned.

26. Carbon dioxide is formed from its elements as follows:

$$C(\text{graphite}) + O_2(g) \longrightarrow CO_2(g) + 94.1 \text{ kcal}$$

How much energy is produced when $88\overline{0}$ g of $CO_2(g)$ are formed?

27. Barium peroxide can be used as a solid propellant for rockets. When it is added to hydrochloric acid, the following reaction takes place:

$$BaO_2(s) + 2HCl(g) \longrightarrow BaCl_2(s) + H_2O_2(l)$$

Barium
peroxide

(*a*) Using Table 10.2, calculate ΔH for the reaction.
(*b*) Calculate the number of grams of BaO_2 that must react to produce $50\overline{0}$ kcal of heat.

28. Liquid water is broken down into its elements by electrolysis. This is an endothermic reaction requiring 68.3 kcal per mole of liquid water decomposed. How much energy is needed to produce $64\overline{0}$ g of oxygen gas?

29. Given the reaction

$$P_2O_5(s) + 3H_2O(l) \longrightarrow 2H_3PO_4(aq) + 228.5 \text{ kcal}$$

calculate ΔH_f^0 for P_2O_5 (use Table 10.2).

30. When you burn 1.20 g of carbon (graphite) in the presence of oxygen, carbon dioxide gas is produced. The heat generated by this reaction can raise the temperature of 470.5 g of water from 20.0°C to 40.0°C. Calculate ΔH_f^0 for carbon dioxide from this information, and compare it with the value given in Table 10.2.

31. One *calorie* is the heat needed to raise the temperature of 1 g of water by 1°C. When you burn 3.2 g of sulfur in the presence of oxygen, sulfur trioxide (SO_3) is produced. The heat generated by this reaction can raise the temperature of 1,890 g of water from 25°C to 30°C. Calculate ΔH_f^0 for sulfur trioxide from this information and compare it with the value given in Table 10.2.

32. Calculate the heat released when 188.4 g of $K_2O(s)$ are produced from its elements.

33. Eric Cottell, a British-born inventor, has found a method of increasing the burning efficiency of oil and gasoline. He does this simply by mixing the oil or gasoline with *water*. Yes, water! Cottell has built a device called the Cottell Ultrasonic Reactor. This device can actually mix oil and water by breaking down the oil and water droplets into extremely fine particles. By using a mixture containing three parts oil to one part water, Cottell has decreased the fuel consumption in his home furnace by 25%. He claims that cars can be made to run almost pollution-free on a mixture of 18% water and 82% gasoline.

One of the ingredients of home heating oil is $C_{16}H_{34}$. Given the following equation, determine ΔH_f^0 for this ingredient.

$$C_{16}H_{34}(s) + 24\tfrac{1}{2}O_2(g) \longrightarrow$$
$$16CO_2(g) + 17H_2O(l) + 2{,}560 \text{ kcal}$$

35. Calculate the heat released when 12.8 g of $SO_2(g)$ are produced from its elements.

37. Calculate the heat absorbed when 51.2 g of $HI(g)$ are produced from its elements.

39. Using Table 10.2, calculate ΔH for the reaction

$$2SO_2(g) + O_2(g) \longrightarrow 2SO_3(g)$$

41. Hydrogen peroxide (H_2O_2) is sometimes used in diluted form as an antiseptic. Upon standing, hydrogen peroxide decomposes according to the reaction

$$2H_2O_2(l) \longrightarrow 2H_2O(l) + O_2(g)$$

(a) Calculate ΔH for this reaction.
(b) Calculate how much heat is produced when 204 g of H_2O_2 decompose.

43. Calculate ΔH for the decomposition of potassium chlorate:

$$2KClO_3(s) \longrightarrow 2KCl(s) + 3O_2(g)$$

[ΔH_f^0 for $KClO_3(s)$ is -93.50 kcal, and ΔH_f^0 for $KCl(s)$ is -104.18 kcal.]

45. The following equation represents the formation of phosphoric acid from phosphorus pentachloride:

$$PCl_5(s) + H_2O(l) \longrightarrow H_3PO_4(aq) + HCl(aq)$$

(a) Balance the equation.
(b) Calculate ΔH for this reaction. [ΔH_f^0 for $PCl_5(s)$ is -95.35 kcal, and ΔH_f^0 for $HCl(aq)$ is -40.02 kcal.]
(c) How much heat is produced from this reaction if 41.70 g of $PCl_5(s)$ are used?

34. Calculate the heat released when 7.30 g of $HCl(g)$ are produced from its elements.

36. Calculate the heat released when 1.618 kg of $HBr(g)$ are produced from its elements.

38. Acetylene gas, C_2H_2, burns in oxygen to produce carbon dioxide gas and liquid water. For every mole of acetylene burned, 312.0 kcal of heat are released. From this information, calculate ΔH_f^0 for acetylene, and compare your answer with the value given in Table 10.2.

40. Use the following equation and Table 10.2 to determine ΔH_f^0 for $Ca(OH)_2(s)$.

$$CaO(s) + H_2O(l) \longrightarrow Ca(OH)_2(s) + 15.6 \text{ kcal}$$

42. Calculate the amount of heat released when 2.000 kg of $BaCl_2(s)$ are produced from $Ba(s)$ and $Cl_2(g)$. Use Table 10.2.

44. When 0.230 g of $Na(s)$ react with $Cl_2(g)$, $NaCl(s)$ is produced. The heat generated by this reaction can raise the temperature of 24.55 g of water from 10.0°C to 50.0°C. Using this information, calculate ΔH_f^0 for $NaCl(s)$.

46. Calculate the amount of heat absorbed when 5.00 moles of $CO_2(g)$ are formed from the reaction

$$C(\text{graphite}) + 2H_2O(g) \longrightarrow$$
$$CO_2(g) + 2H_2(g) - 31.39 \text{ kcal}$$

*47. Calculate ΔH_f^0 for $CuO(s)$ from the following information:

$$3CuO(s) + 2NH_3(g) \longrightarrow$$
$$3Cu(s) + 3H_2O(l) + N_2(g) \qquad \Delta H = -71.1 \text{ kcal}$$

[ΔH_f^0 for $NH_3(g)$ is -11.04 kcal.]

48. Calculate the amount of heat released when 5.00 moles of $CO_2(g)$ are formed from the reaction

$$CO(g) + \tfrac{1}{2}O_2(g) \longrightarrow CO_2(g) + 67.64 \text{ kcal}$$

Learning Goal 8: Using Hess's Law

49. Using Hess's Law, calculate the heat of reaction for the chemical equation

$$H_2(g) + \tfrac{1}{2}O_2(g) \longrightarrow H_2O(l)$$

Use the following information (*do not* use Table 10.2).

$$C(\text{graphite}) + \tfrac{1}{2}O_2(g) \longrightarrow$$
$$CO(g) + 26.4 \text{ kcal} \quad (1)$$

$$C(\text{graphite}) + 2H_2O(g) \longrightarrow$$
$$CO_2(g) + 2H_2(g) - 31.39 \text{ kcal} \quad (2)$$

$$H_2O(g) \longrightarrow$$
$$H_2O(l) + 10.52 \text{ kcal} \quad (3)$$

$$CO(g) + \tfrac{1}{2}O_2(g) \longrightarrow$$
$$CO_2(g) + 67.64 \text{ kcal} \quad (4)$$

50. For the reaction $CaO(s) + CO_2(g) \longrightarrow CaCO_3(s)$, calculate ΔH, using Hess's Law and the following equations.

$$CaO(s) + 151.9 \text{ kcal} \longrightarrow Ca(s) + \tfrac{1}{2}O_2(g) \quad (1)$$
$$CO_2(g) + 94.1 \text{ kcal} \longrightarrow C(s) + O_2(g) \quad (2)$$
$$Ca(s) + C(s) + 1\tfrac{1}{2}O_2(g) \longrightarrow$$
$$CaCO_3(s) + 288.5 \text{ kcal} \quad (3)$$

51. Use Hess's Law to calculate the heat of reaction for the chemical equation

$$Ca(s) + C(s) + \tfrac{3}{2}O_2(g) \longrightarrow CaCO_3(s)$$

Use the following information (*do not* use Table 10.2).

$$Ca(s) + \tfrac{1}{2}O_2(g) \longrightarrow CaO(s) + 151.9 \text{ kcal} \quad (1)$$
$$C(s) + O_2(g) \longrightarrow CO_2(g) + 94.1 \text{ kcal} \quad (2)$$
$$CaO(s) + CO_2(g) \longrightarrow CaCO_3(s) + 42.5 \text{ kcal} \quad (3)$$

52. Using Hess's Law and the following equations, calculate ΔH_f^0 for $NH_4Br(s)$.

$$NH_4Br(s) \longrightarrow$$
$$HBr(g) + NH_3(g) - 44.87 \text{ kcal} \quad (1)$$
$$\tfrac{1}{2}H_2(g) + \tfrac{1}{2}Br_2(g) \longrightarrow HBr(g) + 8.66 \text{ kcal} \quad (2)$$
$$\tfrac{1}{2}N_2(g) + 1\tfrac{1}{2}H_2(g) \longrightarrow NH_3(g) + 11.04 \text{ kcal} \quad (3)$$

53. Use Hess's Law to calculate the heat of reaction for the chemical equation

$$2C(s) + 3H_2(g) + \tfrac{1}{2}O_2(g) \longrightarrow C_2H_6O(l)$$
$$\text{Ethyl alcohol}$$

Use the following information (*do not* use Table 10.2).

$$C(s) + O_2(g) \longrightarrow CO_2(g) + 94.1 \text{ kcal} \quad (1)$$
$$H_2(g) + \tfrac{1}{2}O_2(g) \longrightarrow H_2O(l) + 68.3 \text{ kcal} \quad (2)$$
$$C_2H_6O(l) + 3O_2(g) \longrightarrow$$
$$2CO_2(g) + 3H_2O(l) + 327.6 \text{ kcal} \quad (3)$$

54. Use Hess's Law to calculate the heat of reaction for the chemical equation

$$C(s) + 2H_2(g) + \tfrac{1}{2}O_2(g) \longrightarrow CH_3OH(l)$$

Use the following information (*do not* use Table 10.2).

$$C(s) + O_2(g) \longrightarrow$$
$$CO_2(g) + 94.1 \text{ kcal} \quad (1)$$
$$H_2(g) + \tfrac{1}{2}O_2(g) \longrightarrow$$
$$H_2O(l) + 68.3 \text{ kcal} \quad (2)$$
$$CH_3OH(l) + 1\tfrac{1}{2}O_2(g) \longrightarrow$$
$$CO_2(g) + 2H_2O(l) + 173.7 \text{ kcal} \quad (3)$$

Extra Exercises

55. In a simple calorimeter, a reaction is carried out in a vessel submerged in water. After the reaction occurs, $42\overline{0}$ g of water increase in temperature from $25.0°C$ to $40.0°C$. What is the amount of energy released by the reaction?

56. The heat of reaction for $CaCO_3(s) \longrightarrow CaO(s) + CO_2(g)$ is $\Delta H = 42.5$ kcal. Given that $\Delta H_{CaO}^0 = -151.9$ kcal/mole and $\Delta H_{CO_2}^0 = -94.1$ kcal/mole, calculate $\Delta H_{CaCO_3}^0$. Compare your calculated value with the value given in Table 10.2.

57. Using Table 10.2, determine whether heat is absorbed or released when hydrogen gas reacts at standard-state conditions with the appropriate elements to form each of the following: (a) $C_2H_2(g)$, (b) $HCl(g)$, (c) $H_2O(l)$, (d) $HI(g)$.

58. When burned with excess oxygen, does 1.00 mole of methane (CH_4) or 1.00 mole of acetylene (C_2H_2) produce the greater amount of energy? The reactions are

$$CH_4(g) + 2O_2(g) \longrightarrow CO_2(g) + 2H_2O(g)$$
$$C_2H_2(g) + 2\tfrac{1}{2}O_2(g) \longrightarrow 2CO_2(g) + H_2O(g)$$

59. Energy is required to break chemical bonds. Considering this, determine whether the reaction $N_2O_4(g) \longrightarrow 2NO_2(g)$ would be endothermic or exothermic.

60. The heat of reaction for the burning of carbon is 94 kcal/mole. The coal used in a coal-fired power plant is 70.0% carbon. How much energy does 1.00 kg of coal yield?

61. An energy source that is gaining more attention these days is hydrogen gas. When it combines with oxygen, it forms water and releases 58 kcal/mole. Compare the fuel value of hydrogen gas with that of natural gas, which can be approximated as methane (CH_4), having a heat of reaction of 213 kcal/mole, and that of coal, which would yield 94 kcal/mole if it were pure carbon. [Hint: The best way to make this comparison is to calculate the heat of formation per gram for each substance.]

62. Explain why an exothermic reaction has a negative ΔH and an endothermic reaction has a positive ΔH.

63. Calculate the heat released by the combustion of 50.0 moles of methane.

64. Using Table 10.2, calculate the heat needed to produce the following change in 1.00 mole of water: $H_2O(l) \longrightarrow H_2O(g)$.

11 The Gaseous State: Ideal Behavior

Learning Goals

After you have studied this chapter, you should be able to:

1. Convert pressures in atmospheres to pressures in torr, and vice versa.

2. Use Boyle's Law to calculate changes in pressure and volume at constant temperature.

3. Use Charles's Law to calculate changes in temperature and volume at constant pressure.

4. Convert from Celsius degrees to kelvins, and vice versa.

5. Use the Combined Gas Law to calculate changes in temperature, pressure, or volume.

6. Use Dalton's Law to calculate the partial pressure of a gas in a mixture.

7. Use Avogadro's Principle to find the number of moles of a gas when you are given the volume (at fixed temperature and pressure), and vice versa.

8. Use the Ideal Gas Law to calculate the pressure, volume, temperature, or number of moles of a gas when you are given the other three.

9. Use the Ideal Gas Law to calculate the molecular weight of gas.

10. State and illustrate the kinetic theory of gases.

Introduction

To many people the word *gas* seems to imply mainly the gas we use for heating and cooking. But that's really the least of it. Every time we breathe we inhale gases; in fact, we inhale 35 pounds of them each day. Every time we pass a neon sign, use an aerosol spray, or open a refrigertor, we see examples of gas under pressure. A flying airplane is supported by gases.

Gases have deadly as well as useful aspects. A person in a large city inhales in one day as much toxic, or poisonous, material as he or she would by smoking two packs of cigarettes. More than half of this unhealthy mixture is contributed by the car.

Maybe because many gases are colorless and therefore seem invisible, people took a long time to get around to studying them. Until the late 1500s, air was the only vapor-like substance that scientists had studied. Even though alchemists had come into contact with many vapor-like substances, they had found them so mysterious that they called them spirits and ignored them. They used the word *spirits* for all liquids that vaporized, or turned to gas easily.

Jan Baptista van Helmont (1577–1644), a Flemish physician, was the first person to study vapors other than air. He observed that various vapors behaved like air but had different physical and chemical properties. Like air, these other vapors had no definite shape or volume.

In this chapter, we discuss the basic physical properties of gases: pressure, volume, and temperature.

11.1 Air Pressure and the Barometer

In the early 1600s the Italian mathematician and physicist Evangelista Torricelli discovered that air exerts a measurable pressure on the objects it touches. Torricelli filled a glass tube, which was sealed on one end, with liquid mercury. He then turned the tube upside down and placed it in a bowl, also filled with mercury (Figure 11.1). Torricelli noticed that most of the mercury remained in the upright tube. He concluded that the surrounding air exerted pressure on the surface of the mercury in the bowl, which in turn supported the column of mercury in the tube.

At sea level the height of the mercury supported in the tube averages 760 mm. In honor of Torricelli, the pressure that supports a column of mercury exactly 1 mm high is now called 1 **torr.** Therefore, at sea level, the air pressure averages 760 torr, depending on the weather. (Note that the plural of torr is torr.)

If we take Torricelli's pressure-measuring device, called a **barometer,** to the top of Pike's Peak (which is 14,000 feet above sea level), the height of the mercury column decreases, because the air pressure at the top of the peak is less than it is at sea level. If we take the same barometer to Death Valley (which is 280 feet below sea level), the height of the mercury column increases, because the air pressure below sea level is greater than it is at sea level. We can say that the height of the mercury column is proportional to the atmospheric pressure. In fact, the height of a mer-

Vacuum

h (mm)

Mercury, Hg

Atmospheric pressure

Figure 11.1
A simple barometer

cury column supported by *any* gas is proportional to the pressure of that gas. We can thus express the pressure of any gas as a number of torr.

Because we know that the air pressure at sea level supports a column of mercury 760 mm high, we can also define another unit of pressure, called the **atmosphere** (abbreviated *atm*). We have used this unit before, but now we'll define it in terms of torr:

$$1 \text{ atmosphere} = 760 \text{ mm of mercury} = 760 \text{ torr}$$

Learning Goal 1
Changing atmospheres to
torr, and vice versa

Air Pressure: A Strong Force

The German physicist Otto von Guericke developed a pump that could *evacuate*, or empty, the air from a container. In 1657, intrigued by Torricelli's experiments, he performed an experiment in which he put together two tightly fitted metal hemispheres (Figure 11.2). He then used his pump to evacuate the air from inside them. The hemispheres held together. He attached a team of horses to each hemisphere by means of a rope. The horses pulled in opposite directions, but they could not separate the hemispheres, because air was pressing from the outside but *not* from the inside. The air pressure was stronger than the two teams of horses. When the hemispheres were again filled with air, they easily fell apart.

Gas inlet–outlet

Figure 11.2
von Guericke's hemispheres

11.2 Boyle's Law

Learning Goal 2
Boyle's Law

After hearing of von Guericke's experiment, an English chemist named Robert Boyle decided to test the effect of pressure on the volume of gases. For his experiments in the mid 1600s, Boyle devised a J-shaped tube. He trapped air in the closed end of the tube by pouring mercury into the open end (Figure 11.3). He changed the pressure on the trapped air by changing the amount of mercury in the tube. He found that, if he tripled the pressure on the air, the volume occupied by the air decreased to one-

$5 \text{ cm}^3 \text{ Hg}$ $15 \text{ cm}^3 \text{ Hg}$ $45 \text{ cm}^3 \text{ Hg}$

90 cm^3 air 30 cm^3 air 10 cm^3 air

Figure 11.3
Boyle's experiment

third the original volume. And if he decreased the pressure on the trapped air to one-third the original pressure, the volume occupied by the air tripled.

Boyle failed to realize that the temperature of the air affects this relationship. The French physicist Edme Mariotte, working on his own, also discovered the relationship, and he was the first to state what is now known as **Boyle's Law:** *The volume of a gas increases as its pressure decreases, if its temperature remains constant.* Another way to state this principle is to say that at constant temperature, the volume of a given mass of gas varies inversely with its pressure.

Boyle's Law can also be stated in terms of the ratios of pressures and volumes:

$$\frac{P_i}{P_f} = \frac{V_f}{V_i} \text{ (constant temperature)}$$

or

$$P_i \times V_i = P_f \times V_f \qquad \qquad \textit{Equation 11.1}$$

where P stands for pressure, V stands for volume, i indicates initial (or original) conditions, and f indicates final conditions. Equation 11.1 can be used to find the final pressure or volume of any gas, given a change in the other quantity.

Example 11.1

Suppose a gas has an initial volume V_i of $5\overline{0}$ mL at a pressure of 1.0 atm. What will be its final volume V_f if the pressure of the gas is increased to 2.0 atm? (Assume that the temperature remains constant.)

Solution

First organize the data—that is, the facts that have been given.

$$V_i = 5\overline{0} \text{ mL} \qquad V_f = ?$$

$$P_i = 1.0 \text{ atm} \qquad P_f = 2.0 \text{ atm}$$

Then write Boyle's Law, solve for V_f, and enter the data in the result. (See Supplement A if this calculation seems difficult.)

$$P_i \times V_i = P_f \times V_f \qquad \text{so} \qquad V_f = \frac{P_i V_i}{P_f}$$

$$V_f = \frac{(1.0 \text{ atm})(5\overline{0} \text{ mL})}{2.0 \text{ atm}} = 25 \text{ mL}$$

As expected, increasing the pressure on the gas decreases the volume of the gas. This is what we mean by an *inverse relationship*.

Practice Exercise 11.1   A gas has an initial volume V_i of 500.0 mL at a pressure P_i of 2.50 atm. What will be its final volume V_f if the pressure is increased to 5.00 atm (P_f)? (Assume that the temperature remains constant.)

Example 11.2

Suppose a gas occupies a space of $50\overline{0}$ mL under a pressure of 3.0 atm at $10\overline{0}°C$. What will be the volume of the gas when the pressure is decreased to 1.0 atm? (Assume that the temperature remains constant.)

Solution

First organize the data:

$$V_i = 50\overline{0} \text{ mL} \qquad V_f = ?$$

$$P_i = 3.0 \text{ atm} \qquad P_f = 1.0 \text{ atm}$$

$$t = 10\overline{0}°C \text{ and is constant}$$

Then write Boyle's Law, solve for V_f, and enter the data in the result.

$$P_i \times V_i = P_f \times V_f \qquad \text{so} \qquad V_f = \frac{P_i V_i}{P_f}$$

$$V_f = \frac{(3.0 \text{ atm})(50\overline{0} \text{ mL})}{1.0 \text{ atm}} = 1{,}500 \text{ mL}$$

When the pressure was decreased to one-third of its initial value, the volume increased to three times *its* initial value.

Practice Exercise 11.2   A gas occupies a space of 250.0 mL under a pressure of 4.00 atm at 100.0°C. What will be the volume of the gas when the pressure is decreased to 2.00 atm? (Assume that the temperature remains constant.)

Example 11.3

A gas occupies a volume of 0.100 liter (L) under a pressure of $76\overline{0}$ torr at $2\overline{0}°C$. What pressure will cause the gas to occupy a volume of 50.0 mL at the same temperature?

Solution

Organize the data:

$$V_i = 0.100 \text{ L} = 10\overline{0} \text{ mL} \qquad V_f = 50.0 \text{ mL}$$

$$P_i = 76\overline{0} \text{ torr} \qquad P_f = ?$$

$$t = 2\overline{0}°C \text{ and is constant}$$

Write Boyle's Law, solve for P_f, and enter the data in the result.

$$P_i \times V_i = P_f \times V_f \qquad \text{so} \qquad P_f = \frac{P_i V_i}{V_f}$$

$$P_f = \frac{(76\overline{0} \text{ torr})(1\overline{0}0 \text{ mL})}{50.0 \text{ mL}} = 1{,}520 \text{ torr}$$

Practice Exercise 11.3   A gas occupies a volume of 0.200 L under a pressure of $76\overline{0}$ torr at 50.0°C. What pressure will cause the gas to occupy a volume of 25.0 mL at the same temperature?

11.3 Charles's Law

Learning Goal 3
Charles's Law

More than a hundred years after Boyle's work, the French physicist Jacques Charles discovered a relationship between the volume and the temperature of a gas. He found that a gas expands when its temperature is increased and that it shrinks when its temperature is decreased. A gas originally at 0°C expands 1/273 of its volume for each 1°C increase in temperature, and it shrinks 1/273 of its volume for each 1°C decrease (if the pressure of the gas is held constant). This is the basis of *Charles's Law.* It stands to reason: Most things do shrink as they get cooler and expand as they get hotter. The old tube-type tires used on automobiles are a good example. These tires used to expand as the air inside them got warm, as often happened on a hot day when the car was traveling at high speed. If the tires expanded enough, a blowout would occur.

Theoretically, if we took 1 L of a gas at 0°C and kept cooling it (at constant pressure), the volume of the gas would shrink to zero mL at −273°C (Figure 11.4). (This, of course, has never been accomplished, because all known gases become liquid before reaching that temperature.) The tem-

Figure 11.4
A theoretical plot of gas
volume versus temperature

Figure 11.5

Comparison of kelvins, degrees Celsius, and degrees Fahrenheit

Learning Goal 4
Changing °C to K, and vice versa

perature −273°C is called **absolute zero.** It is the basis of a temperature scale called the **Kelvin scale,** which was named after Lord Kelvin, the British physicist who proposed it. The units of this scale are kelvins, K, not degrees. One kelvin covers the same temperature interval as one Celsius degree (Figure 11.5). The relationship between the two temperature scales is

$$K = °C + 273 \qquad \text{Equation 11.2}$$

Using the Kelvin scale, we can state **Charles's Law** in the following way: *The volume of a gas varies directly with its temperature in kelvins if its pressure remains constant.* Charles's Law can also be stated in symbols:

$$\frac{V_i}{V_f} = \frac{T_i}{T_f} \text{ (constant pressure)} \qquad \text{Equation 11.3}$$

where T is the temperature in kelvins (a lower-case t stands for degrees Celsius).

You must remember to change degrees Celsius to kelvins when you work with the formula for Charles's Law. The direct relationship between the temperature and the volume of a gas, as expressed in Equation 11.3, applies *only* to temperatures measured in kelvins. To convert from one scale to the other, use the formula

$$T = t + 273$$

Example 11.4

A gas at 27°C occupies 1,$\overline{0}00$ mL of space. What will be its volume at 127°C? (Assume that the pressure remains constant.)

Solution

First organize the data and change degrees Celsius to kelvins.

$$V_i = 1,\overline{0}00 \text{ mL} \qquad\qquad V_f = ?$$

$$T_i = t_i + 273 \qquad\qquad\qquad T_f = t_f + 273$$

$$= 27 + 273 = 3\overline{0}0 \text{ K} \qquad = 127 + 273 = 4\overline{0}0 \text{ K}$$

Then write Charles's Law, solve for V_f, and enter the data in the result.

$$\frac{V_i}{V_f} = \frac{T_i}{T_f} \qquad \text{so} \qquad V_f = \frac{V_i T_f}{T_i}$$

$$V_f = \frac{(1\overline{0}00 \text{ mL})(4\overline{0}0 \text{ K})}{3\overline{0}0 \text{ K}}$$

$$= 1,333 \text{ mL or } 1,330 \text{ mL (three significant figures)}$$

As expected, increasing the temperature increases the volume of the gas. This is what we mean by a *direct relationship*.

Practice Exercise 11.4   A gas at 127°C occupies 2,5$\overline{0}0$ mL of space. What will be its volume at 27.0°C? (Assume that the pressure of the gas remains constant.)

Example 11.5

When a gas is cooled at constant pressure, it contracts from 50.0 mL to 45.0 mL. Its final temperature is 8$\overline{0}$°C. What was its initial temperature in degrees Celsius?

Solution

First organize the data and change degrees Celsius to kelvins.

$$V_i = 50.0 \text{ mL} \qquad V_f = 45.0 \text{ mL}$$

$$t_i = ? \qquad\qquad T_f = t_f + 273$$

$$T_i = ? \qquad\qquad = 8\overline{0} + 273 = 353 \text{ K}$$

Then write Charles's Law, solve for T_i, and enter the data in the result.

$$\frac{V_i}{V_f} = \frac{T_i}{T_f} \qquad \text{so} \qquad T_i = \frac{V_i T_f}{V_f}$$

$$T_i = \frac{(50.0 \text{ mL})(353 \text{ K})}{45.0 \text{ mL}} = 392 \text{ K}$$

Finally, in order to find the answer in degrees Celsius, proceed as follows:

$$T_i = t_i + 273$$

$$t_i = T_i - 273$$

$$t_i = 392 - 273 = 119°C$$

Practice Exercise 11.5   When a certain gas is heated at constant pressure, it expands from 50.0 mL to 60.0 mL. Its final temperature is 75.0°C. What was its initial temperature in degrees Celsius?

11.4 The Combined Gas Law

Learning Goal 5
Combined Gas Law

Boyle's Law and Charles's Law can be combined into a single law that includes all three variables—pressure, volume, and temperature. The single law can be used to determine how changes in any two of these variables affect the third variable. This **Combined Gas Law** can be stated as follows:

$$\frac{P_i V_i}{T_i} = \frac{P_f V_f}{T_f} \qquad\qquad Equation\ 11.4$$

Equation 11.4 incorporates both Charles's Law and Boyle's Law, so it is the only equation you need to memorize. The following examples show why.

Example 11.6

A certain gas at 2°C and 1.00 atm pressure fills a 4.00-L container. What volume will the gas occupy at $10\overline{0}$°C and $78\overline{0}$ torr pressure?

Solution
To begin, organize the data, change degrees Celsius to kelvins, and convert all the pressures to the same units.

$$P_i = 1.00 \text{ atm } (76\overline{0} \text{ torr}) \qquad P_f = 78\overline{0} \text{ torr}$$

$$V_i = 4.00 \text{ L} \qquad\qquad V_f = ?$$

$$T_i = t_i + 273 \qquad\qquad T_f = t_f + 273$$

$$= 2 + 273 = 275 \text{ K} \qquad = 100 + 273 = 373 \text{ K}$$

Then write the Combined Gas Law, solve for V_f, and enter the data in the result.

$$\frac{P_i V_i}{T_i} = \frac{P_f V_f}{T_f} \quad \text{so} \quad V_f = \frac{V_i P_i T_f}{P_f T_i}$$

$$V_f = \frac{(4.00 \text{ L})(76\overline{0} \text{ torr})(373 \text{ K})}{(78\overline{0} \text{ torr})(275 \text{ K})} = 5.29 \text{ L}$$

Practice Exercise 11.6   A certain gas at 25.0°C and 1.00 atm pressure fills a container of 5.00 L. What volume will the gas occupy at 50.0°C and 800.0 torr pressure?

Example 11.7

Suppose a gas is kept in a container (that is, at constant volume), but the pressure of the gas is increased from $76\overline{0}$ torr to $86\overline{0}$ torr. This is done by changing the temperature of the gas from 0°C to some higher temperature. What is this higher temperature? (Report the answer in degrees Celsius.)

Solution

To begin, organize the data and change degrees Celsius to kelvins. Note that the volume remains constant.

$$P_i = 76\overline{0} \text{ torr} \qquad\qquad P_f = 86\overline{0} \text{ torr}$$

$$T_i = t_i + 273 \qquad\qquad t_f = ?$$

$$= 0 + 273 = 273 \text{ K} \qquad T_f = ?$$

$$V \text{ is constant; this means} \qquad V_i = V_f$$

Then write the Combined Gas Law:

$$\frac{P_i V_i}{T_i} = \frac{P_f V_f}{T_f}$$

But, because V_i and V_f are equal, they cancel each other, and the equation becomes

$$\frac{P_i}{T_i} = \frac{P_f}{T_f}$$

Then solve the equation for T_f and enter the data in the result.

$$T_f = \frac{P_f T_i}{P_i} = \frac{(86\overline{0} \text{ torr})(273 \text{ K})}{76\overline{0} \text{ torr}} = 309 \text{ K}$$

Finally, change kelvins to degrees Celsius.

$$T_f = t_f + 273 \quad \text{so} \quad t_f = T_f - 273$$

$$t_f = 309 - 273 = 36°C$$

Practice Exercise 11.7 A gas is kept in a container at constant volume. The pressure of the gas is decreased from 850.0 torr to 750.0 torr. The temperature of the gas at the start of the experiment is 50.0°C. What is its final temperature in degrees Celsius?

Example 11.8

A certain gas occupies a volume of 3 L at 25°C and 4 atm pressure. What will be its volume at 25°C and 2 atm pressure?

Solution

First organize the data and change degrees Celsius to kelvins. Note that the temperature remains constant.

$$P_i = 4 \text{ atm} \qquad P_f = 2 \text{ atm}$$

$$V_i = 3 \text{ L} \qquad V_f = ?$$

$$t_i = t_f = 25°C \ (298 \text{ K})$$

Then write the Combined Gas Law.

$$\frac{P_i V_i}{T_i} = \frac{P_f V_f}{T_f}$$

T_i and T_f cancel each other because they are equal (remember that T is constant), and the equation becomes

$$P_i V_i = P_f V_f$$

Finally, solve this equation for V_f and enter the data in the result.

$$V_f = \frac{P_i V_i}{P_f} = \frac{(4 \text{ atm})(3 \text{ L})}{2 \text{ atm}} = 6 \text{ L}$$

Practice Exercise 11.8 A gas occupies a volume of 5.00 L at 298 K and 1.50 atm pressure. What will be its volume at 25°C and 3.00 atm pressure?

11.5 Standard Temperature and Pressure (STP)

For work with gases, scientists have defined a set of standard reference conditions. **Standard temperature and pressure (STP)** are defined as 273 K (0°C) and 1 atm pressure (760 torr). Whenever volumes of gases are reported, they may be corrected to STP so that comparisons may be made. That is, the Combined Gas Law is used to determine what the volume of the gas would be at STP.

Example 11.9

A gas collected at 127°C has a volume of 10 L and a pressure of 800 torr. What would be its volume at STP?

Solution

To begin, organize the data and change degrees Celsius to kelvins.

$$P_i = 800 \text{ torr} \qquad\qquad P_f = 760 \text{ torr}$$

$$V_i = 10 \text{ L} \qquad\qquad V_f = ?$$

$$T_i = t_i + 273 \qquad\qquad T_f = t_f + 273$$

$$= 127 + 273 = 400 \text{ K} \qquad = 0 + 273 = 273 \text{ K}$$

Then write the Combined Gas Law, solve for V_f, and enter the data in the result.

$$\frac{P_i V_i}{T_i} = \frac{P_f V_f}{T_f} \quad \text{so} \quad V_f = \frac{P_i V_i T_f}{P_f T_i}$$

$$V_f = \frac{(800 \text{ torr})(10 \text{ L})(273 \text{ K})}{(760 \text{ torr})(400 \text{ K})} = 7.2 \text{ L}$$

Practice Exercise 11.9 A gas collected at 25.0°C has a volume of 250.0 mL and a pressure of 780.0 torr. What would be its volume at STP?

11.6 Dalton's Law of Partial Pressures

Learning Goal 6
Dalton's Law

In the laboratory, gases are usually collected by the water-displacement method (Figure 11.6), because many gases are relatively insoluble in water. This is a convenient method for obtaining a sample of a gas, except for one thing: Gas collected by this method always has some water vapor in it. Therefore the pressure in the collection bottle (P_{total}) is due to both the pressure of the gas (P_{gas}) and the pressure of the water vapor (P_{water}). To find out how much gas has been collected, we need to know the pressure of the gas alone—that is, its partial pressure P_{gas}. The **partial pressure** of any gas in a mixture of gases is the pressure that the gas *would* exert at the temperature of the mixture, if it were the *only* gas in the container.

About 1800, John Dalton discovered a principle that is useful in this situation. **Dalton's Law of Partial Pressures** states that *the total pressure of a mixture of gases is the sum of the partial pressures of all the gases in the mixture.* This means that we can simply subtract the pressure of the water vapor from the total pressure to find the pressure exerted by the gas. In other words,

Figure 11.6

Collection of a gas over water. In this example, hydrochloric acid is added, through the thistle tube, to a piece of zinc in the bottle. Hydrogen gas travels out of the bottle through the rubber tube and is collected in the flask. As the hydrogen gas fills the flask, the water that was originally in the flask is pushed out.

Gas and water vapor

$$P_{total} = P_{gas} + P_{water} \quad \text{(Dalton's Law)}$$

so

$$P_{gas} = P_{total} - P_{water}$$

The partial pressure of water vapor is known at any given temperature. Table 11.1 gives this value for various temperatures.

Example 11.10

Suppose we collect a sample of oxygen gas over water at 27°C and at a pressure of $75\overline{0}$ torr. The gas fills a $50\overline{0}$-mL container but has water vapor mixed in with it. What would be the volume of the *dry* gas at STP?

Solution

First find the initial pressure of the dry gas (P_{O_2}). To do this, subtract the pressure of the water vapor (P_{H_2O}) from the total pressure in the collection bottle (P_{total}). Look up the vapor pressure of water at 27°C and find that it is 26.74 torr. Then

$$P_{O_2} = P_{total} - P_{H_2O}$$

$$P_{O_2} = 75\overline{0} \text{ torr} - 26.74 \text{ torr} = 723 \text{ torr}$$

Therefore the initial pressure of the dry oxygen gas (P_i) is 723 torr. The next step is to find its volume at STP.

Organize the data and change degrees Celsius to kelvins.

$P_i = 723$ torr	$P_f = 76\overline{0}$ torr
$V_i = 50\overline{0}$ mL	$V_f = ?$
$T_i = t_i + 273$	$T_f = t_f + 273$
$= 27 + 273 = 30\overline{0}$ K	$= 0 + 273 = 273$ K

Then write the Combined Gas Law, solve for V_f, and enter the data in the result.

$$\frac{P_i V_i}{T_i} = \frac{P_f V_f}{T_f} \qquad so \qquad V_f = \frac{P_i V_i T_f}{P_f T_i}$$

$$V_f = \frac{(723 \text{ torr})(500 \text{ mL})(273 \text{ K})}{(760 \text{ torr})(300 \text{ K})} = 433 \text{ mL}$$

The dry oxygen would have a volume of 433 mL at STP.

Practice Exercise 11.10   A sample of oxygen gas is collected over water at 55.0°C and at a pressure of 800.0 torr. The gas occupies a volume of 250.0 mL. What would be the volume of the *dry* gas at STP?

Table 11.1
Pressure of Water Vapor, P_{H_2O}, at Various Temperatures

Temperature (°C)	Pressure (torr)	Temperature (°C)	Pressure (torr)
0	4.58	32	35.66
5	6.54	33	37.73
10	9.21	34	39.90
15	12.79	35	42.18
16	13.63	36	44.56
17	14.53	37	47.07
18	15.48	38	49.69
19	16.48	39	52.44
20	17.54	40	55.32
21	18.65	45	71.88
22	19.83	50	92.51
23	21.07	55	118.04
24	22.38	60	149.38
25	23.76	65	187.54
26	25.21	70	233.7
27	26.74	75	289.1
28	28.35	80	355.1
29	30.04	85	433.6
30	31.82	90	525.8
31	33.70	95	633.9
		100	760.0

Example 11.11

Suppose 2.0 L of hydrogen gas is collected over water at 45°C and at a pressure of 700 torr. What would be the volume of the dry hydrogen gas at STP?

Solution

Again, first find the pressure of the dry hydrogen gas (P_{H_2}). To do this, subtract the pressure of the water vapor (P_{H_2O}) from the total pressure in the collection bottle (P_{total}). The vapor pressure of water at 45°C is 71.88 torr, so

$$P_{H_2} = P_{total} - P_{H_2O}$$

$$P_{H_2} = 700 \text{ torr} - 71.88 \text{ torr} = 628 \text{ torr}$$

Therefore the initial pressure of the dry hydrogen gas (P_i) is 628 torr. Next organize the data and change degrees Celsius to kelvins.

$$P_i = 628 \text{ torr} \qquad P_f = 760 \text{ torr}$$

$$V_i = 2.0 \text{ L} \qquad V_f = ?$$

$$T_i = t_i + 273 \qquad T_f = t_f + 273$$

$$= 45 + 273 = 318 \text{ K} \qquad = 0 + 273 = 273 \text{ K}$$

Write the Combined Gas Law, solve for V_f, and enter the data in the result.

$$\frac{P_i V_i}{T_i} = \frac{P_f V_f}{T_f} \quad \text{so} \quad V_f = \frac{P_i V_i T_f}{P_f T_i}$$

$$V_f = \frac{(628 \text{ torr})(2.0 \text{ L})(273 \text{ K})}{(760 \text{ torr})(318 \text{ K})} = 1.4 \text{ L}$$

Practice Exercise 11.11   A sample of hydrogen gas is collected over water at 20.0°C and 715 torr. It occupies a volume of 5.50 L. What would be the volume of the *dry* gas at STP?

Joseph Louis Gay-Lussac
(Reprinted with permission from
Torchbearers of Chemistry, by Henry
Monmouth Smith, © Academic Press,
Inc., 1949)

11.7 Gay-Lussac's Law of Combining Volumes

In the early 1800s the French chemist Joseph Louis Gay-Lussac (pronounced "Gay-lew-**sack**") investigated the reaction in which hydrogen gas and oxygen gas form water vapor. He found that if the gases were at the same temperature and pressure, 2 L of hydrogen gas plus 1 L of oxygen gas produced 2 L of water vapor. He also studied the reaction between hydrogen and chlorine. He found that 1 L of hydrogen gas plus 1 L of chlorine gas reacted to form 2 L of hydrogen chloride gas.

Hydrogen gas + oxygen gas $\longrightarrow$ water vapor
2 L   1 L   2 L

Hydrogen gas + chlorine gas $\longrightarrow$ hydrogen chloride
1 L   1 L   2 L

What Gay-Lussac had discovered is now called **Gay-Lussac's Law of Combining Volumes,** which states that *when gases at the same temperature and pressure combine, the volumes of the reactants and products are in the same ratio as the coefficients of the balanced equation for the reaction.*

The Law of Combining Volumes states, for example, that two volumes of hydrogen combine with one volume of oxygen to form two volumes of water vapor

$$2H_2(g) + O_2(g) \longrightarrow 2H_2O(g)$$
2 volumes   1 volume   2 volumes

whether "volume" means liter, cubic inch, gallon, or even coffee-can-full, as long as all three gases are at the same temperature and pressure.

This provides an easy way to work stoichiometric problems involving gases.

Example 11.12

How many liters of carbon dioxide can be produced with 10 L of carbon monoxide? How much oxygen is required (provided that all three gases are at the same temperature and pressure)?

Solution

First write the balanced equation and organize the data.

$$2CO(g) + O_2(g) \longrightarrow 2CO_2(g)$$
10 L   ? L   ? L

Then use the coefficients of the balanced equation, just as in Chapter 9 (except that here liters can be substituted for moles), to calculate the amount of O_2 needed and the amount of CO_2 formed.

$$? \text{ L of } CO_2 = (10 \text{ L } CO)\left(\frac{2 \text{ L } CO_2}{2 \text{ L } CO}\right)$$

$$= 10 \text{ L of } CO_2$$

$$? \text{ L of } O_2 = (10 \text{ L } CO)\left(\frac{1 \text{ L } O_2}{2 \text{ L } CO}\right)$$

$$= 5 \text{ L of } O_2$$

> **Practice Exercise 11.12**   How many liters of hydrogen fluoride gas can be produced from 20.0 L of hydrogen gas and 20.0 L of fluorine gas? (Assume that all gases are at the same temperature and pressure.)

11.8 Molar Gas Volume and Avogadro's Principle

Learning Goal 7
Avogadro's Principle

In 1811 the Italian physicist Amedeo Avogadro, whose name we associate with Avogadro's number, advanced one of the most important hypotheses in chemistry. In an attempt to explain Gay-Lussac's Law of Combining Volumes, Avogadro said that *at the same temperature and pressure, equal volumes of all gases contain the same number of molecules.* This is called **Avogadro's Principle.** To get an idea of its meaning, consider the following experiment.

Suppose exactly 1 mole of hydrogen gas is placed in a container whose volume can be changed and whose temperature can be kept at 0°C (273 K). The volume of the container is adjusted so that the pressure of the hydrogen gas is exactly 1 atm. We find that the volume of the container (and therefore of the gas) is 22.4 L.

We now remove the hydrogen gas and repeat the experiment with exactly 1 mole of oxygen. Again we find that the volume of the container (and the oxygen) is 22.4 L. And every time the experiment is repeated with exactly 1 mole of a different gas, we find that its volume is 22.4 L at standard temperature and pressure (Figure 11.7).

Our experiment enables us to state the following conclusion: *Exactly one mole of any gas at STP has a volume of 22.4 L.* This volume is called **the molar volume of a gas,** and this conclusion can be used to solve these types of problems.

Figure 11.7
Molar gas volume

1 mole H₂ at STP	1 mole He at STP	1 mole N₂ at STP	1 mole O₂ at STP

1 mole H$_2$ at STP
V = 22.4 liters

1 mole He at STP
V = 22.4 liters

1 mole N$_2$ at STP
V = 22.4 liters

1 mole O$_2$ at STP
V = 22.4 liters

Example 11.13

A certain gas occupies a volume of 179.2 L at STP. How many moles of gas are in this sample?

Solution

One mole of any gas has a volume of 22.4 L at STP. Therefore,

$$? \text{ moles of gas} = (179.2 \text{ L})\left(\frac{1 \text{ mole}}{22.4 \text{ L}}\right) = 8.00 \text{ moles of gas}$$

Practice Exercise 11.13   How many moles of gas occupy a volume of 448 mL at STP?

Example 11.14

Determine the volume of 5.00 moles of a gas at STP.

Solution

Again, use the idea that 1 mole of a gas has a volume of 22.4 L at STP. Therefore

$$? \text{ L} = (5.00 \text{ moles})\left(\frac{22.4 \text{ L}}{1 \text{ mole}}\right) = 112 \text{ L}$$

Practice Exercise 11.14   What is the volume of 0.200 mole of a gas at STP?

11.9 The Ideal Gas Law

Learning Goal 8
Ideal Gas Law

We can now combine Boyle's Law, Charles's Law, and Avogadro's Principle into a relationship among the gas variables (*P, V, T*) and the number *n* of moles of gas. This relationship, called the **Ideal Gas Law,** is represented by the equation

$$PV = nRT \qquad \textit{Equation 11.5}$$

Pressure in atm, Volume in liters, Ideal gas constant, Number of moles, Temperature in K

If this equation is used with the pressure expressed in atmospheres, the volume in liters, the temperature in kelvins, and the amount of gas in moles, the value for *R*, the ideal gas constant, is

$$0.0820 \frac{(\text{L})(\text{atm})}{(\text{K})(\text{mole})}$$

The following examples show how the ideal gas equation is used.

Example 11.15

Two moles of a gas exert a pressure of 1,520 torr at a temperature of 127°C. What volume does the gas occupy under these conditions?

Solution

First organize the data. Change the 1,520 torr to atmospheres and the degrees Celsius to kelvins. To convert torr to atmospheres, remember that 1 atm = $76\overline{0}$ torr.

$$? \text{ atm} = (1,520 \text{ torr})\left(\frac{1 \text{ atm}}{76\overline{0} \text{ torr}}\right) = 2.00 \text{ atm}$$

$$P = 1,520 \text{ torr} \ (2.00 \text{ atm})$$

$$V = ?$$

$$n = 2.00 \text{ moles}$$

$$R = 0.0820 \ \frac{\text{(L)(atm)}}{\text{(K)(mole)}}$$

$$T = t + 273 = 127 + 273 = 4\overline{00} \text{ K}$$

Next write the ideal gas equation, solve for V, and enter the data in the result.

$$PV = nRT \qquad \text{so} \qquad V = \frac{nRT}{P}$$

$$V = \frac{(2.00 \text{ moles})\left(0.0820 \ \frac{\text{(L)(atm)}}{\text{(K)(mole)}}\right)(4\overline{00} \text{ K})}{2.00 \text{ atm}} = 32.8 \text{ L}$$

Practice Exercise 11.15   Three moles of a gas exert a pressure of 1,555 torr at a temperature of 27.0°C. What volume does the gas occupy under these conditions?

Example 11.16

A gas has a pressure of $38\overline{0}$ torr at a temperature of $5\overline{0}$°C. The volume of the gas is $8\overline{00}$ mL. How many moles of gas are present?

Solution

Organize the data. Change torr to atmospheres and degrees Celsius to kelvins.

$$? \text{ atm} = (38\overline{0} \text{ torr})\left(\frac{1 \text{ atm}}{76\overline{0} \text{ torr}}\right) = 0.500 \text{ atm}$$

$$P = 0.500 \text{ atm}$$

$$V = 8\overline{0}0 \text{ mL } (0.800 \text{ L})$$

$$n = ?$$

$$R = 0.0820 \frac{\text{(L)(atm)}}{\text{(K)(mole)}}$$

$$T = t + 273 = 5\overline{0} + 273 = 323 \text{ K}$$

Write the ideal gas equation, solve for n, and enter the data in the result.

$$PV = nRT \quad \text{so} \quad n = \frac{PV}{RT}$$

$$n = \frac{(0.500 \text{ atm})(0.800 \text{ L})}{\left(0.0820 \frac{\text{(L)(atm)}}{\text{(K)(mole)}}\right)(323 \text{ K})} = 0.0150 \text{ mole}$$

Practice Exercise 11.16   A gas exerts a pressure of 444 torr at a temperature of 55.0°C. The volume of the gas is 1.00 L. How many moles of gas are present?

11.10 Determining Molecular Weight with the Ideal Gas Law

Learning Goal 9
Ideal Gas Law and molecular weight

The number of moles n of a sample of a gas is equal to the mass of the sample (g) in grams divided by the molecular weight of the gas (MW).

$$n = \frac{g}{MW}$$

If g/MW is used in place of n in the Ideal Gas Law, we have a valuable tool for determining the molecular weights of gases:

$$PV = \frac{gRT}{MW} \qquad \text{Equation 11.6}$$

For an unknown gas, we can weigh a sample of it into a container, measure its pressure, volume, and temperature, and use this form of the Ideal Gas Law to calculate its molecular weight.

Example 11.17

What is the molecular weight of the gas for which the following experimental data apply?

$$\text{Mass} = 1.00 \text{ g} \qquad V = 82.0 \text{ mL } (0.0820 \text{ L})$$

$$P = 1.5 \text{ atm} \qquad T = 3\overline{00} \text{ K}$$

Solution

Write Equation 11.6, solve for MW, and enter the data in the result.

$$PV = \frac{gRT}{MW} \qquad \text{so} \qquad MW = \frac{gRT}{PV}$$

$$MW = \frac{(1.00 \text{ g})\left(0.0820 \dfrac{(\text{L})(\text{atm})}{(\text{K})(\text{mole})}\right)(3\overline{00} \text{ K})}{(1.5 \text{ atm})(0.0820 \text{ L})} = 2\overline{00} \text{ g/mole}$$

Practice Exercise 11.17 Determine the molecular weight of the gas to which the following experimental data apply:

$$\text{Mass} = 2.50 \text{ g}, \quad V = 100.0 \text{ mL}, \quad P = 2.00 \text{ atm}, \quad T = 400.0 \text{ K.}$$

11.11 Kinetic Theory of Gases: A Review of This Chapter

. . . .
Learning Goal 10
Kinetic theory of gases

We can explain all the properties and laws of gases that we have discussed in this chapter with a series of statements called the **kinetic theory of gases.** The kinetic theory is an attempt to account for properties of gases in terms of the forces that exist between the gas molecules and the energy of these molecules. The kinetic theory states:

1. Nearly all gases are composed of molecules. (The so-called noble gases are composed of atoms, not molecules. Remember, atoms of noble gases don't combine readily with other atoms.)

2. The forces of attraction between gas molecules increase as the molecules move closer together. At normal atmospheric pressure and room temperature, the distances between gas molecules are large compared to the size of the molecules.

3. Gas molecules are always in motion. They often collide with other gas molecules or with their container. The molecules do not stick together after collisions occur, and they do not lose any energy on account of the collisions.

4. Gas molecules move faster when the temperature rises and slower when the temperature drops.

5. All gas molecules (heavy as well as light) have the same average kinetic energy—that is, energy of motion—at the same temperature.

If the gas laws are considered in light of these five statements, it is clear why gases behave as they do. For example, the kinetic theory explains

Figure 11.8
A sample of gas in a small container. The pressure gauge shows the pressure that is exerted by the gas.

Figure 11.9
The same sample of gas in a container twice the size of that shown in Figure 11.8

Boyle's Law. Remember, Boyle's Law says that the volume of a gas increases as its pressure decreases and that the pressure of a gas increases as its volume decreases. When we assume that gases are composed of molecules that are constantly in motion, colliding with each other and with the walls of their container, it seems reasonable that:

(a) Keeping these molecules in a smaller space causes them to strike the walls of their container more often; this means that more force is exerted against the walls of the container; therefore the pressure is greater (Figure 11.8).

(b) Allowing these molecules more space to move around in causes them to strike the walls less often; therefore the pressure is decreased (Figure 11.9).

The kinetic theory also explains Charles's Law, which says that the volume of a gas increases as its temperature increases and that it decreases as its temperature decreases. Consider a balloon filled with air (Figure 11.10). The molecules of air constantly move around, hitting the walls of the balloon. Suppose a candle is held near the balloon, which warms the air inside. As the molecules of air get warmer they move faster, hitting the walls of the balloon with more force. Because the balloon is made of flexible material, it expands. As the molecules of air cool to room temperature, they move more slowly, hitting the walls of the balloon more gently, so the balloon contracts.

Summary

In the early 1600s, Torricelli devised a pressure-measuring device called a barometer and discovered that air exerts pressure on objects it touches. The units of pressure we use are atmospheres and torr, and air exerts a pressure of about 760 torr (1 atm) at sea level.

The experiments of Robert Boyle in the mid-1600s led to his formulation of Boyle's Law, which states that, at constant temperature, the vol-

Figure 11.10
The balloon experiment shows the temperature–volume relationship expressed by Charles's Law.

ume of a given mass of gas varies inversely with the pressure. In the 1780s Jacques Charles discovered a relationship between the volume and the temperature of a gas. Charles's Law states that, at constant pressure, the volume of a given amount of gas varies directly with its temperature in kelvins. On the Kelvin scale, the temperature in kelvins is equal to the Celsius temperature plus 273. Boyle's Law and Charles's Law can be combined into a single law, the Combined Gas Law, which shows the interrelationship of changes in the temperature, pressure, and volume of a gas.

Standard temperature and pressure (STP) for gases are defined as 273 K (0°C) and 1 atm pressure (760 torr). The Combined Gas Law may be used to determine what the volume of a gas would be at STP. Dalton's Law of Partial Pressures states that the total pressure exerted by a mixture of gases is the sum of the partial pressures that each of the gases would exert if it were alone at the same temperature as the mixture.

In the early 1800s, Gay-Lussac discovered the Law of Combining Volumes: When gases at the same temperature and pressure combine, the volumes of the reactants and products are in the same ratio as the coefficients of the balanced equation for the reaction. In an attempt to explain the Law of Combining Volumes, Avogadro hypothesized that equal volumes of all gases at the same temperature and pressure contain the same number of molecules. From the hypothesis, called Avogadro's Principle, we conclude that 1 mole of any gas at STP has a volume of 22.4 L. This is called the molar volume of a gas.

Boyle's Law, Charles's Law, and Avogadro's Principle can be combined into a single relationship among the pressure P, temperature T, volume V, and number of moles n of a gas. This relationship, called the Ideal Gas Law, is represented by the equation $PV = nRT$, where R is the ideal gas constant and is equal to 0.082 (L)(atm)/(K)(mole). The Ideal Gas Law can be used to determine the molecular weight of a gas when n is replaced by the fraction that represents mass/molecular weight (g/MW) and the equation is rearranged.

The kinetic theory of gases is concerned with the motion of gas molecules and the forces of attraction between molecules. It can be used to explain the various physical properties of gases.

Career Sketch

Veterinarians

Veterinarians are doctors of veterinary medicine. They treat diseases and injuries of animals and help prevent the spread of animal diseases (many of which can be transmitted to humans).

Veterinarians treat diseases and injuries of animals and help prevent the spread of animal diseases.
(American Veterinary Medicine Association)

Veterinarians treat animals in hospitals and clinics, as well as on farms and ranches. They perform surgery on sick and injured animals and prescribe and administer medicines and vaccines.

Veterinary medicine offers a variety of practice specialties. More than one-third of all veterinarians treat only small animals or pets. They diagnose illnesses in pets, spay or neuter them, give shots, and set broken bones. About another third treat both large and small animals. A large number specialize in the health and breeding of farm animals: cattle, poultry, sheep, pigs, and horses. Many veterinarians inspect meat and poultry as part of public-health programs. Still others teach in colleges of veterinary medicine, do research related to animal diseases, foods, and drugs, or work as part of a medical research team to learn about prevention and treatment of diseases in humans.

The training of veterinarians involves many of the same courses required of students in other branches of medicine.

Career Sketch

Fire Control and Safety Technician

Recognizing fire hazards, applying technical knowledge, and performing services to control and prevent fires are some of the responsibilities of fire control and safety technicians. These men and women protect life and property, conduct safety inspections, and plan fire protection systems.

Each year fire causes thousands of deaths and destroys more than a billion dollars' worth of property. Fire control and safety technicians work with local fire departments, fire insurance companies, industry, government agencies, business firms, and consulting companies.

The requirements for this occupation are a knowledge of the chemistry of how things burn and of what makes some things more likely than others to catch fire or to generate fumes, heat, and smoke. Students study basic fire science and mathematics and must understand combustion, much special equipment, and many materials, structures, and devices.

Fire safety technicians work in government, business, and industry to help prevent fires.
(Photograph courtesy of Arco)

Knowledge of gas density, and of the chemical and physical properties of gases, is also important. They may specialize in the chemistry of fire extinguishing or in the chemistry of materials and combustion. Over 100 technical institutes and community colleges offer programs in this relatively new field.

Self-Test Exercises

Learning Goal 1: Changing Atmospheres to Torr, and Vice Versa

1. Express the following pressures in atmospheres: (a) 250 torr, (b) 1,550 torr, (c) 7,600 torr.

2. Express the following pressures in atmospheres: (a) 350 torr, (b) 2,500 torr, (c) 1,960 torr.

3. Express the following pressures in torr: (a) 3.00 atm, (b) 0.300 atm, (c) 0.150 atm.

4. Express the following pressures in torr: (a) 2.00 atm, (b) 0.250 atm, (c) 0.123 atm.

5. Express the following pressures in atmospheres: (a) 76 torr, (b) 2,660 torr, (c) 3.80 torr.

6. Express the following pressures in atmospheres: (a) 152 torr, (b) 304 torr, (c) 0.152 torr.

7. Express the following pressures in torr: (a) 0.0100 atm, (b) 12.0 atm, (c) 5.50 atm.

8. Express the following pressures in torr: (a) 0.200 atm, (b) 15.2 atm, (c) 7.6 atm.

Learning Goal 2: Boyle's Law

9. Given the following data for a gas, use Boyle's Law to determine V_f: $P_i = 0.60$ atm, $V_i = 30.0$ L, $P_f = 0.10$ atm.

10. Given the following data for a gas, use Boyle's Law to determine V_f: $P_i = 0.500$ atm, $V_i = 40.0$ L, $P_f = 0.200$ atm.

11. Given the following data for a gas, use Boyle's Law to determine V_f: $P_i = 3.00$ atm, $V_i = 500$ mL, $P_f = 380$ torr.

12. Given the following data for a gas, use Boyle's Law to determine V_f: $P_i = 2.00$ atm, $V_i = 250$ mL, $P_f = 400$ torr.

13. Given the following data for a gas, use Boyle's Law to determine P_f: $P_i = 1,900$ torr, $V_i = 250$ mL, $V_f = 50.0$ mL.

14. Given the following data for a gas, use Boyle's Law to determine P_f: $P_i = 1,520$ torr, $V_i = 455$ cm³, $V_f = 75.0$ cm³.

Learning Goal 3: Charles's Law

15. Given the following data for a gas, use Charles's Law to determine t_f (that is, express your answer in °C): $t_i = 77.0$°C, $V_i = 2.00$ L, $V_f = 4.00$ L.

16. Given the following data for a gas, use Charles's Law to determine t_f (that is, express your answer in °C): $t_i = 90.0$°C, $V_i = 4.00$ L, $V_f = 5.50$ L.

17. Given the following data for a gas, use Charles's Law to determine t_f (that is, express your answer in °C): $t_i = -73$°C, $V_i = 8.00$ L, $V_f = 32.0$ L.

18. Given the following data for a gas, use Charles's Law to determine t_f (that is, express your answer in °C): $t_i = 85.0$°C, $V_i = 7.50$ L, $V_f = 4.50$ L.

19. Given the following data for a gas, use Charles's Law to determine V_f: $t_i = 327°C$, $V_i = 20\overline{0}$ L, $t_f = -173°C$.

20. Given the following data for a gas, use Charles's Law to determine V_f: $t_i = 400.0°C$, $V_i = 45\overline{0}$ L, $t_f = -73°C$.

Learning Goal 4: Changing °C to K, and Vice Versa

21. Change the following temperatures to °C: (a) $50\overline{0}$ K, (b) 4 K.

22. Change the following temperatures to °C: (a) 353 K, (b) 3 K.

23. Change the following temperatures to kelvins: (a) 10°C, (b) −10°C, (c) −73°C.

24. Change the following temperatures to kelvins: (a) 15°C, (b) −15°C, (c) −50°C.

25. Change the following temperatures to °C: (a) 0 K, (b) $10\overline{0}$ K, (c) $20\overline{0}$ K, (d) $30\overline{0}$ K.

26. Change the following temperatures to °C: (a) 5 K, (b) 150 K, (c) 250 K, (d) 350 K.

27. Change the following temperatures to kelvins: (a) 20°C, (b) 0°C, (c) $-10\overline{0}$°C, (d) 212°F.

28. Change the following temperatures to kelvins: (a) 25°C, (b) −5°C, (c) −75°C, (d) 424°F.

Learning Goal 5: Combined Gas Law

29. Given the following data for a gas, use the Combined Gas Law to determine P_f:

$$P_i = 0.5 \text{ atm} \qquad V_f = 5 \text{ L}$$
$$V_i = 10 \text{ L} \qquad t_f = 327°C$$
$$t_i = 27°C$$

30. Given the following data for a gas, use the Combined Gas Law to determine P_f:

$$P_i = 0.20 \text{ atm} \qquad V_f = 10.0 \text{ L}$$
$$V_i = 8.0 \text{ L} \qquad t_f = 35\overline{0}°C$$
$$t_i = 43°C$$

31. A tire is filled with air at a pressure of 1,520 torr at 20.0°C. What will be the pressure in the tire after the car is driven and the temperature of the air in the tire rises to 34.0°C? (Assume that the volume of the tire remains constant, which is the case in a tubeless tire.)

32. A tubeless tire (whose volume remains constant) is filled with air at a pressure of $76\overline{0}$ torr at 30.0°C. What will be the pressure in the tire after the car is driven and the temperature of the air in the tire rises to 35.0°C?

33. A balloon is filled with a gas to a volume of 400 mL, at a pressure of 1.00 atm and a temperature of 25.0°C. The temperature is increased to 50.0°C and the pressure to $78\overline{0}$ torr. What will be the volume of the balloon?

34. A balloon is filled with air to a volume of 350.0 mL, at a pressure of 1.20 atm and a temperature of 28.0°C. The temperature is increased to 55.0°C and the pressure is increased to 2.00 atm. What is the new volume of the balloon?

35. Suppose that you collect a gas over water at 30.0°C. The volume of the gas is $80\overline{0}$ mL, and the total pressure is $78\overline{0}$ torr. What is the volume of *dry* gas at STP?

36. A gas is collected over water at 28.0°C. The volume of the gas is 750.0 mL, and the total pressure is 790.0 torr. What is the volume of the *dry* gas at STP?

37. Assume that your lungs can hold about $50\overline{0}$ mL of air at STP. You take a deep breath and hold it, then dive into the ocean to a depth at which the temperature is 10.0°C and the pressure is $1,0\overline{0}0$ torr. What is the volume of the air in your lungs now?

38. A gas is collected over water at 35.0°C. The volume of the gas is 1.00 L, and the total pressure is 800.0 torr. What is the volume of the *dry* gas at STP?

39. The Gulf of Mexico puffer-fish can live in deep water, where the water pressures are enormous. These fish have the ability to compress and expand their bodies to compensate for changes in pressure. They do this by holding in or letting out air. If a fish of this type decided to move from an area in which the water pressure was 4.0 atm to an area in which it was 1.5 atm, by what ratio would the volume of its body increase?

40. Assume that the Gulf of Mexico puffer-fish described in Exercise 39 moves from an area in which the water pressure is 2.00 atm to an area in which it is 4.50 atm. What happens to the volume of its body?

41. An airtight chamber is filled with a gas at a pressure of 25.0 atm and a temperature of 50.0°C. The temperature inside the chamber is increased to $50\overline{0}$°C. What happens to the pressure of the gas? (The volume of the chamber remains constant.)

42. An airtight chamber is filled with a gas to a pressure of 15.0 atm and a temperature of 350.0°C. If the temperature inside the chamber is decreased to 50.0°C, what will happen to the pressure of the gas? (The volume of the chamber remains constant.)

43. A gas has a volume of $40\overline{0}$ mL at a pressure of 3.00 atm and a temperature of 127°C. What would its volume be at STP?

44. A gas has a volume of 40.0 mL at a pressure of 2.50 atm and a temperature of 100.0°C. What would its volume be at STP?

45. A sample of oxygen gas is collected over water. The volume of gas collected is $30\overline{0}$ mL, at a pressure of $75\overline{0}$ torr and a temperature of 15.0°C. What is the volume of the *dry* gas at STP?

46. A sample of hydrogen gas is collected over water. The volume of gas collected is 250.0 mL, at a pressure of 815 torr and a temperature of 27.0°C. What is the volume of the dry gas at STP?

47. Complete the following table for each data set.

Data set	P	V	n	T
1	Constant	Increased	Constant	?
2	?	Decreased	Constant	Constant
3	Decreased	?	Increased	Constant

48. Complete the following table for each data set.

Data set	P	V	n	T
1	decrease	?	constant	constant
2	?	increase	constant	constant
3	increase	constant	constant	?

Learning Goal 6: Dalton's Law

49. Find the total pressure in an enclosed vessel that contains the following gases: (*a*) nitrogen at $2\overline{0}0$ torr, (*b*) hydrogen at $3\overline{0}0$ torr, (*c*) oxygen at $5\overline{0}$ torr, (*d*) helium at $15\overline{0}$ torr.

50. Find the total pressure in an enclosed vessel that contains the following gases: (*a*) helium at $2\overline{0}0$ torr, (*b*) hydrogen at $15\overline{0}$ torr, (*c*) nitrogen at $9\overline{0}$ torr, (*d*) oxygen at $15\overline{0}$ torr.

51. Nitrogen gas makes up about $8\overline{0}$ volume percent of the atmosphere. What is the pressure exerted by this nitrogen gas when the total atmospheric pressure is $76\overline{0}$ torr?

52. Suppose we collect a $75\overline{0}$-mL sample of oxygen gas over water at 22.0°C and $77\overline{0}$ torr. What would be the volume of the *dry* gas at STP?

Learning Goal 7: Avogadro's Principle

53. Methane (CH_4) burns in air to produce carbon dioxide (CO_2) and water vapor. How many liters of water vapor are produced from 12 L of CH_4?

54. Ethane (C_2H_6) burns in air to produce carbon dioxide and water vapor. How many liters of water vapor are produced from 15.0 L of ethane?

55. Determine the volume of 4.00 moles of oxygen gas at STP.

57. You are given the reaction

$$2Na + 2H_2O \longrightarrow 2NaOH + H_2$$

How many grams of sodium would be needed to release 10.0 L of hydrogen gas at STP?

*59. Suppose that $8\overline{0}$ mL of hydrogen gas react with $8\overline{0}$ mL of oxygen gas. How many milliliters of oxygen gas remain uncombined? How many milliliters of water vapor are formed?

61. Sulfur dioxide can combine with oxygen in the air to produce sulfur trioxide:

$$SO_2(g) + O_2(g) \longrightarrow SO_3(g)$$

(a) Balance the equation. (b) How many liters of oxygen gas are needed to combine with $15\overline{0}$ L of SO_2? (c) How many liters of SO_3 are produced?

63. Butane burns in air to produce carbon dioxide and water vapor:

$$C_4H_{10}(g) + O_2(g) \longrightarrow CO_2(g) + H_2O(g)$$

(a) Balance the equation. (b) How many liters of oxygen gas are needed to burn 15.0 L of butane? (c) How many liters of carbon dioxide and water are produced?

*65. Nitrogen gas can combine with oxygen gas to produce nitrogen dioxide:

$$N_2(g) + O_2(g) \longrightarrow NO_2(g)$$

(a) Balance the equation. (b) Suppose that 75.0 mL of nitrogen gas react with 75.0 mL of oxygen gas. How many milliliters of nitrogen gas remain uncombined? (c) How many milliliters of $NO_2(g)$ are formed?

56. What is the volume of 3.00 moles of hydrogen gas at STP?

*58. Consider the reaction

$$2Li + 2H_2O \longrightarrow 2LiOH + H_2$$

How many grams of lithium are needed to release 10.0 L of hydrogen gas at STP?

*60. Suppose that 1.00 L of hydrogen gas reacts with 1.00 L of oxygen gas at STP. How many grams of oxygen gas remain uncombined? How many grams of water vapor are formed?

62. Nitric oxide can combine with oxygen gas in the air to produce nitrogen dioxide:

$$NO(g) + O_2(g) \longrightarrow NO_2(g)$$

(a) Balance the equation. (b) How many liters of oxygen gas are needed to combine with 10.0 L of NO? (c) How many liters of NO_2 are formed?

64. Propane burns in air to produce carbon dioxide and water vapor:

$$C_3H_8(g) + O_2(g) \longrightarrow CO_2(g) + H_2O(g)$$

(a) Balance the equation. (b) How many liters of oxygen gas are needed to burn 25.0 L of propane? (c) How many liters of carbon dioxide gas are formed?

*66. Nitrogen gas can combine with hydrogen gas to produce ammonia:

$$N_2(g) + H_2(g) \longrightarrow NH_3(g)$$

(a) Balance the equation. (b) Suppose that 120.0 mL of N_2 can react with 120.0 mL of H_2 gas. How many milliliters of nitrogen gas remain uncombined? (c) How many milliliters of ammonia gas are formed?

Learning Goal 8: Ideal Gas Law

67. Determine the number of grams of nitrogen gas present in a 3.00-L container at a temperature of 27°C and a pressure of 1,520 torr.

69. Part of Avogadro's Principle is that 1.00 mole of any gas at STP has a volume of 22.4 L. Using this information, solve the ideal gas equation for R.

71. Determine the volume of 0.500 mole of $N_2(g)$ at STP.

68. Determine the number of grams of hydrogen gas present in a 2.00-L container at a temperature of 127°C and a pressure of $2,28\overline{0}$ torr.

70. What volume will 0.75 mole of hydrogen gas occupy at STP?

72. How many grams of chlorine gas are present in a 1,500.0-mL flask at a temperature of 150.0°C and a pressure of 1.50 atm?

73. Determine the number of grams of NO_2 gas present in a $50\overline{0}$-mL flask, at a temperature of 227°C and a pressure of $3,04\overline{0}$ torr.

74. How many moles of fluorine gas are present in a 2.00-L flask at a temperature of 127°C and a pressure of 1.00 atm?

75. Determine the pressure of 12.20 moles of H_2 gas occupying a volume of $2,00\overline{0}$ mL at a temperature of −73°C.

76. What is the pressure of 4.00 moles of oxygen gas that occupy a volume of 5.00 L at a temperature of 0°C?

77. A 128-g sample of SO_2 has a pressure of $38\overline{0}$ torr and a volume of $50\overline{0}$ mL. What is the temperature of this gas?

78. A 92.0-g sample of NO_2 gas has a pressure of $77\overline{0}$ torr and a volume of $95\overline{0}$ mL. What is the temperature of this gas?

Learning Goal 9: Ideal Gas Law and Molecular Weight

*79. A gaseous compound is made up of 83% carbon and 17% hydrogen by weight. When 1.2 g of this gas are placed in a $50\overline{0}$-mL container at 25°C, the pressure of the gas is found to be 1.0 atm. (a) What is the empirical formula of the gas? (b) What is the molecular weight of the gas? (c) What is the molecular formula of the gas?

80. Given that 4.00 g of a certain gas have a volume of 500.0 mL at a pressure of 2,005 torr and a temperature of 27.0°C, calculate the molecular weight of the gas.

81. Calculate the molecular weight of a gas, given that 2.0 g of the gas have a volume of $40\overline{0}$ mL at a pressure of 3,040 torr and a temperature of 127°C.

*82. Using the Ideal Gas Law equation and remembering that density equals mass divided by volume, calculate the density of hydrogen gas at STP. (Report your answer in grams per liter.)

*83. Using the ideal gas equation and remembering that density equals mass divided by volume $(D = m/V)$, calculate the density of oxygen gas at STP. (Report the answer in grams per liter.)

84. A 20.0-g sample of a gas has a pressure of 4.00 atm and a volume of 1,500.0 mL at a temperature of 30.0°C. What is the molecular weight of the gas?

85. A 14.63-g sample of a gas has a pressure of 6.00 atm and a volume of $2,50\overline{0}$ mL at a temperature of 27°C. What is the molecular weight of this gas?

86. Given that 150.0 g of a certain gas have a volume of 5.00 L at a pressure of $2,00\overline{0}$ torr and a temperature of 150.0°C, calculate the molecular weight of the gas.

*87. A gaseous compound is 69.57% oxygen and 30.43% nitrogen by weight. When 1.44 g of this gas are placed in a $25\overline{0}$-mL flask at 20.0°C, the pressure of the gas is found to be 1.50 atm. (a) What is the empirical formula of the gas? (b) What is the molecular weight of the gas? (c) What is the molecular formula of the gas?

88. Given that 75.0 g of a certain gas have a volume of 4.50 L at a pressure of $1,56\overline{0}$ torr and a temperature of 127°C, calculate the molecular weight of the gas.

Learning Goal 10: Kinetic Theory of Gases

89. State the five basic principles of the kinetic theory of gases. Explain how each of these principles is related to the gas laws discussed in this chapter.

90. Explain how the kinetic theory of gases accounts for the properties of gases in terms of the forces between gas molecules and the energy these molecules have.

Extra Exercises

91. Explain Boyle's Law and Charles's Law, using the kinetic theory of gases.

92. One-half of a 10.0-L tank of hydrogen gas at 25°C and 10.0 atm is released into a balloon at constant temperature. The atmospheric pressure is $76\overline{0}$ torr. Determine the volume of the balloon.

93. The balloon in Problem 92 is released. It rises 10.0 km into the atmosphere, where it bursts. What was its final volume, given that the temperature was −50.0°C and the pressure was $23\overline{0}$ torr?

94. An industrial process for the production of ammonia involves reacting hydrogen gas with nitrogen gas. How many liters of ammonia gas could be produced from 1.00 kg of nitrogen gas, given that the process operates at a temperature of 30.0°C and a pressure of $75\overline{0}$ torr?

95. Change the following temperatures to kelvins: (*a*) $4\overline{0}$°F, (*b*) $18\overline{0}$°F, (*c*) 212°F.

96. Calculate the volume occupied by 1.00 mole of oxygen gas at STP (density = 1.429 g/L).

97. Calculate the density of CH_4 gas at STP.

98. Here is a very interesting experiment (see if your instructor will do it): Boil a small amount of water in a tin can. While the water is boiling, seal the can and remove the heat. As the can cools, it collapses. Explain this in terms of the gas laws you are now familiar with.

99. A volume of 272 mL of oxygen is collected at 27.0°C. What volume would the oxygen occupy at $20\overline{0}$°C, assuming that the pressure is held constant?

100. The following data are collected on a gas sample: P = 1.00 atm, V = 2.00 L, t = 27.0°C, mass = 2.56 g. What is the molecular weight of the gas? Suggest an element or compound that the gas might be if it is known to be diatomic.

12 Hydrogen, Oxygen, and Water

Learning Goals

After you have studied this chapter, you should be able to:

1. Describe the laboratory methods of preparing oxygen, and write equations for them.
2. Describe the commercial methods of preparing oxygen.
3. List the physical properties of oxygen.
4. Write equations for some important reactions of oxygen with other elements.
5. Describe the common sources of hydrogen.
6. Write equations for the laboratory and commercial preparation of hydrogen.
7. List the physical properties of hydrogen.
8. Write equations for some important chemical reactions of hydrogen.
9. List the main uses of hydrogen.
10. List some chemical and physical properties of water.
11. Write equations for the reaction of water with oxides of metals and nonmetals.
12. Explain what a hydrate is, give examples of hydrates, and calculate the percentage of water in a hydrated salt.
13. Define the terms *anhydrous, hygroscopic, deliquescence,* and *efflorescence.*
14. Define the term *potable* and list sources of potable water.
15. List some common water pollutants.

Joseph Priestley
(Reprinted with permission from
Torchbearers of Chemistry, *by Henry
Monmouth Smith,* © *Academic Press,
Inc., 1949)*

Introduction

Water covers more than two-thirds of the earth's surface, composes about 70% of our bodies, and accounts for nearly all of the weight of plant and animal life on earth. We accumulate it in our oceans, rivers, and lakes. It seeps into the ground and is also found in the air we breathe. It composes glaciers and ice caps. In fact, there are about 1.4 billion cubic kilometers of water on earth.

All living organisms need this vital substance in order to survive. The normal human diet requires more than 2.5 quarts of water each day, which come from foods, beverages, and the products of digestion.

We begin this chapter by examining oxygen and hydrogen, which are the two elements that compose water, and then take a closer look at the chemistry of this important compound.

12.1 Oxygen

In 1774 the English chemist Joseph Priestley heated a certain orange-red compound, which decomposed to yield mercury metal plus an unknown gas (Figure 12.1). He noted that this gas had very unusual properties. A burning object would burn more brightly in this gas than in air. And a glowing wood splint would burst into flames when placed in a test tube filled with the gas. This unknown substance was oxygen.

Oxygen is by far the most abundant element (by weight) on earth. About one-fifth of the air we breathe is oxygen, O_2. Because this oxygen

Figure 12.1
Preparation of oxygen in the laboratory

Table 12.1
The Elements As Percentages (By Weight) of Earth's Crust

Element	Percentage	Element	Percentage
Oxygen	49.5	Sodium	2.6
Silicon	25.7	Potassium	2.4
Aluminum	7.5	Magnesium	1.9
Iron	4.7	Hydrogen	0.9
Calcium	3.4	Others	1.4

is not combined with any other element, it is called *free oxygen*. About half of the outer crust of the earth is also composed of oxygen (Table 12.1). Because this oxygen is combined with other elements, it is called **combined oxygen.**

Laboratory Preparation of Oxygen

Learning Goal 1
Laboratory preparation of oxygen

Joseph Priestley originally prepared oxygen by heating mercury(II) oxide.

$$2HgO \xrightarrow{\text{Heat}} 2Hg + O_2$$

A more common method of preparing oxygen in the laboratory is by decomposing potassium chlorate.

$$2KClO_3 \xrightarrow{\text{Heat}} 2KCl + 3O_2$$

Or we can decompose water in the laboratory by passing a direct electric current through it to produce hydrogen and oxygen. This is called **electrolysis** (Figure 12.2).

$$2H_2O \xrightarrow{\text{Electricity}} 2H_2 + O_2$$

Commercial Preparation of Oxygen

Learning Goal 2
Commercial preparation of oxygen

When very pure oxygen is needed, as in hospitals, it is prepared commercially by the electrolysis of water. Most of the time, however, oxygen is obtained from the air. Air is 21% oxygen and 78% nitrogen; the other 1% (by volume) consists of noble gases, carbon dioxide, and water. Companies that prepare oxygen commercially do it by freeing air of carbon dioxide and water and liquefying what is left. Then they allow the liquid air to evaporate. The nitrogen and noble gases evaporate first because they have lower boiling points than oxygen. So oxygen is left behind. This method of preparation can produce oxygen that is 99.5% pure.

Physical Properties of Oxygen

At room temperature, oxygen is a colorless, odorless, tasteless gas that is a little denser than air and slightly soluble in water. This slight solubility provides enough oxygen for fish and aquatic plants to live on. (The solubility of oxygen in water at 20°C is about 9 mg of oxygen per liter of water.)

The small amount of oxygen that dissolves in water also helps to decontaminate it by chemically changing organic pollutants to harmless material. In many sewage treatment plants, oxygen or air is bubbled through sewage sludge to aerate it, so that the oxygen is available for microorganisms to break down the waste materials. This is called the activated sludge process of sewage treatment. If it weren't for this process, many rivers and lakes around the world would be polluted with sewage.

Chemical Properties of Oxygen

Oxygen combines with most elements—metals as well as nonmetals—to form oxides, as shown in the following examples.

$$2Mg + O_2 \longrightarrow 2MgO$$

$$C + O_2 \longrightarrow CO_2$$

Many compounds also react with oxygen. For example, gasoline and other fuels burn in oxygen to produce heat energy.

$$CH_4(g) + 2O_2(g) \longrightarrow CO_2(g) + 2H_2O(l) + 212.8 \text{ kcal}$$

Methane
(natural gas)

$$C_8H_{18}(l) + 12\tfrac{1}{2}O_2(g) \longrightarrow 8CO_2(g) + 9H_2O(l) + 1,307.5 \text{ kcal}$$

Octane

If there isn't enough oxygen the combustion is incomplete, and some carbon or carbon monoxide is produced in addition to carbon dioxide. This is what happens in a car engine. It is one reason why automotive engineers are trying to design an engine in which gasoline and oxygen combine to produce pure carbon dioxide and water.

Uses of Oxygen

For human life to exist, there must be oxygen. Where there isn't enough of it naturally, as in a plane flying at high altitude, we have to find some way to supply it.

Figure 12.2
Electrolysis of water

Water

H_2
10 mL — 10 mL

20 mL — 20 mL

$e^- \quad e^-$
Platinum   Platinum
cathode   anode
$\ominus$ dc battery $\oplus$

The hyperbaric chamber introduces pressurized oxygen to divers suffering from the bends.
(University of Southern California)

Oxygen

Acetylene

Oxygen   Acetylene

Figure 12.3
The oxyacetylene torch

People hospitalized with respiratory or heart problems may need pure oxygen. Prematurely born infants often need it, and some people with emphysema or other lung diseases are kept alive by daily sessions with their breathing machines, which increase the amount of oxygen reaching the lungs.

The use of the oxygen hyperbaric chamber (high-pressure chamber) has become an important life-saving technique in medicine. The hyperbaric chamber operates with oxygen pressures between 3 and 4 atm, which makes it possible for nearly 20 times the normal amount of oxygen to be dissolved in the blood. This is an advantage to people recovering from heart attack, stroke, gas gangrene, or various illnesses. Babies born with hyaline membrane conditions of the lung can often be saved by treatments in hyperbaric chambers.

Oxygen has many uses in industry. For example, the oxyacetylene torch makes use of oxygen and acetylene (a flammable gas) to produce extremely high temperatures (around 4,000°C). This torch (Figure 12.3) is used for cutting and welding metal. The reaction is

$$2C_2H_2(g) + 5O_2(g) \longrightarrow 4CO_2(g) + 2H_2O(g) + 600 \text{ kcal}$$
Acetylene

Steel refiners use oxygen to burn off impurities, such as carbon, phosphorus, and sulfur, that would otherwise spoil the steel. And, in the aerospace industry, liquid oxygen (LOX) is used to power rockets.

12.2 Hydrogen

In 1766 the English chemist Henry Cavendish investigated the properties of a gas that was produced when certain metals reacted with acids. He noticed that this gas was very flammable. Eventually he designed an experiment to find out what products were formed when this gas was burned in air. The result of the experiment was astounding: Water was produced! On hearing of this experiment, the French chemist Antoine Lavoisier named this gas **hydrogen,** which means "water-former" (Figure 12.4).

Free hydrogen is found in only very small quantities in the earth's atmosphere, although it is sometimes found in volcanic gases. The H_2 molecule is scarce because it is so light that it escapes from the earth's gravitational field. However, in the universe as a whole, free hydrogen is the most abundant element, and it seems to be the basic building block of the stars. For example, our sun is composed of 75% (by weight) hydrogen.

On earth hydrogen is commonly found in the combined form. For example, it is hydrogen, in combination with either carbon or nitrogen, that makes up the cells of plants and animals. Hydrogen accounts for about 10% of the weight of the human body. Two of the most common sources of combined hydrogen are water and natural gas (CH_4).

Learning Goal 5
Common sources of hydrogen

Laboratory Preparation of Hydrogen

Chemists usually prepare pure hydrogen by the reaction of an active metal and an acid. For example,

Learning Goal 6
Laboratory and commercial preparation of hydrogen

$$Mg(s) + 2HCl(aq) \longrightarrow MgCl_2(aq) + H_2(g).$$

H_2 burning with O_2 in air

H_2O droplets being formed

Hydrogen generator (Zn metal plus HCl)

Figure 12.4
Combustion of hydrogen in air

Commercial Preparation of Hydrogen

Industries that use hydrogen manufacture it cheaply by passing steam over heated carbon.

$$C(s) + H_2O(g) \xrightarrow{1,000°C} CO(g) + H_2(g)$$

The hydrogen cannot easily be separated from the carbon monoxide. However, for certain uses, the H_2 and CO do not have to be separated. The mixture can be used as a chemical fuel called **water gas.**

An inexpensive way of preparing pure hydrogen involves the reaction between iron and steam.

$$3Fe(s) + 4H_2O(g) \longrightarrow Fe_3O_4(s) + 4H_2(g)$$

Because the iron oxide is a solid, there is no problem in separating the products.

We can also use the electrolysis of water to produce pure hydrogen.

$$2H_2O(l) \xrightarrow{\text{Electricity}} 2H_2(g) + O_2(g)$$

However, this is the most expensive method of producing it.

Physical Properties of Hydrogen

Learning Goal 7
Physical properties of hydrogen

At room temperature, hydrogen is a colorless, odorless, tasteless gas. It is the lightest of all elements, and it has the lowest density of all known substances.

Chemical Properties of Hydrogen

Learning Goal 8
Chemical reactions of hydrogen

Hydrogen reacts vigorously with most nonmetallic elements:

$$2H_2(g) + O_2(g) \longrightarrow 2H_2O(g) + 116 \text{ kcal}$$

$$H_2(g) + Cl_2(g) \longrightarrow 2HCl(g) + 44 \text{ kcal}$$

$$H_2(g) + F_2(g) \longrightarrow 2HF(g) + 130 \text{ kcal}$$

These reactions produce a great deal of heat energy. For example, the reaction between hydrogen and oxygen in an oxyhydrogen torch produces temperatures close to 2,500°C.

The commercially important reaction of hydrogen and nitrogen to produce ammonia (called the *Haber process,* as we noted earlier) requires high temperature and pressure.

$$N_2(g) + 3H_2(g) \xrightarrow[\text{1,000 atm}]{500°C} 2NH_3(g)$$

Hydrogen also reacts with some of the Group IA and IIA metals to produce **hydrides**—compounds composed of hydrogen and one other element.

$$2Na(s) + H_2(g) \longrightarrow 2NaH(s)$$

Sodium
hydride

$$Ca(s) + H_2(g) \longrightarrow CaH_2(s)$$

Calcium
hydride

(Note that in these compounds, hydrogen is present as H^{1-}.)

Hydrogen reacts with many compounds in what are called **hydrogenation** reactions. These are reactions in which hydrogen is added to a compound. For example,

$$CO(g) + 2H_2(g) \longrightarrow CH_3OH(l)$$

Methyl (wood)
alcohol

Methyl alcohol is important to industry. It serves as a starting material for making many organic compounds, such as formaldehyde and perfumes. It also has the potential to be used as a component of automobile fuel in the future.

Uses of Hydrogen

Learning Goal 9
Uses of hydrogen

Under normal conditions, hydrogen is less dense than air. Taking advantage of this fact, the Germans in the 1920s and 1930s pioneered the use of hydrogen to lift dirigibles or zeppelins (giant blimps), in which they could fly over the Atlantic. One year a dirigible made ten crossings. However, hydrogen's combustibility made it extremely dangerous to use, so much so that passengers were forbidden to smoke. You may have heard of the *Hindenburg* disaster at Lakehurst, New Jersey, in May of 1937. The *Hindenburg*—which was the most modern dirigible of its time—was 804 feet long and carried more than 7 million cubic feet of hydrogen, but only 50 passengers. It was making a landing at Lakehurst after a transatlantic crossing when a spark ignited the hydrogen and blew the dirigible to pieces, causing the loss of 36 lives. In 1939 the Germans finally decided to stop using such airships, and other nations followed suit. Hydrogen obviously could not be tamed for such uses.

The use of hydrogen to form ammonia is extremely important in the production of synthetic fertilizers. Without synthetic fertilizers, U.S. farmers would not be able to produce enough food to feed the American population, let alone enough food to export to other countries.

Another example of hydrogen's contribution to food is that, through hydrogenation, liquid oils react with hydrogen to produce solid fats such as margarine.

Scientists also use hydrogen in *cryogenics,* the science that is concerned with extremely low temperatures. They have found hydrogen to be useful in cooling other substances to temperatures near absolute zero, because it remains liquid (does not freeze) at these temperatures.

12.3 Water

Water is the world's most important liquid. You might be able to live a month without food, but you could survive only two or three days without water. All vegetation disappears without water: In West Africa, an area the size of the United States is becoming a desert because there has been almost no rain for some years. The human body is composed of about two-thirds water, and many fruits and vegetables are as much as 80 or 90% water. Milk, one of our basic foods, is more than 90% water.

Yet the supply of water is *not* unlimited. Today the entire population of the United States has only about the same quantity of drinkable water that was available when Columbus landed in 1492. Americans use an estimated 2,000 gallons per person per day, and the trend is to use still more. (This figure of 2,000 gallons includes water used by industry, by power plants, and so forth.)

Physical Properties of Water

Learning Goal 10
Chemical and physical properties of water

Pure water is a tasteless, odorless, colorless substance that has a melting point of 0°C and a boiling point of 100°C at 1 atm pressure. These two temperatures serve as reference points on the Celsius temperature scale.

The density of water is another reference standard. It relates metric units of volume (milliliters) to metric units of mass (grams). At 4°C, water has a density of 1 gram per milliliter (1 g/mL). This means that 1 gram of water at 4°C occupies a volume of 1 milliliter. If we measured out exactly 1 mL of 4°C water, we would find that it had a mass of 1 g (Figure 12.5).

Water has intrigued humanity since the days of the ancient Greeks. One of its most interesting properties is the way in which its volume changes when its temperature is changed. Suppose we start with a certain amount of water at room temperature. If we warm it, the volume of the water increases. If we cool the water below room temperature, its volume at first decreases. This is exactly the behavior we would expect from a liquid. However, as we cool the water below 4°C, the volume of the water starts to *increase.* And its volume continues to increase until it freezes at 0°C. This behavior is very unusual, because the volumes of most liquids decrease continuously as the liquids are cooled.

Figure 12.5

Establishing the density of water (at 4°C). The experiment shows that 1 mL of water has a mass of 1 g (in other words, the density of water is 1 g/mL).

One result of this strange behavior of water is that the density of ice is less than the density of water. Remember that density D = mass/volume. At 4°C, the volume of, say, 10 g of water is 10 mL. But at 0°C the volume of the ice (solidified water) has increased to 10.9 mL. Therefore

$$D \text{ (at 4°C)} = 10 \text{ g/10 mL} = 1 \text{ g/mL}$$

$$D \text{ (at 0°C)} = 10 \text{ g/10.9 mL} = 0.92 \text{ g/mL}$$

Because the density of ice is less than the density of water, ice floats on water. This is not true of most other substances. These substances reach their minimum volume (and maximum density) at the freezing point, not above it. This causes the solid form to sink in the liquid form.

It is fortunate for us that ice floats on water. If it didn't, when lakes, rivers, and seas froze and water turned to ice in the winter, the ice would sink to the bottom. The water above the ice would protect it from the sun in the summer, so the ice would never melt. Eventually, the entire body of water would become frozen, and all life forms in it would be killed. And, if it is true that life on earth originated in the sea, then human beings would never have evolved. However, ice *does* float on water, and the layer of floating ice insulates the rest of the water and prevents it from freezing. This explains why an entire body of water does not freeze, even though the air temperature above it stays below freezing for many months (Figure 12.6).

Water also has a much higher freezing point and boiling point than would normally be expected. The unusual behavior of water results from its bonding and the three-dimensional shape of the water molecule. If we examine a single water molecule, we find that:

1. The bonds between the hydrogen and oxygen atoms are polar. Each hydrogen atom seems to have a slight positive charge, and the oxygen atom seems to have a slight negative charge.

2. The bond angle between the hydrogen atoms is about 105° (Figure 12.7).

Figure 12.6

Floating ice insulates the lake water below.

Figure 12.7

The structure of a water molecule

The slightly positively charged hydrogen atom of one water molecule can attract the slightly negatively charged oxygen atom of another water molecule. The two molecules then are held together by a force called a **hydrogen bond** (Figure 12.8). Thousands of water molecules may bond together in this way to form long chains (Figure 12.9). It is the hydrogen bonds, which hold the water molecules together, that give water its unusually high boiling point.

But how do we explain the fact that water expands as it freezes? If we looked at the three-dimensional structure of a crystal of ice, it would seem to be hexagonal, or six-sided (Figure 12.10). To attain this type of geometry while maintaining their 105° bond angle, the molecules have to move farther apart in ice than they are in liquid water. Therefore water that is in the form of ice has a greater volume than water that is in the form of a liquid.

Figure 12.8

Hydrogen bonding between two water molecules

Figure 12.9

Hydrogen bonding among many water molecules. (Note: True bond angles are not shown.)

Figure 12.10
The crystal structure of ice.
(Note: Not all the hydrogen
atoms are shown, because
some of them would be
behind the plane of the
paper.)

Chemical Properties of Water

Water is one of the most stable compounds—so stable that for many years
it was considered one of the basic *elements* of matter. Even when heated as
high as 2,000°C, water shows little decomposition. However, if an electric
current is passed through it, water breaks down.

$$2H_2O \xrightarrow{\text{Electricity}} 2H_2 + O_2$$

Water reacts with oxides of *non*metals to produce *acids*. For example, in
the case of sulfur trioxide,

$$SO_3 + H_2O \longrightarrow H_2SO_4$$
Sulfuric
acid

and in the case of sulfur dioxide,

$$SO_2 + H_2O \longrightarrow H_2SO_3$$
Sulfurous
acid

Water reacts with oxides of metals to produce *bases*. For example,

$$CaO + H_2O \longrightarrow Ca(OH)_2$$

$$Na_2O + H_2O \longrightarrow 2NaOH$$

Learning Goal 11
Reactions of water

Many salts combine with water to form compounds called *hydrates*. A
hydrate is a crystalline compound that contains a definite proportion of
water molecules loosely held. Some common hydrates are

$Na_2B_4O_7 \cdot 10H_2O$	Borax	$MgSO_4 \cdot 7H_2O$	Epsom salts
$CaSO_4 \cdot 2H_2O$	Gypsum	$Na_2CO_3 \cdot 10H_2O$	Washing soda
$KAl(SO_4)_2 \cdot 12H_2O$	Alum	$CuSO_4 \cdot 5H_2O$	Blue vitriol

The centered dot between the formula for the salt and the formula for the
water is not a decimal point. It is a way of showing that the water
molecules maintain their integrity in the crystal structure. When a hydrated salt is heated, the water is driven off. For example,

Learning Goal 12
Hydrates and water of
hydration

$$MgSO_4 \cdot 7H_2O \xrightarrow{\text{Heat}} MgSO_4 + 7H_2O$$

$$CuSO_4 \cdot 5H_2O \xrightarrow{\text{Heat}} CuSO_4 + 5H_2O$$

A blue   A white
compound   compound
(hydrated salt)   (anhydrous salt)

The salt is left in its **anhydrous** form—the form in which the water of
hydration is no longer attached to the salt.

Example 12.1

Anhydrous nickel chloride, a golden-yellow salt, is used as an absorbent for ammonia in gas masks. When the anhydrous salt picks up water, it changes color to form nickel chloride hexahydrate, which is green. The reaction is as follows:

$$NiCl_2 + 6H_2O \longrightarrow NiCl_2 \cdot 6H_2O$$
$$\text{golden-yellow} \qquad\qquad\qquad \text{green}$$

Calculate the percentage of water (by weight) in nickel chloride hexahydrate.

Solution

Calculate the molecular weight of the hydrated salt as follows:

$$NiCl_2 \cdot 6H_2O$$

$$58.7 + 71.0 + 6(18.00)$$

$$129.7 + 108.00 = 237.7$$

Now determine the percentage of water (by weight) by dividing the weight of the water molecules in the hydrated salt ($6H_2O$) by the molecular weight of the hydrated salt ($NiCl_2 \cdot 6H_2O$). Don't forget to multiply by 100 to express the answer as a percentage.

$$\text{Percent } H_2O \text{ (by weight)} = \frac{\text{weight of water's contribution}}{\text{weight of hydrated salt}} \times 100$$

$$= \frac{108.0}{237.7} \times 100$$

$$= 45.44\%$$

Practice Exercise 12.1   Pale green anhydrous chromium(III) nitrate also exists in a deep violet nonahydrate form, as shown in the following reaction:

$$Cr(NO_3)_3 + 9H_2O \longrightarrow Cr(NO_3)_3 \cdot 9H_2O$$
$$\text{pale green} \qquad\qquad\qquad \text{deep violet}$$

Calculate the percentage of water (by weight) in chromium(III) nonahydrate.

Learning Goal 13
Definitions of anhydrous, hygroscopic, deliquescence, and efflorescence

On exposure to air, some anhydrous salts (as well as other substances) can absorb water from the atmosphere. Such a substance is said to be **hygroscopic.** Some substances are so hygroscopic that they absorb enough water from the atmosphere to form a solution. These substances are said

Chemical samples being dried or kept dry

Anhydrous calcium chloride

Figure 12.11

A desiccator is used to dry chemical samples and to keep them in a moisture-free atmosphere.

Learning Goal 14
Potable water

to be **deliquescent.** One such substance is sodium hydroxide (NaOH). A small piece of solid sodium hydroxide, if left in an open beaker, soon absorbs enough water from the atmosphere to form a solution.

Some hygroscopic substances, calcium chloride ($CaCl_2$), for example, are used in the laboratory as drying agents. Such substances are called **desiccants.** For example, if you want to keep some chemicals in a dry atmosphere, you can place them in a container called a **desiccator** (Figure 12.11). Anhydrous calcium chloride in the desiccator ensures that the atmosphere in the desiccator remains moisture-free. (Of course, in time, the calcium chloride absorbs as much water as it can, and it must be changed.)

Desiccant pellets are sometimes placed in vitamin containers. The pellets absorb any moisture that might accumulate in the containers and cause the vitamins to decompose.

We have described hygroscopic substances as those that can spontaneously absorb water from the atmosphere. There are also substances that can do the opposite. A few hydrated salts can spontaneously lose their water of hydration when exposed to the atmosphere. This process is known as **efflorescence.** For example,

$$Na_2SO_4 \cdot 10H_2O \xrightarrow[\text{temperature}]{\text{Room}} Na_2SO_4 + 10H_2O$$

12.4 Water Pollution

Water—a vital substance for human survival—must be free of contamination from deadly chemicals and bacteria in order to be potable (drinkable). In the United States, each American uses over 340 liters (90 gallons) of water a day for domestic purposes, and statistics show that the demand is increasing. Between 1900 and 1980 the U.S. population rose by 320%, whereas water consumption increased by nearly 1,250%.

The world's supply of water is enormous, and water is found as vapor, liquid, and ice. Over 97% of the earth's water is located in the oceans, however, and is unfit to drink. Just 3% of all water is found as fresh water fit for human consumption.

Our sources of potable water include lakes, rivers, springs, and wells. As our wells and springs are rapidly drained, many large cities are turning to rivers and lakes as sources of potable water. To be considered drinkable, water must meet standards set by the U.S. Public Health Service (Table 12.2). Because of the pollutants that have been dumped into them over the years, very few of our rivers and lakes can now meet these standards.

In the past, a nearby river was commonly used as the discharge point for waste water. In the United States today, despite legislation to control

Table 12.2
U.S. Public Health Service Standards for Potable Water

Contaminating ion(s)	Maximum concentration (milligrams/liter)	Contaminating ion(s)	Maximum concentration (milligrams/liter)
Arsenic	0.05	Lead	0.05
Barium	1.00	Manganese	0.05
Cadmium	0.01	Nitrate	45
Chloride	250	Organics	0.20
Chromium	0.05	Selenium	0.01
Copper	1.00	Silver	0.05
Cyanide	0.20	Sulfate	250
Fluoride	2.00	Zinc	5.00
Iron	0.30	Total dissolved solids	500

Learning Goal 15
Water pollutants

this form of pollution, millions of gallons of raw, untreated waste are still being dumped into our waterways.

There are many ways in which water pollution occurs today. Nearly half of the water withdrawn from rivers, lakes, and estuaries each year in the United States is used to cool electric power plants. When the water is returned it contains no pollutants, but its temperature is increased significantly, which can adversely affect the ecological relationships in the body of water.

Another widespread problem called **eutrophication** occurs when erosion and run-off introduce plant nutrients from the land into lakes. The algae normally present in a healthy lake feed on phosphates and nitrates, nutrients that originated in agricultural or urban centers. However, populations of algae grow out of control when the lake becomes overloaded with phosphates and nitrates derived from fertilizers, animal waste, or sewage treatment plants. The vital oxygen content of the lakes then decreases, and the water tastes bad and has a foul odor. It is the decay of the resulting algae that causes eutrophication.

The process of eutrophication has destroyed the natural balance of life in the Great Lakes. The effect on Lake Erie has been particularly serious. Located near major population centers, the Great Lakes are sinks for large amounts of phosphates and toxic chemicals. However, as a result of pollution control measures, conditions have continued to improve at the Great Lakes for more than a decade.

Ground water contamination is yet another problem of growing proportions. Accidental chemical spills, leaking sewer pipelines, leaks from the hundreds of thousands of underground storage tanks and abandoned hazardous waste sites, and agricultural and industrial run-off—all contribute to ground water pollution. This is the most difficult type of pollution to control, and there are just a few thousand people trained in the ground water field who can work effectively on the problem.

Oil spills constitute a continuing water pollution hazard. Besides natural seepage of crude petroleum from deposits below the ocean bottom, accidents, urban run-off, and ruptures of offshore wells contribute to the problem.

When water pollutants get into the bodies of birds, fish, and shellfish, their life cycles can be altered or death can occur. Once these substances enter the food chain, their effects can be far-reaching. When humans consume fish that have ingested harmful pollutants, for example, these unwanted substances build up in human tissues, creating potential health problems (Figure 12.12).

Legislation has been enacted in an effort to put an end to our waste water contamination. America's two major water pollution control laws are the Federal Water Pollution Act of 1972 and the Clean Water Act of 1977 (which was amended in 1981). These laws required the Environmental Protection Agency to set a system of standards for waste water leaving a sewage treatment or industrial plant. They also required all municipalities to treat sewage biologically by 1988. In addition, goals were set so that by 1983, all U.S. waters would be safe for fishing and swimming. Elimination of the discharge of 129 highly toxic pollutants into U.S. waters was to end by 1985, and the discharge of any pollutants into waterways would be allowed by EPA permit only.

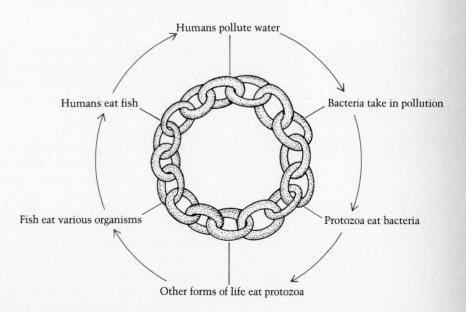

Figure 12.12

The path of water pollution through the food chain

Because of this legislation, the amount of pollution dumped into our waterways by industry was reduced by one-half between 1972 and 1980. Problems still exist today; for example, 37 states could not meet the goal of making all bodies of water safe for fishing and swimming by 1983. Much pollution from agricultural and urban run-off could not be effectively controlled, and the slow pace at which bureaucratic machinery turns, construction delays, and lack of funds made compliance impossible in many areas. Even so, and despite the fact that our waterways continue to be polluted, conditions are much better than they were a decade or two ago.

Summary

By weight, oxygen is the most abundant element on earth. About one-fifth of air is the diatomic molecule O_2, called free oxygen because it is not combined with other substances. And about half of the outer crust of the earth is composed of oxygen combined with other elements. Oxygen can be prepared in the laboratory by heating mercury(II) oxide or potassium chlorate or via the electrolysis of water. Commercially, oxygen is prepared by freeing air of carbon dioxide and water and then liquefying what is left. The nitrogen and inert gases evaporate from the liquid air, leaving 99.5% pure oxygen behind. Oxygen is a colorless, odorless, tasteless gas, a little more dense than air and slightly soluble in water. Oxygen combines with most elements—metals as well as nonmetals—to form oxides. Many compounds also react with oxygen.

Free hydrogen is found in only very small quantities in the earth's atmosphere, but it is the most abundant free element in the universe. On earth, it is usually found in combination with other elements such as carbon or nitrogen. In the laboratory, hydrogen is usually prepared via the reaction of an active metal and an acid. Commercially, hydrogen is prepared by passing steam over heated carbon, through the reaction of iron and steam, or via the electrolysis of water. At room temperature, hydrogen is a colorless, odorless, tasteless gas. It is the lightest of all elements, and it has the lowest density of all known substances. Hydrogen reacts with most nonmetallic elements. Under high temperature and pressure, hydrogen reacts with nitrogen to produce ammonia. Hydrogen also reacts with some of the Group IA and IIA metals to produce compounds called hydrides.

Water is the world's most important liquid. Pure water is a colorless, odorless, tasteless substance that has a melting point of 0°C and a boiling point of 100°C at 1 atm pressure. At 4°C, water has a density of 1 g/mL. Because of the way in which hydrogen and oxygen atoms bond to form water, and because of the three-dimensional shape of the water molecule, the volume of water *increases* as water is cooled below 4°C. For this reason, ice is less dense than liquid water and floats on water. Water is a polar molecule: The positively charged hydrogen atom of one water molecule can attract the negatively charged oxygen atom of another water molecule to form a *hydrogen bond*. Such bonds can hold together long strings of water molecules, which accounts for the high boiling point of water. Water reacts with oxides of nonmetals to produce acids and with oxides of metals to produce bases. Water combines chemically with many salts to form hydrates.

Water pollution is a serious problem that must be solved soon if we are to continue to have enough drinking water. The pollution of our lakes, rivers, springs, and wells continues to be a significant problem. Erosion, run-off, chemical spills, leaks, and the use of water for cooling electric power plants contribute to the problem. Legislation has resulted in significant increases in water quality, but there is still room for a great deal of improvement.

Career Sketch

Medical Laboratory Workers

There are three kinds of medical laboratory workers: medical technologists, technicians, and assistants. They perform tests under the general direction of pathologists and other physicians or under that of scientists who specialize in clinical chemistry, microbiology, or other biological sciences. Medical laboratory workers analyze body tissues, blood, and other body fluids by using precision techniques and instruments.

Medical technologists require four years of post-high-school training. They perform complicated chemical, microscopic, and bacteriological tests. For example, they test cholesterol levels and examine blood under a microscope to detect the presence of diseases. They also make cultures of body fluid or tissue to determine the presence of bacteria, parasites, or other microorganisms; and they analyze the samples for chemical content or reaction. They may also type blood samples.

Technologists in small laboratories perform many types of tests. Those in large laboratories usually specialize in such areas as microbiology, parasitology, biochemistry, hematology (the study of blood cells), and nuclear medical technology.

Medical technologists conduct tests related to the treatment of patients and are called on to display independent judgment. Some do research, develop laboratory techniques, teach, or perform administrative duties.

Medical laboratory technicians generally require two years of post-high-school training. They perform a wide range of laboratory procedures and

Medical laboratory workers perform a variety of vital functions while working under the direction of pathologists, physicians, and scientists who specialize in clinical chemistry, microbiology, and other biological diseases.
(Centers for Disease Control)

tests that require a high level of skill, but not the in-depth knowledge of highly trained medical technologists. Like technologists, they may work in several areas or specialize in one area.

Medical laboratory assistants generally have a year of formal training after high school. They help medical technologists and technicians in routine tests and related work that can be learned in a relatively short time. In large laboratories, they may concentrate in one area. For example, they may identify abnormal blood cells on slides. In addition to performing routine tests, assistants may store and label plasma, clean and sterilize laboratory equipment, prepare solutions following standard procedures, keep records of tests, and identify specimens.

Career Sketch

Waste Water Treatment Chemist

Waste water treatment chemists analyze samples of streams, examine raw and treated waste water, and look at sludge and other by-products of waste water treatment. They work to determine the efficiency of the processes used in waste water treatment plants and to ensure that local, state, and federal standards are met. They test water for chemical, bacteriological, and physical components, and they monitor and regulate the discharge of waste into sewer and treatment systems. In addition, they make surcharge assessments for waste water treatment plants.

The proper treatment of waste water is an essential factor in preserving our supply of clean water.
(Steve Delaney/EPA)

Most waste water treatment chemists have a high school diploma and graduated from a four-year college, majoring in chemistry or environmental science. Two-year programs are also available for waste water treatment technicians.

Self-Test Exercises

Learning Goals 1 and 2: Laboratory and Commerical Preparation of Oxygen

1. List some source of (*a*) free oxygen, (*b*) combined oxygen.

2. What is the difference between free oxygen and combined oxygen?

3. Write a chemical equation for (*a*) the laboratory preparation of oxygen, (*b*) the commercial preparation of oxygen.

4. Why is there a difference between the laboratory preparation and the commercial preparation of oxygen?

5. Complete the following equations, and then balance them.

(a) $HgO \xrightarrow{\text{Heat}}$

(b) $KClO_3 \xrightarrow{\text{Heat}}$

(c) $H_2O \xrightarrow{\text{Electricity}}$

6. What compounds are used to form oxygen in the following reactions?

(a) $? \xrightarrow{\text{Electricity}} H_2(g) + O_2(g)$

(b) $? \xrightarrow{\text{Heat}} KCl(s) + O_2(g)$

Learning Goal 3: Physical Properties of Oxygen

7. Misguided people often boil water before putting it in a fish tank. The following day the fish are usually found dead. Why?

8. What is the activated-sludge process of sewage treatment?

9. How can molecular oxygen be removed from water?

10. How do you think pure oxygen would be prepared for use in a sewage-treatment plant?

*11. List some of the problems we would have if the composition of our atmosphere were changed to 20% nitrogen and 80% oxygen (by volume).

12. A tank of water at 20°C contains 4.00 L of water. How much oxygen is present in this water?

13. What is the chemical meaning of the term LOX?

14. Which is more dense, oxygen or air?

Learning Goal 4: Important Reactions of Oxygen

15. Describe a simple test to detect oxygen.

16. Complete the following equation for the combustion of ethane, and then balance it.

$$C_2H_6(g) + O_2(g) \longrightarrow ? + ?$$

17. When a substance burns in air, it combines with the oxygen in the air. Therefore the substance should gain weight. Yet when many substances burn in air—for example, candles—they seem to lose weight. Explain.

18. Complete the following equation, and then balance it.

$$Ca + O_2 \longrightarrow ?$$

19. What products are formed as a result of the complete combustion of hydrocarbon fuels?

20. What products are formed as a result of the *incomplete* combustion of hydrocarbon fuels?

21. Complete the following equations, and then balance them.

(a) $Zn + O_2 \longrightarrow$

(b) $C_2H_6 + O_2 \longrightarrow$

(c) $Al + O_2 \longrightarrow$

22. The reaction that occurs in the oxyacetylene torch is

$C_2H_2(g) + 2\frac{1}{2}O_2(g) \longrightarrow$
Acetylene $\qquad 2CO_2(g) + H_2O(g) + \overline{30}0$ kcal

How much heat is produced in the combustion of 2.86 g of acetylene?

*23. The oxyacetylene torch produces $\overline{30}0$ kcal for every mole of acetylene burned.

$C_2H_2(g) + 2\frac{1}{2}O_2(g) \longrightarrow$
Acetylene $\qquad 2CO_2(g) + H_2O(g) + \overline{30}0$ kcal

How much heat is produced in the burning of 143 g of acetylene?

24. Complete the following equation, and then balance it.

$$Sr + O_2 \longrightarrow ?$$

Learning Goal 5: Common Sources of Hydrogen

25. Describe some common sources of hydrogen.

26. What celestial bodies contain most of the hydrogen in our universe?

Learning Goal 6: Laboratory and Commercial Preparation of Hydrogen

27. Write a chemical equation for (*a*) the commercial preparation of hydrogen, (*b*) the laboratory preparation of hydrogen.

28. Complete the following equations for the preparation of hydrogen gas.

(*a*) $C(s) + H_2O(g) \xrightarrow{1,000°C} ? + ?$

(*b*) $Zn(s) + HCl(aq) \longrightarrow ? + ?$

Learning Goal 7: Physical Properties of Hydrogen

29. A balloon filled with hydrogen floats on air.
(*a*) Does this mean that the hydrogen has a negative mass? Explain.
(*b*) How would you measure the mass of a sample of hydrogen?

30. Determine the density of hydrogen in grams per liter, given the following data: mass of sample, 0.18 g; volume, 2,000 mL.

Learning Goal 8: Chemical Reactions of Hydrogen

31. Describe a simple test to detect hydrogen.

32. Match each gas on the left with its description on the right
(*a*) hydrogen gas
(*b*) oxygen gas

(1) This gas supports combustion.
(2) This gas is flammable and can explode when ignited by a spark in the presence of air.

33. Predict the products formed when hydrogen combines with (*a*) I_2, (*b*) Li.

34. Predict the products formed when hydrogen combines with (*a*) Cl_2, (*b*) Ca.

35. Complete the following equations, and then balance them.
(*a*) $H_2 + N_2 \longrightarrow$
(*b*) $H_2 + Br_2 \longrightarrow$
(*c*) $H_2 + O_2 \longrightarrow$

36. Complete the following equations, and then balance them.
(*a*) $CO + H_2 \longrightarrow ?$
(*b*) $Li + H_2 \longrightarrow ?$
(*c*) $Mg + H_2 \longrightarrow ?$

Learning Goal 9: Uses of Hydrogen

37. List some major uses of hydrogen.

38. Why is hydrogen so useful in the field of cryogenic research?

Learning Goal 10: Chemical and Physical Properties of Water

39. Why is it incorrect to say that rainwater is "pure" water?

40. The density of ice at 0°C is 0.92 g/cm³. What is the volume of a cube of ice that has a mass of 18.4 g?

41. A glass is filled to capacity with ice cubes and water. When the ice cubes melt, will the water overflow? (Assume that there is no evaporation.) Explain.

42. The density of water at 4°C is 1.00 g/cm³. What is the mass of 535 cm³ of water?

43. The density of ice at 0°C is 0.92 g/cm³. What is the mass of an ice cube that measures 4.0 cm on each side?

44. What would happen to our rivers, lakes, and seas if ice did not float on water?

45. Draw a picture of a hydrogen bond between two water molecules.

46. How large is the bond angle between the hydrogen atoms in a water molecule?

47. The density of water at 4°C is 1.00 g/cm³. What is the volume of 1.00 kg of water?

48. What is a hydrogen bond?

Learning Goal 11: Reactions of Water

49. Predict the products formed when water combines with (*a*) MgO, (*b*) SO₃, (*c*) Na.

50. Predict the products formed when water combines with: (*a*) CaO, (*b*) SO₂, (*c*) K.

51. Complete the following equations, and then balance them.
(*a*) $SrO + H_2O \longrightarrow$
(*b*) $SO_2 + H_2O \longrightarrow$
(*c*) $Li + H_2O \longrightarrow$

52. Complete the following equations, and then balance them.
(*a*) $Na_2O + H_2O \longrightarrow ?$
(*b*) $CO_2 + H_2O \longrightarrow ?$
(*c*) $Rb + H_2O \longrightarrow ?$

Learning Goal 12: Hydrates and Water of Hydration

53. Determine the percentage of water (by weight) in $CuSO_4 \cdot 5H_2O$.

54. Determine the percentage of water (by weight) in $CaSO_4 \cdot 2H_2O$.

55. A commercial for a well-known cleaning powder says it "shakes out white, then turns blue." What type of reaction is occurring? What compound could be used to produce this effect?

56. What is the difference between a hydrated salt and an anhydrous salt?

57. Determine the percent composition by weight of water in each of the following hydrated salts.
(*a*) $MgSO_4 \cdot 7H_2O$
(*b*) $Na_2CO_3 \cdot 10H_2O$
(*c*) $KAl(SO_4)_2 \cdot 12H_2O$

58. Determine the percentage by weight of water in each of the following hydrated salts.
(*a*) $Ca(ClO_3)_2 \cdot 2H_2O$
(*b*) $CaCO_3 \cdot 6H_2O$
(*c*) $CaCl_2 \cdot 6H_2O$

59. Complete the following equations, and then balance them.
(*a*) $CaSO_4 \cdot 2H_2O \xrightarrow{\text{Heat}}$
(*b*) $KAl(SO_4)_2 \cdot 12H_2O \xrightarrow{\text{Heat}}$

60. Complete the following equations, and then balance them.
(*a*) $Mg(C_2H_3O_2)_2 \cdot 4H_2O \xrightarrow{\text{Heat}} ? + ?$
(*b*) $MgBr_2 \cdot 6H_2O \xrightarrow{\text{Heat}} ? + ?$

*61. There are three hydrated forms of the salt $CoSO_4$. They are $CoSO_2 \cdot 1H_2O$, $CoSO_4 \cdot 6H_2O$, and $CoSO_4 \cdot 7H_2O$. From the following data, determine which form of $CoSO_4 \cdot nH_2O$ is being analyzed: When a chemist heats 52.58 g of hydrated salt, 21.60 g of water are produced.

*62. The two hydrated forms of the salt $Mg(NO_3)_2$ are $Mg(NO_3)_2 \cdot 2H_2O$ and $Mg(NO_3)_2 \cdot 6H_2O$. From the following data, determine which of the hydrated salts is being analyzed: When 25.64 g of hydrated salt are heated, 10.80 g of water are produced.

Learning Goal 13: Definitions of Anhydrous, Hygroscopic, Deliquescence, and Efflorescence

63. Define and give an example of each of the following terms: (*a*) Anhydrous salt, (*b*) hygroscopic substance, (*c*) deliquescent substance, (*d*) substance that undergoes efflorescence.

64. Complete the following sentences.
(*a*) A salt that has lost its water of hydration is called a(n) _____ salt.
(*b*) A substance that can absorb water from the atmosphere is _____ .
(*c*) A substance that absorbs enough water from the atmosphere to form a solution is _____ .
(*d*) A salt that can spontaneously lose its water of hydration to the atmosphere is a(n) _____ salt.

Learning Goal 14: Potable Water

65. Define the term *potable*, and list four sources of potable water.

66. Is water from the ocean classified as potable water?

Learning Goal 15: Water Pollutants

67. List some common water pollutants. How do they get into our lakes and rivers?

68. Name three sources of water pollution.

69. Discuss the importance of the food chain as it pertains to water pollution.

*70. How can a body of water clean itself?

Learning Goals 1 to 15

71. Name each substance from its description.
(*a*) This is the most abundant element (by weight) on earth. It composes about one-fifth of the air we breathe.
(*b*) This substance was once considered a basic *element* of matter. Even when heated to 2,000°C, it shows little decomposition.
(*c*) The name of this very flammable substance means "water-former."

72. Name each substance from its description.
(*a*) This substance becomes less dense when it freezes.
(*b*) A burning object burns more brightly in this gas than in air.
(*c*) This substance is the basic building block of the stars.

Extra Exercises

73. A 5.00-g sample of magnesium sulfate hydrate is heated until all the water is driven off. The remaining salt weighs 2.44 g. Determine the formula of the hydrate.

74. Calculate the percentage of water in barium chloride dihydrate.

75. Complete the following reactions, and then balance them.
(*a*) $K + H_2O \longrightarrow$
(*b*) $MgO + H_2O \longrightarrow$

76. Both hygroscopic and deliquescent substances absorb water. How could you distinguish one from the other?

77. Ozone (O_3), a form of oxygen, can be formed as a pollutant in photochemical smog. The observed reaction is

$$CO + 2O_2 \longrightarrow CO_2 + O_3$$

Determine how much O_3 can be produced with 1.00 mole of carbon monoxide at STP, assuming that excess oxygen is available.

78. Both hydrogen and oxygen are colorless, odorless, and tasteless. Suppose you were given a sample of one of these gases. What could you do to determine which gas your sample represented?

79. When water is formed from its elements, the reaction is highly *exothermic*. Does this mean that energy would be needed to decompose water into its elements, hydrogen and oxygen?

80. Even though water is a relatively small molecule, it has a very high boiling point compared to molecules of similar size. Can you explain this in terms of the structure of water?

81. Calcium chloride, $CaCl_2$, is often used as a desiccant (drying agent) for wet chemicals. Explain.

82. Certain weather indicators operate by color changes. For example, one indicator turns red if the air is moist and turns blue in dry weather. Explain how this happens.

13   *The Liquid and Solid States*

Learning Goals

After you've studied this chapter, you should be able to:

1. Discuss the properties that distinguish gases, liquids, and solids from one another.
2. Apply the kinetic theory to describe the behavior of liquids and solids.
3. Define and illustrate the terms *equilibrium vapor pressure, evaporation,* and *boiling point.*
4. Explain what happens as a solid changes to liquid and the liquid changes to gas.
5. Explain the processes of distillation and fractional distillation, and define the relevant terms.
6. Describe what is meant by *crystalline* solids and *amorphous* solids.
7. Discuss the three kinds of crystalline solids and the forces of attraction in each.
8. Define and give an example of a *crystal lattice* and a *unit cell.*

Introduction

Matter can exist in any of three physical states: gas, liquid, and solid. The early chemists—the alchemists—had trouble dealing with gases, and it was only in the seventeenth century that research into gases really began. The alchemists were, however, able to study liquids and solids extensively, and most of what we know of their work is related to these two states of matter.

In Chapter 11 we discussed the special properties of gases, and in this chapter we do the same for liquids and solids. We begin by using the kinetic theory (which was applied to gases in Section 11.11) to explain some of the differences among the three states.

13.1 The Kinetic Theory Extended to Liquids and Solids

· · · · · · · · · · · · · ·
Learning Goal 1
Properties of gases, liquids, and solids

· · · · · · · · · · · · · ·
Learning Goal 2
Kinetic theory of liquids and solids

Table 13.1 lists three physical properties of gases, liquids, and solids. In Chapter 11 we explained these properties of gases according to the kinetic theory of gases. We noted that gas molecules are relatively far apart, are always in motion, and thus show little attraction for each other. It is for these reasons that a gas has no shape of its own, will move through and fill any volume, and is easily compressed.

In a liquid, however, the molecules are very close to each other. Therefore, according to the kinetic theory, the forces of attraction between liquid molecules are much greater than those between gas molecules. The liquid molecules tend to "stick together" and have a definite volume, but they do not maintain a specific shape. And, because the molecules in a liquid are so close together, liquids tend to be incompressible. When an ocean liner goes through the locks of a canal, it does not crush the water down. Water molecules can't be squeezed any closer together than they already are.

The particles composing a solid are not only close together but are also held together in a fixed position in relation to one another. Later in this chapter we'll discuss the forces that hold the particles together. For now, we'll just say that the fact that the particles of a solid *are held in a fixed position* explains why a solid has a definite shape, has a definite volume, and is practically incompressible.

Table 13.1
Properties of Gases, Liquids, and Solids

Property	Gases	Liquids	Solids
Volume	Have no definite volume (gases expand to fill their containers)	Have a definite volume	Have a definite volume
Shape	Have no definite shape	Have no definite shape (liquids assume the shape of their containers)	Have a definite shape (most solids have a well-defined crystalline arrangement)
Compressibility	Are easily compressed	Tend to be incompressible	Cannot be compressed (they are practically incompressible)

13.2 Evaporation of Liquids

. . . .
Learning Goal 3
Equilibrium vapor pressure,
evaporation, and boiling point

Figure 13.1
A solution of sugar and water

. . . .
Learning Goal 4
Change of state

If a liquid is placed in an open container, it slowly disappears. That is, the liquid **evaporates** as its molecules change from the liquid state to the gaseous (or vapor) state. For any liquid, the speed of evaporation depends on the temperature of the liquid and its exposed surface area.

Before we discuss the process of evaporation, let us perform an experiment: We place a cube of sugar in a container of water and leave the container undisturbed for a few hours. Later, when we examine the container, we see that the cube of sugar has disappeared. It has dissolved in the water. If we taste the solution, we discover that the molecules of sugar have been distributed evenly throughout the water (Figure 13.1). All parts of the solution have the same sweetness. This could happen only if the molecules had moved around. From experiments like this, we conclude that *the molecules of a liquid are constantly in motion*. We can use this knowledge to explain evaporation.

Because liquid molecules are constantly in motion, each must have a specific amount of energy. And when the molecules collide, energy can be transferred from one molecule to another. If a molecule on the surface of a liquid collides with a molecule that is on its way up toward the surface, the surface molecule may obtain enough energy from the collision to escape from the liquid entirely (Figure 13.2). That molecule has then changed state, from liquid to vapor. This is similar to what happens in a game of pool. When the cue ball strikes another ball, it passes energy to that other ball. The struck ball then starts to move. And if it is struck hard enough, it can leave the table entirely.

A Dynamic Equilibrium

If we place a cover on a container that is partly filled with water, the evaporation process begins, but it seems to stop after a time. Here is what happens.

Figure 13.2
How liquids evaporate

● Liquid molecule on surface

⬤ Liquid molecule on its way up to surface

Figure 13.3
Liquid molecules escaping
from the surface of the liquid

Figure 13.4
A dynamic equilibrium

Figure 13.5
The higher the temperature,
the more molecules escape.

As Figure 13.3 shows, water molecules begin to escape from the surface of the liquid. These molecules occupy the space above the water in the partly filled container. As evaporation continues, more and more escaping molecules occupy this same space. The molecules are now closer together, they collide with one another more often, and some of them are bumped back into the water. At first, many molecules escape and few return to the water. But after a while, as the concentration of molecules in the vapor state increases, the number of molecules returning to the liquid also increases. Eventually, the number of molecules escaping from the water equals the number of molecules returning to it (Figure 13.4). The net result is that no additonal liquid seems to be evaporating from the container. But, in reality, there is a *dynamic equilibrium* between evaporation and return to the liquid state. *These two opposing processes are occurring at the same rate.*

Equilibrium Vapor Pressure

The molecules that have escaped the liquid in Figure 13.4 exert a pressure on the container, the cover, and the surface of the liquid. This pressure is called the *equilibrium vapor pressure*. We can define **equilibrium vapor pressure** as *the pressure exerted by a vapor when it is in equilibrium with its liquid at a given temperature.*

Every liquid has its own characteristic equilibrium vapor pressure, and this equilibrium presssure depends on the temperature. To see this, consider what happens when we raise the temperature of the water in Figure 13.4. Because we are adding heat to raise the temperature, we are also supplying additional energy to the liquid molecules. As a result, a greater number of molecules obtain sufficient energy to escape from the surface of the liquid and become vapor (Figure 13.5). The equilibrium that had existed is upset, but a new equilibrium is soon established at the new temperature. And, because there are now more molecules in the space above the liquid, the new equilibrium vapor pressure is higher than the original one. That is, the higher the temperature, the greater the equilibrium vapor pressure for any substance.

The more easily a liquid evaporates, the more **volatile** it is said to be. Liquids that have high equilibrium vapor pressures tend to evaporate easily and so are considered volatile. Some examples are alcohol, gasoline, and ether. Liquids that have low equilibrium vapor pressures tend to evaporate less easily. Examples include motor oil, water, and glycerine.

13.3 Boiling of Liquids

If a beaker of water is heated over a flame, bubbles soon form on the bottom of the beaker. These first bubbles are merely dissolved air (oxygen, nitrogen, and so forth) being driven out of the water by the heat.

As the temperature rises, however, the water molecules become more active. Soon other bubbles form at the bottom of the beaker and begin to rise. These bubbles are water vapor (steam). They form on the *bottom* of the beaker because the water is hottest there. At first these steam bubbles disappear when they reach the cooler water toward the top of the beaker (Figure 13.6). After a while, however, they begin to rise all the way to the surface. At this point we say that the water is *boiling*. What we mean is that the vapor pressure exerted by the bubbles of steam has become equal to the atmospheric pressure (Figure 13.7). So the bubbles of steam are able to rise up and out of the water.

The **boiling point** of a liquid is *the temperature at which the equilibrium vapor pressure of the liquid equals the atmospheric pressure.* This means that the boiling point depends on the atmospheric pressure: Increasing the atmospheric pressure increases the boiling point, and decreasing the atmospheric pressure decreases the boiling point. So, when we discuss the boiling points of liquids, we mean the *normal* boiling points, at the *normal* pressure of 1 atm, unless we state otherwise.

The temperature of a liquid remains constant as it boils at constant pressure. For example, if you heat a container of water, the temperature of the water increases to 100°C (at 1 atm pressure), and the water then starts to boil. But the temperature of the liquid does not increase beyond 100°C *no matter how much heat you add.* What happens is this: The heat you add goes first to raising the temperature of the water to 100°C. After that, all the heat that is added is used to vaporize the water molecules. That is, energy is required for the water molecules to move fast enough to escape the liquid, and this energy is supplied by the heat. The more heat you add, the faster the water is vaporized. But the temperature of the water remains at the boiling point until the last molecule escapes.

Figure 13.6
The first bubbles of water vapor

Figure 13.7
Boiling water

13.4 Freezing of Liquids

Learning Goal 4
Change of state

When heat is removed from a liquid, its temperature decreases. At a certain temperature, the liquid begins to **freeze,** or change to a solid. This temperature is called the **freezing point** of the liquid. As more heat (or energy) is removed, more and more of the liquid freezes *at this same temperature.* Once all the liquid has been transformed to the solid state, its temperature will decrease if more heat is removed.

For example, the freezing point of pure water is 0°C. As liquid water is cooled, its temperature decreases. At 0°C the liquid begins to turn to ice—or solid water. As more heat is removed by cooling, more of the liquid turns to ice, but the temperature stays at 0°C. Finally, when all the water has been frozen, the temperature of the ice will drop below 0°C if it is cooled further.

When a solid is heated, its temperature increases until its freezing point is reached. At that temperature, the solid begins to **melt,** or change to a liquid. (For this reason, the freezing point of a substance is also called its **melting point.**) As more heat is added, more of the solid turns to liquid *at this same temperature.* Once all the solid has become liquid, its temperature will increase if more heat is added. Figure 13.8 shows the complete process—solid to liquid to gas—for ice at 1 atm pressure.

Some substances can change directly from the solid state to the gaseous state without first becoming a liquid. This process is called **sublimation.** An example is dry ice, which is solid carbon dioxide. It goes directly from the solid state to the gaseous state at 1 atm pressure.

Figure 13.8

How water at 1 atm pressure changes from a solid to a liquid to a gas. Below 0°C, water is in the solid state—ice. In this diagram, we begin with ice at −10°C. Energy (heat) is supplied so that the temperature of the ice rises. When it reaches 0°C, the ice starts to melt into water. As more energy is supplied, more ice melts into water. This water remains at 0°C until all the ice melts. After all the ice has melted, the energy supplied starts to warm the water. This continues until the water temperature reaches 100°C. At this point, the water starts to vaporize into steam. The temperature of the water remains constant at 100°C as the liquid is turned into gas.

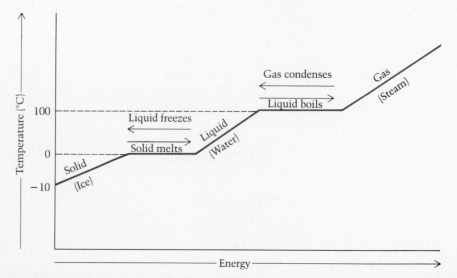

13.5 The Process of Distillation

· · · · · · · · · · · · · ·
Learning Goal 5
Process of distillation

The fact that liquids boil gives rise to an interesting method for separating one liquid from another, or even separating a solid from a liquid in which it is dissolved. This process is called *distillation*.

Figure 13.9 shows a simple distillation apparatus. It works this way: Suppose you have a solution of sodium chloride and water. You want to separate the sodium chloride from the water and at the same time collect the purified water. You place the sodium chloride–water solution in the distillation flask and heat it to boiling. As the solution boils, water in the form of vapor fills the flask and travels toward the distilling head. From there the hot water vapor travels down the double-jacketed, water-cooled condenser. As the vapor travels down within the inner walls of the condenser, it is cooled by tap water running between the outer walls of the condenser. The water vapor **condenses,** or changes back to its liquid form. The condensed liquid, known as the *distillate,* is collected in a receiving flask. Because only the water, and not the sodium chloride, vaporizes and passes through the condenser, the distillate contains only pure water. The sodium chloride remains behind in the distillation flask.

Miscible *substances are those that form solutions when mixed.* Two or more miscible liquids that have different boiling points may also be separated

Figure 13.9
A distillation apparatus

by the process of distillation. For example, when a mixture of ethyl alcohol (boiling point, 78.5°C) and water (boiling point, 100°C) is heated, the component with the lower boiling point (ethyl alcohol) vaporizes more readily than the component with the higher boiling point (water). Therefore the distillate is richer in ethyl alcohol than in water. And it can be distilled again to yield even purer ethyl alcohol. If this process is repeated enough times, a distillate of essentially pure ethyl alcohol is obtained.

This technique, called *fractional distillation*, is extremely useful in petroleum refining. The different components of crude petroleum all have different boiling points, so they can be separated by distillation. Petroleum companies use huge distilling columns, known as *fractionating columns*, to separate the various petroleum products (Figure 13.10). When the crude

Figure 13.10
A fractionating column for processing crude petroleum

oil is heated, the components (or fractions) with lower boiling points rise to the top of the column, those with higher boiling points rise to the middle of the column, and solids remain at the bottom of the column. As the components rise up the column and condense, they are drawn off, to be purified further if necessary.

13.6 Solids

Learning Goal 6
Crystalline and amorphous solids

Solids have two main characteristics that distinguish them from gases and liquids: *fixed shape* and *fixed volume*. Most solids owe these unique properties to their well-defined crystal structure (Figure 13.11). (We'll explain the term *crystal* in a moment.) However, a few solids, such as glass and paraffin, do not have a well-defined crystal structure, although they do have definite shape and volume. *Such solids are called* **amorphous,** *which means that they have no definite internal structure or form.* This makes them interesting to research scientists. However, we shall concentrate on crystalline solids.

Particles That Make Up Solids

Learning Goal 7
The kinds of crystalline solids

Solids can be composed of three types of particles: atoms, molecules, and ions. Solids such as carbon, silicon, copper, and gold are composed of atoms. Solids such as dry ice (CO_2), iodine, and water (ice) are composed of molecules. Solids such as sodium chloride, calcium chloride, and sodium bromide are composed of ions. The three types of solids have very different properties. Table 13.2 lists some of the differences.

Figure 13.11
Examples of different types of crystalline solids

Diamond
Atomic solid
(atoms occupy the lattice positions)

Carbon dioxide
Molecular solid
(molecules occupy the lattice positions)

Sodium chloride
Ionic solid
(ions occupy the lattice positions)

Table 13.2
Some Properties of Crystalline Solids

Type	Melting point	Hardness	Conductivity	Example
Ionic	High	Hard and brittle	Nonconductors of electricity	Sodium chloride, NaCl
Molecular	Low	Soft	Nonconductors of electricity	Dry ice, CO_2
Atomic				
Metallic	High	Varying hardness	Good conductors of electricity	Copper, Cu
Nonmetallic	High	Hard and brittle	Nonconductors of electricity	Diamond, C

What Is a Crystal?

Learning Goal 8
Crystal lattice and unit cell

A **crystal** is *a solid that has a fixed, regularly repeating, symmetrical internal structure. This symmetrical structure, formed by the particles in a crystal, is called the* **crystal lattice.** A crystal lattice is not an actual object, but a *structure* or *pattern.* The particles that make up the pattern are different in different types of crystals, but in each case they occupy specific *lattice positions.* In an atomic solid, *atoms* occupy the lattice positions; in a molecular solid, *molecules* occupy them; and in an ionic solid, *ions* occupy them (Figure 13.11). It is this internal structure that results in the plane (flat) surfaces exhibited by crystals (see Figure 13.12).

Figure 13.12
A quartz crystal
(Courtesy of the American Museum of Natural History)

Unit cell of magnesium (Mg)

Figure 13.13
The unit cell of magnesium

Table 13.3
Six Types of Crystals

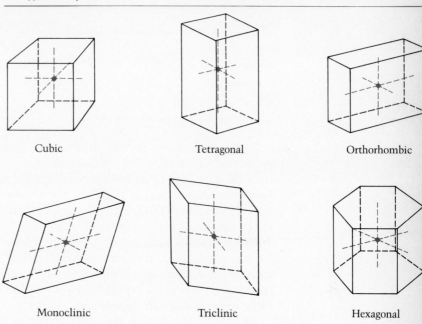

Cubic	Tetragonal	Orthorhombic

Monoclinic	Triclinic	Hexagonal

Figure 13.14
Sodium chloride crystals
(Courtesy Morton Salt)

Each repeated unit in a crystal lattice is called a **unit cell** (Figure 13.13), and the unit cell is repeated over and over again in the crystal. Because the whole crystal is made up of repetitions of the lattice structure, we know the three-dimensional pattern of the crystal when we know the lattice. Table 13.3 shows six different types of crystals—each precise, orderly, and capable of existing side by side with exact copies of itself. Figure 13.14 is a photograph taken with a microscope, showing lattice structure.

What Holds a Crystal Together?

The properties of a crystalline solid are partly the result of the forces that bind together the particles that make up the crystal. Let's look briefly into the bonding forces within the three types of crystals (ionic, molecular, and atomic).

In an *ionic* crystal, the attractive forces between oppositely charged ions hold the crystal together. This kind of bonding force is very strong, so ionic solids have very high melting points. For example, table salt (NaCl) has a melting point of 800°C. A great deal of heat energy is required to force those ions apart.

Table 13.4
Melting Points of Some Atomic Crystals

Substance	Melting point
Iron	1,535°C
Copper	1,083°C
Diamond	3,500°C
Silicon	1,410°C

Molecular crystals, on the other hand, are usually held together by weak electrical forces. Only molecular crystals whose molecules are polar have binding forces that have some degree of strength. The forces binding most molecular crystals together are so weak that these crystals frequently have much lower melting points than ionic crystals. For example, ice (a molecular crystal) melts at 0°C.

Atomic crystals have strong binding forces—Table 13.4 shows just *how* strong. When the atoms are nonmetallic, they are held together by covalent bonds. An example is diamond, made up of carbon atoms that bond covalently. It takes a temperature of 3,500°C to break the bonds that unite carbon atoms in diamond.

In atomic crystals that are composed of metallic atoms, it is the attraction of opposite charges that makes the bonds so strong. Scientists think of the metallic lattice as composed of positive ions surrounded by a cloud of electrons. The electrons are given off by the metallic atoms, but they are considered part of the entire crystal—electrons at large, you might say. Because the electrons are free to wander throughout the crystal, they are able to conduct electricity. This is why metals are good conductors of electricity.

Summary

Gases have no definite shape or volume and are easily compressed. Liquids have a definite volume but no definite shape, and they tend to be incompressible. Solids have a definite volume and shape and are practically incompressible.

The kinetic theory helps explain the properties of liquids and solids. The forces of attraction between liquid molecules and between solid molecules are much greater than those between gas molecules. As a result, the molecules of a liquid are very close to each other. This gives the liquid a definite volume but no specific shape. The particles composing a solid are not only close together but are also held in a fixed position. This gives the solid both a definite volume and a definite shape.

The molecules of a liquid are constantly in motion. When molecules collide, energy is transferred from one molecule to another. A molecule near the surface of the liquid can obtain enough energy in a collision to evaporate, or escape from the liquid entirely. Molecules that evaporate (or vaporize) have changed from the liquid state to the gaseous state.

When the liquid is in a covered container, the molecules that have escaped the liquid remain above its surface. Some of these molecules return to the liquid and others escape it. Eventually, a dynamic equilibrium is attained, in which the rate of escape equals the rate of return to the liquid. The pressure exerted by the escaped molecules at this point is called the equilibrium vapor pressure of the liquid.

The boiling point of a liquid is the temperature at which its equilibrium vapor pressure equals the atmospheric pressure. A heated liquid will continue to boil at its boiling point until it has evaporated completely. The boiling phenomenon is used in the process of distillation, in which liquids are separated by being evaporated in turn.

The melting point (or freezing point) of a substance is the temperature at which it begins to change from the solid state to the liquid state (or from the liquid state to the solid state). The temperature of a substance remains at its melting point as long as it is melting or freezing.

Most solids owe their fixed shape and volume to their crystal structure, although a few solids, such as glass and paraffin, are amorphous (do not have a definite internal structure) rather than crystalline. A crystal is a solid that has a definite, repeating, symmetrical internal structure. The form of this internal structure is called the crystal lattice.

In ionic crystals, ions occupy the fixed lattice positions, and the attractive forces between oppositely charged ions hold the crystal together. Ionic solids have high melting points because these attractive forces are so strong. Molecular crystals, in which molecules occupy the lattice positions, are usually held together by weak electrical forces. Most molecular crystals have much lower melting points than ionic crystals. Only molecular crystals whose molecules are polar have binding forces that are fairly strong. Atomic crystals, in which atoms occupy the lattice positions, have strong binding forces. These crystals can be composed of either nonmetallic or metallic ions.

Career Sketch

Dietitians

Dietitians plan well-balanced and appetizing meals to help healthy people stay healthy or sick people recover their good health. There are three subgroups: clinical (the largest group), administrative, and research dietitians.

Dietitians plan well-balanced and appetizing meals to help healthy people stay healthy and to help sick people recover good health.
(San Pedro Peninsula Hospital, California)

Administrative dietitians supervise the planning and preparation of meals for large numbers of people, for such institutions as hospitals, universities, and schools. They also select, train, and direct food-service workers; purchase food, equipment, and supplies; enforce sanitary and safety regulations; and prepare records and reports. Dietitians who are directors of a dietetic department also coordinate dietetic service with the activities of other departments. They are responsible for the budget of the dietetic department, which in large organizations may amount to millions of dollars annually.

Clinical dietitians, sometimes called *therapeutic dietitians*, plan diets and supervise the service of meals to patients in hospitals, nursing homes, or clinics. Clinical dietitians confer with doctors about patients' nutrition, instruct patients and their families on diet requirements, and suggest ways for patients to follow their diets after they leave the hospital or clinic.

Research dietitians conduct research aimed at improving the nutrition of both healthy and sick people. Some specialize in nutrition science and education, some in food management, and some in food service systems and equipment. Some research dietitians study the way the body uses food. Their projects may include investigating the nutritional needs of the aging, of persons with a chronic disease, or of space travelers. Research dietitians usually work in medical centers or in colleges or universities, but they may also work in community health programs.

Self-Test Exercises

Learning Goal 1: Properties of Gases, Liquids, and Solids

1. (*a*) How is the boiling point of a liquid affected by an increase in atmospheric pressure?
(*b*) How is the melting point of a solid affected by an increase in atmospheric pressure?

3. How is the boiling point of a liquid affected by a decrease in atmospheric pressure?

5. State whether each of the following statements refers to a solid, liquid, or gas.
(*a*) It has definite volume but no definite shape.
(*b*) It has no definite volume or shape.
(*c*) It has both definite volume and definite shape.

2. True or False: A solid has a definite shape but no definite volume.

4. True or False: Gases are easily compressed, but solids and liquids are not.

6. True or False: Solids, liquids, and gases all have a definite volume.

Learning Goal 2: Kinetic Theory of Liquids and Solids

7. Explain each of the following properties of liquids in terms of the kinetic theory: (*a*) no definite shape, (*b*) definite volume, (*c*) incompressibility.

9. A man heats a pot of potatoes. When the water starts to boil, he turns up the heat "to cook the potatoes faster." Is he wasting gas, or will the potatoes actually cook faster? Explain.

11. It takes 15 minutes to cook a hard-boiled egg at sea level. Explain why it takes more time to cook a hard-boiled egg on a mountain top.

*13. Adding salt to water raises the boiling point of water. Give a possible explanation for this phenomenon.

15. As liquid evaporates from an open container, the temperature of the remaining liquid drops. Give a possible explanation for this occurrence.

17. Will an equilibrium vapor pressure be reached in the space above a liquid in an open container (for example, a half-filled glass of water)? Explain.

19. Describe an experiment that you could perform to show that the molecules of a liquid are constantly in motion.

8. Explain each of the following properties of solids in terms of the kinetic theory: (*a*) definite shape, (*b*) definite volume, (*c*) incompressibility.

*10. The directions on a box of fudge brownie mix say, "High-Altitude Directions (3,500 to 6,500 feet): Stir two tablespoons of all-purpose flour into brownie mix (dry). Increase baking time to 35 to 40 minutes." Why are these high-altitude directions necessary? If you wanted to bake cake-like brownies, no adjustment would be necessary. Why?

12. A kettle full of water is boiling on a stove. If you turn up the heat, will the water get hotter? Explain.

14. You have a bottle of rubbing alcohol on the shelf and you forget to put the cap on the bottle. What will happen to the alcohol in the bottle?

16. A house heated by a forced-hot-air system has air that is very dry. How could the situation be remedied?

18. Name a liquid that evaporates slowly.

20. A thermometer containing mercury, a toxic liquid metal, breaks. Why is it dangerous to collect the mercury and let it sit in an open container?

21. If you place equal amounts of gasoline and water in separate glasses, the gasoline evaporates first. Explain.

23. Adding salt to water lowers its freezing point. Give an explanation for this occurrence.

22. Which system would have a higher equilibrium vapor pressure: a covered jar half-filled with water at 25°C or a covered jar half-filled with water at 75°C?

24. Have you ever been inside a vehicle that is being filled with gasoline and smelled the fumes from the gasoline? If a quart of oil were being added to this same vehicle, you would not smell any fumes from the oil. Why is this so?

Learning Goal 3: Equilibrium Vapor Pressure, Evaporation, and Boiling Point

25. Define the following terms: (a) normal boiling point, (b) equilibrium vapor pressure, (c) evaporation.

27. Define the term *dynamic equilibrium* as it applies to vapor pressure.

26. Which liquid do you think would boil first, ethyl alcohol or salad oil? Explain your answer.

28. What happens to the boiling point of a liquid if the weather changes from sunny and dry to rainy? (*Hint:* Keep in mind that sunny and dry weather usually means a high atmospheric pressure system is in the area, whereas rainy weather usually means a low atmospheric pressure system is in the area.)

29. Explain the difference between the terms *boiling point* and *normal boiling point*.

30. A beaker of water is boiling at 1 atm pressure. The temperature of the water is 100°C. What is the temperature of the water vapor at the boiling point?

Learning Goal 4: Change of State

31. Explain what happens to molecules of water as they go from solid (ice) to liquid (water) to gas (steam).

*32. When dry ice (solid carbon dioxide) is placed in a beaker of water, vigorous bubbling is seen and a white vapor leaves the beaker. Explain what is happening.

Learning Goal 5: Process of Distillation

33. Explain the following processes: (a) distillation, (b) fractional distillation.

35. Define the following terms: (a) condensation, (b) distillate, (c) distilling head, (d) condenser.

34. Water and isopropyl alcohol are mixed and distilled. Which liquid is collected first?

36. What is the difference between distillation and fractional distillation?

Learning Goal 6: Crystalline and Amorphous Solids

37. Solids that have no well-defined crystalline structure are said to be _____ .

39. Crystalline solids are said to exist in a _____ structure.

38. Name two solids with a well-defined crystalline structure.

40. Name two amorphous solids.

Learning Goal 7: The Kinds of Crystalline Solids

41. List the properties associated with: (*a*) solids composed of oppositely charged ions, (*b*) solids composed of atoms, (*c*) solids composed of molecules.

43. The compound naphthalene is a soft solid substance with a low melting point. What type of solid (ionic, molecular, or atomic) is naphthalene? Explain.

45. The compound cesium chloride has a melting point of 646°C. It is a brittle substance. What type of crystalline solid is it?

47. For each of the following statements, classify the solid as ionic, molecular, or atomic (metallic or nonmetallic).
(*a*) The solid has a high melting point and is a good conductor of electricity.
(*b*) The solid has a high melting point, does not conduct electricity, and can be decomposed into simpler substances.

49. Fill in the blanks.
(*a*) In atomic solids, _____ occupy the lattice positions.
(*b*) In molecular solids, _____ occupy the lattice positions.
(*c*) In ionic solids, _____ occupy the lattice positions.

51. Describe the forces that hold the crystals together in (*a*) atomic solids, (*b*) molecular solids, (*c*) ionic solids.

42. Compare the melting points of ionic solids, molecular solids, and atomic solids.

44. True or False: Ionic solids are hard and brittle with low melting points.

46. True or False: Molecular solids are nonconductors of electricity.

48. Which types of solids can be good conductors of electricity? Which types can be good nonconductors? What determines whether a solid is a conductor or a nonconductor?

50. Which types of solids are hard and brittle?

52. A certain compound is a soft solid with a low melting point and does not conduct electricity. Describe the particles that occupy its lattice positions.

Learning Goal 8: Crystal Lattice and Unit Cell

53. Trace the unit cell of sodium chloride from the drawing in Figure 13.11, and show how several unit cells can build up into a crystalline structure.

55. Define the following terms: (*a*) crystal lattice, (*b*) unit cell.

54. Name six types of crystals.

56. Which type of crystal structure is exhibited by sodium chloride?

Extra Exercises

57. Explain, in terms of the kinetic theory, why the temperature of a pure substance remains constant during a phase change.

58. Compare and contrast the characteristics of the three types of solids.

59. In what way is a *unit cell* like a molecule?

60. Is it possible for a substance *not* to have a normal boiling point? Explain.

61. Compare the freedom of motion of a molecule of a gas, with that of a molecule of a liquid, and that of a molecule of a solid.

62. Using Figure 13.8, explain why steam at 100°C produces more severe burns than liquid water at 100°C.

63. Explain what is happening when a glass rod is heated and begins to sag.

64. State whether each of the following statements refers to a solid, a liquid, or a gas. (*Note:* Each statement may refer to more than one state of matter.) (*a*) Has definite volume. (*b*) Has definite shape. (*c*) Has no definite volume.

65. Describe an experiment that you could perform to show that the molecules of a gas are constantly in motion.

66. Do liquids that evaporate easily at a particular temperature have *higher* or *lower* equilibrium vapor pressures than liquids that do not evaporate easily at the same temperature? Explain.

Cumulative Review / Chapters 10–13

After you have studied Chapters 10, 11, 12, and 13, answer these questions to review what you have learned.

1. Distinguish between endothermic and exothermic reactions.

2. How much heat is needed to raise the temperature of 500.0 g of water from 15.0°C to 35.0°C?

3. How much heat is required to raise the temperature of 25.0 g of copper from 30.0°C to 90.0°C?

4. How much heat is required to raise the temperature of 25.0 mL of ethyl alcohol from 25.0°C to 50.0°C? (The density of ethyl alcohol is 0.800 g/mL.)

5. Why must we use standard state conditions when calculating changes in heat?

6. How much heat is needed to raise the temperature of 1.000 kg of water from 50.0°C to 75.0°C?

7. Write the thermochemical equation for the formation of one mole of water from hydrogen gas and oxygen gas.

8. Use Table 10.2 to determine ΔH for the reaction

$$H_2O(l) + SO_3(g) \longrightarrow H_2SO_4(aq)$$

9. Calculate the heat released when 254 g of $SO_2(g)$ are produced from its elements.

10. Calculate the amount of heat released when 10.2 g of $H_2O_2(l)$ decompose to form $H_2O(l)$ and $O_2(g)$.

11. A gas has an inital volume of 55 mL at a pressure of 2.0 atm. What will be its final volume if the pressure of the gas is increased to 4.5 atm and the temperature remains constant?

12. Suppose a gas occupies a space of 15.05 mL under a pressure of 2.50 atm at 110.0°C. What will be the volume of the gas when the pressure is decreased to 2.00 atm? (Assume the temperature remains constant.)

13. A gas occupies a volume of 0.200 L under a pressure of 770 torr at 50.0°C. What pressure will cause the gas to occupy a volume of 250.0 mL at the same temperature?

14. A gas occupies 1,500.0 mL of space at 27°C. What will be its volume at 227°C, assuming its pressure remains constant?

15. When a gas is cooled at constant pressure, it contracts from 100.0 mL to 55.0 mL. Its final temperature is 180.0°C. What was its initial temperature?

16. A gas at 50.5°C and 1.50 atm pressure fills a 2.50-L container. What volume will the gas occupy at 250.0°C and 775 torr?

17. A gas collected at 127°C has a volume of 10.5 L and a pressure of 775 torr. What will be its volume at STP?

18. Suppose a sample of oxygen is collected over water at 20.0°C and a pressure of 775 torr. The gas fills a 400-mL container but has water vapor mixed in with it. What would be the volume of the *dry* gas at STP?

19. Suppose 5.00 L of hydrogen gas are collected over water at 28.0°C and a pressure of 725 torr. What would be the volume of the *dry* hydrogen gas at STP?

20. A gas occupies a volume of 358.4 L at STP. How many moles of gas are in this sample?

21. Determine the volume of 15.0 moles of a gas at STP.

22. Twenty (20.0) moles of a gas exert a pressure of 2,280 torr at a temperature of 227°C. What volume does the gas occupy under these conditions?

23. A gas exerts a pressure of 550.0 torr at a temperature of 155°C. If the volume of the gas is 2.00 L, how many moles of gas are present?

24. Determine the molecular weight of a gas, given the following experimental data: mass = 3.55 g, volume = 255 mL, pressure = 1,520 torr, and T = 227°C.

25. When ethane (C_2H_6) reacts with oxygen, carbon dioxide and water vapor are formed. How many liters of carbon dioxide are produced from 5.00 L of ethane?

26. Using the Ideal Gas Law and the formula for density $(D = m/V)$, calculate the density of chlorine gas at STP. Report your answer in grams per liter.

27. Once released, a weather balloon travels to the upper atmosphere. Why does its volume increase there?

28. An old tube tire has a volume of 2.55 L at 25°C, and 1.39 atm pressure. During a trip, the temperature

of the tire increases by 12°C and the pressure increases to 1.41 atm. What is the new volume of the tire?

29. A 25.0-g sample of a gas has a pressure of 4.50 atm and a volume of 1.45 L at a temperature of 25.0°C. What is the molecular weight of the gas?

30. Examining the physical states of the halogens at room temperature, we find that fluorine is a gas, bromine a liquid, and iodine a solid. Use the kinetic theory of gases to explain these properties.

31. Is the oxygen found on the earth's crust composed predominantly of free oxygen or combined oxygen?

32. When oxygen is prepared for use in hospitals, which method of preparation is employed?

33. Describe the steps involved in separating oxygen from air.

34. State the solubility of oxygen in water at 20.0°C. How is this dissolved oxygen important to marine life?

35. A balloon is filled with 4.0 L of air at standard pressure. What will happen to the volume of air in the balloon if it is placed in a hyperbaric chamber at 3.5 atm pressure? (Assume the temperature doesn't change.)

36. A sample of intestinal gas is analyzed and found to have the following approximate composition in percent by volume: 42.0% CO_2, 40.0% H_2, 16.9% N_2, and 0.00300% CH_4. If the total pressure in the intestine is 815 torr, what is the partial pressure of each gas?

37. Complete the following chemical equations.
(a) $Li(s) + H_2(g) \longrightarrow ?$
(b) $Sr(s) + H_2(g) \longrightarrow ?$

38. Complete the following statement. Hydrogen reacts with nitrogen under high temperature and pressure to produce _____ in the _____ process.

39. Complete and balance the following chemical equation. $CO(g) + H_2(g) \longrightarrow ?$

40. Match each of the following reactions with its description.
(a) $SO_2 + H_2O \longrightarrow H_2SO_3$ 1. electrolysis of water
(b) $CaO + H_2O \longrightarrow Ca(OH)_2$ 2. metal oxide + water
(c) $2H_2O \longrightarrow 2H_2 + O_2$ 3. nonmetal oxide + water

41. A 20.8-g sample of calcium carbonate is heated until all the water is driven off. The remaining salt weighs 10.00 g. Determine the formula of the hydrate.

42. Calculate the percentage of water in calcium chloride hexahydrate.

43. Explain how DDT found in a polluted body of water can enter the food chain.

44. Describe the process of eutrophication in a body of water.

45. Calculate the percentage of water in calcium sulfate dihydrate.

46. A body of water is tested and found to have a chloride ion concentration of 400 milligrams per liter. According to the U.S. Public Health Service standards for potable water, is this water safe to drink?

47. A hydrate of barium chloride is analyzed and found to contain 14.7% water by weight. Determine the formula of this hydrate.

48. Calculate the formula for the hydrate of sodium carbonate that is 63% water.

49. How many grams of water are needed to produce 25.00 g of oxygen by electrolysis?

50. Match each of the following formulas with the name of the proper hydrate.
(a) $MgSO_4 \cdot 7H_2O$ 1. alum
(b) $KAl(SO_4)_2 \cdot 12H_2O$ 2. gypsum
(c) $CaSO_4 \cdot 2H_2O$ 3. epsom salts

51. Match each of the following properties with the correct term.
(a) Has no definite shape but has a definite volume   1. solid
(b) Is easily compressed   2. liquid
(c) Has definite shape and volume   3. gas

52. Eggs are being cooked in boiling water. The heat is increased. Which of the following statements is true in this situation?
(a) The eggs cook faster.
(b) The water evaporates more quickly.

53. On a sunny day following a winter snowstorm, the volume of snow decreases, yet the temperature has not allowed melting to take place. What is the most probable explanation for this occurrence?

54. A crystalline solid has a low melting point and is soft. Is it more likely to be a molecular or an atomic solid?

55. A crystalline solid has a high melting point and is hard and brittle. It is more likely to be an ionic or a molecular solid?

56. Is a solid with no definite shape more likely to be a crystalline or an amorphous solid?

57. Using the kinetic theory, explain what happens to water as it changes from ice to vapor.

58. Benzene evaporates more easily than water. Which liquid has the higher equilibrium vapor pressure?

59. Why are some solids capable of conducting electricity whereas others are not?

60. Why are atomic crystals that are composed of metallic atoms so strong?

14 *The Chemistry of Solutions*

Learning Goals

After you have studied this chapter, you should be able to:

1. Define the terms *solution*, *solute*, and *solvent*.

2. Predict solubilities using the idea that "like dissolves like."

3. Calculate concentrations of solutions in percentage by weight, percentage by volume, and percentage by weight–volume.

4. Calculate the molarity, number of moles, or volume of a solution when you are given the other two.

5. Explain the concept of equivalents for an acid or base, and calculate the number of equivalents when you are given the number of grams or moles.

6. Calculate the normality, number of equivalents, or volume of a solution when you are given the other two.

7. Calculate concentrations of normal and molar solutions after they have been diluted.

8. Explain what electrolytic and nonelectrolytic solutions are, and give examples of each.

9. Define the terms *ion* and *ionization*.

10. Explain what is meant by the molality of a solution.

11. Calculate the molality of a solution, given mass of solute and mass of solvent.

12. Explain how a solute affects the boiling point and freezing point of a solution.

13. Define the following terms: *colligative properti* (K_f), and *boiling-point elevation* (K_b).

14. Calculate the boiling and freezing points of a so given the molality of the solution, its K_f, and its $K_$

330

Introduction

Solutions: We drink them, we breathe them, we swim in them, we are even composed of them. Every time you drink a cup of tea or a soda, you swallow a solution. Each time you breathe, you inhale a solution—air. When you swim in the ocean, you're swimming in a solution of salt in water. Even your blood is a solution.

There is a lot more to solutions than just two things put together. In this chapter we look at the types of solutions, some of their properties, and some special units that apply to solutions.

14.1 Like Dissolves Like

Learning Goal 1
Definitions of solution, solute, and solvent

In Chapter 3 we defined a **solution** as a homogeneous mixture. A solution can contain two substances, three substances, or more. The most common types of solutions are made by dissolving a solid in a liquid—for example, a salt in water. However, solutions can be made up from combinations of all three states of matter. Table 14.1 shows six possible types of solutions and gives an example of each.

Because solutions are always composed of at least two substances, we need to be able to identify the role that each substance plays. The **solute** is *the substance that is being dissolved*, whereas the **solvent** is *the substance that is doing the dissolving*. For example, in a solution of salt water, salt is the solute and water is the solvent.

Table 14.1
Different Types of Solutions

	Solid	Liquid	Gas
Solid	Copper metal dissolved in silver metal (for example, coins)	———	———
Liquid	Salt dissolved in water	Ethyl alcohol dissolved in water	———
Gas	Hydrogen dissolved in platinum metal	Carbon dioxide dissolved in water (soda water)	Oxygen gas dissolved in nitrogen gas

When a solution is composed of two substances in the same state (as is a liquid–liquid solution), it is difficult to establish which substance is the solute and which is the solvent. One rule of thumb is that the substance present in the larger amount is the solvent. In a solution of 10 mL of ethyl alcohol and 90 mL of water, the ethyl alcohol is the solute and the water is the solvent.

Example 14.1

Determine which is the solute and which is the solvent in each of the following solutions.
(a) Sugar and water
(b) Hydrogen chloride gas and water
(c) 75.0 mL of ethyl alcohol and 25.5 mL of water
(d) 80 mL of nitrogen and 20 mL of hydrogen
(e) Soda water

Solution

(a) Sugar is the solute; water is the solvent.
(b) Hydrogen chloride is the solute; water is the solvent.
(c) Water is the solute; ethyl alcohol is the solvent.
(d) Hydrogen is the solute; nitrogen is the solvent.
(e) Carbon dioxide is the solute; water is the solvent.

Practise Exercise 14.1   Which is the solute and which is the solvent in each of the following solutions?
(a) Salt and water
(b) Sulfur trioxide gas and water
(c) $3\overline{0}$ mL of isopropyl alcohol and 25 mL of water
(d) 40 mL of hydrogen gas and 50 mL of oxygen gas
(e) Carbonated mineral water

Learning Goal 2
Substances that form
solutions

Before we go on, we should note that some pairs of substances do not form solutions when mixed; such substances are said to be **immiscible.** For example, gasoline and water do not readily form a solution. The same holds true for oil and vinegar, as in salad dressing (Figure 14.1). There are also substances that do mix, but only to a slight degree.

How can we determine whether two substances will form a solution? The answer is found in the statement "like dissolves like." To understand its significance, recall that a molecule may be either polar or nonpolar. "Like dissolves like" means that polar substances dissolve other polar substances and that nonpolar substances dissolve other nonpolar substances. But a polar and a nonpolar substance generally do not form a solution. Alcohol and water mix because both are composed of polar molecules.

Figure 14.1
Two immiscible liquids

Gasoline and water do not mix, because gasoline is composed of nonpolar molecules and water is composed of polar molecules.

Example 14.2

State whether the following pairs of substances form solutions. [*Hint:* Refer to Chapter 7.]
(a) Methane (CH$_4$) and water
(b) HBr and water
(c) *Trans*-dichloroethene and carbon tetrachloride

Solution
(a) Methane is nonpolar, whereas water is polar. They do not form a solution.
(b) Both HBr and water are polar. They do form a solution.
(c) *Trans*-dichloroethene is nonpolar, and so is carbon tetrachloride. They do form a solution.

Practice Exercise 14.2   State whether the following pairs of substances form solutions. (*Hint:* You may want to refer to Chapter 7.)
(a) Oil and water
(b) Hydrogen chloride gas and water
(c) Benzene and carbon tetrachloride

14.2 Concentrations of Solutions By Percentage

Learning Goal 3
Percentage concentrations

Because solutions are mixtures, their components can be present in different ratios. For example, we can make many different salt-and-water solutions, each with a different **concentration,** or ratio of solute to solvent. The concentrations or relative quantities of solutions can vary widely, so we must have a way of describing them. (It is important to note, however, that we can increase the concentration of the solute in a solution only to a certain point. This concentration, called the **saturation point,** is the point at which no more solute dissolves to form a stable solution.)

One method of defining the concentrations of solutions is based on the percentage of solute in the solution. This method can cause confusion, because there can be three types of percentage concentrations:

1. Percentage by weight

2. Percentage by volume

3. Percentage by weight–volume

A 30%-by-weight solution of alcohol and water is not the same as a 30%-by-volume solution of alcohol and water. Therefore a label that reads "30% alcohol in water" tells us nothing. The solution could be a 30%-by-weight solution, a 30%-by-volume solution, or a 30%-by-weight–volume solution. For this reason chemists seldom use the percentage method of defining concentrations. However, biologists and medical workers do use percentage by weight–volume, so we will discuss all three types.

Percentage-By-Weight Solutions

To find percentage by weight, we divide the *mass of the solute* by the *total mass of the solution* and multiply the result by 100. (The mass of the solution equals the mass of the solute plus the mass of the solvent.) For example, suppose we made a solution of alcohol in water by mixing 30 g of alcohol with enough water to make 100 g of solution (Figure 14.2). The concentration of our solution would be 30% alcohol by weight.

$$\text{Percent alcohol by weight} = \frac{\text{weight of alcohol}}{\text{total weight of solution}} \times 100$$

$$= \frac{30 \text{ g}}{100 \text{ g}} \times 100 = 30\%$$

Figure 14.2

Preparing a 30%-by-weight alcohol–water solution

Example 14.3

Find the concentrations of the following solutions in percentage by weight.

(a) $2\bar{0}$ g of salt with enough water to make $6\bar{0}$ g of solution
(b) $5\bar{0}$ g of sugar with enough water to make $35\bar{0}$ g of solution

Solution

(a) Percent salt by weight $= \dfrac{\text{weight of salt}}{\text{total weight of solution}} \times 100$

$$= \frac{2\bar{0} \text{ g}}{6\bar{0} \text{ g}} \times 100 = 33\%$$

(b) Percent sugar by weight $= \dfrac{\text{weight of sugar}}{\text{total weight of solution}} \times 100$

$$= \frac{5\bar{0} \text{ g}}{35\bar{0} \text{ g}} \times 100 = 14\%$$

Practice Exercise 14.3 Find the concentrations of the following solutions in percentage by weight.
(a) 40.0 g of salt with enough water to make 90.0 g of solution
(b) 75.0 g of sugar with enough water to make 250.0 g of solution

Percentage-By-Volume Solutions

To find percentage by volume, we divide the *volume of the solute* by the *total volume of the solution* and multiply the result by 100. For example, suppose we mixed 30 mL of alcohol with enough water to make 100 mL of solution (Figure 14.3). The concentration of the solution would be 30% alcohol by volume.

Figure 14.3

Preparing a 30%-by-volume alcohol–water solution

100 mL

30 mL alcohol

Step 1
Measure 30 mL of alcohol

Step 2
Add enough water to make 100 mL of solution

$$\text{Percent alcohol by volume} = \frac{\text{volume of alcohol}}{\text{total volume of solution}} \times 100$$

$$= \frac{30 \text{ mL}}{100 \text{ mL}} \times 100 = 30\%$$

Example 14.4

Find the concentrations of the following solutions in percentage by volume.
(a) 50.0 mL of alcohol with enough water to make $\overline{4}00$ mL of solution
(b) 10 mL of benzene in enough carbon tetrachloride to make 50 mL of solution

Solution

(a) Percent alcohol by volume $= \dfrac{\text{volume of alcohol}}{\text{total volume of solution}} \times 100$

$$= \frac{50.0 \text{ mL}}{400 \text{ mL}} \times 100 = 12.5\%$$

(b) Percent benzene by volume $= \dfrac{\text{volume of benzene}}{\text{total volume of solution}} \times 100$

$$= \frac{10 \text{ mL}}{50 \text{ mL}} \times 100 = 20\%$$

Practice Exercise 14.4   Find the concentrations of the following solutions in percentage by volume.
(a) 60.0 mL of alcohol with enough water to make 300.0 mL of solution
(b) 50.0 mL of benzene in enough carbon tetrachloride to make 90.0 mL of solution

---

Percentage-By-Weight–Volume Solutions

To find the percentage by weight–volume, we divide the *mass of the solute* in grams by the *volume of the solution* in milliliters and multiply the result by 100. For example, suppose that we mix $\overline{3}0$ g of alcohol with enough water to make $1\overline{0}0$ mL of solution (Figure 14.4). The concentration of the solution would be $\overline{3}0\%$ alcohol by weight–volume.

Figure 14.4

Preparing a
30%-by-weight–volume
alcohol–water solution

30 g alcohol

30-g mass

Water

Step 1
Measure 30 g of
alcohol

Step 2
Add water to make
100 mL of solution

Alcohol

$$\text{Percent alcohol by weight–volume} = \frac{\text{weight of alcohol}}{\text{volume of solution}} \times 100$$

$$= \frac{30\ \text{g}}{100\ \text{mL}} \times 100 = 30\%$$

Example 14.5

Find the concentrations of the following solutions in percentage by weight–volume.
(a) 2.0 g of iodine in enough carbon tetrachloride to make 80 mL of solution
(b) 130 g of sugar in enough water to make 650 mL of solution

Solution

(a) Percent iodine by weight–volume $= \dfrac{\text{weight of iodine}}{\text{volume of solution}} \times 100$

$$= \frac{2.0\ \text{g}}{80\ \text{mL}} \times 100 = 2.5\%$$

(b) Percent sugar by weight–volume $= \dfrac{\text{weight of sugar}}{\text{volume of solution}} \times 100$

$$= \frac{130\ \text{g}}{650\ \text{mL}} \times 100 = 20.0\%$$

Practice Exercise 14.5 Find the concentrations of the following solutions in percentage by weight–volume:
(a) 4.5 g of *trans*-dichloroethene in enough carbon tetrachloride to make 100.0 mL of solution
(b) 150.0 g of sucrose in enough water to make 750.0 mL of solution

14.3 Molarity

Chemists have a better way of measuring concentration. They have developed a concentration concept called *molarity*. The **molarity** of a solution is *the number of moles of solute per liter of solution*. Molarity (abbreviated M) is equal to

. . . .
Learning Goal 4
Calculations involving
molarity

$$\frac{\text{moles of solute}}{\text{liter of solution}} \quad \text{so that} \quad M = \frac{\text{moles}}{\text{liter}}$$

There can be no confusion when the concentrations of solutions are expressed in these terms. A solution made by dissolving 117 g (2.00 moles) of NaCl in enough water to obtain 1 L of solution (Figure 14.5) would be a 2-molar NaCl solution (usually written as $2M$ NaCl). (For a review of moles, see Chapter 3.) To calculate the molarity of this solution, we would write

$$M = \frac{\text{moles}}{\text{liter}} \quad ?M = \frac{2 \text{ moles}}{1 \text{ liter}} = 2M$$

Chemists run many reactions in solutions. That's why they need a concentration unit that expresses the *number of particles of solute* present in a given amount of solution. This is because most reactions take place between the solute *particles*.

117 g of NaCl ——   ——117-g mass

Step 1
Weigh out 2 moles of NaCl

H_2O ——

1 liter   1 liter

Figure 14.5
Preparing 1 L of a $2M$ NaCl
solution

Step 2
Transfer NaCl to a 1-liter flask

Step 3
Add water and fill to 1 liter

Example 14.6

Calculate the molarity of each of the following solutions.
(a) 3.6 g of HCl in enough water to make 500 mL of solution
(b) 160 g of NaOH in enough water to make 6.0 L of solution
(c) 36.8 g of ethyl alcohol (C_2H_6O) in enough water to make 1,600 mL of solution

Solution

(a) We first calculate that

$$3.6 \text{ g of HCl} = 0.10 \text{ mole}$$

$$500 \text{ mL of solution} = 0.500 \text{ L}$$

Then,

$$M = \frac{\text{moles}}{\text{liter}} \qquad ?M = \frac{0.10 \text{ mole}}{0.500 \text{ L}} = 0.20M$$

(b) We have 6.0 L of solution, and we first calculate that

$$160 \text{ g of NaOH} = 4.0 \text{ moles}$$

Then,

$$M = \frac{\text{moles}}{\text{liter}} \qquad ?M = \frac{4.0 \text{ moles}}{6.0 \text{ L}} = 0.67M$$

(c) We first calculate that

$$36.8 \text{ g of } C_2H_6O = 0.800 \text{ mole}$$

$$1,600 \text{ mL of solution} = 1.600 \text{ L}$$

Then,

$$M = \frac{\text{moles}}{\text{liter}} \qquad ?M = \frac{0.800 \text{ mole}}{1.600 \text{ L}} = 0.500M$$

Practice Exercise 14.6   Calculate the molarity of each of the following solutions.
(a) 9.80 g of H_2SO_4 in enough water to make 0.500 L of solution
(b) 4.60 g of KOH in enough water to make 5.00 L of solution
(c) 73.6 g of ethyl alcohol (C_2H_6O) in enough water to make 1.500 L of solution

Example 14.7

Suppose that you have a supply of 3.0M NaOH solution in your laboratory. You take 200 mL of this solution and evaporate it to dryness. How many grams of *solid* NaOH remain?

Solution

First find the number of moles of NaOH in $\overline{2}00$ mL (0.200 L) of a 3.0*M* solution, and then change it to the number of grams by the usual method.

$$M = \frac{moles}{liter} \qquad so \qquad moles = (M)(liter)$$

$$? \text{ moles of NaOH} = (3.0M)(0.200 \text{ L}) = \left(3.0 \frac{moles}{\cancel{liter}}\right)(0.200 \cancel{liter})$$

$$= 0.60 \text{ mole of NaOH}$$

$$? \text{ grams of NaOH} = (0.60 \cancel{mole})\left(\frac{40 \text{ g}}{1 \cancel{mole}}\right) = 24 \text{ g of NaOH}$$

Practice Exercise 14.7   Suppose that you have a supply of 2.00*M* $Ca(OH)_2$ solution in your laboratory. You evaporate 250.0 mL of this solution to dryness. How many grams of *solid* $Ca(OH)_2$ do you obtain?

Example 14.8

Suppose that you have a supply of 0.50*M* $Ca(NO_3)_2$ solution in your laboratory. How many milliliters of this solution must be evaporated to obtain 32.8 g of solid $Ca(NO_3)_2$?

Solution

Change the 32.8 g of $Ca(NO_3)_2$ to moles, and then solve for the number of liters of 0.50*M* $Ca(NO_3)_2$ solution needed to obtain this number of moles.

$$32.8 \text{ g of } Ca(NO_3)_2 = 0.200 \text{ mole}$$

$$M = \frac{moles}{liter} \qquad so \qquad liters = \frac{moles}{M}$$

$$? \text{ L} = \frac{0.200 \text{ mole}}{0.50M} = \frac{0.200 \cancel{mole}}{0.50 \cancel{mole}/L} = 0.40 \text{ L}$$

You would need to evaporate 0.40 L (400 mL) of a 0.50*M* $Ca(NO_3)_2$ solution to obtain 32.8 g of solid $Ca(NO_3)_2$.

Practice Exercise 14.8   Suppose that you have a supply of 0.200*M* $Mg(NO_3)_2$ solution in your laboratory. How many milliliters of this solution must be evaporated to obtain 1.48 g of solid $Mg(NO_3)_2$?

14.4 Normality

Chemists have developed another important concentration concept: **normality.** The definition of normality is similar to that of molarity, but normality is easier to use in certain chemical calculations. It is most often applied to solutions of acids and bases. The **normality** of a solution *is the number of equivalents of solute per liter of solution.* Normality (abbreviated *N*) is equal to

$$\frac{\text{equivalents of solute}}{\text{liter of solution}} \quad \text{so that} \quad N = \frac{\text{equivalents}}{\text{liter}}$$

Learning Goal 5
Equivalents of acids and bases

Of course, the question "What is an equivalent?" immediately comes to mind. Because we'll use normality for calculations involving acids and bases, we will define an equivalent in terms of an acid or a base. An **equivalent of an acid** is *the mass of the acid that contains Avogadro's number of hydrogen ions.* An **equivalent of a base** is *the mass of the base that contains Avogadro's number of hydroxide ions.*

To find the number of grams in 1 equivalent of an acid or base, first obtain the molecular weight of the substance. Next divide the molecular weight in grams by the number of replaceable hydrogen ions (for an acid) or the number of replaceable hydroxide ions (for a base). For example, sulfuric acid (H_2SO_4) has a molecular weight of 98. Because H_2SO_4 contains two replaceable hydrogen ions, there are 98/2, or 49, grams per equivalent of sulfuric acid.

Learning Goal 6
Calculations involving normality

$$\text{For } H_2SO_4: \quad \frac{49 \text{ g}}{1 \text{ equivalent}}$$

This means that a 1*N* solution of H_2SO_4 contains 49 g of H_2SO_4 per liter of solution.

Table 14.2 lists the weights of 1 mole and of 1 equivalent for some common acids and bases.

Example 14.9

What is the normality of a solution formed by dissolving 196 g of H_2SO_4 in enough water to make 800 mL of solution? What is the molarity of the solution?

Solution

First convert the grams of H_2SO_4 to equivalents, and then use the normality formula. H_2SO_4 has

$$\frac{49 \text{ g}}{1 \text{ equivalent}} \quad \text{or} \quad \frac{1 \text{ equivalent}}{49 \text{ g}}$$

Therefore

$$? \text{ equivalents of } H_2SO_4 = (196 \text{ g})\left(\frac{1 \text{ equivalent}}{49 \text{ g}}\right)$$

$$= 4.0 \text{ equivalents of } H_2SO_4$$

$$N = \frac{\text{equivalents}}{\text{liter}} \qquad ?N = \frac{4 \text{ equivalents}}{0.80 \text{ L}} = 5.0N$$

To find the molarity of this solution, we first convert grams of H_2SO_4 to moles, and then find the number of moles per liter. (The molecular weight of H_2SO_4 is 98.)

$$? \text{ moles of } H_2SO_4 = (196 \text{ g})\left(\frac{1 \text{ mole}}{98 \text{ g}}\right) = 2.0 \text{ moles of } H_2SO_4$$

$$M = \frac{\text{moles}}{\text{liter}} \qquad ?M = \frac{2.0 \text{ moles}}{0.80 \text{ L}} = 2.5M$$

Practice Exercise 14.9   What is the normality of the solution formed by dissolving 60.0 g of $HC_2H_3O_2$ in enough water to make 500.0 mL of solution? What is the molarity of the solution?

Table 14.2
The Weight of 1 Mole and of 1 Equivalent for Some Common Acids and Bases

Substance	Weight of 1 mole	Weight of 1 equivalent	Number of replaceable H^+ or OH^- ions
HCl (hydrochloric acid)	$\dfrac{36 \text{ g}}{1 \text{ mole}}$	$\dfrac{36 \text{ g}}{1 \text{ equivalent}}$	1
$HC_2H_3O_2$ (acetic acid)	$\dfrac{60 \text{ g}}{1 \text{ mole}}$	$\dfrac{60 \text{ g}}{1 \text{ equivalent}}$	1
H_2SO_4 (sulfuric acid)	$\dfrac{98 \text{ g}}{1 \text{ mole}}$	$\dfrac{49 \text{ g}}{1 \text{ equivalent}}$	2
H_3PO_4 (phosphoric acid)	$\dfrac{98 \text{ g}}{1 \text{ mole}}$	$\dfrac{32.7 \text{ g}}{1 \text{ equivalent}}$	3
NaOH (sodium hydroxide)	$\dfrac{40 \text{ g}}{1 \text{ mole}}$	$\dfrac{40 \text{ g}}{1 \text{ equivalent}}$	1
$Ca(OH)_2$ (calcium hydroxide)	$\dfrac{74 \text{ g}}{1 \text{ mole}}$	$\dfrac{37 \text{ g}}{1 \text{ equivalent}}$	2

Example 14.10

What are the normality and the molarity of the solution formed by dissolving 19.6 g of H_3PO_4 in enough water to make $30\overline{0}$ mL of solution?

Solution

Change the 19.6 of H_3PO_4 to equivalents, and then use the normality formula. H_3PO_4 has

$$\frac{32.7 \text{ g}}{1 \text{ equivalent}} \quad \text{or} \quad \frac{1 \text{ equivalent}}{32.7 \text{ g}}$$

Therefore

$$? \text{ equivalents of } H_3PO_4 = (19.6 \text{ g})\left(\frac{1 \text{ equivalent}}{32.7 \text{ g}}\right)$$

$$= 0.599 \text{ equivalent of } H_3PO_4$$

$$N = \frac{\text{equivalents}}{\text{liter}} \qquad ?N = \frac{0.599 \text{ equivalent}}{0.300 \text{ L}} = 2.00N$$

To find the molarity of the solution, change grams of H_3PO_4 to moles, and then find the number of moles per liter.

$$? \text{ moles of } H_3PO_4 = (19.6 \text{ g})\left(\frac{1 \text{ mole}}{98 \text{ g}}\right) = 0.20 \text{ mole of } H_3PO_4$$

$$M = \frac{\text{moles}}{\text{liter}} \qquad ?M = \frac{0.20 \text{ mole}}{0.300 \text{ L}} = 0.67M \text{ solution of } H_3PO_4$$

Practice Exercise 14.10   What are the normality and the molarity of the solution formed by dissolving 7.20 g of HCl in enough water to make 450.0 mL of solution?

Example 14.11

How many grams of $Ca(OH)_2$ are there in $5\overline{0}$ mL of a 3.0N solution?

Solution

Solve the normality formula for the number of equivalents of solid $Ca(OH)_2$ dissolved in $5\overline{0}$ mL of a 3.0N solution. Then change the number of equivalents of $Ca(OH)_2$ to grams.

$$N = \frac{\text{equivalents}}{\text{liter}} \qquad \text{so} \qquad \text{equivalents} = (N)(\text{liter})$$

? equivalents of $Ca(OH)_2 = (3.0N)(0.050 \text{ L})$

$$= \left(3.0 \frac{\text{equivalents}}{\cancel{L}}\right)(0.050 \cancel{L})$$

$$= 0.15 \text{ equivalent of } Ca(OH)_2$$

$Ca(OH)_2$ has

$$\frac{37 \text{ g}}{1 \text{ equivalent}}$$

Therefore

$$? \text{ g of } Ca(OH)_2 = (0.15 \cancel{\text{equivalent}})\left(\frac{37 \text{ g}}{1 \cancel{\text{equivalent}}}\right)$$

$$= 5.6 \text{ g of } Ca(OH)_2$$

Practice Exercise 14.11   How many grams of NaOH are there in 75.0 mL of a 4.00N NaOH solution?

14.5 Dilution of Solutions

Learning Goal 7
Calculations for dilutions of
solutions

Solutions of acids and bases are usually purchased from chemical supply houses. For the sake of economy, the suppliers furnish most of these solutions in highly concentrated form. (This is like buying concentrated soups and adding water to them.) Chemists have to know how to dilute concentrated solutions to the strengths they want. If the concentration is expressed as a normality, the formula for doing this is

$$N_c V_c = N_d V_d$$

where N_c is the normality of the concentrated solution, V_c is the volume of the concentrated solution, N_d is the normality of the dilute solution, and V_d is the volume of the dilute solution. If the concentration is expressed in terms of molarity, the formula becomes

$$M_c V_c = M_d V_d$$

The volume V can be expressed in any convenient unit. But the units used, such as liters or milliliters, must be the same on both sides of the equation. This formula can be applied when solving practical laboratory problems.

Example 14.12

How would you prepare 2 L of $3M$ HCl from a $12M$ HCl solution?

Solution

You need to find the amount V_c of concentrated HCl that, when diluted with water, will give 2 L of $3M$ HCl.

$$M_c V_c = M_d V_d \qquad \text{so} \qquad V_c = \frac{M_d V_d}{M_c}$$

$$= \frac{(3\cancel{M})(2\text{ L})}{12\cancel{M}} = 0.5\text{ L}$$

Therefore, to make the final solution, measure 0.5 L of concentrated HCl into a flask and dilute with water until you have 2 L. (But see the cautionary note in Example 14.13.)

Practice Exercise 14.12   How would you prepare 3.00 L of a 4.00M HCl solution from a stock solution of 6.00M HCl?

Example 14.13

How would you prepare 300 mL of a 0.6N H_2SO_4 solution from a 36N solution?

Solution

You need the amount V_c of concentrated H_2SO_4 that, when diluted with water, will give 300 mL of 0.6N H_2SO_4.

$$N_c V_c = N_d V_d \qquad \text{so} \qquad V_c = \frac{N_d V_d}{N_c}$$

$$= \frac{(0.6\cancel{N})(300\text{ mL})}{36\cancel{N}} = 5\text{ mL}$$

Therefore, to make the final solution, measure 5 mL of concentrated H_2SO_4 into a flask and dilute with water to 300 mL. This will be a 0.6N H_2SO_4 solution. (A cautionary note: *Always add acid to water, never the reverse. In the case of H_2SO_4, adding water to the concentrated acid will make the acid react violently. In the problem we just solved, it would be wise to put most of the water in the flask before adding the H_2SO_4.*)

Practice Exercise 14.13   How would you prepare 500.0 mL of a 0.400N H_3PO_4 solution from a stock solution of 12.0N H_3PO_4?

14.6 Ionization in Solutions

Learning Goal 8
Electrolytic and
nonelectrolytic solutions

Learning Goal 9
Ions and ionization

For centuries, scientists have tried to find out what happens when one substance dissolves in another. The chemists of the 1800s were puzzled by the fact that some solutions would conduct electric current and others would not. The English physicist Michael Faraday tried to explain this phenomenon. He classified solutions generally in the following way: **Electrolytic solutions** are *solutions that conduct electric current*. **Nonelectrolytic solutions** are *solutions that do not conduct electric current*.

Faraday said that electrolytic solutions—such as sodium chloride in water—contain charged particles. He called these particles *ions* (from a Greek word that may be translated as "wanderer"). He said that ions wander through the solution carrying electric current (Figure 14.6). But questions arose about the nature of these ions. For example, why are there ions in a salt-and-water solution but no ions in a sugar-and-water solution? It was left to Arrhenius to come up with the answer.

In 1884, in his Ph.D. thesis, Svante Arrhenius advanced his ideas about ions. He suggested that Faraday's ions were really simple atoms (or groups of atoms) carrying a positive or a negative charge. He said that when some substances dissolve in solution, they break up into ions:

$$NaCl(s) \xrightarrow{\text{H}_2\text{O}} Na^{1+}(aq) + Cl^{1-}(aq)$$

Figure 14.6
An electrolytic solution conducts electricity.

This process is what we call *ionization* (Figure 14.7). To be more exact, we should say the solid sodium chloride *dissociates* when placed in water, because actually it is composed of ions. When placed in water, the ions break apart and are free to move about in the solution. Solutions that contain ions are electrolytic because they contain charged particles that can carry electric current. Substances that ionize in solution are called *electrolytes*.

On the other hand, some substances dissolve in solution and *do not ionize*. They simply break up into their neutral molecules and become surrounded by the molecules of solvent:

$$C_6H_{12}O_6(s) \xrightarrow{\text{H}_2\text{O}} C_6H_{12}O_6(aq)$$
$$\text{(glucose)}$$

● Sodium ion, Na^{1+}
● Chloride ion, Cl^{1-}
Ω Water molecule

Figure 14.7
Dissociation of NaCl in water

These solutions are nonelectrolytic because they contain no charged particles to carry electric current (Figure 14.8). Substances that do not ionize in solution are called *nonelectrolytes*.

Electrolytes and nonelectrolytes acting as solutes have some very interesting effects on the boiling points and freezing points of solutions. However, before we can examine these effects, we must discuss another way of expressing the concentration of solutes in solutions.

14.7 Molality

We can express the concentration of a solute in a solution in terms of the *number of moles of solute per kilogram of solvent*. This concentration unit is known as *molality (m)*.

Learning Goal 10
Definition of molality

$$\text{Molality } (m) = \frac{\text{number of moles of solute}}{\text{kilogram of solvent}}$$

We can determine the molality of a solution if we know either the number of moles or the mass of a solute dissolved in a given mass of solvent.

Figure 14.8
Sugar dissolves in water without ionizing.

Example 14.14

Calculate the molality of each of the following solutions.
(a) 36 g of glucose ($C_6H_{12}O_6$) dissolved in $50\overline{0}$ g of water
(b) 1.03 g of sodium bromide (NaBr) dissolved in $25\overline{0}$ g of water
(c) 3.40 g of ammonia (NH_3) dissolved in $30\overline{0}$ g of water

Solution
(a) First change the 36 g of $C_6H_{12}O_6$ (molecular weight = 180) to moles.

$$? \text{ moles} = (36 \text{ g})\left(\frac{1 \text{ mole}}{180 \text{ g}}\right) = 0.20 \text{ mole}$$

Next change $50\overline{0}$ g of water to kilograms.

$$? \text{ kg} = (50\overline{0} \text{ g})\left(\frac{1 \text{ kg}}{1{,}000 \text{ g}}\right) = 0.500 \text{ kg}$$

Finally, substitute these numbers into the molality formula.

$$m = \frac{\text{number of moles of solute}}{\text{kilogram of solvent}}$$

$$? \, m = \frac{0.20 \text{ mole of glucose}}{0.500 \text{ kg of water}} = 0.40m$$

Learning Goal 11
Calculation of molality

(b) First change 1.03 g of NaBr (molecular weight = 103) to moles.

$$? \text{ moles} = (1.03 \text{ g})\left(\frac{1 \text{ mole}}{103 \text{ g}}\right) = 0.0100 \text{ mole}$$

Next change $25\overline{0}$ g of water to kilograms.

$$? \text{ kg} = (25\overline{0} \text{ g})\left(\frac{1 \text{ kg}}{1{,}000 \text{ g}}\right) = 0.250 \text{ kg}$$

Finally, substitute these numbers into the molality formula.

$$m = \frac{\text{number of moles of solute}}{\text{kilogram of solvent}}$$

$$? \, m = \frac{0.0100 \text{ mole of NaBr}}{0.250 \text{ kg of H}_2\text{O}} = 0.0400m$$

(c) First change the 3.40 g of NH_3 (molecular weight = 17.0) to moles.

$$? \text{ moles} = (3.40 \text{ g})\left(\frac{1 \text{ mole}}{17 \text{ g}}\right) = 0.200 \text{ mole}$$

Next change the $\overline{300}$ g of water to kilograms.

$$? \text{ kg} = (3\overline{00} \text{ g})\left(\frac{1 \text{ kg}}{1,000 \text{ g}}\right) = 0.300 \text{ kg}$$

Finally, substitute these numbers into the molality formula.

$$m = \frac{\text{number of moles of solute}}{\text{kilogram of solvent}}$$

$$? \, m = \frac{0.200 \text{ mole of NH}_3}{0.300 \text{ kg of H}_2\text{O}} = 0.667m$$

Practice Exercise 14.14   Calculate the molality of each of the following solutions.
(a) 72 g of glucose ($C_6H_{12}O_6$) dissolved in $4\overline{00}$ g of water
(b) 1.17 g of NaCl dissolved in $5\overline{00}$ g of water
(c) 0.680 g of NH_3 dissolved in $25\overline{0}$ g of water

Chemists find that molality is a useful concept when they must deal with the effect of a solute on the boiling point and freezing point of a solution. Let's see what this means.

14.8 Colligative (Collective) Properties of Solutions

Certain properties of solutions depend more on the number of solute particles present than on the type of solute particles. In other words, these properties are the same regardless of what substance is the solute. *These properties, called* **colligative properties,** *depend mainly on the number of solute particles present in a given quantity of solvent.*

One important colligative property of solutions is related to the vapor pressure of the solution. When a solute is dissolved in a solvent, the vapor

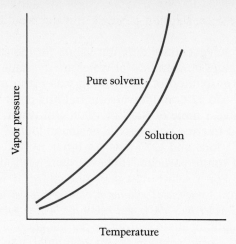

Figure 14.9
At a given temperature, the vapor pressure of the pure solvent is always greater than that of the solution of a nonvolatile solute.

pressure of the solution *decreases*. In other words, if we compared the vapor pressure of pure water at 50°C with the vapor pressure of a sugar–water solution at 50°C, we would find that the sugar–water solution had a lower vapor pressure than the pure water (Figure 14.9).

Two other colligative properties of solutions result from this vapor-pressure-reduction property:

Learning Goal 12
Effect of solute on boiling and freezing points

1. The boiling point of a solution is *higher* than the boiling point of the pure solvent alone. This is called the **boiling-point-elevation** property.

2. The freezing point of a solution is *lower* than the freezing point of the pure solvent alone. This is called the **freezing-point-depression** property.

For example, when salt is added to water, the boiling point of the water is raised, and food cooks in the water at a higher temperature. And, by adding salt to the ice that is used to freeze ice cream, we lower its freezing point. We both raise the boiling point and lower the freezing point of auto radiator water when we add antifreeze/coolant.

14.9 Boiling-Point Elevation, Freezing-Point Depression, and Molality

The amount by which the boiling point of a solution is elevated or the freezing point lowered depends on (1) the solvent and (2) the concentration of the solute particles. For water, the **boiling-point-elevation constant** K_b is 0.52°C per mole of solute particles per kilogram of water. That is, a one-molal ($1m$) solution of sugar in water raises the boiling point of the water from 100°C to 100.52°C. The **freezing-point-depression constant** K_f for water is 1.86°C per mole of solute particles per kilogram of

Learning Goal 13
Definition of colligative properties, K_f and K_b

water. So a $1m$ solution of sugar in water would have a freezing point of $-1.86°C$.

Other solvents have different values for the constants K_b and K_f.

The values of K_f and K_b for water are the same for all non-ionic solutes. Thus a $1m$ solution of glycerine in water also raises the boiling point of water to $100.52°C$ and lowers the freezing point to $-1.86°C$. (However, if we dissolve 1 mole of NaCl in 1,000 g of water, the effect on the boiling point and freezing point would be about double, because the NaCl ionizes into Na^{1+} ions and Cl^{1-} ions. In other words, 1 mole of NaCl yields 2 moles of solute particles. However, here we'll discuss only non-ionizing solutes.) For a solute that does not ionize in water, we can make this statement: *A one-molal solution of non-ionizing solute lowers the freezing point of 1 kg of water by 1.86°C and raises the boiling point by 0.52°C.*

Learning Goal 14
Calculations involving boiling
and freezing points of
solutions

Each additional mole of solute per kilogram of solvent lowers the freezing point of the solution by an additional $1.86°C$ and raises the boiling point of the solution by an additional $0.52°C$. We can state this mathematically as

$$\Delta t = mK_f \qquad \text{(for freezing-point depression)}$$

$$\Delta t = mK_b \qquad \text{(for boiling-point elevation)}$$

where Δt is the boiling-point elevation or freezing-point depression in Celsius degrees.

Example 14.15

Calculate the boiling point of each of the following solutions.
(a) 36 g of glucose dissolved in $50\overline{0}$ g of water
(b) 9.2 g of glycerol ($C_3H_8O_3$) in $25\overline{0}$ g of water

Solution

(a) First we need to determine the molality of a solution of 36 g of glucose in $50\overline{0}$ g of water. If you look back at Example 14.14(a), you'll see that we calculated this molality to be $0.40m$.

Now use the formula for boiling-point elevation to calculate the change in temperature Δt.

$$\Delta t = mK_b = (0.40\,\cancel{m})(0.52°C/\cancel{m}) = 0.21°C$$

If pure water boils at $100°C$, then this $0.40m$ solution boils at $100.21°C$.

(b) First determine the molality of a solution of 9.2 g of $C_3H_8O_3$ (molecular weight = 92) in $25\overline{0}$ g of water.

$$? \text{ moles } C_3H_8O_3 = (9.2\,\cancel{g})\left(\frac{1 \text{ mole}}{92\,\cancel{g}}\right) = 0.10 \text{ mole}$$

$$? \text{ kg water} = (25\overline{0} \text{ g}) \left(\frac{1 \text{ kg}}{1{,}000 \text{ g}} \right) = 0.250 \text{ kg}$$

$$? \text{ } m = \frac{0.10 \text{ mole}}{0.250 \text{ kg}} = 0.40m$$

Now use the formula for boiling-point elevation to calculate the change in temperature Δt.

$$\Delta t = mK_b = (0.40m)(0.52°\text{C}/m) = 0.21°\text{C}$$

If pure water boils at 100°C, then this 0.40*m* solution boils at 100.21°C.

Practice Exercise 14.15 Calculate the boiling point of each of the following solutions.
(a) 72.0 g of glucose ($C_6H_{12}O_6$) dissolved in $25\overline{0}$ g of water
(b) 1.84 g of glycerol ($C_3H_8O_3$) dissolved in $50\overline{0}$ g of water

Example 14.16

Calculate the freezing point of each of the solutions in Example 14.15.

Solution

(a) We already know that the molality of this solution is 0.40*m*. We also know that K_f for water is 1.86°C. Therefore

$$\Delta t = mK_f = (0.40m)(1.86°\text{C}/m) = 0.74°\text{C}$$

Because pure water freezes at 0°C, this 0.40*m* solution freezes at $-0.74°\text{C}$.
(b) We already know that the molality of this solution is 0.40*m*. Therefore

$$\Delta t = mK_f = (0.40m)(1.86°\text{C}/m) = 0.74°\text{C}$$

This 0.40*m* solution freezes at $-0.74°\text{C}$.

Practice Exercise 14.16   Calculate the freezing point of each of the solutions in Practice Exercise 14.15.

Summary

A solution is a homogeneous mixture composed of at least two substances. The solute is the substance that is dissolved, and the solvent is the substance in which it is dissolved. Pairs of substances that do not form solutions when mixed are said to be immiscible. In general, polar substances dissolve polar substances, and nonpolar substances dissolve nonpolar substances.

Because solutions are mixtures, they may have various ratios of solvent to solute. These ratios, or concentrations, may be measured as percentage by weight, percentage by volume, or percentage by weight–volume. In addition, the concentration of a solution may be measured as its molarity: the number of moles of solute per liter of solution. Concentration may also be measured as normality: the number of equivalents of solute per liter of solution.

An equivalent of any acid is the mass of that acid that contains Avogadro's number of hydrogen ions. An equivalent of a base is the mass of that base that contains Avogadro's number of hydroxide ions. The number of grams in one equivalent of an acid or base is computed by dividing the molecular weight of the substance by the number of replaceable hydrogen ions (for an acid) or replaceable hydroxide ions (for a base) in one molecule of the substance.

Electrolytic solutions are solutions that conduct electricity, and nonelectrolytic solutions are solutions that do not conduct electricity. Electrolytic solutions are capable of conducting electricity because they contain ions, which are charged atoms or groups of atoms. Substances that ionize, or break up into ions, when they dissolve are called electrolytes.

The molality of a solution is the number of moles of solute per kilogram of solvent. Molality is useful in dealing with colligative properties of solutions, which are properties that depend mainly on the number of solute particles present (in a given quantity of solvent), rather than on the type of solute particles. Vapor-pressure reduction, boiling-point elevation, and freezing-point depression are examples of colligative properties. In general, a solution exhibits a lower vapor pressure, a higher boiling point, and a lower freezing point than the pure solvent alone.

Career Sketch

Pharmacists

Pharmacists dispense medicines prescribed by physicians and dentists. They also supply many medicines that people can buy without prescriptions. Pharmacists must understand the composition and effect of drugs, especially because their customers often turn to them for advice about nonprescription medicines. Pharmacists also often test drugs for purity and strength. They may maintain files on patient medication and advise physicians on the selection and use of medicines. Compounding—the actual mixing of ingredients—is now only a small part of pharmacists' prac-

Pharmacists dispense prescription medication as well as carry out research to develop and test new drugs. (American Pharmaceutical Association)

tice, because most medicines today are produced by manufacturers in the form that will be used by the patient.

Many pharmacists in community drugstores also have other duties. Besides dispensing medicines, they often buy and sell nonpharmaceutical merchandise, hire and supervise personnel, and oversee the general operation of the pharmacy. Some pharmacists, however, operate prescription pharmacies that dispense only medicines, medical supplies, and health accessories.

Pharmacists in hospitals and clinics dispense prescriptions and advise the medical staff on the selection and effects of drugs. They also make sterile solutions, buy medical supplies, teach in schools of nursing and allied health professions, and perform administrative duties. An increasing number of pharmacists work as consultants in hospitals, nursing homes, and other health care facilities.

. . . .

Self-Test Exercises

Learning Goal 1: Definitions of Solution, Solute, and Solvent

1. Define the following terms and give an example of each: (*a*) solute, (*b*) solvent, (*c*) solution.

2. True or False: A solution is a heterogeneous mixture.

3. Which is the solute and which the solvent in each of the following solutions?
(a) 5%-by-weight sodium hydroxide in water
(b) 3%-by-volume ether in carbon tetrachloride
(c) 1%-by-weight–volume iodine in alcohol

5. Give an example of: (a) a solid–solid solution, (b) a solid–liquid solution, (c) a solid–gas solution, (d) a liquid–liquid solution.

4. True or False: A coin composed of two metals that were melted, mixed together, and solidified is an example of a solution.

6. Which is the solute and which the solvent in each of the following solutions?
(a) 10%-by-weight sodium chloride in water
(b) 5%-by-weight benzene in carbon tetrachloride
(c) 10%-by-weight–volume methyl alcohol in ethyl alcohol

Learning Goal 2: Substances That Form Solutions

7. True or False: "Like dissolves like." What does this statement refer to in terms of molecular structure?

9. The point at which no more solute dissolves in a solution is called the _____ point.

8. You are served a cup of coffee in a restaurant. Assuming that no sugar has been added to the coffee, you add two teaspoons of sugar, stir well, and notice, at the bottom of the cup, a layer of sugar that won't dissolve. What type of solution have you produced?

10. What are the meanings of the terms *miscible* and *immiscible*?

Learning Goal 3: Percentage Concentrations

11. What is the concentration in percentage by weight of each of the following solutions?
(a) 40 g of sodium nitrate with enough water to make 200 g of solution
(b) 3.0 g of ethyl alcohol with enough water to make 200 g of solution
(c) 30 g of potassium chloride *plus* 170 g of water

13. How many grams of KOH are there in 200 g of 10%-by-weight KOH solution?

15. What are the percentages by volume of the solutes in the following solutions?
(a) 200 mL of ether in enough carbon tetrachloride to make 1.0 L of solution
(b) 5 mL of alcohol in enough water to make 80 mL of solution

*17. Find the volume of alcohol present in 100 g of 10.0%-by-weight alcohol–water solution. (*Hint:* The density of alcohol is 0.800 g/mL.)

12. What is the concentration in percentage by weight of each of the following solutions?
(a) 80.0 g of sodium chloride with enough water to make 250 g of solution
(b) 7.50 g of methyl alcohol with enough water to make 400 g of solution
(c) 50.0 g of potassium nitrate *plus* 150 g of water

14. How many grams of NaOH are there in 500.0 g of a 15.0%-by-weight NaOH solution?

16. What is the percentage by volume of the solute in each of the following solutions?
(a) 300.0 mL of ether in enough carbon tetrachloride to make 2.00 L of solution
(b) 25.0 mL of alcohol in enough water to make 100.0 mL of solution

*18. Find the volume of alcohol present in 250.0 g of a 10.0%-by-weight alcohol–water solution. (*Hint:* The density of alcohol is 0.800 g/mL.)

19. How many milliliters of chloroform are there in $5\overline{0}0$ mL of a 4.0%-by-volume chloroform–ether solution?

20. How many milliliters of carbon tetrachloride are there in 250.0 mL of a 5.00%-by-weight–volume carbon tetrachloride–benzene solution?

21. What are the percentages by weight–volume of the solutes in the following solutions?
(a) $5\overline{0}$ g of sodium chloride in 1.5 L of solution
(b) $1\overline{0}$ g of ammonia dissolved in $2\overline{0}0$ g of solution (Assume that the density of the solution is 1.00 g/cm^3.)

22. What is the percentage by weight–volume of the solutes in the following solutions?
(a) 25.0 g of sodium nitrate in 1.00 L of solution
(b) 15.0 g of sodium hydroxide dissolved in 100.0 g of solution (Assume that the density of the solution is 1.00 g/cm^3.)

23. How many grams of sugar are there in $1\overline{0}0$ mL of a 15%-by-weight–volume sugar–water solution?

24. How many grams of lactose (milk sugar) are there in 150.0 mL of a 10%-by-weight–volume lactose–water solution?

*25. If a person's blood-sugar level falls below 60 mg per 100 mL, insulin shock can occur. What is the percentage by weight of sugar in the blood at this level? (Assume that the density of blood is 1.2 g/mL.)

*26. A person's blood-sugar level is calculated to be 120.0 mg per $10\overline{0}$ mL blood. What is the percentage by weight of sugar in the blood at this level? (The density of blood is 1.20 g/mL.)

27. What are the concentrations in percentage by weight of the following solutions?
(a) 25.0 g of NaCl in enough water to make $2\overline{0}0$ g of solution
(b) 8.00 g of sugar in enough water to make $16\overline{0}$ g of solution

28. What is the concentration in percentage by weight of each of the following solutions?
(a) 55.0 g of Na_2CO_3 in enough water to make 400.0 g of solution
(b) 14.5 g of sugar in enough water to make 290.0 g of solution

29. How many grams of KCl are there in $5\overline{0}0$ g of a 6.0%-by-weight KCl solution?

30. How many grams of KNO_3 are there in 750.0 g of a 7.50%-by-weight KNO_3 solution?

31. What are the percentages by volume of the solutes in the following solutions?
(a) $15\overline{0}$ mL of ethyl alcohol in enough water to make 3.0 L of solution
(b) $5\overline{0}0$ mL of methyl alcohol (dry gas) in enough gasoline to make 60.0 L of solution. (This represents a can of dry gas in a nearly full tank of gasoline in a standard-size automobile.)

32. What is the percentage by volume of the solute in each of the following solutions?
(a) 175 mL of isopropyl alcohol in enough water to make 4.00 L of solution
(b) 600.0 mL of methyl alcohol (dry gas) in enough gasoline to make 60.0 L of solution

33. Determine the volume of ether present in $5\overline{0}0$ g of a 4.0%-by-weight ether–carbon tetrachloride solution. (*Hint:* The density of ether is 0.714 g/cm^3.)

*34. Determine the volume of ether in 750.0 g of a 10.0%-by-weight ether–carbon tetrachloride solution. (*Hint:* The density of ether is 0.714 g/cm^3.)

35. How many milliliters of ethyl alcohol are there in 2.00 L of a 10.0%-by-volume ethyl alcohol–water solution?

36. How many milliliters of isopropyl alcohol are there in 4.50 L of a 20.0%-by-volume isopropyl alcohol–water solution?

37. What are the percentages by weight–volume of the solutes in the following solutions?
(a) 25.0 g of sodium nitrate in $8\overline{0}0$ mL of solution
(b) 0.800 g of potassium chloride in 16.0 mL of solution

38. What is the percentage by weight–volume of the solutes in the following solutions?
(a) 50.0 g of sodium carbonate in 500.0 mL of solution
(b) 0.160 g of potassium nitrate in 80.0 mL of solution

39. How many grams of sodium chloride are there in $25\overline{0}$ mL of a 5.00%-by-weight–volume solution of sodium chloride and water?

40. How many grams of sodium chloride are there in 100.0 mL of a 9.00%-by-weight–volume solution of sodium chloride and water?

Learning Goal 4: Calculations Involving Molarity

41. What is the molarity of each of the following solutions?
(a) 5.4 g of HCl in enough water to make $\overline{50}0$ mL of solution
(b) 117 g of sodium chloride in enough water to make 4 L of solution

*43. What is the molarity of a 10.0%-by-weight HCl solution? (Assume that the density of the solution is 1.00 g/mL.)

45. How many grams of KOH are obtained by evaporating to dryness $\overline{5}0$ mL of a 3.0M KOH solution?

*47. The Merck Manual reports that a blood-alcohol level of $15\overline{0}$ to $20\overline{0}$ mg per $1\overline{0}0$ mL produces intoxication and that a level of $30\overline{0}$ to $40\overline{0}$ mg per $1\overline{0}0$ mL produces unconsciousness. At a blood-alcohol level above $50\overline{0}$ mg per $1\overline{0}0$ mL, a person may die. What is the molarity of the blood with respect to alcohol at the level of $50\overline{0}$ mg per $1\overline{0}0$ mL? (*Hint:* The molecular weight of alcohol is 46.)

*49. Salt is a necessary ingredient of our diets, and the body has a system to maintain a delicate balance of salt. In certain illnesses, this salt balance can be lost, and a doctor or nurse must administer salt intravenously, using a 0.85% solution. What is the molarity of a 0.85%-by-weight–volume sodium chloride solution?

51. What is the molarity of each of the following solutions?
(a) 24.5 g of H_2SO_4 in enough water to make 1.50 L of solution
(b) 10.1 g of KNO_3 in enough water to make $\overline{50}0$ mL of solution

*53. What is the molarity of a 5.00%-by-weight magnesium chloride solution? (Assume that the density of the solution is 1.00 g/cm³.)

55. How many grams of Na_3PO_4 are obtained by evaporating to dryness $25\overline{0}$ mL of a 0.400M Na_3PO_4 solution?

42. What is the molarity of each of the following solutions?
(a) 1.18 g of HCl in enough water to make 2.00 L of solution
(b) 23.40 g of sodium chloride in enough water to make 5.00 L of solution

*44. What is the molarity of a 5.00%-by-weight HCl solution? (Assume that the density of the solution is 1.00 g/cm³.)

46. How many grams of NaOH are obtained by evaporating to dryness 25.0 mL of a 2.00M NaOH solution?

*48. A person has been drinking alcohol. What is the molarity of the alcohol in this person's blood if the alcohol concentration is $40\overline{0}$ mg per $1\overline{0}0$ mL of blood?

50. How many milliliters of water are needed to make a 2.00M NaCl solution if 5.85 g of NaCl are present?

52. What is the molarity of each of the following solutions?
(a) 49.0 g of H_2SO_4 in enough water to make 2.00 L of solution
(b) 2.02 g of KNO_3 in enough water to make 200.0 mL of solution

*54. What is the molarity of a 7.50%-by-weight sodium nitrate solution? (Assume that the density of the solution is 1.00 g/cm³.)

56. How many grams of $NaC_2H_3O_2$ are obtained by evaporating to dryness 400.0 mL of a 0.500M $NaC_2H_3O_2$ solution?

Learning Goal 5: Equivalents of Acids and Bases

57. Determine the number of equivalents in each of the following samples: (a) 4.9 g of H_2SO_4, (b) 7.2 g of HCl, (c) 32.1 g of $Fe(OH)_3$.

58. Determine the number of equivalents in each of the following samples: (a) 9.8 g of H_2SO_4, (b) 3.6 g of HCl, (c) 6.42 g of $Fe(OH)_3$.

59. For each of the following acids and bases, determine the number of grams per mole and the number of grams per equivalent.
(*a*) H_2SO_3 (*b*) $Fe(OH)_3$ (*c*) $Fe(OH)_2$ (*d*) H_3BO_3

61. Determine the number of equivalents in:
(*a*) 1.23 g of H_2SO_3,   (*b*) 26.7 g of $Fe(OH)_3$.

63. Determine the number of grams in: (*a*) 25.0 equivalents of $Fe(OH)_2$,   (*b*) 0.450 equivalent of H_3BO_3.

60. For each of the following acids and bases, determine the number of grams per mole and the number of grams per equivalent: (*a*) $HC_2H_3O_2$,   (*b*) NaOH, (*c*) $Mg(OH)_2$,   (*d*) HNO_3.

62. Determine the number of equivalents in each of the following: (*a*) 246 g of H_2SO_3,   (*b*) 4.00 g of NaOH.

64. Determine the number of grams in each of the following: (*a*) 0.250 equivalents of $Mg(OH)_2$, (*b*) 2.00 equivalents of $HC_2H_3O_2$.

Learning Goal 6: Calculations Involving Normality

65. Determine the normality of each of the following solutions.
(*a*) $97\overline{0}$ g of H_3PO_4 in enough water to make 10.0 L of solution
(*b*) 3.7 g of $Ca(OH)_2$ in enough water to make 800 mL of solution
(*c*) Calculate the molarities of the solutions in (*a*) and (*b*).

67. How many grams of $Ca(OH)_2$ are there in $80\overline{0}$ mL of a 2.0*N* $Ca(OH)_2$ solution?

69. How many milliliters of a 0.60*N* H_3PO_4 solution do you need to get 4.9 g of H_3PO_4?

71. Determine the normality of each of the following solutions.
(*a*) 11.23 g of $Fe(OH)_2$ in enough water to make $25\overline{0}$ mL of solution
(*b*) 4.12 g of H_3BO_3 in enough water to make $40\overline{0}$ mL of solution
(*c*) Calculate the molarities of the solutions in (*a*) and (*b*).

73. How many grams of $Al(OH)_3$ are there in $25\overline{0}$ mL of a 0.500*N* $Al(OH)_3$ solution?

75. How many milliliters of 0.100*N* H_2SO_4 solution do you need to obtain 2.94 g of H_2SO_4?

66. Determine the normality of each of the following solutions.
(*a*) 465 g of H_3PO_4 in enough water to make 8.00 L of solution
(*b*) 740.0 g of $Ca(OH)_2$ in enough water to make 740.0 mL of solution
(*c*) Calculate the molarities of the solutions in (*a*) and (*b*).

68. How many grams of $Mg(OH)_2$ are there in $75\overline{0}$ mL of a 3.00*N* $Mg(OH)_2$ solution?

70. How many milliliters of a 0.450*N* H_3PO_4 solution do you need to get 98.0 g of H_3PO_4?

72. Determine the normality of each of the following solutions.
(*a*) 2.246 g of $Fe(OH)_2$ in enough water to make 500.0 mL of solution
(*b*) 16.48 g of H_3BO_3 in enough water to make 500.0 mL of solution
(*c*) Calculate the molarities of the solutions in (*a*) and (*b*).

74. How many grams of $Fe(OH)_3$ are there in 500.0 mL of a 0.250*N* $Fe(OH)_3$ solution?

76. How many milliliters of a 0.200*N* H_3PO_4 solution do you need to obtain 29.40 g of H_3PO_4?

Learning Goal 7: Calculations for Dilutions of Solutions

77. How would you make $50\overline{0}$ mL of 4.0*N* H_2SO_4 from a 16*N* solution?

79. How would you make 3 L of a 0.1*N* NaOH solution from a 6*N* solution?

78. How would you make 250.0 mL of 8.00*N* H_3PO_4 from a 10.0*N* solution?

80. How would you make 4.00 L of 0.200*N* $Ca(OH)_2$ from a 2.00*N* solution?

81. How would you prepare a $25\overline{0}$ mL of a 5.00M HCl solution from a 12.0M solution?

82. How would you prepare 450.0 mL of a 7.50M HNO$_3$ solution from an 11.0M solution?

83. How would you prepare 5.00 L of 0.0500N H$_2$SO$_4$ from a 3.00M solution?

84. How would you prepare 4.00 L of a 0.0200N HC$_2$H$_3$O$_2$ solution from a 2.00N solution?

Learning Goals 8 and 9: Electrolytic and Nonelectrolytic Solutions, Ions and Ionization

85. Define the terms *ion* and *ionization*, and give examples of each.

86. True or False: (*a*) Ions are groups of charged atoms. (*b*) Charged elements are not ions.

87. Explain why some solutions conduct electric current and others do not.

88. Solution A conducts electricity but solution B does not. If solutions A and B are mixed and no chemical reaction occurs, will the resulting solution conduct electricity?

89. Define and give an example of an electrolytic solution and of a nonelectrolytic solution.

90. Which will conduct electricity, a glass of salt water or a cup of tea with sugar?

Learning Goal 10: Definition of Molality

91. Define *molality*. What information is necessary to calculate the molality of a solution?

92. How is *molality* different from *molarity*?

Learning Goal 11: Calculation of Molality

*93. Calculate the molality of the intravenous salt solution in Self-Test Exercise 49. (Assume that there is 0.85 g of NaCl per $10\overline{0}$ g of water.)

94. A solution contains 0.373 g of KCl per $10\overline{0}$ g of water. What is its molality?

95. Determine the molality of each of the following solutions.
(*a*) 6.4 g of methanol (CH$_3$OH) in $25\overline{0}$ g of water,
(*b*) 90.0 g of glucose (C$_6$H$_{12}$O$_6$) in 1,50$\overline{0}$ g of water

96. Determine the molality of each of the following solutions.
(*a*) 0.128 g of CH$_3$OH in $50\overline{0}$ g of water
(*b*) $45\overline{0}$ g of C$_6$H$_{12}$O$_6$ in 2,00$\overline{0}$ g of water

Learning Goal 12: Effect of Solute on Boiling and Freezing Points

97. Explain how a solute affects the boiling point and the freezing point of a solution.

98. What would happen to the engine of an automobile if antifreeze were not added to the water in the radiator during a cold winter?

Learning Goal 13: Definition of Colligative Properties, K_f and K_b

99. Define the following terms, and give an example of each: (*a*) colligative property, (*b*) K_f, (*c*) K_b.

100. What are the values of K_f and K_b for water? What do they mean?

Learning Goal 14: Calculations Involving Boiling and Freezing Points of Solutions

101. You can winterize a typical car radiator system by adding 1.00 gal of antifreeze to 1.00 gal of water. At what temperature does this water–antifreeze system begin to freeze? (Assume that the antifreeze is pure ethylene glycol, $C_2H_6O_2$, which has a density of 1.10 g/cm³.) [*Hint:* You must change the 1.00 gal of ethylene glycol solute to grams and then to moles. Also, the 1.00 gal of water must be changed to kilograms.]

102. Calculate the boiling point of a solution prepared by dissolving 5.40 g of glucose in $50\overline{0}$ g of water.

103. Calculate the boiling point of the solution in Self-Test Exercise 101.

104. Calculate the boiling point of a solution prepared by dissolving 0.184 g of glycerol ($C_3H_8O_3$) in $1,00\overline{0}$ g of water.

105. Calculate the boiling point and the freezing point of a solution prepared by dissolving 27.0 g of glucose, $C_6H_{12}O_6$, in $10\overline{0}$ g of water.

106. Calculate the freezing point of the solution in Self-Test Exercise 104.

Extra Exercises

*107. Use the kinetic theory to explain how and why a solute affects the boiling and freezing points of a solution.

108. Calculate the molarity of a solution of HCl that has a hydrogen-ion concentration of $10^{-2}M$.

109. Calculate the molarity and the normality of a solution of sulfuric acid that has a $[H^{1+}]$ of $10^{-2}M$.

110. How many grams of H_2SO_4 are necessary to make 10.0 L of a 5.00%-by-weight sulfuric acid solution? (Assume that the density of the solution is 1.00 g/mL.)

111. Find the molarity and the normality of the solution in Extra Exercise 110.

112. How does a solution differ from a mixture?

113. In Chicago salt is used on roads, at times, to keep the roads free of ice. In northern Minnesota, however, people use sand to avoid slipping on ice. Why don't the Minnesota people try to use salt to get rid of their ice?

114. Which of the following would lower the freezing point of $1,00\overline{0}$ g of water by the greatest amount? Explain your choice.
(a) $10\overline{0}$ g of glucose ($C_6H_{12}O_6$)
(b) $10\overline{0}$ g of sucrose ($C_{12}H_{22}O_{11}$)
(c) $10\overline{0}$ g of ethanol (C_2H_6O)

*115. Concentrated hydrochloric acid is delivered as 37.00% HCl by weight. The density of this solution is 1.19 g/mL. What volume of this acid is needed to produce 1.00 L of a 0.500M HCl solution?

116. You dissolve a 15.0-g sample of an unknown substance (nonelectrolyte) in $10\overline{0}$ g of water. The solution freezes at $-1.86°C$. What is the molecular weight of the substance?

117. A chemist prepares a solution by taking 1.00 L of water and adding 2.00 moles of solute. Is this solution a 2.00M solution or a 2.00m solution? Explain.

*118. You add a 10.0-g sample of a substance to $1,00\overline{0}$ g of water. You find that the freezing point of the solution is $-5.00°C$. What is the molecular weight of the substance?

119. You add 5.85 g of NaCl to $1,00\overline{0}$ g of water. What is the approximate boiling point of the solution?

120. Determine the number of equivalents in: (a) 72.0 g of HCl, (b) 98.0 g of H_2SO_4.

121. Determine the number of equivalents in: (a) 1.00 mole of HCl, (b) 1.00 mole of H_2SO_4.

15 Acids, Bases, and Salts

Learning Goals

After you have studied this chapter, you should be able to:

1. Define the terms *acid* and *base* in the way Arrhenius did.

2. Define the terms *acid* and *base* according to the Brønsted–Lowry concept.

3. List the important characteristics of acids and bases.

4. Write equations for the preparation of acids and bases.

5. Write equations for important chemical reactions of acids and bases.

6. Explain what is meant by strong acid and strong base, and give some examples of each.

7. Explain what is meant by weak acid and weak base, and give some examples of each.

8. Explain what is meant by dynamic equilibrium.

9. Write an equation for the ionization of water.

10. Define pH and explain the different values on the pH scale.

11. Determine the pH of a solution, given its hydrogen-ion or hydroxide-ion concentration in moles per liter.

12. State the relationship between the hydrogen-ion concentration and the hydroxide-ion concentration in aqueous solutions.

13. Explain what the term *salt* means, and give some examples.

14. Cite the most important property of salts.

15. List methods and equations for the preparation of various salts.

16. Explain what is meant by titration.

Introduction

Three important groups of compounds—acids, bases, and salts—are the subject of this chapter. You are probably familiar with many compounds found in each of these groups. Acetic acid is the substance that gives vinegar its characteristic taste and odor, and citric acid is responsible for the tartness of lemons and grapefruits. Many popular household cleansers contain the base ammonia. Bases are the substances that make soaps and cleansers feel slippery. In this chapter we also discuss titration, a process that enables the chemist to determine the strengths of acids and bases very accurately. And finally we examine salts, which are widely used in both home and industry.

15.1 Acids

Three different definitions of acids and bases have been advanced in the last 100 years, and each has its place in explaining observed behavior. Two of these definitions will serve as the basis for our discussion of the properties, reactions, and uses of acids.

The Arrhenius Definition

In 1884 a Swedish scientist named Svante Arrhenius described an acid as a substance that releases hydrogen ions (H^{1+}) when it is dissolved in water.

Let's look at how an acid dissolves in water. In Chapter 7 we saw that water is a polar molecule wherein the oxygen end is negatively charged and the hydrogen end is positively charged. When we bubble hydrogen chloride gas (HCl) into a beaker of water, the gas seems to dissolve in the water, but actually much more is happening. The HCl gas is reacting with the water (Figure 15.1). The covalently bonded hydrogen chloride molecules are being broken apart (ionized) by the water into hydrogen ions (H^{1+}) and chloride ions (Cl^{1-}), and the hydrogen ions are adding on to the negative end of the water molecules, forming a new type of ion. This new ion, $(H_3O)^{1+}$, is called a **hydronium ion.** The reaction is

$$HCl + H_2O \longrightarrow (H_3O)^{1+} + Cl^{1-}$$

Figure 15.1

The positive end of the HCl molecule is attracted by the negative end of the water molecule. The attraction is strong enough to break the bond between the H and the Cl, forming a hydronium ion, $(H_3O)^{1+}$, and a chloride ion, Cl^{1-}.

Table 15.1
Some Common Arrhenius Acids

Name of acid	Formula
Hydrochloric acid	HCl
Nitric acid	HNO_3
Sulfuric acid	H_2SO_4
Phosphoric acid	H_3PO_4
Acetic acid	$HC_2H_3O_2$

Similar reactions occur when the other acids listed in Table 15.1 are dissolved in water. For example,

$$H_2SO_4 + 2H_2O \longrightarrow 2(H_3O)^{1+} + (SO_4)^{2-}$$

Learning Goal 1
Arrhenius acids

According to Arrhenius, HCl is an acid because it undergoes the following reaction when placed in water:

$$HCl + H_2O \longrightarrow (H_3O)^{1+} + Cl^{1-} \qquad \text{or, in a simpler form,}$$

$$HCl \xrightarrow{\text{Water}} H^{1+} + Cl^{1-}$$

Table 15.1 lists some additional Arrhenius acids. When we confine ourselves to the Arrhenius definition of an acid, then, our discussion is limited to aqueous solution and to substances that liberate hydrogen ions in solution. Because of these limitations, other definitions for the term *acid* were proposed. However, the Arrhenius definition is adequate for most of our purposes.

The Brønsted–Lowry Definition

In 1923 two chemists, J. N. Brønsted and T. M. Lowry, working independently, proposed a much broader definition of the term *acid*. They described an acid as any substance that is a proton donor. According to this definition, an acid is any substance that can donate a proton (an H^{1+} ion) to another substance. This definition includes all substances that qualify as acids under the Arrhenius definition, and many other compounds and ions as well (Table 15.2). For example, the ammonium ion is a Brønsted–Lowry acid because

$$(NH_4)^{1+} + H_2O \longrightarrow NH_3 + (H_3O)^{1+} \qquad \text{or, more simply,}$$

Learning Goal 2
Brønsted–Lowry acids

$$\underset{\text{Acid}}{(NH_4)^{1+}} \xrightarrow{\text{Water}} NH_3 + H^{1+}$$

Table 15.2
Some Common Brønsted–Lowry Acids

Name of acid	Formula
Hydrochloric acid	HCl
Nitric acid	HNO_3
Sulfuric acid	H_2SO_4
Phosphoric acid	H_3PO_4
Acetic acid	$HC_2H_3O_2$
Hydronium ion	$(H_3O)^{1+}$
Hydrogen sulfate ion	$(HSO_4)^{1-}$
Ammonium ion	$(NH_4)^{1+}$
Water	H_2O

Properties of Acids

Learning Goal 3
Characteristics of acids

The acids listed in Table 15.1 have many characteristics in common.

1. *Each has a sour taste.* (The word *acid* comes from the Latin *acidus*, which means "sour.")

2. *Each can change the color of certain dyes.* For example, blue litmus dye or blue litmus paper turns red in the presence of an acid. These dyes are called **indicators** because they indicate whether a substance is an acid or a base (Figure 15.2).

3. *Each can react with certain active metals, such as zinc and magnesium, to produce hydrogen gas.*

$$Zn + 2HCl \longrightarrow ZnCl_2 + H_2(g)$$

$$Zn + H_2SO_4 \longrightarrow ZnSO_4 + H_2(g)$$

4. *Each can neutralize bases.* (We will discuss this property later.)

But the most important characteristic of acids is the one we used in the definition: *They dissolve in water to produce hydrogen ions* (H^{1+}). These ions are responsible for the other properties of acids.

Figure 15.2
The reaction of an acid–base indicator. Hydrochloric acid turns blue litmus red.

Preparation of Acids

There are many different ways of preparing acids. One of the most important ways involves the reaction of a nonmetal oxide with water.

$$SO_3 \; + H_2O \longrightarrow \; H_2SO_4$$

Sulfur   Sulfuric
trioxide   acid

$$N_2O_5 \; + H_2O \longrightarrow \; 2HNO_3$$

Nitrogen(V)   Nitric
oxide   acid

The acids that result from some of these reactions are called *mineral* acids (because they are prepared from minerals). Other acids, called *organic* acids, are obtained from living things such as citrus fruits or fermented grapes or are found in animal matter.

Reactions of Acids

Acids undergo many types of reactions, two of which are important to beginning chemistry students:

1. The reactions of acids with metals

2. The reactions of acids with bases

We have already discussed the reaction of an acid with a metal. You learned in Chapter 8 that this type of reaction is a *single-replacement reaction* involving the replacement of hydrogen atoms in an acid by atoms of a metal (which must lie above hydrogen in the activity series, Table 8.1). For example,

$$Mg + H_2SO_4 \longrightarrow \; MgSO_4 + H_2$$

$$Mg + 2HBr \longrightarrow \; MgBr_2 + H_2$$

Hydrobromic
acid

We have also already discussed the reaction of an acid with a base (in Section 8.4). We'll come back to this type of reaction, sometimes called a neutralization reaction, later in this chapter when we cover the reactions of bases.

Uses of Some Common Acids

Let us now survey some of the modern uses of acids.

Hydrochloric acid, a strong mineral acid known commercially as muriatic acid, is used industrially to clean metals that are about to be coated or plated. (This cleaning is known as *pickling.*) Hydrochloric acid is also used to clean brick and cement.

Nitric acid, another mineral acid, is highly reactive. When a fair-skinned person happens to spill some on his or her skin, it turns the skin yellow. This is because of a reaction between the nitric acid and the protein in skin. Nitric acid is used mainly in the production of fertilizers but is also used to make dyes, plastics, and explosives.

Sulfuric acid, also a very strong mineral acid, is another one that is used in fertilizers. In addition, it is used to make smokeless powder (a type of gunpowder) and nitroglycerine. Automobile batteries contain a diluted form of sulfuric acid. Such batteries work because there is a reaction between their metal plates and the acid, creating a movement of electrons and therefore an electric current. The Russians are manufacturing sulfuric acid from the oxides of sulfur that are created when fossil fuels are burned to make electricity. In other words, sulfuric acid can be made as a by-product of electric power plants.

Phosphoric acid is a weak mineral acid. Its uses range from flavoring soft drinks to producing superphosphates for fertilizers. And the false teeth of millions of people are held in place by dental cements containing phosphoric acid.

Acetic acid is found in vinegar. According to the pure-food laws of the United States, vinegar must contain no less than 4% acetic acid. Concentrated acetic acid is used in making synthetic fibers called *acetates*. It is also used in manufacturing various plastics. (Remember, though, that acetic acid—unlike the other acids we have just discussed—is an organic acid.)

15.2 Bases

As in the case of acids, three different definitions of bases have been advanced in the last 100 years. Again, we will look at the two most useful definitions—those of Arrhenius and of Brønsted and Lowry.

The Arrhenius Definition

Arrhenius's definition of an acid applies only to aqueous solutions, and the same is true of his definition of a base. A base is a substance that releases hydroxide ions, $(OH)^{1-}$, when it is dissolved in water. According to Arrhenius, NaOH is a base because it undergoes the following reaction when placed in water:

. . . .
Learning Goal 1
Arrhenius bases

$$NaOH \xrightarrow{\text{Water}} Na^{1+} + (OH)^{1-}$$

Table 15.3 lists some additional Arrhenius bases.

Incidentally, there is another word that people often use for a base; it is **alkali.** An alkali is any substance that dissolves to yield a basic solution.

Table 15.3
Some Common Arrhenius Bases

Name of base	Formula
Lithium hydroxide	LiOH
Sodium hydroxide	NaOH
Potassium hydroxide	KOH
Calcium hydroxide	$Ca(OH)_2$
Magnesium hydroxide	$Mg(OH)_2$
Iron(III) hydroxide	$Fe(OH)_3$

Many of the common bases found in the laboratory are composed of metallic ions plus hydroxide ions—for example, sodium hydroxide (NaOH).

The Brønsted–Lowry Definition

Just as Brønsted and Lowry defined an acid as a proton donor, they defined a base as a proton acceptor. In other words, according to Brønsted and Lowry, a base is any substance that accepts protons (H^{1+} ions). This definition includes all substances that qualify as bases under the Arrhenius definition, and many other compounds and ions as well (Table 15.4). For example, the compound ammonia is a Brønsted–Lowry base because

$$NH_3 + H^{1+} \longrightarrow (NH_4)^{1+}$$

Learning Goal 2
Brønsted–Lowry bases

Properties of Bases

Learning Goal 3
Characteristics of bases

The bases listed in Table 15.3 have many characteristics in common:

1. *Each has a bitter taste.*

Table 15.4
Some Common Brønsted–Lowry Bases

Name of base	Formula	Name of base	Formula
Ammonia	NH_3	Chloride ion	Cl^{1-}
Lithium hydroxide	LiOH	Carbonate ion	$(CO_3)^{2-}$
Sodium hydroxide	NaOH	Sulfate ion	$(SO_4)^{2-}$
Potassium hydroxide	KOH	Acetate ion	$(C_2H_3O_2)^{1-}$
Calcium hydroxide	$Ca(OH)_2$	Hydroxide ion	$(OH)^{1-}$
Magnesium hydroxide	$Mg(OH)_2$	Water	H_2O

Figure 15.3

The reaction of an acid–base indicator. Sodium hydroxide turns red litmus blue.

Sodium hydroxide

Red litmus dye

Blue

2. *Each can change the color of certain dyes.* For example, red litmus turns blue in the presence of a base (Figure 15.3).

3. *Each feels slippery or soapy on the fingers.*

4. *Each can neutralize acids.*

But the most important characteristic of bases is the one we used in the definition: *They dissolve in water to produce hydroxide ions,* $(OH)^{1-}$. These hydroxide ions are responsible for the other properties of bases.

How do bases dissolve in water to produce hydroxide ions? Consider, for example, solid sodium hydroxide (NaOH). Sodium hydroxide is an ionic compound made up of sodium ions and hydroxide ions, or Na^{1+} and $(OH)^{1-}$. The ions are held together by the attraction of their opposite charges. When we add sodium hydroxide to water, the negative end of each water molecule (the oxygen end) attracts the sodium ion, and the positive end (the hydrogen end) attracts the hydroxide ion. This attraction breaks the bonds between the sodium ions and the hydroxide ions (Figure 15.4). The result is that there are now free sodium ions and free hydroxide ions in solution. The reaction can be represented by the equation

Figure 15.4

The dissociation of NaOH

$$NaOH \xrightarrow{\text{Water}} Na^{1+} + (OH)^{1-}$$

Preparation of Bases

Learning Goal 4
Preparation of bases

Bases can be prepared in many ways. One major method involves the reaction of a metal oxide with water.

$$Na_2O + H_2O \longrightarrow 2NaOH$$
Sodium
oxide

$$CaO + H_2O \longrightarrow Ca(OH)_2$$
Calcium
oxide

Reactions of Bases

Bases undergo many types of reactions. Two that are important to the beginning chemistry student are

1. The reactions of bases with nonmetal oxides
2. The reactions of bases with acids

· · · · · · · · · · · · · ·
Learning Goal 5
Chemical reactions of bases

When a base reacts with a nonmetal oxide, it produces a salt and water:

$$2NaOH + CO_2 \longrightarrow Na_2CO_3 + H_2O$$

$$Ca(OH)_2 + CO_2 \longrightarrow CaCO_3 + H_2O$$

When a base reacts with an acid, it again produces a salt and water:

$$NaOH + HCl \longrightarrow NaCl + H_2O$$

$$2NaOH + H_2SO_4 \longrightarrow Na_2SO_4 + 2H_2O$$

$$3Ca(OH)_2 + 2H_3PO_4 \longrightarrow Ca_3(PO_4)_2 + 6H_2O$$

These latter reactions are sometimes called *neutralization reactions* because, when we slowly add an acid to a base, the H^{1+} and $(OH)^{1-}$ ions neutralize each other by forming water.

Uses of Some Common Bases

Like acids, bases play an important role in our daily lives, and they are just as dangerous to our skin and clothes. They are used in our homes, in industrial processes, and in manufacturing drugs and medicines.

For example, *sodium hydroxide* (NaOH) is a base used in drain cleaners, or in what we call lye. It is also used to make soaps and cellophane. The petroleum industry uses sodium hydroxide to neutralize acids in the refining process.

Calcium hydroxide, $Ca(OH)_2$, sometimes called slaked lime, is important in the construction industry. It is used in mortar, plaster, and cement, and in many other building and paving materials.

Ammonia (NH_3) dissolved in water is the household cleaner we know as ammonia water, and it has an intense, almost suffocating odor. Ammonia is the chief ingredient of smelling salts. It is excellent for removing stains, and it is also used in manufacturing textiles (such as rayon), plastics, and fertilizers.

Magnesium hydroxide, $Mg(OH)_2$, dispersed in water, commonly called milk of magnesia, has many uses as a medicine. In small doses (a few hundred milligrams), magnesium hydroxide acts as an antacid (bases counteract acid just as reliably in the stomach as in a laboratory). In large doses (two to four grams), it acts as a laxative.

Table 15.5
Strong Acids and Bases

	Name	Formula
Acids	Hydrochloric acid	HCl
	Sulfuric acid	H_2SO_4
	Nitric acid	HNO_3
Bases	Sodium hydroxide	NaOH
	Potassium hydroxide	KOH

15.3 The Strengths of Acids and Bases

Learning Goal 6
Strong acids and bases

Acids and bases can both be subdivided into two categories: (1) *strong* acids and *strong* bases and (2) *weak* acids and *weak* bases. The strong acids and bases are those that essentially ionize completely. For example, hydrochloric acid is a strong acid. One mole of hydrogen chloride gas dissolved in 1 L of water contains essentially 1 mole of hydrogen ions and 1 mole of chloride ions. There are no hydrogen chloride molecules to speak of. Table 15.5 gives some examples of strong acids and bases. Weak acids and bases are those that do not ionize completely. Acetic acid is a weak acid. If you were to analyze a solution containing 1 mole of acetic acid ($HC_2H_3O_2$) in 1 L of solution, you would find that most of the acetic acid stays in the form of molecules. Only about 0.4% ionizes to form hydrogen ions and acetate ions. We can represent the dissociation of this weak acid in the following manner:

Learning Goal 7
Weak acids and bases

$$HC_2H_3O_2 \; \underset{\text{Water}}{\rightleftharpoons} \; H^{1+} + (C_2H_3O_2)^{1-}$$

The *double arrow* indicates that, in a given solution of acetic acid, an *equilibrium* (or balance) exists between the acetic acid molecules, the hydrogen ions, and the acetate ions. This equilibrium lies in the direction of (provides more of) the undissociated $HC_2H_3O_2$. That's why the arrow that points to the left is longer than the arrow that points to the right. (If the situation were the other way around, the arrow pointing to the right would be longer.) In weak acids and weak bases, the equilibrium is always in the direction of the undissociated species. Table 15.6 offers some examples of weak acids and bases. (Also see Figure 15.5.)

Figure 15.5
One difference between a weak acid and a strong acid

No reaction visible

Vigorous reaction
$Zn + H_2SO_4 \rightarrow ZnSO_4 + H_2(g)$

Table 15.6
Weak Acids and Bases

	Name	Formula
Acids	Acetic acid	$HC_2H_3O_2$
	Carbonic acid	H_2CO_3
	Boric acid	H_3BO_3
Bases	Ammonia (in water)	NH_3
	Zinc hydroxide	$Zn(OH)_2$

Learning Goal 8
Dynamic equilibrium

15.4 Weak Acids and Bases and Dynamic Equilibrium

The equilibrium that exists in a solution of a weak acid or weak base is not a static equilibrium but rather a *dynamic* equilibrium. That is, some molecules are always dissociating to form ions, and some ions are always recombining to form molecules. But these dissociations and recombinations take place at the same rate, so the proportion of molecules and ions remains constant—they remain in equilibrium.

As an analogy, consider a popular restaurant during the busy lunch hour. The restaurant can seat 200 people, and it is full at noon. But some people are leaving the restaurant and others are entering. A quarter of an hour later, the situation is the same: two hundred people in the restaurant, some people entering, and others leaving. And again, 15 minutes later, the same situation prevails. During the lunch hour, those who enter the restaurant take the places of those who have eaten and left, maintaining a population equilibrium of 200 diners. But the equilibrium is dynamic (changing), because individual diners are constantly being replaced by other diners.

A chemical equilibrium is dynamic in the same sense. For example, in a solution of acetic acid, there are some hydrogen ions (H^{1+}), some acetate ions ($(C_2H_3O_2)^{1-}$), and many acetic acid molecules ($HC_2H_3O_2$). But at any given time, some hydrogen ions are combining with acetate ions to form acetic acid molecules. And some acetic acid molecules are ionizing to form hydrogen ions and acetate ions. As long as these two processes go on at the same rate, the concentrations of the various species don't change, and a dynamic equilibrium is maintained.

Chemists use various methods to measure the extent of dissociation of acids and bases. There are also quantitative ways of describing the extent of dissociation.

15.5 The Ionization of Water

Learning Goal 9
Ionization of water

Like weak acids and bases, water has a tendency to ionize. One way to represent the ionization of water is as follows:

$$H\text{---}OH + H_2O \; \rightleftharpoons \; (H_3O)^{1+} \;+\; (OH)^{1-} \quad \textit{(Equation 15.1)}$$

Hydronium   Hydroxide
ion   ion

Here the formula for the first water molecule is written as H—OH so that we can see how the ionization occurs. Equation 15.1 shows that a hydronium ion and a hydroxide ion are formed.

We can also represent the ionization of water in this simpler way:

$$H_2O \; \rightleftharpoons \; H^{1+} + (OH)^{1-} \quad \textit{(Equation 15.2)}$$

This equation shows that a hydrogen ion and a hydroxide ion are formed. Even though Equation 15.1 is more accurate (technically), Equation 15.2 is easier to use and gets the point across. However, as the double arrow indicates, the ionization of water occurs only to a slight extent. In fact, in exactly 1 L of water, which is 55.6 moles of water, only 0.0000001 mole (which can be written as 10^{-7} mole) actually dissociates into hydrogen and hydroxide ions. In other words, there are 10^{-7} mole of hydrogen ions and 10^{-7} mole of hydroxide ions present in 1 L of water. Because there are equal amounts of hydrogen ions and hydroxide ions in pure water, it is said to be neutral.

15.6 The pH Scale

Learning Goal 10
pH and pH scale

The pH scale is used to measure and express the acidity or basicity (acid or base strength) of a solution. The **pH** of a solution is *equal to the negative of the logarithm of its hydrogen-ion concentration in moles per liter.* This equation is written as

$$pH = -\log [H^{1+}]$$

where the brackets, [ ], mean *concentration in moles per liter.* The pH scale runs from 0 to 14 (Figure 15.6).

Recall, from our discussion of the ionization of water, that the concentration of hydrogen ions in pure water is 10^{-7} mole per liter. It happens that this concentration corresponds to a pH of 7. Moreover, we noted that pure water is neutral (neither acidic nor basic). Therefore any solution with a pH of 7 is considered neutral.

Figure 15.6
The pH scale

Table 15.7
The pH Values of Some Common Substances

Substance	pH	Substance	pH
1M HCl	0.1	Rainwater	6.2
0.5M H$_2$SO$_4$	0.3	Milk	6.5
Lemons	2.3	Pure water	7.0
Vinegar	2.8	Sea water	8.5
Soft drinks	3.0	Milk of magnesia	10.0
Oranges	3.5	0.1M NH$_4$OH	11.1
Tomatoes	4.2	1M NaOH	14.0

A solution that has a pH between 0 and 2 is considered very acidic. A solution that has a pH between 2 and 5 is moderately acidic, and one with a pH between 5 and 7 is slightly acidic. A solution with a pH of 7 is, as we saw, neutral. A solution that has a pH between 7 and 9 is slightly basic, one with a pH between 9 and 11 is moderately basic, and one with a pH between 11 and 14 is very basic. Table 15.7 lists the pH values of some common substances.

We can determine the pH of a solution by many methods. One simple method is to place a few drops of a chemical dye or a piece of paper tape that contains the dye in the solution. The dye or paper turns a color that depends on the pH of the solution. There is also available a device called a pH meter, which directly reads the pH of a solution.

Learning Goal 11
pH, hydrogen-ion, and hydroxide-ion concentrations in solutions

It is easy to find the pH of a solution when its hydrogen-ion concentration (in moles per liter) is given as a power of 10. For example, for pure water, with a hydrogen-ion concentration of 10^{-7} mole/liter,

$$pH = -\log [H^{1+}] = -\log 10^{-7} = -(-7) = 7$$

Note that the logarithm of a power of 10 is simply that power of 10. For example, $\log 10^{-2} = -2$.

Example 15.1

Determine the pH of each of the following solutions. Then use Figure 15.6 to determine how acidic or basic the solution is.
(a) A solution whose [H^{1+}] is 10^{-3}
(b) A solution whose [H^{1+}] is 10^{-13}
(c) A solution whose [H^{1+}] is 10^{0}

Solution
pH = $-\log$ [H^{1+}]. Therefore,

(a) pH = $-\log 10^{-3}$ = 3, which is moderately acidic.
(b) pH = $-\log 10^{-13}$ = 13, which is very basic.
(c) Since 10^0 = 1, pH = $-\log 10^0$ = 0, which is very acidic.

Practice Exercise 15.1 Determine the pH of each of the following solutions. Then use Figure 15.6 to determine how acidic or basic the solution is.
(a) A solution whose $[H^{1+}]$ is 10^{-5}
(b) A solution whose $[H^{1+}]$ is 10^{-10}
(c) A solution whose $[H^{1+}]$ is 10^{-1}

15.7 Relationship Between the Hydrogen-Ion and Hydroxide-Ion Concentrations

Learning Goal 12
Relationship between hydrogen-ion and hydroxide-ion concentrations

There is an interesting relationship between the hydrogen-ion concentration and the hydroxide-ion concentration in aqueous (water) solutions: The product of $[H^{1+}]$ and $[OH^{1-}]$ is always equal to 10^{-14}. In other words,

$$[H^{1+}][OH^{1-}] = 10^{-14}$$

Table 15.8
The Relationship Between Hydrogen-Ion Concentration and Hydroxide-Ion Concentration

$[H^{1+}]$	*pH*	$[OH^{1-}]$
$10^0 = 1$	0	10^{-14}
$10^{-1} = 0.1$	1	10^{-13}
$10^{-2} = 0.01$	2	10^{-12}
$10^{-3} = 0.001$	3	10^{-11}
$10^{-4} = 0.0001$	4	10^{-10}
$10^{-5} = 0.00001$	5	10^{-9}
$10^{-6} = 0.000001$	6	10^{-8}
$10^{-7} = 0.0000001$	7	10^{-7}
$10^{-8} = 0.00000001$	8	10^{-6}
$10^{-9} = 0.000000001$	9	10^{-5}
$10^{-10} = 0.0000000001$	10	10^{-4}
$10^{-11} = 0.00000000001$	11	10^{-3}
$10^{-12} = 0.000000000001$	12	10^{-2}
$10^{-13} = 0.0000000000001$	13	10^{-1}
$10^{-14} = 0.00000000000001$	14	10^0

Remember that pure (neutral) water has an H^{1+} concentration of $10^{-7}M$ and an $(OH)^{1-}$ concentration of $10^{-7}M$. The product of these two numbers is $10^{-14}M^2$:

$$[H^{1+}][OH^{1-}] = (10^{-7}M)(10^{-7}M) = 10^{-14}M^2$$

The value 10^{-14} is the *dissociation constant* of water. It is usually given the symbol K_w. Therefore,

$$[H^{1+}][OH^{1-}] = K_w = 10^{-14}$$

(We generally don't bother writing the M^2 after 10^{-14}, but remember that it is there all the same.) A solution that has a hydrogen-ion concentration of $10^{-5}M$ has to have a hydroxide-ion concentration of $10^{-9}M$, because

$$(10^{-5}M)[OH^{1-}] = 10^{-14}M^2 \quad \text{so} \quad [OH^{1-}] = \frac{10^{-14}\,M^2}{10^{-5}\,M} = 10^{-9}M$$

Table 15.8 summarizes the relationships among hydrogen-ion concentration, hydroxide-ion concentration, and pH.

Example 15.2

Determine the pH of each of the following strong electrolytes.
(a) A $0.001M$ HCl solution
(b) A $0.01M$ HNO_3 solution
(c) A solution whose $[OH^{1-}]$ is 10^{-3}
(d) A solution whose $[OH^{1-}]$ is 10^{-11}
(e) A $0.0001M$ NaOH solution

Solution

(a) A $0.001M$ HCl solution releases 0.001 mole of hydrogen ions per liter of solution, so its $[H^{1+}]$ is 10^{-3}. Therefore its pH is 3.

$$pH = -\log [H^{1+}] = -\log 10^{-3} = 3$$

(b) A $0.01M$ HNO_3 solution releases 0.01 mole of hydrogen ions per liter of solution, so its $[H^{1+}]$ is 10^{-2}. Therefore its pH is 2.

$$pH = -\log [H^{1+}] = -\log 10^{-2} = 2$$

(c) A solution whose $[OH^{1-}]$ is 10^{-3} has $[H^{1+}] = 10^{-11}$ (see Table 15.8). Therefore its pH is 11.

$$pH = -\log [H^{1+}] = -\log 10^{-11} = 11$$

(d) A solution whose $[OH^{1-}]$ is 10^{-11} has $[H^{1+}] = 10^{-3}$ (see Table 15.8). Therefore its pH is 3.

$$pH = -\log [H^{1+}] = -\log 10^{-3} = 3$$

(e) A $0.0001M$ NaOH solution releases 0.0001 mole of hydroxide ions per liter of solution, so its $[OH^{1-}]$ is 10^{-4}. A solution whose

$[OH^{1-}]$ is 10^{-4} has $[H^{1+}] = 10^{-10}$ (see Table 15.8). Therefore its pH is 10:

$$pH = -\log [H^{1+}] = -\log 10^{-10} = 10$$

Practice Exercise 15.2   Determine the pH of each of the following solutions.
(a) A 0.0001M HCl solution
(b) A 0.001M HCl solution
(c) A solution whose $[OH^{1-}]$ is 10^{-5}
(d) A solution whose $[OH^{1-}]$ is 10^{-11}
(e) A 0.10M KOH solution

15.8 Salts

Definition of Salt

We have seen that acids and bases react with each other in what are called neutralization reactions. The products of these reactions are salts and water.

$$HCl + NaOH \longrightarrow \underset{\substack{\text{Sodium} \\ \text{chloride}}}{NaCl} + H_2O$$

$$HCl + NH_4OH \longrightarrow \underset{\substack{\text{Ammonium} \\ \text{chloride}}}{NH_4Cl} + H_2O$$

$$H_2SO_4 + 2NH_4OH \longrightarrow \underset{\substack{\text{Ammonium} \\ \text{sulfate}}}{(NH_4)_2SO_4} + 2H_2O$$

In these reactions, the hydrogen ions from the acid combine with the hydroxide ions from the base to produce water. At the same time, the positive ions of the base combine with the negative ions of the acid to produce a compound that we call a salt (Figure 15.7). A **salt** is a compound composed of the positive ion of a base and the negative ion of an acid.

Most salts are composed of a metal ion combined with a nonmetal (or polyatomic) ion that has a negative charge. Examples of such salts are sodium chloride (NaCl), which is table salt, silver bromide (AgBr), potassium sulfate (K_2SO_4), and iron(III) phosphate ($FePO_4$). Some salts are composed of a polyatomic ion that has a positive charge combined with a nonmetal (or polyatomic) ion that has a negative charge. Examples of such salts are ammonium chloride (NH_4Cl) and ammonium nitrate (NH_4NO_3).

Learning Goal 13
Definition of salt

Figure 15.7
A neutralization reaction

$$HCl + NaOH \longrightarrow NaCl + H_2O$$

Properties, Preparations, and Some Uses of Salts

. . . .
Learning Goal 14
Important properties of salts

Different salts have different properties, but *all salts are ionic substances* (which means that they are bonded ionically). Their solubilities in water may vary, but their aqueous solutions always contain ions. For example, sodium chloride dissociates into sodium ions and chloride ions when it is dissolved in water.

$$NaCl \xrightarrow{\text{Water}} Na^{1+} + Cl^{1-}$$

. . . .
Learning Goal 15
Preparation of salts

There are many ways to prepare salts. However, we'll limit our discussion to three methods: (1) direct combination of two elements, (2) reaction of an active metal with an acid, and (3) neutralization of an acid by a base.

1. When we heat a piece of sodium metal in the presence of an excess of chlorine gas, a vigorous reaction takes place. The product is pure sodium chloride.

$$2Na + Cl_2 \longrightarrow 2NaCl$$

Many salts can be prepared in this manner—simply by heating the elements together.

2. In Chapter 12 we noted that hydrogen could be produced by the reaction of an active metal with an acid. The other product formed in such a reaction is a salt. We can use this type of reaction to produce the salt we want by choosing the appropriate metal and acid. If we want to prepare magnesium chloride, for example, we use magnesium metal and hydrochloric acid.

$$Mg + 2HCl \longrightarrow MgCl_2 + H_2$$

An important fact to keep in mind is that the metal we choose must be able to replace the hydrogen in the acid. Table 8.1, the activity series, reveals which metals can replace hydrogen.

3. To form a salt by a neutralization reaction, we choose a base that contains the positive ion of the salt, and an acid that contains the negative ion of the salt. For example, if we want to prepare potassium bromide, we can use the reaction between potassium hydroxide and hydrobromic acid:

$$KOH + HBr \longrightarrow KBr + H_2O$$

Salts have many different uses. Many salts, such as sodium chloride, are necessary for life itself. Calcium phosphate is the main ingredient of our bones. Other salts, such as calcium phosphate dihydrate [$Ca_3(PO_4)_2 \cdot 2H_2O$], are used for building materials. The variety of uses for salts is matched only by the number of salts known.

15.9 Titration

When explaining double-replacement reactions in Section 8.4, we discussed neutralization as one kind of double-replacement reaction. **Neutralization** is *the reaction between an acid and a base to form water and a salt.* A neutralization reaction is often represented by the general equation

$$HX + MOH \longrightarrow MX + H_2O$$

wherein HX (an acid) and MOH (a base) form a salt (MX) and water.

The most common use that chemists make of neutralization reactions is called titration. **Titration** is *an analytical procedure that enables the chemist to find out how much of a substance is present in solution.* In the laboratory, titration is sometimes used to find the concentrations of chemicals in urine, blood, plasma, and drinking water. When a neutralization titration is performed, the amount of acid or base can be determined.

In an acid–base titration, a measured volume of a solution of acid or base of known concentration (called the titrant) is added to a measured volume of a solution of acid or base whose concentration is to be determined. A device called a burette is used for dispensing the measured quantities of acid or base (Figure 15.8). The burette enables the chemist to stop adding titrant by means of a stopcock at the bottom of the burette.

Learning Goal 16
Titration

Figure 15.8
A titration set-up. A measured volume of one solution is placed in the flask. The burette above it contains another solution. The second solution is added gradually until the end point is reached. The stopcock controls the rate of flow of the solutions into the flask.

Burettes

Burette clamps

Stopcock

Erlenmeyer flask

The stopcock can be closed to stop the flow of solution at any time. Calibration marks etched onto the surface of the burette reveal exactly how much titrant has been added. The titrant used is generally a *standard solution* of acid or base, which is a solution whose concentration is accurately known.

How does the chemist know when enough titrant has been added? In other words, when is it evident that the neutralization is complete? *When equal numbers of equivalents of acid and base have been added to the reaction vessel (which is usually an Erlenmeyer flask), the* **equivalence point** *has been reached.* At this point, the volume and concentration of the titrant that has been added can be used to calculate the concentration of the acid or base being titrated.

In order to determine whether the equivalence point has been reached, the chemist must know (1) the pH of a solution of the salt formed by complete neutralization of the particular acid and base involved in the titration and (2) its concentration.

In general, the pH at the equivalence point varies with the salt formed by the particular acid and base used in the neutralization reaction, and with the concentration of the salt formed. If we are neutralizing a solution of a strong base, such as sodium hydroxide, with a standard solution of a strong acid, such as nitric acid, we begin with a solution of sodium hydroxide of unknown concentration and quite high pH. As the nitric acid is added, the pH decreases, because OH^{1-} is removed by the reaction $H^{1+} + (OH)^{1-} \rightarrow H_2O$. When the equivalence point is reached, the numbers of equivalents of acid and base are the same and there are equal amounts of H^{1+} and OH^{1-}. The pH at the equivalence point should be 7, because when $[H^{1+}] = [OH^{1-}]$, each value is $10^{-7}M$. When a strong acid and a strong base take part in a titration reaction, the equivalence point is reached when the pH reaches 7. At that point, neutralization is complete and no more titrant should be added. The salt sodium nitrate is neutral.

What happens when a strong acid and a weak base take part in a titration? At the equivalence point, when the numbers of equivalents of acid and base are the same, is the pH equal to 7? No, it is not, because when the salt of a strong acid and a weak base forms, it further reacts with water to produce a solution with a pH less than 7. For example, a solution of NH_4Cl is slightly acidic.

Similarly, when a strong base and a weak acid take part in a titration, at the equivalence point the numbers of equivalents of acid and base are the same, but the solution is slightly basic. When the salt of a strong base and a weak acid forms, it further reacts with water to produce a solution with a pH greater than 7. Sodium acetate is an example of a salt that gives a slightly basic solution.

Chemicals called indicators can be carefully selected to show the point at which all the available hydrogen ions have reacted with all the available hydroxide ions in a solution. One must, however, take into account the pH of the solution of salt formed.

15.10 Indicators

An **indicator** is *a dye whose color depends on its pH.* Paper saturated with an indicator is called indicator paper. Litmus is the most common indicator. It turns blue when the solution in which it is placed is basic; it turns pink when the solution is acidic. *In general, indicators change color within a certain pH range, and the range of pH values within which the color change takes place depends on the particular indicator.* For example, litmus changes from pink to blue within the pH range of 5.5 to 7.5. It is blue above 7.5 and red below 5.5. Another common indicator, phenolphthalein, is colorless in acid solution and red-violet in alkaline solution. It changes color within the pH range of 8.0 to 9.8. Table 15.9 lists common indicators and the pH ranges within which their color changes take place.

Table 15.9
Table of Common Indicators

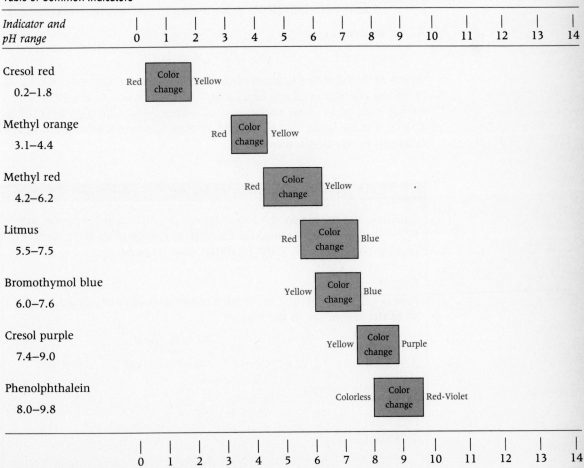

When an indicator is added to the solution that is being titrated and the indicator just changes color, what we call the **end point** has been reached. To do a titration, chemists estimate the pH at the equivalence point, select an indicator that changes color in the appropriate pH range, and add a few drops of that indicator to the solution being titrated.

15.11 Acid–Base Titrations

In the laboratory, units of normality are used to simplify calculations. Equal numbers of equivalents of acid and base react with each other. That is,

Number of equivalents of acid = number of equivalents of base

Remember, normality (N) is the number of equivalents of acid or base present in one liter of solution, so

Number of equivalents = normality (N) × volume (liters)

At the equivalence point,

$$N_a \times V_a = N_b \times V_b$$

where N_a is the concentration of the acid, V_a is the volume of the acid, N_b is the concentration of the base, and V_b is the volume of the base. Liters do not have to be used as the unit of volume. Any unit may be used as long as it is the same for both V_a and V_b. This equation is often used in calculations involving titrations.

Example 15.3

Calculate the normality (or molarity; see the alternative solution that follows) of an HNO_3 solution if 25.0 mL of the solution are neutralized by 100.0 mL of 0.200N (0.100M) $Ca(OH)_2$ solution.

Solution

First organize the data. Let N_a refer to the normality of the acid and N_b to the normality of the base.

$$N_a = ? \qquad N_b = 0.200N$$

$$V_a = 25.0 \text{ mL} \qquad V_b = 100.0 \text{ mL}$$

Rewrite the formula

$$N_a \times V_a = N_b \times V_b$$

in terms of N_a:

$$N_a = \frac{N_b V_b}{V_a}$$

Then solve that equation.

$$N_a = \frac{(0.200N)(100.0 \text{ mL})}{25.0 \text{ mL}} = 0.800N$$

The concentration of the acid is 0.800N.

Alternative Solution Using Molarity

Some instructors prefer to have students perform titration calculations by using only molarity. Here's how it's done. First calculate how many moles of $Ca(OH)_2$ solution it took to neutralize the HNO_3 solution.

$$M_{Ca(OH)_2} = 0.100 \frac{\text{mole}}{\text{L}} \qquad V_{Ca(OH)_2} = 100.0 \text{ mL, or } 0.1000 \text{ L}$$

$$M = \frac{\text{moles}}{\text{liter}}$$

$$\text{moles} = (M)(\text{liter})$$

$$? \text{ moles} = \left(0.100 \frac{\text{mole}}{\text{L}}\right)(0.1000 \text{ L})$$

$$= 0.0100 \text{ mole } Ca(OH)_2$$

Next write the balanced equation for the reaction between the HNO_3 and the $Ca(OH)_2$.

$$2HNO_3 + Ca(OH)_2 \longrightarrow Ca(NO_3)_2 + 2H_2O$$

The balanced equation tells us that 2 moles of HNO_3 are neutralized by 1 mole of $Ca(OH)_2$. Use this information and the factor-unit method to calculate the moles of HNO_3 neutralized by 0.0100 mole of $Ca(OH)_2$.

$$? \text{ mole } HNO_3 = (0.0100 \text{ mole } Ca(OH)_2) \left(\frac{2 \text{ moles } HNO_3}{1 \text{ mole } Ca(OH)_2}\right)$$

$$= 0.0200 \text{ mole } HNO_3$$

Finally, calculate the molarity of the HNO_3 solution from this information and the volume of the HNO_3 (25.0 mL, or 0.0250 L).

$$M = \frac{\text{mole}}{\text{liter}}$$

$$?M = \frac{0.0200 \text{ mole}}{0.0250 \text{ L}} = 0.800M$$

Practice Exercise 15.3   Calculate the normality (or molarity) of an HCl solution if 150 mL of the solution are neutralized by 50 mL of 1.0N (1.0M) NaOH solution.

Example 15.4

A quality control laboratory is analyzing a sample of vinegar to determine its acetic acid content. The technicians find that it requires 14.0 mL of 1.25N (1.25M) NaOH solution to neutralize a 25.0-mL sample of vinegar. What is the normality (or molarity) of the vinegar solution?

Solution
Organize the data.

$$N_a = ? \qquad\qquad N_b = 1.25N$$

$$V_a = 25.0 \text{ mL} \qquad V_b = 14.0 \text{ mL}$$

Substitute the values into the equation and solve it for N_a.

$$N_a \times V_a = N_b \times V_b$$

$$N_a = \frac{N_b V_b}{V_a} = \frac{(1.25N)(14.0 \text{ mL})}{25.0 \text{ mL}} = 0.700N$$

Alternative Solution Using Molarity
First calculate how many moles of NaOH it took to neutralize the vinegar.

$$M_{\text{NaOH}} = 1.25 \frac{\text{mole}}{\text{L}} \qquad V_{\text{NaOH}} = 14.0 \text{ mL, or } 0.0140 \text{ L}$$

$$M = \frac{\text{mole}}{\text{liter}}$$

$$\text{moles} = (M)(\text{liter})$$

$$? \text{ moles} = \left(1.25 \frac{\text{mole}}{\text{L}}\right)(0.0140 \text{ L})$$

$$= 0.0175 \text{ mole NaOH}$$

Next write the balanced equation for the reaction between the $HC_2H_3O_2$ and the NaOH.

$$HC_2H_3O_2 + NaOH \longrightarrow NaC_2H_3O_2 + H_2O$$

The balanced equation reveals that 1 mole of $HC_2H_3O_2$ is neutralized by 1 mole of NaOH. Use this information and the factor-unit method to calculate the moles of $HC_2H_3O_2$ neutralized by the 0.0175 mole of NaOH.

$$? \text{ moles } HC_2H_3O_2 = (0.0175 \text{ mole } HC_2H_3O_2)\left(\frac{1 \text{ mole NaOH}}{1 \text{ mole } HC_2H_3O_2}\right)$$

$$= 0.0175 \text{ mole NaOH}$$

Finally, calculate the molarity of the vinegar from this information and the volume of the vinegar solution (25.0 mL, or 0.0250 L).

$$M = \frac{\text{moles}}{\text{liter}}$$

$$?M = \frac{0.0175 \text{ mole}}{0.0250 \text{ L}} = 0.700M$$

Practice Exercise 15.4 Calculate the normality (or molarity) of a solution of acetic acid if 43.50 mL of vinegar were neutralized by 25.00 mL of 0.15N (0.15M) NaOH.

Example 15.5

A 50.00-mL sample of $Mg(OH)_2$ is titrated with 65.00 mL of 0.500N (0.500M) HCl. What is the normality (or molarity) of the $Mg(OH)_2$ solution?

Solution
Organize the data.

$$N_a = 0.500N \qquad N_b = ?$$

$$V_a = 65.00 \text{ mL} \qquad V_b = 50.00 \text{ mL}$$

Substitute the values into the equation and solve it for N_b.

$$N_a \times V_a = N_b \times V_b$$

$$N_b = \frac{N_a V_a}{V_b} = \frac{(0.500N)(65.00 \text{ mL})}{50.00 \text{ mL}} = 0.650N \text{ Mg(OH)}_2$$

Alternative Solution Using Molarity
First calculate how many moles of HCl it took to neutralize the $Mg(OH)_2$ solution.

$$M_{\text{HCl}} = 0.500 \frac{\text{mole}}{\text{L}} \qquad V_{\text{HCl}} = 65.00 \text{ mL, or } 0.06500 \text{ L}$$

$$M = \frac{\text{mole}}{\text{liter}}$$

$$\text{mole} = (M)(\text{liter})$$

$$? \text{ moles} = \left(0.500 \frac{\text{mole}}{\text{L}}\right)(0.06500 \text{ L})$$

$$= 0.0325 \text{ mole HCl}$$

Next write the balanced equation for the reaction between the HCl and the $Mg(OH)_2$.

$$2HCl + Mg(OH)_2 \longrightarrow MgCl_2 + 2H_2O$$

The balanced equation reveals that 2 moles of HCl are neutralized by 1 mole of $Mg(OH)_2$. Use this information and the factor-unit method to calculate the number of moles of $Mg(OH)_2$ neutralized by the 0.0325 mole of HCl.

$$? \text{ moles } Mg(OH)_2 = (0.0325 \text{ mole HCl}) \left(\frac{1 \text{ mole } Mg(OH)_2}{2 \text{ moles HCl}} \right)$$

$$= 0.0163 \text{ mole } Mg(OH)_2$$

Finally, calculate the molarity of the $Mg(OH)_2$ solution from this information and the volume of the $Mg(OH)_2$ solution (50.00 mL, or 0.05000 L).

$$M = \frac{\text{mole}}{\text{liter}}$$

$$?M = \frac{0.0163 \text{ mole}}{0.05000 \text{ L}} = 0.326M \text{ } Mg(OH)_2$$

Practice Exercise 15.5   A 25.65-mL sample of NaOH solution is titrated with 100.0 mL of 0.105N (0.105M) lactic acid solution. What is the normality (or molarity) of the NaOH solution?

Example 15.6

A chemistry student was determining the normality of an $Fe(OH)_3$ solution. It required 30.00 mL of a 2.00N (2.00M) HCl solution to neutralize 10.00 mL of the $Fe(OH)_3$ solution. Calculate the normality (or molarity) of the $Fe(OH)_3$ solution.

Solution

Organize the data.

$$N_a = 2.00N \qquad N_b = ?$$

$$V_a = 30.00 \text{ mL} \qquad V_b = 10.00 \text{ mL}$$

Substitute the values into the equation and solve it for N_b.

$$N_a \times V_a = N_b \times V_b$$

$$N_b = \frac{N_a V_a}{V_b} = \frac{(2.00N)(30.00 \text{ mL})}{10.00 \text{ mL}} = 6.00N$$

Alternative Solution Using Molarity

First calculate how many moles of HCl it took to neutralize the $Fe(OH)_3$ solution.

$$M_{HCl} = 2.00 \frac{mole}{L} \qquad V_{HCl} = 30.00 \text{ mL, or } 0.03000 \text{ L}$$

$$M = \frac{mole}{liter}$$

$$mole = (M)(liter)$$

$$? \text{ moles} = \left(2.00 \frac{mole}{\cancel{L}}\right)(0.03000 \; \cancel{L})$$

$$= 0.0600 \text{ mole HCl}$$

Next write the balanced equation for the reaction between HCl and $Fe(OH)_3$.

$$3HCl + Fe(OH)_3 \longrightarrow AlCl_3 + 3H_2O$$

The balanced equation reveals that 3 moles of HCl are neutralized by 1 mole of $Fe(OH)_3$. Use this information and the factor-unit method to calculate the number of moles of $Fe(OH)_3$ neutralized by the 0.0600 mole of HCl.

$$? \text{ mole } Fe(OH)_3 = (0.0600 \; \cancel{\text{mole HCl}})\left(\frac{1 \text{ mole } Fe(OH)_3}{3 \; \cancel{\text{moles HCl}}}\right)$$

$$= 0.0200 \text{ mole } Fe(OH)_3$$

Finally, calculate the molarity of the $Fe(OH)_3$ solution from this information and the volume of the $Fe(OH)_3$ solution (10.00 mL, or 0.01000 L).

$$M = \frac{mole}{liter}$$

$$?M = \frac{0.0200 \text{ mole}}{0.01000 \text{ L}} = 2.00M$$

Practice Exercise 15.6 In a titration, it took 20.4 mL of $0.1330N$ ($0.0665M$) sulfuric acid solution to neutralize 26.5 mL of potassium hydroxide solution of unknown concentration. Calculate the normality (or molarity) of the potassium hydroxide solution.

Summary

Arrhenius defined an acid as a substance that releases hydrogen ions, H^{1+}, when dissolved in water. Brønsted and Lowry defined an acid as a proton donor. Some common Arrhenius acids are HCl, HNO_3, H_2SO_4, and $HC_2H_3O_2$. These are also Brønsted–Lowry acids, as are such substances as $(H_3O)^{1+}$, $(HSO_4)^{1-}$, $(NH_4)^{1+}$, and H_2O. Generally, acids taste sour, change the color of certain dyes, react with certain metals to produce hydrogen gas, and neutralize bases. Most important, they dissolve in water to increase the hydrogen ion concentration. Acids are commonly prepared by the reaction of a nonmetal oxide with water. Acids react with metals as well as with bases. They are used in the manufacturing of a variety of products.

Bases, like acids, can be defined in more than one way. Arrhenius defined a base as a substance that releases hydroxide ions, $(OH)^{1-}$, when dissolved in water. Brønsted and Lowry defined a base as a proton acceptor. Some common Arrhenius bases are $LiOH$, KOH, $NaOH$, $Ca(OH)_2$, $Mg(OH)_2$, and NH_3. These are also Brønsted–Lowry bases, as are such substances as chloride ions, carbonate ions, sulfate ions, acetate ions, hydroxide ions, and water. Generally, bases taste bitter, change the color of certain dyes, feel soapy or slippery, and neutralize acids. Most important, they dissolve in water to produce hydroxide ions. Bases can be produced by the reaction of a metal oxide with water. They react with nonmetal oxides and with acids. Like acids, they have a number of uses in industry and in our daily lives.

Strong acids and bases are those that essentially ionize completely in a solution. Weak acids and bases are those that don't ionize completely. In solution, the ions and molecules of a weak acid or a weak base are in dynamic equilibrium.

Water ionizes partially, just like a weak acid or a weak base. When a water molecule ionizes, a hydronium ion $(H_3O)^{1+}$ and a hydroxide ion are formed. In 1 L of water, only 10^{-7} mole of water actually dissociates, forming 10^{-7} mole of hydronium ions and 10^{-7} mole of hydroxide ions.

The pH scale provides a means of measuring the acid or base strength of solutions. The pH of any solution is equal to the negative of the logarithm of its hydrogen-ion concentration in moles per liter. The pH of pure water and that of any other neutral solution is 7. An acidic solution has a pH below 7, and a basic solution has a pH above 7. The product of the hydrogen-ion concentration and the hydroxide-ion concentration in an aqueous solution is always 10^{-14}.

A salt is a compound composed of the positive ion of a base and the negative ion of an acid. All salts are bonded ionically, and their aqueous solutions always contain ions. Salts can be prepared by direct combination of two elements, by reaction of an active metal with an acid, or by neutralization of an acid by a base.

Titration is an analytical procedure that enables the chemist to determine the concentration of a substance present in solution. A measured volume of a solution of unknown concentration is found. Acid–base titrations involve a neutralization reaction.

. . . .

Career Sketch

Radiologic (X-Ray) Technologists

The people who operate x-ray equipment and take x-ray pictures (known as radiographs) are called *radiologic technologists*. They usually work under the supervision of radiologists—physicians who specialize in the use of x rays.

There are three specialties in the field of radiologic technology. The most widely known is x-ray technology, which involves taking x-ray pictures of parts of the human body. A radiologist then studies the x rays to

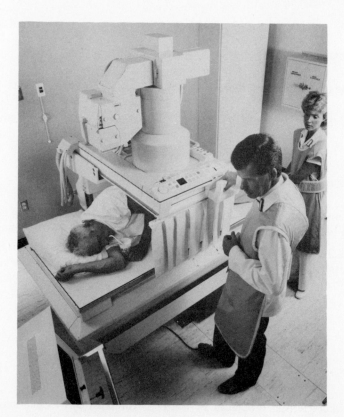

X-ray technicians administer a barium swallow before performing fluoroscopic x-ray imaging procedures.
(GE Medical Systems Group)

diagnose the patient's problem. X rays are also taken of the chest during routine medical checkups to detect the presence of lung diseases.

The other two specialties are nuclear medicine technology and radiation therapy. In nuclear medicine technology, the radiologic technologist applies radioactive material to help radiologists diagnose or treat illnesses or injuries. In radiation therapy, radiation-producing machines are used to administer therapeutic treatments recommended by radiologists. This method is often used in treating various kinds of cancer. Radiologic technologists study chemistry, anatomy and physiology, radiologic physics, radiation biology, and radiography. They generally have two-year or four-year college degrees and work in hospitals, private physicians' offices, clinics, x-ray manufacturing firms, and x-ray supply companies. In most states two-year graduates who complete 2,000 hours of clinical practice may take the American Registry Examination. They are then prepared to take state certification licenses.

. . . .

Self-Test Exercises

Learning Goal 1: Arrhenius Acids and Bases

1. Using the Arrhenius definition, define: (a) an acid, (b) a base.

3. What is the acid found in vinegar?

2. Name two Arrhenius acids and two Arrhenius bases.

4. What are the limitations of the Arrhenius definition of acids and bases?

Learning Goal 2: Brønsted–Lowry Acids and Bases

5. Using the Brønsted–Lowry definition, define (a) an acid, (b) a base.

*7. Complete the following equation, which shows why ammonia (NH_3) is a Brønsted–Lowry base.

$$NH_3 + H_2O \rightleftharpoons \underline{\qquad}$$

6. Why is water both an acid and a base according to the Brønsted–Lowry definitions?

*8. Complete the following equations, which show why each substance is a Brønsted–Lowry acid or base:

(a) $(HSO_4)^{1-} \longrightarrow ? + ?$
    Acid

(b) $Cl^{1-} + ? \longrightarrow ?$
    Base

Learning Goal 3: Characteristics of Acids and Bases

9. What are the characteristic properties of acids, bases, and salts?

10. What do we call a substance that is used to reveal whether a chemical is an acid or a base?

11. What is the difference between hydrogen chloride gas and hydrochloric acid?

13. Why do people store acids in containers made of nonmetal materials?

15. The two types of acids are mineral acids and _____ acids.

17. Place the words *red* and *blue* in the proper spaces.
(a) _____ litmus paper turns _____ in acid.
(b) _____ litmus paper turns _____ in base.

19. Write a reaction showing the formation of a hydronium ion from HBr.

21. Write a reaction showing the formation of hydroxide ions when calcium hydroxide dissolves in water.

12. What is a hydronium ion?

14. A substance that dissolves in water to produce hydroxide ions is a(n) _____ .

16. A substance that dissolves in water to produce hydrogen ions is a(n) _____ .

18. True or False: An acid feels soapy or slippery.

20. Write an equation for the formation of hydronium ions from HNO_3.

22. Write an equation for the formation of hydroxide ions when magnesium hydroxide dissolves in water.

Learning Goal 4: Preparation of Acids and Bases

23. Write a reaction for the preparation of each of the following bases from a metal oxide and water: (a) KOH, (b) $Mg(OH)_2$.

25. Write a reaction for the preparation of each of the following acids from a nonmetal oxide and water: (a) H_2SO_3 (sulfurous acid), (b) H_2CO_3 (carbonic acid).

24. Write a reaction for the preparation of each of the following bases from a metal oxide and water: (a) LiOH, (b) $Ca(OH)_2$.

26. Write a reaction for the preparation of each of the following acids from a nonmetal oxide and water: (a) HNO_3, (b) H_3PO_4.

Learning Goal 5: Chemical Reactions of Acids and Bases

27. Complete and balance the following equations.
(a) $H_3PO_4 + Mg(OH)_2 \longrightarrow$
(b) $P_2O_5 + H_2O \longrightarrow$
(c) $K_2O + H_2O \longrightarrow$
(d) $Mg(OH)_2 + CO_2 \longrightarrow$
(e) $Zn + H_3PO_4 \longrightarrow$

29. Suppose you spill some sodium hydroxide on your clothes. What could you do to keep this chemical from ruining them?

31. Do vinegar and lye react? If a reaction does take place, write a chemical equation for it.

28. Complete and balance the following equations.
(a) $HC_2H_3O_2 + Ca(OH)_2 \longrightarrow$
(b) $Na_2O + H_2O \longrightarrow$
(c) $Mg + HCl \longrightarrow$
(d) $Sr(OH)_2 + CO_2 \longrightarrow$
(e) $N_2O_5 + H_2O \longrightarrow$

*30. Suppose that you accidentally drank a liquid containing sodium hydroxide. What should you do to neutralize its effect?

*32. Power plants that burn coal with a high sulfur content send huge quantities of SO_2 into the atmosphere. When the SO_2 reacts with the moisture in the clouds, acid rain results. Write an equation for the formation of acid rain.

33. Paper mills emit huge quantities of sulfur dioxide (SO_2) into the atmosphere at the mill site. What can happen to exposed metal machinery when rain falls near an operating paper mill?

34. Write a balanced chemical equation for each of the following word equations.
(a) Magnesium + hydrochloric acid $\longrightarrow$? + ?
(b) Sulfuric acid + sodium hydroxide $\longrightarrow$? + ?
(c) ? + ? $\longrightarrow$ magnesium sulfate + hydrogen gas
(d) ? + ? $\longrightarrow$ zinc phosphate + water

35. Write balanced chemical equations for the following word equations.
(a) Zinc + sulfuric acid $\longrightarrow$
zinc sulfate + hydrogen gas
(b) Hydrochloric acid +
lithium hydroxide $\longrightarrow$? + ?
(c) ? + ? $\longrightarrow$ calcium sulfate + H_2
(d) ? + ? $\longrightarrow$ calcium phosphate + water

36. Complete and balance the following equations.
(a) $Zn + H_3PO_4 \longrightarrow$
(b) $Zn + HCl \longrightarrow$
(c) $Zn + H_2SO_4 \longrightarrow$

37. Complete and balance the following equations.
(a) $Zn + HCl \longrightarrow$
(b) $HCl + Ca(OH)_2 \longrightarrow$
(c) $Mg(OH)_2 + CO_2 \longrightarrow$

38. All of the reactions in Exercise 36 can be classified as _____ reactions.

Learning Goal 6: Strong Acids and Bases

39. Define and give an example of (a) a strong acid, (b) a strong base.

40. What is the difference between a strong acid and a strong base?

41. Write the formulas for the following acids: (a) nitric acid, (b) hydrofluoric acid, (c) hydroiodic acid.

42. Write formulas for the following acids: (a) phosphoric acid, (b) sulfuric acid, (c) boric acid.

43. Give names for the following bases: (a) NaOH, (b) $Ca(OH)_2$, (c) $Fe(OH)_2$, (d) $Al(OH)_3$.

44. Give names for the following bases: (a) $Mg(OH)_2$, (b) $Zn(OH)_2$, (c) KOH, (d) LiOH.

45. Give names for the following acids: (a) HNO_3, (b) H_3PO_4, (c) HI, (d) H_2SO_4.

46. Name the following acids: (a) HCl, (b) H_2SO_3, (c) H_2CO_3, (d) $HC_2H_3O_2$.

47. Write the formulas for the following bases: (a) lithium hydroxide, (b) barium hydroxide, (c) iron(II) hydroxide, (d) iron(III) hydroxide.

48. Write formulas for the following bases: (a) potassium hydroxide, (b) sodium hydroxide, (c) zinc hydroxide, (d) calcium hydroxide.

Learning Goal 7: Weak Acids and Bases

49. Define and give an example of (a) a weak acid, (b) a weak base.

50. True or False: A weak acid or a weak base ionizes almost completely.

Learning Goal 8: Dynamic Equilibrium

*51. Explain what is meant by dynamic equilibrium. Show how it is related to weak acids and bases.

*52. Explain how a dynamic equilibrium comes to exist when carbon dioxide is dissolved in water to produce carbonic acid, a weak acid.

Learning Goal 9: Ionization of Water

53. Write an equation for the ionization of water.

54. How many moles of hydrogen ions and how many moles of hydroxide ions are present in 2 L of water?

Learning Goal 10: pH and pH Scale

55. State how acidic or basic each of the following substances is (very acidic, moderately acidic, very basic, and so on).
(a) Vinegar, pH = 2.8   (b) Sea water, pH = 8.5
(c) Blood, pH = 7.4   (d) Saliva, pH = 7.0

56. State how acidic or basic each of the following substances is (very acidic, moderately acidic, very basic, and so on).
(a) oranges, pH = 3.5   (b) milk, pH = 6.5
(c) lemons, pH = 2.3   (d) rainwater, pH = 6.2.

57. Why is it important to measure the pH of a swimming pool?

58. Why must the water in a fish tank be the proper pH?

*59. The pH scale is a logarithmic scale. This means that a difference of one pH unit indicates a multiple of 10 between the actual acidities or basicities of two substances.
(a) The pH of grapefruit juice is 3; the pH of beer is 5. How many times more acidic is grapefruit juice than beer?
(b) The pH of a lemon is 2; the pH of pure water is 7. How many times more acidic is lemon than water?

*60. The pH of soft drinks is 3.0 and the pH of pure water is 7.0. How many times more acidic is the soft drink than the water?

61. In your own words, give a working definition of pH.

62. What is "pH-balanced shampoo"?

Learning Goal 11: pH, Hydrogen-Ion, and Hydroxide-Ion Concentrations in Solutions

63. Determine the pH of each of the following aqueous solutions.
(a) $[H^{1+}] = 10^{-3}$ (b) $[H^{1+}] = 10^{-4}$
(c) $[H^{1+}] = 10^{-2}$ (d) $[H^{1+}] = 10^{-11}$
(e) $[OH^{1-}] = 10^{-4}$ (f) $[OH^{1-}] = 10^{-9}$

64. Determine the pH of each of the following aqueous solutions.
(a) $[H^{1+}] = 10^{-5}$ (b) $[H^{1+}] = 10^{-10}$
(c) $[H^{1+}] = 10^{-8}$ (d) $[OH^{1-}] = 10^{-5}$
(e) $[OH^{1-}] = 10^{-10}$ (f) $[OH^{1-}] = 10^{-8}$

65. Determine the $[H^{1+}]$ and $[OH^{1-}]$ for aqueous solutions with the following pH's.
(a) pH = 5   (b) pH = 9
(c) pH = 2   (d) pH = 12

66. Determine the $[H^{1+}]$ and $[OH^{1-}]$ of each of the following aqueous solutions.
(a) pH = 4   (b) pH = 7
(c) pH = 1   (d) pH = 13

Learning Goal 12: Relationship Between Hydrogen-Ion and Hydroxide-Ion Concentrations

67. For each of the following aqueous solutions, determine the hydrogen-ion concentration and the pH.
(a) $[OH^{1-}] = 10^{-5}$ (b) $[OH^{1-}] = 10^{-8}$
(c) $[OH^{1-}] = 10^{-10}$ (d) $[OH^{1-}] = 10^{-2}$

*68. For each of the following aqueous solutions, determine the hydrogen-ion concentration and the pH.
(a) $[OH^{1-}] = 10^{-4}$ (b) $[OH^{1-}] = 10^{-11}$
(c) $[OH^{1-}] = 10^{-9}$ (d) $[OH^{1-}] = 10^{-3}$

*69. Determine the OH^{1-} concentration for each of the following aqueous solutions.
(a) $[H^{1+}] = 10^{-3}$ (b) $[H^{1+}] = 10^{-4}$
(c) $[H^{1+}] = 10^{-8}$ (d) pH = 9
(e) pH = 6   (f) pH = 2

*70. Determine $[OH^{1-}]$ for each of the following aqueous solutions.
(a) $[H^{1+}] = 10^{-2}$ (b) $[H^{1+}] = 10^{-9}$
(c) $[H^{1+}] = 10^{-5}$ (d) pH = 3
(e) pH = 4   (f) pH = 7

Learning Goal 13: Definition of Salt

71. Name each of the following salts: (a) Na_2S, (b) KBr,   (c) Na_2SO_4,   (d) KNO_3,   (e) $CaCO_3$, (f) BaI_2.

72. Name each of the following salts: (a) Na_2CO_3, (b) KCl,   (c) $BaCl_2$,   (d) $NaNO_3$,   (e) $Ca(NO_3)_2$, (f) CaI_2.

73. Write the formula for each of the following salts: (a) silver nitrate,   (b) cobalt(II) chloride,   (c) copper(I) sulfite,   (d) mercury(II) phosphate.

74. Write the formula for each of the following salts: (a) silver chloride,   (b) iron(III) nitrate, (c) iron(II) nitrate,   (d) copper(I) chloride.

Learning Goal 14: Important Properties of Salts

75. List at least two important properties of salts.

76. What kind of bonding is found in all salts?

Learning Goal 15: Preparation of Salts

77. List at least two ways in which salts can be prepared by chemical reactions.

78. How would you prepare the salt silver nitrate?

79. Write a reaction showing the preparation of the salt lithium sulfate (Li_2SO_4), using an acid–base reaction.

80. Write a reaction showing the preparation of the salt lithium nitrate ($LiNO_3$), using an acid–base reaction.

81. Write a reaction showing the preparation of the salt zinc bromide ($ZnBr_2$), using a single-replacement reaction.

82. Write a reaction showing the preparation of the salt zinc chloride ($ZnCl_2$), using a single-replacement reaction.

83. Write a reaction showing the preparation of each of the following salts, using a single-replacement reaction: (a) magnesium chloride, (b) zinc sulfate.

84. Write a reaction showing the preparation of each of the following salts, using a single-replacement reaction: (a) magnesium bromide, (b) copper(II) sulfate.

85. Write a reaction showing the preparation of each of the following salts, using an acid–base reaction: (a) potassium phosphate, (b) calcium nitrate.

86. Write a reaction showing the preparation of each of the following salts, using an acid–base reaction: (a) lithium phosphate, (b) calcium chloride.

87. Write a reaction for the preparation of each of the following salts by the direct combination of their elements: (a) KCl, (b) $CaBr_2$.

88. Write a reaction for the preparation of each of the following salts by the direct combination of their elements: (a) KI, (b) $CaCl_2$.

Learning Goal 16: Titration

89. Calculate the normality of an HNO_3 solution if 30.0 mL of the solution are neutralized by 75.0 mL of 0.25N NaOH.

90. Calculate the normality and the molarity of an HCl solution if 125 mL of the solution are neutralized by 25.0 mL of 1.5N NaOH.

91. Calculate the normality and the molarity of a solution of acetic acid if 45.00 mL of the solution are neutralized by 50.00 mL of 0.2500N (0.2500M) NaOH.

92. A 35.0-mL sample of NaOH solution is titrated with 75.0 mL of 0.150N lactic acid. Calculate the normality of the NaOH solution.

93. A 25.0-mL sample of vinegar is titrated with 30.00 mL of 0.150N (0.150M) NaOH solution. Calculate the normality and the molarity of the vinegar.

94. A 35.00-mL sample of KOH solution required 25.00 mL of 1.000N H_2SO_4 to neutralize it. Calculate the concentration of the KOH solution.

95. If 50.0 mL of 1.00M HCl solution neutralize 75.0 mL of $Ca(OH)_2$ solution, what is the molarity of the $Ca(OH)_2$ solution?

96. Calculate the molarity of an NaOH solution if 25.0 mL of 0.150M HCl are needed to titrate 30.0 mL of the NaOH solution.

Extra Exercises

97. Consider the reaction PH_3 + NaH $\longrightarrow$
$$NaPH_2 + H_2.$$
(*a*) Identify PH_3 as a Brønsted–Lowry acid or base.
(*b*) Identify NaH as a Brønsted–Lowry acid or base.

98. Consider the reaction NH_3 + HCl $\longrightarrow$
$$(NH_4)^{1+} + Cl^{1-}.$$
(*a*) Identify NH_3 as a Brønsted–Lowry acid or base.
(*b*) Identify HCl as a Brønsted–Lowry acid or base.

99. Consider the reaction $(HSO_4)^{1-}$ + $(H_3O)^{1+}$ $\longrightarrow$
$$H_2SO_4 + H_2O.$$
(*a*) Identify $(HSO_4)^{1-}$ as a Brønsted–Lowry acid or base.
(*b*) Identify $(H_3O)^{1+}$ as a Brønsted–Lowry acid or base.

100. Calculate the pH and $[OH^{1-}]$ of a 0.001M solution of NaOH.

101. Complete and balance the following equations.
(*a*) HCl + NaOH $\longrightarrow$
(*b*) BaO + HCl $\longrightarrow$
(*c*) Zn + H_2SO_4 $\longrightarrow$

102. Show by a reaction with water that each of the following is a base.
(*a*) $(OH)^{1-}$ (*b*) NH_3 (*c*) $(HSO_4)^{1-}$

103. Show by a reaction with water that each of the following is an acid.
(*a*) H_2O (*b*) $(HCO_3)^{1-}$ (*c*) HNO_3

104. Many midwestern industries give off large amounts of sulfur trioxide, SO_3, through their smokestacks. Explain why the pH of lakes in New England is decreasing.

105. Explain why phosphorus trichloride, PCl_3, might be a Brønsted–Lowry base.

106. Suggest a way to prepare each of the following via a correctly balanced equation.
(*a*) Na_2SO_4 (*b*) $ZnCl_2$ (*c*) KCl (*d*) H_2SO_3
[*Hint:* Choose your reactants, and then use them in a combination reaction, single-replacement reaction, or double-replacement reaction.]

16 Chemical Kinetics and Chemical Equilibrium

Learning Goals

After you have studied this chapter, you should be able to:

1. Define the terms *chemical kinetics* and *chemical equilibrium*, and state LeChatelier's principle.

2. Define the terms *reaction mechanism*, *activated complex*, and *activation energy*.

3. List the factors that affect the rate of a chemical reaction and those that affect chemical equilibrium, and describe the effect of each.

4. Explain how the rates of the forward and reverse reactions change as a chemical reaction approaches equilibrium.

5. Explain what is meant by a reversible reaction.

6. Write the equilibrium expression for a chemical reaction and calculate the equilibrium constant K_{eq}, given the necessary information.

7. Calculate the ionization constant K_a for a weak acid or K_b for a weak base, given the appropriate data.

8. Calculate the equilibrium concentrations of reactants and products, given the ionization constant K_a or K_b and the initial concentrations of the reactants.

9. Write the expression for the solubility product constant K_{sp} for a slightly soluble salt, and calculate K_{sp}, given the appropriate data.

10. Calculate the solubility of a salt when you are given its solubility product constant K_{sp}.

11. Explain what is meant by the term *buffer*.

12. Explain how the carbonic acid/bicarbonate buffer system maintains the pH of the blood at a fairly constant level.

Introduction

In water, hydrogen ions, H^{1+}, react with hydroxide ions, $(OH)^{1-}$, to form water, H_2O. This reaction occurs in less than a millionth of a second. Water is also formed when hydrogen gas, H_2, reacts with oxygen gas, O_2. But this reaction is so slow that if we mixed the two gases together today and left them, we could come back in the year 2000 and find hardly any water. Obviously, these two reactions occur at vastly different rates.

As another example of very different reaction rates, compare the oxidation of gunpowder (an explosion) with the oxidation of iron (rusting). Gunpowder explodes almost instantaneously, but the rusting of iron is a slow reaction that takes place over a long period of time.

In this chapter we discuss several ideas that are related to chemical reaction rates. We look at the process by which reactants combine to form products. We discuss the dynamic equilibrium that exists between reactants and products in many chemical reactions. And we talk about the factors that affect both this equilibrium and the rate at which the reaction takes place.

16.1 Chemical Kinetics: Rates of Reaction and the Reaction Mechanism

Learning Goal 1
Chemical kinetics

Learning Goal 2
Mechanism of reactions

Chemical kinetics is *the study of how chemical reactions occur and the rates at which they occur.* One theory of chemical kinetics explains how molecules and atoms react. According to this **collision theory,** a chemical reaction can occur only if reactant molecules collide with each other. However, a simple collision is not enough. The molecules must possess a certain amount of energy. We can say that there is an energy barrier—a certain minimum amount of energy—that is necessary for a reaction to take place. As Figure 16.1 shows, this energy barrier is larger than the energy of the reactant molecules or the energy of the product molecules.

Thus the reactant molecules—if they are to react—must pick up some extra energy. Molecules that have this extra energy are said to be **activated.** And the total energy needed to overcome the energy barrier is called the **activation energy.** When two activated reactant molecules (say, one activated molecule X_2^* and one activated molecule Y_2^*) collide, they join to form an activated complex, which may be represented by the symbol $X_2^* Y_2^*$. This activated complex is not very stable (that is, it does not last long). It is simply part of the path of a chemical reaction: **the reaction mechanism.** However, the molecules that make up the activated complex do possess the required energy of activation. So when the activated complex breaks down, it can form the reaction product XY, or the reverse, the reactants X_2 and Y_2 again (Figure 16.2).

Sometimes two colliding reactant molecules have enough energy to form an activated complex, but they do not have the proper *orientation.*

Figure 16.1
Reactant molecules must attain enough energy to overcome this energy barrier. Once the energy barrier is overcome, reaction can take place in either direction.

That is, the wrong parts of the molecules are touching, so the X–Y bond cannot form. In that case, there is no activated complex formed that can lead to the product XY. A collision that does lead to a chemical reaction is called an *effective* collision.

On the basis of this collision theory, we can see that several factors influence the rate at which a chemical reaction occurs. Of course, the *reactants* themselves affect their reaction rate. That is, different sets of reactants react at different rates, depending on their physical and chemical

Learning Goal 3
Factors that affect the rate of a chemical reaction

Energy level of activated complex

$X_2^* Y_2^*$

Activation
energy

$X_2 + Y_2$
Energy level
of reactants

2XY

Energy level of products

Progress of the reaction ⟶

Figure 16.2
The activation energy is the amount of energy needed to form the activated complex through which the forward or reverse reaction occurs.

properties. But, for a given set of reactants, the *concentration*, the *temperature*, and the *presence of catalysts* influence the rate at which the reaction proceeds.

Concentration

The greater the *concentration* of a reactant, the more molecules are present and the more often collisions take place. Therefore, increasing the concentration increases the reaction rate. We can speed up a slow reaction by adding more of one reactant. This makes it easier for collisions to occur. Collisions between molecules and atoms may be likened to the conditions in a crowded roller-skating rink. When many skaters are present on the rink, collisions occur often. When just a few skaters are present, collisions are less likely to occur.

Temperature

By increasing the *temperature* at which a reaction takes place, we increase the energy of the reactant particles. This has three main effects. (1) It causes the reactant particles to move about more rapidly so collisions occur more frequently. (2) It causes more reactant particles to attain activation energy. (3) It causes more product particles to attain activation energy. For these reasons, more collisions occur in a given time and more of them are effective collisions, so the reaction rate increases. (*It has been found that an increase of about* 10°C *often doubles the rate at which a reaction takes place.*) Returning to our skating-rink analogy, we note that when the skaters are moving about on the rink at greater speeds, they are more likely to collide with one another.

Presence of Catalysts

A *catalyst* is a substance that increases the rate at which a reaction occurs. A catalyst remains unchanged at the end of a reaction. It does not change the concentrations or the temperature of the reactants. It does, however, present a different path for a reaction, a path with a lower energy barrier. It makes possible a new activated complex whose lower activation energy increases the chance of effective collisions (Figure 16.3). Because more reactant molecules can attain this lower activation energy, the reaction rate increases.

The rate of reaction between hydrogen gas and oxygen gas, for example, has a very high activation energy; it proceeds very slowly. The addition of a catalyst consisting of finely divided platinum lowers the activation energy and allows the reaction to take place at a much higher rate. The platinum is unchanged when the reaction ends.

Figure 16.3
The presence of a proper catalyst presents a new path with a lower activation energy.

16.2 Reversible Chemical Reactions

Learning Goal 5
Reversible reactions

In previous chapters, we have considered mainly irreversible reactions. An **irreversible reaction** is *a chemical reaction in which one or more reactants form one or more products—but the products cannot then react to form the reactants.* An irreversible reaction stops when one of the reactants is used up. For example, when a raw egg is cooked and has turned into a hard-boiled egg, or when ingredients are mixed and baked to form a cake, an irreversible reaction has occurred. It is not possible to unboil the egg or unbake the cake. When an irreversible reaction stops, we say that it has *gone to completion.*

But many reactions are reversible. A **reversible reaction** is *a chemical reaction in which the products, once they are formed, can react to yield the original reactants.* We can represent the general case of a reversible reaction as

$$A + B \; \rightleftharpoons \; C + D$$

where A and B are the original reactants, and C and D are the original products formed. The reaction $A + B \longrightarrow C + D$ is called the *forward* reaction, and $C + D \longrightarrow A + B$ is called the *reverse* reaction. The double arrow tells us that the reaction is reversible. The reaction *does not* go to completion, because products C and D are reacting to form reactants A and B at the same time that A and B are reacting to form C and D.

However, the system eventually attains an equilibrium condition. In a **chemical equilibrium system,** *the forward reaction and the reverse reaction occur at the same rate.* Figure 16.4 shows how this equilibrium condition comes about. The forward reaction starts off at a high rate, because the concentration of reactant particles is high. However, as the reaction proceeds, products are formed and the concentration of reactants decreases, so the forward reaction rate decreases. At first the concentration of prod-

Learning Goal 1
Chemical equilibrium

Figure 16.4
When the forward and reverse reactions are proceeding at the same rate, equilibrium has been achieved.

ucts is zero, so the reverse reaction rate is zero. As products are formed by the forward reaction, the concentration of products increases, so the reverse reaction rate increases. Eventually (at equilibrium), the two reaction rates become and remain equal (Figure 16.5).

In a chemical equilibrium, both products and reactants are being formed, so a chemical equilibrium is a *dynamic* equilibrium. However, equilibrium does not mean that equal *amounts* of products and reactants are present—only that they are being formed at equal *rates*.

Learning Goal 4
Rates of reaction and equilibrium

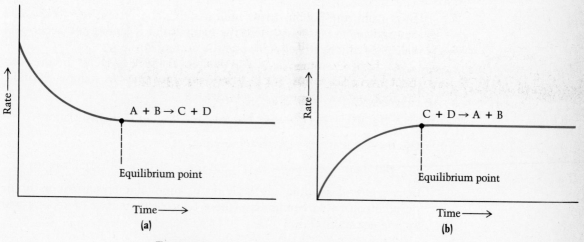

Figure 16.5
Rates of formation of products and reactants before and at equilibrium. (a) In the forward reaction, the reactants combine to form products. Note that the forward reaction starts out fast but slows down as equilibrium is approached. (b) In the reverse reaction the products recombine to form reactants. Note that the reverse reaction starts out slowly but increases in speed as equilibrium is approached.

In Chapter 15 we explain dynamic equilibrium by using the analogy of the people eating in a restaurant during a busy lunch hour. The restaurant can seat 200 people, and all the seats are taken by noon. All the seats are still taken 15 minutes later, but people are constantly entering and leaving the restaurant. This same situation continues over the next two hours. In other words, during lunchtime at this restaurant, there is a population equilibrium. As a given number of people finish eating and leave the restaurant, the same number of hungry people enter it.

As a specific example of a reversible reaction, consider a solution containing 1 mole of acetic acid ($HC_2H_3O_2$) in 1 liter of solution at 25°C. Analysis of the solution would show that most of the acetic acid is in the form of molecules and that only about 0.4% is in the form of hydrogen ions and acetate ions. That is, the forward reaction, in which the acetic acid dissociates into ions, is only about 0.4% complete.

$$HC_2H_3O_2 \xrightarrow{H_2O} H^{1+} + (C_2H_3O_2)^{1-} \quad \text{(about 0.4\% complete)}$$

If you begin with $1M$ ions of H^{1+} and $(C_2H_3O_2)^{1-}$, however, the reaction is 99.6% complete.

$$H^{1+} + (C_2H_3O_2)^{1-} \xrightarrow{H_2O} HC_2H_3O_2 \quad \text{(about 99.6\% complete)}$$

Thus an equilibrium system exists. It can be represented by a single equation:

$$HC_2H_3O_2 \rightleftharpoons H^{1+} + (C_2H_3O_2)^{1-}$$

Once again, the double arrow indicates that the system is *reversible* and that an *equilibrium* exists. Here the equilibrium is among the acetic acid molecules, the hydrogen ions, and the acetate ions.

We have examined a few reversible chemical systems in previous chapters. For example, we discussed the *dissociation* of sodium chloride. The forward reaction may be represented as

$$NaCl(s) + H_2O \longrightarrow Na^{1+}(aq) + Cl^{1-}(aq) \quad \text{(Dissociation)}$$

The reverse reaction may be represented as

$$Na^{1+}(aq) + Cl^{1-}(aq) \longrightarrow NaCl(s) + H_2O \quad \text{(Crystallization)}$$

In a *saturated* solution of NaCl, dissociation and crystallization occur simultaneously. At equilibrium, these two reactions may be represented by one equation:

$$NaCl(s) \underset{H_2O}{\rightleftharpoons} Na^{1+}(aq) + Cl^{1-}(aq)$$

An example of a reversible reaction that is *visible* is the conversion of nitrogen dioxide (NO_2), a colored gas, to dinitrogen tetraoxide (N_2O_4), a colorless gas. This reaction is very sensitive to temperature. When pure NO_2 gas (which is yellow-brown) is sealed in a container, the intensity of

Mixture of
NO_2 and N_2O_4
(lightly colored
gas mixture)

NO_2
(yellow-brown
gas)

30°C   50°C

Figure 16.6
The interconversion of
colored NO_2 to colorless N_2O_4
is temperature-sensitive.

its color changes with temperature, indicating that different amounts of
$NO_2(g)$ and $N_2O_4(g)$ exist in equilibrium (Figure 16.6). The equilibrium
reaction may be represented as follows:

$$2NO_2(g) \underset{\text{Heating}}{\overset{\text{Cooling}}{\rightleftharpoons}} N_2O_4(g)$$

Colored   Colorless

16.3 The Equilibrium Constant

Learning Goal 6
Equilibrium expression and
equilibrium constant

In 1862 Marcellin Berthelot and Jean de Saint Gilles were experimenting
with the reaction in which ethyl alcohol and acetic acid combine to produce ethyl acetate and water:

Ethyl alcohol + acetic acid $\longrightarrow$ ethyl acetate + water

The two scientists mixed various amounts of the reactants and then measured the concentrations of the reactants and products. (We now know
that they were measuring *equilibrium concentrations*.) They found that, in
every case, no matter how much of each reactant they started with, the
concentrations of the products and reactants had the same relation to
each other. They then mixed ethyl acetate and water and found that the
reaction was reversible. And again they found the same kind of relation
among the concentrations of the reactants and products.

Berthelot and Saint Gilles had discovered that, for an equilibrium system, the ratio of the product of the *product concentrations* to the product of
the *reactant concentrations* is a constant. That is,

$$\frac{\text{Product of } product \text{ concentrations}}{\text{Product of } reactant \text{ concentrations}} = \text{constant } K_{eq} \qquad \text{(At temperature } T)$$

regardless of the initial concentrations of the reactants. The concentrations are usually expressed in moles per liter, and the constant K_{eq} is called the **equilibrium constant** for the reaction.

For the general reversible reaction

$$A + B \rightleftharpoons C + D$$

we would write the equilibrium constant as

$$K_{eq} = \frac{[C][D]}{[A][B]}$$

where, as usual, the brackets mean "concentration in moles per liter." This expression for K_{eq} is called the **equilibrium expression.** If we measured the concentrations of the reactants and products at equilibrium at a given temperature, we could substitute them into the equilibrium expression and find the value of K_{eq} for that temperature. (The equilibrium constant changes with the temperature but is not affected by the initial concentrations of the reactants.)

Suppose we wanted to find the equilibrium expression for the reaction

$$H_2(g) + I_2(g) \rightleftharpoons 2HI(g)$$

How would we handle the coefficient 2 on the right side of the equation? One way to do so is to note that we can write the chemical equation without coefficients:

$$H_2(g) + I_2(g) \rightleftharpoons HI(g) + HI(g)$$

Then, as above, we can write the equilibrium expression as

$$K_{eq} = \frac{[HI][HI]}{[H_2][I_2]}$$

Now, because $[HI][HI]$ is $[HI]^2$, we write

$$K_{eq} = \frac{[HI]^2}{[H_2][I_2]}$$

There really is no need to go through these intermediate steps. Instead, we simply write the coefficients of the balanced equation as exponents in the equilibrium expression. Thus, for the general reversible reaction

$$aA + bB \rightleftharpoons cC + dD$$

the equilibrium expression is

$$K_{eq} = \frac{[C]^c[D]^d}{[A]^a[B]^b}$$

Example 16.1

Write the equilibrium expression for the reaction

$$N_2(g) + 3H_2(g) \rightleftharpoons 2NH_3(g)$$

Solution
Apply the method we just discussed, being sure to place the product concentration in the numerator of the ratio and the reactants in the denominator.

$$K_{eq} = \frac{[NH_3]^2}{[N_2][H_2]^3}$$

Practice Exercise 16.1   Write the equilibrium expression for the reaction

$$CO(g) + Cl_2(g) \rightleftharpoons COCl_2(g)$$

Example 16.2

Calculate the equilibrium constant for the reaction

$$2NO(g) + O_2(g) \rightleftharpoons 2NO_2(g)$$

given that the equilibrium molar concentrations (at a particular temperature) are as follows: $[NO_2] = 8.8$, $[NO] = 1.2$, and $[O_2] = 1.6$ mole/L.

Solution
First write the equilibrium expression for the reaction from the balanced equation.

$$K_{eq} = \frac{[NO_2]^2}{[NO]^2[O_2]}$$

Next enter the concentrations and calculate K_{eq}.

$$K_{eq} = \frac{(8.8)^2}{(1.2)^2(1.6)} = 34$$

Practice Exercise 16.2   When carbohydrates are broken down in the body, one important step is the conversion of glucose-1-phosphate to glucose-6-phosphate. The reaction may be represented as follows:

$$\text{Glucose-1-phosphate} \rightleftharpoons \text{glucose-6-phosphate}$$

Calculate K_{eq} for this reaction, given that the concentrations at equilibrium at body temperature are as follows: glucose-1-phosphate, 0.0050 mole/liter; and glucose-6-phosphate, 0.10 mole/liter.

Example 16.3

Calculate K_{eq} for the reaction

$$PCl_5(g) \rightleftharpoons PCl_3(g) + Cl_2(g)$$

given that the equilibrium concentrations at 25°C are as follows: PCl_5, 0.0096 mole/liter; Cl_2, 0.0247 mole/liter; and PCl_3, 0.0247 mole/liter.

Solution
First write the equilibrium expression for the reaction from the balanced equation.

$$K_{eq} = \frac{[PCl_3][Cl_2]}{[PCl_5]}$$

Next enter the concentrations and find K_{eq}.

$$K_{eq} = \frac{(0.0247)(0.0247)}{0.0096} = 0.064$$

Practice Exercise 16.3   Use your result from Practice Exercise 16.2 to calculate the concentration of glucose-6-phosphate at equilibrium, if the concentration of glucose-1-phosphate is 0.10 mole/liter. (Assume that the temperature does not change.)

The magnitude of the equilibrium constant K_{eq} is a good indication of how far the reaction goes toward completion. A large equilibrium constant means that the concentrations of the products are much larger than the concentrations of the reactants at equilibrium. Therefore the reaction tends toward completion. In other words, the forward reaction is favored. A small equilibrium constant means that the concentrations of the products are much smaller than the concentrations of the reactants. Therefore the reverse reaction is favored. In Example 16.2, K_{eq} is relatively large. This means that the forward reaction is favored. In Example 16.3, K_{eq} is relatively small. This means that the reverse reaction is favored. The larger the equilibrium constant at a given temperature, the more highly favored the formation of products. The smaller the equilibrium constant, the more highly favored the formation of reactants.

16.4 Equilibrium Concentrations

Chemical equilibrium can occur only in a reaction that does not go to completion. There is a point in time, called the equilibrium point, at which the rate of the forward reaction is equal to the rate of the reverse reaction. This means that products are being formed at the same rate as reactants. It also means that products are being formed as fast as they are used up to yield reactants. And reactants are being formed as fast as they are used up to yield products. So the concentrations of products and the concentrations of reactants do not change as long as the equilibrium condition is maintained. *Again, we stress the fact that this does not mean that these concentrations are equal. It means only that the concentrations remain unchanged.*

If we know how far a reversible reaction proceeds toward completion, we can determine the equilibrium concentrations of the reactants and products. (The reactants and products are called the *species* taking part in the reaction.) For example, we saw earlier that, at 25°C, a solution of acetic acid molecules is about 0.4% ionized. Then, at equilibrium and 25°C, a $1M$ solution of acetic acid would contain 0.996 mole of $HC_2H_3O_2$ molecules, 0.004 mole of H^{1+} ions, and 0.004 mole of $(C_2H_3O_2)^{1-}$ ions:

$$HC_2H_3O_2 \xrightarrow{H_2O} H^{1+} + (C_2H_3O_2)^{1-} \quad \text{(At equilibrium, 25°C)}$$

$$\underset{\text{0.996 mole}}{} \qquad\qquad \underset{\text{0.004 mole}}{} \quad \underset{\text{0.004 mole}}{}$$

The concentrations are then $0.996M$, $0.004M$, and $0.004M$, respectively. To find these concentrations, we simply find the amounts of the various species that are present at equilibrium, as in the example that follows.

Example 16.4

Into a 1-liter vessel are introduced 1 mole of H_2 gas and 1 mole of I_2 gas at 440°C. These gases are allowed to react to produce HI gas, according to the reaction

$$H_2(g) + I_2(g) \rightleftharpoons 2HI(g)$$

How many moles of each gas are present at equilibrium? The forward reaction is 78.2% complete at equilibrium.

Solution

We started with 1 mole of H_2 and 1 mole of I_2. If the reaction is 78.2% complete at equilibrium, then

? moles of H_2 have reacted = (1.000 mole)(0.782) = 0.782 mole

Because H_2 and I_2 react in the ratio of 1 mole to 1 mole, 0.782 mole of H_2 and 0.782 mole of I_2 have reacted. Then the number of moles of H_2 (and I_2) that *have not reacted* (that is, the number that remain) is $1.000 - 0.782 = 0.218$ mole.

But what about the amount of HI that is produced? According to the balanced equation, for each mole of H_2 and I_2 that reacts, 2 moles of HI are produced. Because 0.782 mole of H_2 and 0.782 mole of I_2 have reacted,

? mole of HI = (2)(0.782) moles = 1.564 moles

So 1.564 moles of HI are produced. We can now summarize the equilibrium condition as follows:

$$H_2(g) \; + \; I_2(g) \; \rightleftharpoons \; 2HI(g) \qquad \text{(At equilibrium)}$$

0.218 mole   0.218 mole   1.564 moles

Practice Exercise 16.4   Into a 1-liter vessel at 400°C are introduced 1 mole of the hypothetical gas A_2 and 1 mole of the hypothetical gas B_2. The gases react to produce AB, according to the reaction

$$A_2(g) + B_2(g) \; \rightleftharpoons \; 2AB(g)$$

How many moles of each gas are present at equilibrium? The forward reaction is 60.0% complete at equilibrium.

16.5 LeChatelier's Principle

Learning Goal 1
LeChatelier's principle

More than a century ago, the French chemist Henri LeChatelier experimented with solutions containing reactants and products at equilibrium. He observed what happened when the equilibrium was disturbed in some way—that is, when a *stress* was applied to the system. He found that the system adjusted itself to a new equilibrium in such a way as to relieve the stress. This observation is now known as **LeChatelier's principle:** *When a stress is applied to a system at equilibrium, the system adjusts to a new equilibrium position, if possible, in such a way as to reduce the effect of the stressful condition.*

A chemical system adjusts its equilibrium only by shifting to new equilibrium concentrations of products and reactants. Depending on the nature of the stress, the formation of products may be favored (the equilibrium shifts to the right through a net forward reaction) or the formation of reactants may be favored (the equilibrium shifts to the left through a net reverse reaction). In each case, a new equilibrium system results. We shall look at the effects of three types of stresses on chemical systems at equilibrium.

Effect of a Change in Concentration

When a system is at equilibrium and we change the concentration of one of its components, a stress is placed on the system. For example, suppose that the system

$$CO_2(g) + H_2(g) \; \rightleftharpoons \; CO(g) + H_2O(g)$$

is at equilibrium. According to LeChatelier's principle, increasing the concentration of one of the reactants (for example, adding CO_2) would shift the equilibrium to the right. That is, the system would act to reduce the stress—here, the excess CO_2—by using it to form more product. The result would be a new equilibrium with greater concentrations of the products and a lower concentration of H_2. The concentration of CO_2 would be greater than the original concentration but lower than the increased concentration that caused the shift in equilibrium.

According to LeChatelier's principle, if the concentration of one of the *products* is increased, the equilibrium shifts to the left. That is, the system acts to relieve the excess of product by using it up to form more reactant. At the new equilibrium point, the concentrations of reactants are higher and the concentrations of products are lower.

Example 16.5

Consider the equilibrium system

$$A(g) + B(g) \rightleftharpoons C(g) + D(g)$$

Use LeChatelier's principle to predict the direction in which the equilibrium shifts when:
(a) The concentration of species A is increased.
(b) The concentration of species B is decreased.
(c) The concentration of species C is decreased.
(d) The concentration of species D is increased.

Solution

(a) When more A is added, the equilibrium shifts to the *right*.
(b) When some B is removed, the equilibrium shifts to the *left*. Decreasing the concentration of a reactant has the same effect as increasing the concentration of a product.
(c) When the concentration of C is decreased, the equilibrium shifts to the *right*. Decreasing the concentration of a product has the same effect as increasing the concentration of a reactant.
(d) When the concentration of D is increased, the equilibrium shifts to the *left*.

Practice Exercise 16.5   Consider the equilibrium system

$$4HCl(g) + O_2(g) \rightleftharpoons 2H_2O(g) + 2Cl_2(g)$$

Use LeChatelier's principle to predict the direction in which the equilibrium shifts when:
(a) The concentration of $HCl(g)$ is decreased.
(b) The concentration of $O_2(g)$ is increased.
(c) The concentration of $H_2O(g)$ is increased.
(d) The concentration of $Cl_2(g)$ is decreased.

Effect of a Change in Pressure or Volume

Recall from your study of the gas laws that the pressure and volume of a gas are inversely related. If we increase the pressure on a system of gas molecules, the volume is decreased. And if we decrease the pressure, the volume is increased. Therefore pressure changes and volume changes should have opposite effects on an equilibrium system composed entirely of gases.

The fewer molecules there are in a given volume of gas, the smaller the stress produced by an increase in pressure. And, according to LeChatelier's principle, an equilibrium system shifts in such a way as to reduce a stress imposed on the system. Therefore, if you *increase* the pressure on a system composed of gas molecules, *the equilibrium shifts in the direction that yields the smaller number of gas molecules*. If you *decrease* the pressure on a system composed of gas molecules, the equilibrium shifts in the direction that yields the greater number of gas molecules. Changes in the volume of a gaseous equilibrium system have the exact opposite effects.

Three additional facts are important in predicting equilibrium shifts due to changes in pressure or volume:

1. The numbers of reactant and product molecules that take part in a reaction (or that are present in the reaction vessel) are proportional to the coefficients of the balanced equation.

2. If the balanced equation for a reaction shows the same number of gas molecules on the right and on the left, then the equilibrium does *not* shift in response to a change in pressure. Shifting either way would do nothing to reduce the stress.

3. In a system consisting of gases along with solids or liquids or both, only the gas molecules are significantly affected by changes in pressure.

Example 16.6

Consider the equilibrium system

$$N_2(g) + 3H_2(g) \rightleftharpoons 2NH_3(g)$$

Predict the direction in which the equilibrium will shift when:
(a) The pressure is increased.
(b) The volume is increased.

Solution

(a) When the pressure is increased, the equilibrium shifts in the direction of *fewer* gas molecules. Because the coefficients on the left side of the balanced equation add up to 4, and the coefficients on the right side add up to 2, the equilibrium will shift to the *right*.

(b) When the volume is increased, the pressure is decreased. And a decrease in pressure shifts the equilibrium in the direction of the *greater* number of gas molecules. Therefore the equilibrium will shift to the *left*.

Practice Exercise 16.6   Consider the equilibrium system

$$CO_2(g) + 2NH_3(g) \rightleftharpoons H_2O(g) + CO(NH_2)_2(s)$$

Predict the direction in which the equilibrium will shift when:
(a) The pressure is increased.
(b) The volume is increased.

Example 16.7

For each of the following equilibrium systems, predict the direction in which the equilibrium will shift when the pressure is increased.
(a) $SiF_4(g) + 2H_2O(g) \rightleftharpoons SiO_2(s) + 4HF(g)$
(b) $H_2(g) + Cl_2(g) \rightleftharpoons 2HCl(g)$

Solution

(a) The equilibrium shifts in the direction of fewer gas molecules. The coefficients of the *gaseous* molecules on the left side of the equation add up to 3. The coefficient for the *gaseous* molecules on the right side of the equation is 4. Therefore the equilibrium will shift to the *left*.
(b) Here the sum of the coefficients of the gases on the left is equal to the "sum" of the coefficients on the right. An equilibrium shift would not reduce the stress due to the increased pressure, and no shift will occur.

Practice Exercise 16.7   For each of the following equilibrium systems, predict the direction in which the equilibrium will shift when the pressure is increased.
(a) $CO(g) + H_2O(g) \rightleftharpoons CO_2(g) + H_2(g)$
(b) $COCl_2(g) \rightleftharpoons CO(g) + Cl_2(g)$

Effect of a Change in Temperature

Chemical reactions that require heat in order to proceed are called *endothermic* reactions. Chemical reactions that produce heat are called *exothermic* reactions (Figure 16.7). When the temperature of an equilibrium system is raised, heat is added to the system. The equilibrium then shifts in the direction that absorbs the added heat—to relieve the

Figure 16.7
Energy diagrams for exothermic and endothermic reactions. (a) An exothermic reaction (energy is released to the environment). (b) An endothermic reaction (energy is absorbed from the environment).

stress caused by the heat. For *endothermic* reactions, an increase in temperature (heat) favors the *forward* reaction. For *exothermic* reactions, an increase in temperature (heat) favors the *reverse* reaction.

An easy way to remember this is to think of heat as a reactant in endothermic reactions or as a product in exothermic reactions. Then an increase in temperature increases the "concentration" of heat. And the equilibrium shifts in such a way as to reduce this concentration.

Example 16.8

For each of the following reactions, predict the direction in which the equilibrium will shift when heat is added.
(a) $C(s) + CO_2(g) + \text{heat} \rightleftharpoons 2CO_2(g)$
(b) $H_2(g) + Cl_2(g) \rightleftharpoons 2HCl(g) + \text{heat}$

Solution

(a) This is an *endothermic* reaction, so the addition of heat will shift the equilibrium to the right. (Increasing the concentration of the *reactant* heat shifts the equilibrium to the right.)
(b) This is an *exothermic* reaction, so the addition of heat will shift the equilibrium to the *left*. (Increasing the concentration of the *product* heat shifts the equilibrium to the left.)

Practice Exercise 16.8   For each of the following reactions, predict the direction in which the equilibrium will shift when heat is added.
(a) $4HCl(g) + O_2(g) \rightleftharpoons 2H_2O(g) + 2Cl_2(g) + heat$
(b) $CO_2(g) + 2NH_3(g) \rightleftharpoons H_2O(g) + CO(NH_2)_2(s) + heat$
(c) $CO_2(g) + H_2(g) + heat \rightleftharpoons CO(g) + H_2O(g)$
(d) $N_2(g) + 3H_2(g) \rightleftharpoons 2NH_3(g) + heat$

Effect of a Catalyst

Recall that catalysts are substances that increase the speed of a chemical reaction but are not themselves chemically changed in the reaction. Experiments have shown that a catalyst has no effect on chemical equilibrium. That is, it does not shift the equilibrium in either direction. Because catalysts change reaction paths, catalysts must increase the forward reaction rate and the reverse reaction rate to exactly the same degree.

Catalysts are used to a great extent, in both the laboratory and commercial preparation of chemicals. In addition, more than 1,000 biological catalysts, or *enzymes*, are known. Enzymes work in our bodies to help us digest our food. They conduct nerve impulses throughout our bodies and break down food components into smaller molecules. If even one enzyme is missing in the body, the results can be disastrous.

An example of a disease caused by a missing enzyme is *phenylketonuria*, or PKU for short. This genetically transmitted disease results in a deficiency of the enzyme phenylalanine hydrolase, which is produced in the liver and which converts the amino acid phenylalanine in protein-containing foods. A baby that is born with PKU may be unusually irritable, may have epileptic seizures, and may vomit often. Phenylacetic acid appears in the urine and sweat and can be detected by its odor. About two-thirds of the babies that have elevated levels of phenylalanine show severe mental retardation. This disease is treated by lowering the amount of phenylalanine in the diet, not by eliminating it altogether. Babies are fed Lofenelac (a formula fortified with vitamins, iron, and other minerals) instead of milk, which contains phenylalanine. Low-protein foods such as fruits, vegetables, and cereals are also included in the diet. With such treatment, mental retardation can be avoided and many other problems associated with this condition can be reversed.

How LeChatelier's Principle Can Be Used Practically

Perhaps the best example of how LeChatelier's principle has been put to practical use in controlling chemical equilibria is the synthesis of ammonia (NH_3). The industrial synthesis of ammonia, which is the fourth-most-important major chemical produced commercially in the United States, is as follows:

$$3H_2(g) + N_2(g) \rightleftharpoons 2NH_3(g) + 22.08 \text{ kcal}$$

If the pressure is increased, the equilibrium shifts to the right, favoring the formation of ammonia. Because ammonia is formed in an exothermic reaction, adding heat would shift the equilibrium to the left, which would *not* favor the formation of ammonia. Therefore, if chemists want to increase the maximum yield of ammonia, the reaction should take place at low temperature and high pressure, in accordance with LeChatelier's principle. Manufacturing plants produce ammonia at pressures of about 1,000 atm and moderate temperatures of about 500°C. At temperatures below 500°C the reaction occurs too slowly. In addition, a catalyst is added to speed up the reaction.

The story of the production of ammonia is an interesting one that illustrates the relationship between science and society. Thomas Malthus theorized that, although food supply grows at a steady rate, population grows many, many times faster. He predicted that population would eventually outstrip agricultural production. In 1890 Sir William Crookes, addressing the British Association for the Advancement of Science, discussed the problem of the *fixed nitrogen* supply. Nitrogen, although it composes 79% of the air, is not useful when in the form of the unreactive N_2 molecule. Atmospheric nitrogen must be *fixed* in order to be used by plants. To *fix* nitrogen means to stabilize it and change it from a gaseous to a solid or liquid form, in which it is bound with another element or with more than one element. Some forms of fixed nitrogen are ammonia, nitrite ions, and nitrate ions. Certain bacteria and algae are capable of fixing atmospheric nitrogen. These organisms are found free in nature. Some live in colonies on the roots of plants called legumes. Crookes maintained that, if people were to have adequate food supplies, scientists must figure out a way to fix atmospheric nitrogen, because nitrogen-containing fertilizers were the only hope for growing enough food. The source of nitrogen used for fertilizers at the time was limited to manures and to sodium nitrate, which is found in Chile.

The Germans were interested in producing fixed nitrogen. A large chemical company called Badische Anilin-und-Sodafabrik appealed to Fritz Haber, a professor of physical chemistry at Karlsruhe, for help. Haber had read a paper published by LeChatelier in 1901, which stated that increased pressure resulted in definite amounts of ammonia being formed from its elements. Haber pursued this path. By 1911 the German chemical industry had built a plant to produce ammonia from its elements by using high pressure, moderate temperatures, and the proper catalyst.

In August 1914, when World War I broke out, the German chemical industry was able to supply ammonium nitrate for use as an explosive. A discovery that was intended for use in fertilizer, to grow food, was used to produce explosives. In 1918 Haber was awarded the Nobel Prize in chemistry for his discovery.

16.6 Ionization Constants

Learning Goal 7
Calculation of K_a and K_b

As we saw in Chapter 15, weak acids and weak bases ionize only partially in solution. An equilibrium is eventually established between the undissociated acid or base (the "reactant") and its ions (the "products"). We can write an equilibrium expression for the dissociation, just as we do for a reaction involving molecules. Then, if we can obtain the equilibrium concentrations of the various species at a particular temperature, we can calculate the equilibrium constant for the acid or base at that temperature. Such an equilibrium constant is called the **ionization constant** or **dissociation constant** of the acid or base. It is usually given the symbol K_a for an acid and K_b for a base.

Example 16.9

Calculate K_a for acetic acid at 25°C if measured equilibrium concentrations are as follows:

$$HC_2H_3O_2 \quad \overset{H_2O}{\rightleftharpoons} \quad H^{1+} \quad + \quad (C_2H_3O_2)^{1-}$$
$$0.09986M \qquad\qquad 0.00134M \qquad 0.00134M$$

Solution

Write the equilibrium expression for the reaction, enter the concentrations of the species, and calculate K_a.

$$K_a = \frac{[H^{1+}][(C_2H_3O_2)^{1-}]}{[HC_2H_3O_2]}$$

$$= \frac{(0.00134)(0.00134)}{0.09986} = 1.80 \times 10^{-5}$$

This small value of K_a tells us that the equilibrium favors the reverse reaction—the formation of acetic acid.

Practice Exercise 16.9   Calculate K_a for nitrous acid, HNO_2, given the following equilibrium concentrations:

$$HNO_2 \quad \rightleftharpoons \quad H^{1+} \quad + \quad NO_2^{1-}$$
$$0.0935M \qquad 6.5 \times 10^{-3}M \quad 6.5 \times 10^{-3}M$$

Learning Goal 8
Calculation of equilibrium concentrations

Table 16.1 lists the ionization constants of some acids and bases at 25°C. Once you know K_a or K_b for a substance at a particular temperature, you can use it to calculate the concentrations of ions and molecules in an equilibrium solution of that substance.

Table 16.1
Ionization Constants of Weak Acids and Bases at 25°C

Substance	Formula	K_a or K_b
Acetic acid	$HC_2H_3O_2$	1.8×10^{-5}
Benzoic acid	$HC_7H_5O_2$	6.3×10^{-5}
Cyanic acid	$HCNO$	2.0×10^{-4}
Formic acid	$HCHO_2$	1.8×10^{-4}
Hydrocyanic acid	HCN	4.0×10^{-10}
Hypochlorus acid	$HClO$	3.5×10^{-8}
Nitrous acid	HNO_2	4.5×10^{-4}
Ammonium hydroxide	NH_4OH	1.8×10^{-5}
Copper(II) hydroxide	$Cu(OH)_2$	1.0×10^{-8}
Zinc hydroxide	$Zn(OH)_2$	4.0×10^{-5}

Example 16.10

Calculate the equilibrium concentrations of H^{1+}, $(C_2H_3O_2)^{1-}$, and $HC_2H_3O_2$ in a 0.50M acetic acid solution at 25°C ($K_a = 1.8 \times 10^{-5}$).

Solution
The reaction is

$$HC_2H_3O_2 \; \rightleftharpoons \; H^{1+} + (C_2H_3O_2)^{1-}$$

The initial concentration of $HC_2H_3O_2$ is 0.50M. Let us suppose that x moles of $HC_2H_3O_2$ dissociate. Then, according to the stoichiometry of the reaction, x moles of H^{1+} ions and x moles of $(C_2H_3O_2)^{1-}$ ions are produced. Summarize this as follows:

	$HC_2H_3O_2$	$\rightleftharpoons$	H^{1+}	+ $(C_2H_3O_2)^{1-}$
Init. conc.	0.50 mole/liter		0 mole/liter	0 mole/liter
Conc. at equil.	(0.50 − x) mole/liter		x moles/liter	x moles/liter

Next derive the equilibrium expression from the balanced equation and enter all known values.

$$K_a = \frac{[H^{1+}][(C_2H_3O_2)^{1-}]}{[HC_2H_3O_2]} \quad \text{so} \quad 1.8 \times 10^{-5} = \frac{(x)(x)}{0.50 - x}$$

At this point we could solve the equation for x. However, we would obtain a quadratic equation. Instead we simplify the mathematics by assuming that x is very small compared to 0.50, so that (0.50 − x) is still about 0.50. (We can safely make this simplifying assumption because the value of K_a, 1.8×10^{-5}, tells us that the acetic acid is only slightly dissociated.) Our expression becomes

$$1.8 \times 10^{-5} = \frac{(x)(x)}{0.50}$$

so that

$$9.0 \times 10^{-6} = x^2 \quad \text{and} \quad 3.0 \times 10^{-3} = x$$

Therefore the equilibrium concentration of H^{1+} is 3×10^{-3} mole/liter. The equilibrium concentration of $C_2H_3O_2^{1-}$ is also 3×10^{-3} mole/liter. The concentration of $HC_2H_3O_2$ is $0.50 - 3 \times 10^{-3}$, which is $0.50 - 0.003$, or 0.497, mole/liter. Rounded to the proper number of significant figures, this is 0.50 mole/liter. The initial assumption was good.

Practice Exercise 16.10   Citric acid is found in citrus fruits and is responsible for their characteristic taste. Calculate the equilibrium concentrations of hydrogen ion, citrate ion, and citric acid for a $1.00M$ citric acid solution ($K_a = 8.4 \times 10^{-4}$).

Example 16.11

Morphine, $C_{17}H_{19}NO_3$, a naturally occurring base, is administered medically to relieve pain. The ionization constant K_b for morphine is 1.60×10^{-6} at 25°C. We can represent the ionization of morphine in water by the following equation (let's abbreviate the formula of morphine as Mor.):

$$\text{Mor}(aq) + H_2O \; \rightleftharpoons \; (\text{HMor})^{1+} + (\text{OH})^{1-}$$

Or, because the concentration of the water changes very little, we can simplify the equation to

$$\text{Mor} \; \overset{H_2O}{\rightleftharpoons} \; (\text{HMor})^{1+} + (\text{OH})^{1-}$$

If the initial concentration of Mor is 0.1000 mole/liter, what is the concentration of each species at equilibrium?

Solution

The initial concentration of Mor is 0.1000 mole/liter. If x moles of Mor dissociate, then according to the stoichiometry of the reaction, x moles of $(\text{HMor})^{1+}$ and x moles of $(\text{OH})^{1-}$ ions are produced. This can be summarized as follows:

	Mor $\overset{H_2O}{\rightleftharpoons}$	$(\text{HMor})^{1+}$ +	$(\text{OH})^{1-}$
Init. conc.	0.1000 mole/liter	0 mole/liter	0 mole/liter
Conc. at equil.	(0.1000 − x) mole/liter	x mole/liter	x mole/liter

Derive the expression K_b from the balanced equation and enter all known values.

$$K_b = \frac{[\text{HMor}^{1+}][\text{OH}^{1-}]}{[\text{HMor}]} \quad \text{so} \quad 1.60 \times 10^{-6} = \frac{(x)(x)}{0.1000 - x}$$

If we assume that x is small compared to 0.1000 (on the basis of the value of K_b), then the expression simplifies to

$$1.60 \times 10^{-6} = \frac{(x)(x)}{0.1000}$$

so that

$$1.60 \times 10^{-7} = x \quad \text{and} \quad 4.00 \times 10^{-4} = x$$

Therefore, the equilibrium concentrations of $(\text{HMor})^{1+}$ and $(\text{OH})^{1-}$ add up to 4.00×10^{-4} mole/liter. The equilibrium concentration of Mor is $0.1000 - 4.00 \times 10^{-4}$, which is $0.1000 - 0.0004$, or 0.0996, mole/liter.

Practice Exercise 16.11   The ionization constant for the reaction

$$\text{NH}_3 + \text{H}_2\text{O} \quad \rightleftharpoons \quad (\text{NH}_4)^{1+} + (\text{OH})^{1-}$$

is $K_b = 1.8 \times 10^{-5}$. The initial concentration of NH_3 is 1.0 mole/liter. Calculate the concentration of each species at equilibrium.

16.7 The Solubility Product

Many salts are only slightly soluble in water. For example, the salt silver chloride has a solubility of 1.3×10^{-5} mole/liter at 25°C. In other words, if we place some silver chloride in a liter of water at 25°C, only 1.3×10^{-5} mole of it will dissociate. The rest of the silver chloride will remain undissolved. However, there will (as usual) be a dynamic equilibrium between the ions and the undissolved salt. We can represent this equilibrium with the equation

$$\text{AgCl}(s) \quad \rightleftharpoons \quad \text{Ag}^{1+} + \text{Cl}^{1-}$$

We could write an expression for the equilibrium constant K_{eq} for this reaction. However, the denominator is constant, so we eliminate the denominator: We define the **solubility product** constant K_{sp} of a slightly soluble salt as the product of the concentrations of its ions in a saturated solution at a particular temperature. For silver chloride, then,

$$K_{sp} = [\text{Ag}^{1+}][\text{Cl}^{1-}]$$

We have already noted that the solubility of AgCl at 25°C is 1.3×10^{-5} mole/liter. Therefore, on the basis of the stoichiometry of the ionization reaction, $[Ag^{1+}]$ is 1.3×10^{-5} and $[Cl^{1-}]$ is 1.3×10^{-5}. From these concentrations, we obtain

$$K_{sp} = [Ag^{1+}][Cl^{1-}]$$
$$= (1.3 \times 10^{-5})(1.3 \times 10^{-5}) = 1.7 \times 10^{-10}$$

Generally speaking, the higher the solubility product constant of a salt, the more soluble the salt is in water. For example, the solubility product constants given in Table 16.2 show that barium sulfate ($BaSO_4$) is about three times as soluble as silver chloride at 25°C. And silver acetate is more than a hundred times as soluble as barium sulfate.

Suppose we have 1 liter of a saturated aqueous solution of silver chloride. We know that the concentration of Ag^{1+} ions in this solution is 1.3×10^{-5} mole/liter and that the concentration of Cl^{1-} ions is the same. What would happen to the concentration of Cl^{1-} ions if we increased the concentration of Ag^{1+} ions to 0.010 mole/liter? Would it change? We can use the solubility product to find out.

Because K_{sp} for silver chloride is 1.7×10^{-10}, we have

$$K_{sp} = [Ag^{1+}][Cl^{1-}]$$

$$1.7 \times 10^{-10} = (0.010)[Cl^{1-}]$$

Then

$$[Cl^{-1}] = \frac{1.7 \times 10^{-10}}{0.010} = 1.7 \times 10^{-8}$$

Therefore the concentration of the chloride ion *decreases* to 1.7×10^{-8} mole/liter. (Remember, before we added the excess Ag^{1+}, the Cl^{1-} concentration was much higher: 1.3×10^{-5} mole/liter.) *This result, in which the concentration of one ion is reduced when the concentration of another ion is increased at equilibrium, is called the* **common ion effect.**

Table 16.2
Solubility Product Constants, K_{sp}, at 25°C

Substance	K_{sp}	Substance	K_{sp}
AgCl	1.7×10^{-10}	BaSO$_4$	1.5×10^{-9}
AgBr	5.0×10^{-13}	CaF$_2$	3.9×10^{-11}
AgI	8.5×10^{-17}	CuS	9.0×10^{-45}
AgC$_2$H$_3$O$_2$	2.0×10^{-3}	Fe(OH)$_3$	6.0×10^{-38}
Ag$_2$CrO$_4$	1.9×10^{-12}	Mn(OH)$_2$	2.0×10^{-13}
BaCrO$_4$	8.5×10^{-11}		

Example 16.12

Write the K_{sp} expressions for the following slightly soluble salts.
(a) AgBr (b) $BaCrO_4$ (c) CaF_2

Solution

First write the equilibrium equation. Then use the equation to write the K_{sp} expression.

(a) $AgBr \rightleftharpoons Ag^{1+} + Br^{1-}$
$K_{sp} = [Ag^{1+}][Br^{1-}]$

(b) $BaCrO_4 \rightleftharpoons Ba^{2+} + CrO_4^{2-}$
$K_{sp} = [Ba^{2+}][CrO_4^{2-}]$

(c) $CaF_2 \rightleftharpoons Ca^{2+} + 2F^{1-}$
$K_{sp} = [Ca^{2+}][F^{1-}]^2$

Practice Exercise 16.12   Write the K_{sp} expressions for the following salts. (a) $PbSO_4$ (b) Ag_2CrO_4 (c) AgI

Example 16.13

Calculate the solubility of the salt $BaSO_4$ in grams per liter, given that $K_{sp} = 1.5 \times 10^{-9}$.

Solution

Write the equilibrium equation. Then write the expression for K_{sp} and insert its value.

$$BaSO_4 \rightleftharpoons Ba^{2+} + (SO_4)^{2-}$$

$$K_{sp} = [Ba^{2+}][(SO_4)^{2-}]$$

$$1.5 \times 10^{-9} = [Ba^{2+}][(SO_4)^{2-}]$$

Learning Goal 10
Calculation of the solubility of a salt

The equilibrium equation tells us that the concentrations of Ba^{2+} and $(SO_4)^{2-}$ in solution are equal. We let x represent these concentrations in *moles per liter of solution.* Then

$$1.5 \times 10^{-9} = [Ba^{2+}][(SO_4)^{2-}] \quad \text{or} \quad 1.5 \times 10^{-9} = (x)(x)$$

Therefore

$$1.5 \times 10^{-9} = x^2 \quad \text{and so} \quad 3.9 \times 10^{-5} = x$$

Therefore 3.9×10^{-5} moles of $BaSO_4$ can dissolve per liter of solution. If we want to express the solubility in *grams per liter,* we have to multiply this result by the molecular weight of $BaSO_4$.

$$\frac{?\text{ g}}{\text{liter}} = \left(\frac{3.9 \times 10^{-5} \text{ mole}}{1 \text{ liter}}\right)\left(\frac{236 \text{ g}}{1 \text{ mole}}\right)$$

$$= 9.2 \times 10^{-3} \text{ g/liter}$$

Practice Exercise 16.13 Calculate the solubility in grams per liter of $PbSO_4$, given that $K_{sp} = 1.3 \times 10^{-8}$ for this salt at 25°C.

16.8 Buffer Solutions and the Control of pH

Learning Goal 11
Buffers

A **buffer solution** is *a solution that resists changes in pH when small amounts of strong acids or bases are added to it or when it is diluted with water.* A buffer solution may be made up of

1. A weak acid and its salt, or

2. A weak base and its salt

Pure water has a pH of 7. But if we added even a small amount of an acid or a base to pure water, its pH would change drastically. The top part to Table 16.3 gives two examples of such changes. However, if the water contained a buffer, the pH would change much less, as shown in the bottom part of Table 16.3.

Learning Goal 12
Carbonic acid/bicarbonate
buffer system

To appreciate how a buffer system works, consider the carbonic acid/bicarbonate buffer system, which maintains the pH of our blood at a fairly constant level: Carbon dioxide and water, which are present in the blood, combine to form carbonic acid.

$$H_2O + CO_2 \rightleftharpoons \underset{\text{Carbonic acid}}{H_2CO_3} \qquad (Equation\ 1)$$

Table 16.3
Changes in pH As a Result of Adding HCl and NaOH to Pure Water and to an Aqueous Acetic Acid–Sodium Acetate Buffer Solution

Solution	pH	Change in pH
Pure H_2O (1 liter)	7	—
Add: 0.01 mole HCl	2	−5
Add: 0.01 mole NaOH	12	+5
Buffer solution (1 liter) 0.5M $HC_2H_3O_2$ + 0.5M $NaC_2H_3O_2$	4.74	—
Add: 0.01 mole HCl	4.72	−0.02
Add: 0.01 mole NaOH	4.76	+0.02

As Equation 1 indicates, an equilibrium exists among the water, carbon dioxide, and carbonic acid. However, some of the carbonic acid ionizes to form an equilibrium with hydrogen ions and bicarbonate ions.

$$H_2CO_3 \rightleftharpoons H^{1+} + (HCO_3)^{1-} \qquad \textit{(Equation 2)}$$

Bicarbonate ion

The system composed of Equations 1 and 2 can be represented as follows:

$$H_2O + CO_2 \rightleftharpoons H_2CO_3 \rightleftharpoons H^{1+} + (HCO_3)^{1-} \qquad \textit{(Equation 3)}$$

Equation 3 indicates that the blood contains water, carbon dioxide, carbonic acid, hydrogen ions, and bicarbonate ions in equilibrium with each other. The fact that they are in equilibrium with each other is very important, because under these conditions LeChatelier's principle applies: *If the concentration of any one of these substances changes, the equilibrium shifts in a direction to minimize this change.* For example, if too many hydrogen ions form in the blood, the equilibrium shifts to the left, reducing the number of hydrogen ions. If too much carbon dioxide is present in the blood, the equilibrium shifts to the right, and the concentration of CO_2 is reduced.

The normal pH of blood is 7.4. However, suppose a person ingests a basic substance or has an internal disorder that causes the concentration of hydroxide ions in the blood to increase (a condition known as *metabolic alkalosis*). If the blood had *no* buffer system, its pH would increase greatly. This could prove fatal to the individual. But, because the blood has the benefit of the carbonic acid/bicarbonate buffer system, the following reaction occurs:

$$H_2CO_3 + OH^{1-} \longrightarrow H_2O + (HCO_3)^{1-}$$

The carbonic acid reacts with the hydroxide ions, producing water and bicarbonate ion and maintaining the pH at about 7.4. The excess bicarbonate ion is eventually eliminated via the kidneys and urine.

Now suppose that the concentration of hydrogen ions in a person's blood were to increase suddenly, as the result of the person's ingesting an acidic substance, experiencing kidney failure, or having diabetes mellitus. (This condition is known as *metabolic acidosis*.) The excess hydrogen ions react with the bicarbonate ions, producing carbonic acid. The excess carbonic acid then decomposes to form water and carbon dioxide, which is eventually eliminated from the blood. The reactions can be summarized in this way:

$$(HCO_3)^{1-} + H^{1+} \longrightarrow H_2CO_3 \longrightarrow H_2O + CO_2$$

Again, the pH of the blood remains almost constant.

Buffer systems work very well—within their limits. It is, of course, possible to overpower a buffer system by adding huge amounts of a strong acid or base. This uses up all the buffer ions and causes a drastic change in pH.

Example 16.14

The blood buffer system can be represented, in part, as follows:

$$H_2O + CO_2 \; \rightleftharpoons \; H_2CO_3 \; \rightleftharpoons \; H^{1+} + (HCO_3)^{1-}$$

In *hypoventilation*, the body does not take in enough oxygen. As a result, the level of CO_2 in the blood increases. Describe the effect of hypoventilation on the blood buffer system. What condition would it lead to?

Solution
Hypoventilation leads to a buildup of CO_2, which shifts the equilibrium to the *right*. This leads to an acidosis condition. (In medical terms, this is known as *respiratory acidosis*.)

Practice Exercise 16.14   In terms of the blood buffer system described in Example 16.14, explain which way the equilibrium would shift if a person *hyperventilated*. (A person who is hyperventilating is breathing at an above-normal rate.) What condition would this lead to?

Summary

A reversible reaction is one in which the products can react to yield the reactants. At first, the formation of products (the forward reaction) proceeds at a high rate. This rate decreases as products are formed and the concentration of reactants decreases. The formation of reactants (the reverse reaction) proceeds slowly at first. However, this rate increases as more product is formed in the forward reaction. Eventually, the forward and reverse reaction rates become equal, and the reaction is said to be at equilibrium. This chemical equilibrium is a dynamic one, in which products and reactants are both being formed. The concentrations of products and reactants remain constant as long as this equilibrium condition is maintained.

According to the collision theory, for two species to react, their molecules must collide. In addition, the colliding molecules must be activated. That is, they must possess a certain amount of energy called the activation energy. If they do, they will form an activated complex. And if the reactant molecules are oriented properly, the activated complex can form product molecules.

Increased concentrations of reactants increase the reaction rate by producing more frequent collisions. Increased temperature increases the reaction rate by increasing the velocity and energy of reactant molecules.

Catalysts increase the reaction rate by offering a new path for reaction, a path that involves a lower activation energy.

According to LeChatelier's principle, if a stress is applied to a chemical system at equilibrium, the equilibrium shifts in the direction that tends to relieve the stress. Thus, for a chemical system at equilibrium, an increase in the concentration of one species shifts the equilibrium in such a way as to use up more of that species. An increase in temperature shifts the equilibrium in such a way as to favor an endothermic reaction. An increase in the pressure on (or a decrease in the volume of) gaseous species shifts the equilibrium in such a way as to favor the production of fewer molecules. The introduction of a catalyst has no effect on the equilibrium.

For any reaction, the equilibrium constant K_{eq} is the ratio of the product of the *product concentrations* to the product of the *reactant concentrations* at equilibrium. The larger the equilibrium constant, the more the formation of products is favored at equilibrium. The value of K_{eq} for a given reaction depends on the temperature, but not on the initial concentrations of the reactants.

The equilibrium constant for a weak acid or base is called its ionization constant, represented as K_a or K_b. The ionization constant indicates the degree of dissociation in aqueous solution. The solubility product constant K_{sp} for a slightly soluble salt is equal to the product concentrations of its ions (in moles/liter) at equilibrium. This constant is useful in computing solubilities.

A buffer system or buffer solution is a solution that resists changes in pH when small amounts of strong acids or bases are added. Buffer solutions are solutions of either a weak acid and its salt or a weak base and its salt.

. . . .

Career Sketch

Geophysicist

Geophysicists study the physics of the earth, including its composition, physical aspects, interior, atmosphere, land, bodies of water, and underground layers. They use geology, mathematics, physics, chemistry, and engineering. The 12,000 geophysicists working in the United States are employed in the petroleum industry and in mining, exploration, consulting, and research.

Geophysicists specialize in a variety of areas. Volcanologists study the formation and eruption of volcanos. Paleomagneticians specialize in analyzing the historical magnetic fields of the earth through the study of rock layers and lava flow. Planetologists study the atmosphere and composi-

A geophysicist at the U.S. Geological Survey's National Earthquake Information Service in Golden, Colorado, checks seismographs, which record seismic waves from earthquakes around the world.
(U.S. Department of the Interior Geological Survey)

tion of the planets, of the moon, and of bodies in outer space. In addition, there are geomagneticians, solid-earth geophysicists, exploration geophysicists, hydrologists, seismologists, and tectonophysicists.

. . . .

Self-Test Exercises

Learning Goal 1: Chemical Kinetics, Chemical Equilibrium, and LeChatelier's Principle

1. Match each term on the left with the proper definition on the right.
(a) Chemical kinetics
(b) Chemical equilibrium
(c) LeChatelier's principle

1. A system under stress will move in a direction to relieve that stress ___
2. The study of the rate of speed at which a chemical reaction occurs ___

2. Define and give an example of a chemical equilibrium.

3. A dynamic state in
which two or more op-
posing processes are
taking place simulta-
neously at the same
rate   ___

3. Explain how LeChatelier's principle is related to chemical equilibrium.

4. Define the term *reaction mechanism.*

Learning Goal 2: Mechanism of Reactions

5. Use a specific example to explain a reaction mechanism. Identify the activated complex.

6. What is meant by the term *energy barrier?*

7. Draw a graph showing the progress of a reaction versus changes in energy for an exothermic reaction. Label the *activation energy.*

8. Define the term *activated complex.*

9. How does a catalyst affect the activation energy of a reaction?

10. What is an effective collision?

Learning Goal 3: Factors That Affect the Rate of a Chemical Reaction

11. Name the four factors that affect the rate of a chemical reaction.

12. How can a change in concentration increase the speed of a reaction?

13. For a reversible reaction, explain what happens to the rate of the forward reaction as the reaction proceeds. Explain what happens to the rate of the reverse reaction.

14. What is the rate of the reverse reaction at the start of a chemical reaction?

15. Explain how temperature affects reaction rate. What is the general rule?

16. What effect does a decrease in temperature have on reaction rate?

*17. For the gas-phase reaction

$$2H_2(g) + 2NO(g) \longrightarrow 2H_2O(g) + N_2(g)$$

the initial rate of reaction at 800°C is 20 torr per minute. What would be the approximate rate at 820°C?

18. Consider the reaction

$$A(g) + B(g) \xrightarrow{\text{catalyst}} AB(g)$$

Explain what would happen to the reaction rate if: (*a*) the concentration of A were doubled, (*b*) the catalyst were removed.

19. Explain how a catalyst affects the rate of a reaction.

20. Explain the following observation: On a windy day a brush fire burns more rapidly than on a calm day.

Learning Goal 4: Rates of Reaction and Equilibrium

21. Explain how the rates of the forward and reverse reactions change as a chemical reaction approaches equilibrium.

22. What is meant by the phrase "reaction goes to completion"?

23. Which of the following statements are correct?
(*a*) In a reaction at equilibrium, the concentrations of the reactants and those of the products are equal.
(*b*) A catalyst increases reaction rates by changing reaction mechanisms.
(*c*) The amount of product available at equilibrium is proportional to how fast equilibrium is attained.
(*d*) A catalyst lowers the activation energy of a reaction by equal amounts for both forward and reverse reactions.

24. How do catalysts affect reaction rates?

Learning Goal 5: Reversible Reactions

25. Write the forward and reverse reactions for the ionization of acetic acid.

26. Write the forward and reverse reactions for the ionization of carbonic acid.

27. Write the forward and reverse reactions for the dissociation of sodium chloride and water.

28. Write the forward and reverse reactions for the dissociation of sodium sulfate and water.

Learning Goal 6: Equilibrium Expression and Equilibrium Constant

29. Write an equilibrium expression for each of the following equations.
(*a*) $2A(g) + 3B(g) \rightleftharpoons 5C(g) + 4D(g)$
(*b*) $H_2(g) + I_2(g) \rightleftharpoons 2HI(g)$
(*c*) $H_2(g) + Cl_2(g) \rightleftharpoons 2HCl(g)$
(*d*) $2NO_2(g) \rightleftharpoons N_2O_4(g)$

30. Write an equilibrium expression for each of the following equations.
(*a*) $A(g) + 2B(g) \rightleftharpoons C(g) + 4D(g)$
(*b*) glucose-1-phosphate $\rightleftharpoons$ glucose-6-phosphate
(*c*) $N_2(g) + 2O_2(g) \rightleftharpoons 2NO_2(g)$
(*d*) $H_2O(g) + CO(g) \rightleftharpoons H_2(g) + CO_2(g)$

31. Write an equilibrium expression for each of the following equations.
(*a*) $N_2(g) + 3H_2(g) \rightleftharpoons 2NH_3(g)$
(*b*) $2NO(g) + O_2(g) \rightleftharpoons 2NO_2(g)$
(*c*) $PCl_5(g) \rightleftharpoons PCl_3(g) + Cl_2(g)$
(*d*) $2CO_2(g) \rightleftharpoons 2CO(g) + O_2(g)$

32. Write an equilibrium expression for each of the following equations.
(*a*) $2SO_2(g) + O_2(g) \rightleftharpoons 2SO_3(g)$
(*b*) $C(s) + H_2O(g) \rightleftharpoons CO(g) + H_2(g)$
(*c*) $A(g) + 3B(g) \rightleftharpoons 2C(g)$
(*d*) $H_2(g) + Br_2(g) \rightleftharpoons 2HBr(g)$

33. Calculate K_{eq} at a given temperature for the reaction

$$2NO(g) + O_2(g) \rightleftharpoons 2NO_2(g)$$

The concentrations at equilibrium are as follows: $NO(g)$, 0.600 mole/liter; $O_2(g)$, 0.800 mole/liter; and $NO_2(g)$, 4.40 moles/liter.

34. Calculate K_{eq} at 400°C for the reaction

$$H_2(g) + I_2(g) \rightleftharpoons 2HI(g)$$

given the following data: equilibrium concentration of $H_2(g)$, 0.436M; equilibrium concentration of $I_2(g)$, 0.436M; and equilibrium concentration of $HI(g)$, 3.128M.

35. Calculate K_{eq} at 440°C for the reaction

$$H_2(g) + I_2(g) \rightleftharpoons 2HI(g)$$

The concentrations at equilibrium are as follows: $H_2(g)$, 0.218 mole/liter; $I_2(g)$, 0.218 mole/liter; and $HI(g)$, 1.564 moles/liter.

36. Consider the reaction

glucose-1-phosphate $\rightleftharpoons$ glucose-6-phosphate

Calculate K_{eq} if the concentrations at equilibrium at a certain temperature are as follows: glucose-6-phosphate, 1.00 mole/liter; and glucose-1-phosphate, 0.050 mole/liter.

Learning Goal 7: Calculation of K_a and K_b

37. Calculate K_a for benzoic acid at 25°C for the reaction

$$HC_7H_5O_2 \rightleftharpoons H^{1+} + (C_7H_5O_2)^{1-}$$

The equilibrium concentrations are as follows: $HC_7H_5O_2$, 0.0975 mole/liter; H^{1+}, 0.0025 mole/liter; and $(C_7H_5O_2)^{1-}$, 0.0025 mole/liter.

39. Calculate K_a for cyanic acid at 25°C for the reaction

$$HCNO(aq) \rightleftharpoons H^{1+}(aq) + (CNO)^{1-}(aq)$$

The equilibrium concentrations are as follows: HCNO, 0.0955 mole/liter; H^{1+}, 0.0045 mole/liter; and $(CNO)^{1-}$, 0.0045 mole/liter.

38. Calculate K_b for NH_3 at 20°C for the reaction

$$NH_3(aq) + H_2O(aq) \rightleftharpoons (NH_4)^{1+}(aq) + (OH)^{1-}(aq)$$

The equilibrium concentrations are as follows: $(NH_4)^{1+}$, 0.0013 mole/liter; $(OH)^{1-}$, 0.0013 mole/liter; and $NH_3(aq)$, 0.0939 mole/liter.

40. Calculate the K_a for acetic acid at 25°C for the reaction

$$HC_2H_3O_2(aq) \rightleftharpoons H^{1+}(aq) + (C_2H_3O_2)^{1-}(aq)$$

The equilibrium concentrations are as follows: $HC_2H_3O_2$, 0.3990 mole/liter; H^{1+}, 0.00268 mole/liter; and $(C_2H_3O_2)^{1-}$, 0.00268 mole/liter.

Learning Goal 8: Calculation of Equilibrium Concentrations

41. Calculate the equilibrium concentrations of H^{1+} and $(C_2H_3O_2)^{1-}$ at 25°C for a 0.100M acetic acid solution ($K_a = 1.80 \times 10^{-5}$).

43. Calculate the equilibrium concentrations of H^{1+} and $(CN)^{1-}$ for a 0.500M hydrocyanic acid solution ($K_a = 4.00 \times 10^{-10}$).

45. Calculate the equilibrium concentrations of $(CuOH)^{1+}$ and $(OH)^{1-}$ for a 0.500M copper(II) hydroxide solution ($K_b = 1.00 \times 10^{-8}$).

47. Calculate the equilibrium concentrations of $(NH_4)^{1+}$ and $(OH)^{1-}$ for a 0.100M ammonium hydroxide solution ($K_b = 1.80 \times 10^{-5}$).

42. Calculate the equilibrium concentrations of H^{1+} and F^{1-} for a 0.400M HF solution if K_a is 8.29×10^{-4}.

44. The vitamin niacin, which we can represent as HNiac, dissociates to form H^{1+} and $(Niac)^{1-}$ ions. K_a for niacin is 1.40×10^{-5}. Calculate the equilibrium concentrations of H^{1+} and $(Niac)^{1-}$ for a 1.00M HNiac solution.

46. Calculate the equilibrium concentrations of Ag^{1+} and $(OH)^{1-}$ for a 1.50M silver hydroxide solution ($K_b = 1.1 \times 10^{-4}$).

48. Calculate the equilibrium concentrations of $(CaOH)^{1+}$ and $(OH)^{1-}$ ions for a 1.00M calcium hydroxide solution at 25°C ($K_b = 3.74 \times 10^{-3}$).

Learning Goal 9: Writing K_{sp} Expressions and Calculating K_{sp}

49. Write K_{sp} expressions for these slightly soluble salts: (*a*) AgI   (*b*) $Fe(OH)_3$ (*c*) Ag_2CrO_4

50. Write K_{sp} expressions for these slightly soluble salts: (*a*) $BaCO_3$ (*b*) BaF_2 (*c*) PbI_2

51. Write K_{sp} expressions for these slightly soluble salts: (*a*) $BaCrO_4$ (*b*) CaF_2 (*c*) CuS

52. Write K_{sp} expressions for these slightly soluble salts: (*a*) ZnS (*b*) $CuCl$ (*c*) $Al(OH)_3$

53. Calculate K_{sp} for $BaSO_4$ at 25°C if the concentrations of Ba^{2+} and $(SO_4)^{2-}$ ions are both 3.9×10^{-5} mole/liter. [*Hint:* The equilibrium equation for the ionization of $BaSO_4$ is $BaSO_4 \rightleftharpoons Ba^{2+} + SO_4{}^{2-}$.]

54. Calculate K_{sp} for silver chloride if the concentrations of the Ag^{1+} and Cl^{1-} ions are both 1.30×10^{-5} mole/liter at 25°C.

55. Calculate K_{sp} for Ag_2CrO_4 at 25°C if the concentration of Ag^{1+} ions is 1.56×10^{-4} mole/liter and the concentration of $(CrO_4)^{2-}$ ions is 7.80×10^{-5} mole/liter. [The equilibrium equation for the ionization of Ag_2CrO_4 is $Ag_2CrO_4 \rightleftharpoons 2Ag^{1+} + (CrO_4)^{2-}$.]

56. Calculate K_{sp} for calcium fluoride if the concentration of Ca^{2+} is 2.14×10^{-4} mole/liter and that of F^{1-} is 4.28×10^{-4} mole/liter at 25°C.

Learning Goal 10: Calculation of the Solubility of a Salt

57. Calculate the solubility of CuS at 25°C in grams per liter, given that $K_{sp} = 9.0 \times 10^{-45}$.

58. Calculate the solubility of AgI at 25°C in grams per liter, given that K_{sp} is 8.5×10^{-17}.

59. Calculate the solubility of Ag_2CrO_4 at 25°C in grams per liter, given that $K_{sp} = 1.9 \times 10^{-12}$.

60. Calculate the solubility of $AgBr$ at 18°C in grams per liter, given that K_{sp} is 4.1×10^{-13}.

Learning Goals 11 and 12: Buffers, Buffer Systems

61. How does one prepare a buffer solution?

62. Write the equation for the equilibrium among $(HCO_3)^{1-}$, H^{1+}, H_2CO_3, and CO_2 in water. What effect would an increase in the $(HCO_3)^{1-}$ concentration have on the hydrogen ion concentration?

63. Write the equation(s) that represent the carbonic acid/bicarbonate buffer system in the blood.

64. What are the effects of hypoventilation and of hyperventilation on the carbonic acid/bicarbonate buffer system in the blood?

65. Given the following reactions at equilibrium,
$$H_2O + CO_2 \rightleftharpoons H_2CO_3 \rightleftharpoons H^{1+} + (HCO_3)^{1-}$$
decide which way the equilibrium shifts when:
(*a*) Excess H^{1+} ion is added
(*b*) Excess CO_2 is added
(*c*) Bicarbonate ion, $(HCO_3)^{1-}$, is removed

66. State whether each of the following statements is true or false, and explain your answers.
(*a*) Carbonic acid is capable of neutralizing any weak base present in the blood.
(*b*) Bicarbonate ion is capable of neutralizing any weak acid that is present in the blood.

67. Show by the use of a chemical equation how the carbonic acid/bicarbonate buffer system in the blood works in the event of a sudden increase in the concentration of H^{1+} ions.

68. A person who hyperventilates and goes into alkalosis is told to breathe into a paper bag. Why? Explain your answer in terms of the carbonic acid/bicarbonate buffer system.

Extra Exercises

69. For each of the reactions at the right, in which direction does the equilibrium shift if the pressure is increased?

(*a*) $H_2(g) + I_2(g) \rightleftharpoons 2HI(g)$
(*b*) $2SO_2(g) + O_2(g) \rightleftharpoons 2SO_3(g)$
(*c*) $N_2(g) + 3H_2(g) \rightleftharpoons 2NH_3(g)$

70. For each of the following reactions, in which direction does the equilibrium shift if the temperature is increased?
(a) $H_2(g) + Cl_2(g) \rightleftharpoons 2HCl(g) + 44.2$ kcal
(b) $2CO_2(g) + 135.2$ kcal $\rightleftharpoons 2CO(g) + O_2(g)$
(c) $H_2(g) + I_2(g) + 12.4$ kcal $\rightleftharpoons 2HI(g)$

71. Write an equilibrium expression for each of the following reactions.
(a) $N_2(g) + O_2(g) \rightleftharpoons 2NO(g)$
(b) $2CO_2(g) \rightleftharpoons 2CO(g) + O_2(g)$
(c) $2SO_2(g) + O_2(g) \rightleftharpoons 2SO_3(g)$

72. For the reaction
$$A(g) + 2B(g) + 3C(g) \rightleftharpoons 4D(g)$$
calculate K_{eq} if the concentrations at equilibrium are: $[A] = 4.0$, $[B] = 3.0$, $[C] = 2.0$, $[D] = 1.0$.

73. Calculate K_a for formic acid, $HCHO_2$, at 25°C, given the following equilibrium concentrations: $[HCHO_2] = 0.0958$, $[H^{1+}] = 0.0042$, and $[(CHO_2)^{1-}] = 0.0042$.

74. Calculate the equilibrium concentration of H^{1+} and $(NO_2)^{1-}$ at 25°C for a $1.0M$ HNO_2 solution $(K_a = 4.5 \times 10^{-4})$.

75. Which of the following indicates the reaction that goes farthest to completion?
(a) $K = 1$, (b) $K = 10$,   (c) $K = 100$, (d) $K = 0.1$

76. Write the K_{sp} expression for each of these salts:
(a) $Mn(OH)_2$ (b) $AgCl$ (c) A_2X_3

77. Calculate the solubility of CaF_2 in grams per liter at 25°C, given that $K_{sp} = 3.9 \times 10^{-11}$.

78. Which compound, Ag_2CrO_4 or $BaCrO_4$, has the greater *molar* solubility? [*Hint:* Check Table 16.2.]

79. Write an equilibrium equation for each of the following systems:
(a) A mixture of ice and water at 0°C
(b) A mixture of water and steam in a closed system at 100°C
(c) Carbonic acid (H_2CO_3) in equilibrium with carbon dioxide and water
(d) Ammonia and water in equilibrium with ammonium ion and hydroxide ion

Cumulative Review / Chapters 14–16

1. Which is the solute and which is the solvent in each of the following solutions?
(a) 75 mL of nitrogen and 35 mL of oxygen
(b) Sulfur dioxide gas and water
(c) Lemon juice, sugar, and water in lemonade

2. State whether each of the following pairs of substances form a solution.
(a) Hydrogen fluoride gas and water
(b) Benzene and *trans*-dichloroethane

3. Find the concentration of each of the following solutions in percentage by weight.
(a) 50.0 g of salt in enough water to make 75.0 g of solution
(b) 175 g of sugar in enough water to make 1,$\overline{5}$00 g of solution

4. Find the concentration of each of the following solutions in percentage by volume.
(a) 25.0 mL of alcohol in enough water to make 455 mL of solution
(b) 35.0 mL of benzene dissolved in enough carbon tetrachloride to make 95.0 mL of solution

5. If 5.0 g of benzene are dissolved in enough carbon tetrachloride to make 75 mL of solution, what is the concentration of the solution in percentage by weight–volume?

6. What is the molarity of the solution formed by dissolving 1.84 g of ethyl alcohol (C_2H_5OH) in enough water to make 1.5 L of solution?

7. When 25$\overline{0}$ mL of 1.33M NaOH solution are evaporated to dryness, how many grams of NaOH remain?

8. How many milliliters of 1.00M $CaCl_2$ solution must one evaporate to obtain 26.0 g of $CaCl_2$?

9. Distinguish between normality and molarity.

10. How many grams of H_3PO_4 are present in 1.00 L of a 2.00N solution of H_3PO_4?

11. Calculate the normality and the molarity of a solution of calcium hydroxide formed by dissolving 18.5 g of calcium hydroxide in enough water to make 5$\overline{0}$0 mL of solution.

12. How would 5.50 L of 1.33N NaOH solution be prepared from a 36.0N solution of NaOH?

13. A photographer is preparing 2.00 L of 0.150M silver nitrate solution. How many grams of silver nitrate are needed?

14. True or False: A solution is a heterogeneous mixture.

15. Calculate the grams of solute in 1$\overline{0}$0 mL of 18.0M H_2SO_4 solution.

16. A solution of sugar in water contains 25.0% sugar, by weight, and has a density of 1.09 g/mL. The molecular formula for sugar is $C_{12}H_{22}O_{11}$.
(a) Calculate the number of grams of sugar present in 2.00 L of solution.
(b) Calculate the molarity of the solution.

17. Calculate the grams of solute present in 100.0 g of 0.90% by weight NaCl solution.

18. To what volume must a solution of 75.0 g of H_2SO_4 in 4$\overline{0}$0 mL of water be diluted to give a 0.200M solution?

19. Calculate the boiling point of a solution prepared by dissolving 100.0 g of ethylene glycol, $C_2H_6O_2$, in 500.0 g of water. K_b for water is 0.520°C/m.

20. A solution is prepared by dissolving 12.8 g of naphthalene, $C_{10}H_8$, in 1.00 L of water. Is this solution 0.100M or 0.100m?

21. State the difference between the Arrhenius and the Brønsted–Lowry definitions of an acid.

22. True or False: A solution of pH 4 is 100 times more acidic than a solution of pH 7.

23. Complete and balance each of the following equations.
(a) $HC_2H_3O_2$ + $Ca(OH)_2 \longrightarrow$?
(b) N_2O_5 + $H_2O \longrightarrow$?

24. Write an equation for the formation of acid rain in the atmosphere.

25. Determine the pH of a solution that has a hydrogen ion concentration of $1.0 \times 10^{-2}M$.

26. Calculate the pH of black coffee if its hydrogen ion concentration is $1.0 \times 10^{-6}M$.

27. How many milliliters of 0.400M HCl can one prepare by diluting 75$\overline{0}$ mL of 12.0M HCl?

28. Calculate the molarity of each of the ions present in a solution prepared by combining 50.0 mL of 2.00M HCl with 50.0 mL of 2.00M NaOH.

29. If 40.0 mL of 1.50M NaOH solution are needed to titrate 50.0 mL of HCl, what is the molarity of the HCl?

30. Calculate the molarity of NaOH if 15.0 mL of 0.150M HCl are needed to titrate 50.00 mL of NaOH.

31. Differentiate between the terms *ionization* and *dissociation*.

32. Which solution is more acidic, 1M H_2SO_4 or 1N H_2SO_4?

33. Which solution is more acidic, 1M HCl or 1M $HC_2H_3O_2$?

34. Calculate the molarity of the sum of all the ions present in 0.020M NaCl if the solution is 100% ionized.

35. Calculate the concentration of Sr^{2+} ions in a solution of $SrCl_2$ that has a Cl^{1-} ion concentration of 0.500M. (Assume 100% ionization.)

36. If 50.00 mL of 0.500M HCl neutralize 100.0 mL of $Ca(OH)_2$ solution, what is the molarity of the $Ca(OH)_2$ solution? The reaction is

$$Ca(OH)_2(aq) + 2HCl(aq) \longrightarrow CaCl_2(aq) + 2H_2O$$

37. Do the terms *strong acid* and *strong base* refer to concentration or ionization?

38. When a mole of $SrCl_2$ is dissolved in water, are there more Sr^{2+} ions or more Cl^{1-} ions present?

39. Calculate the pH of tomato juice whose H^{1+} concentration is 1.0×10^{-4}.

40. What is the $(OH)^{1-}$ concentration of a solution whose H^{1+} concentration is 10^{-8}?

41. Cite four factors that affect the rate of a chemical reaction?

42. As a chemical reaction approaches equilibrium, what happens to the rates of the forward and reverse reactions?

43. Express the following reaction in equation form: Water and water vapor are both present at 100°C in a pressurized system.

44. Is the following reaction endothermic or exothermic?

$$2NO(g) + O_2(g) \longrightarrow 2NO_2(g) + 27 \text{ kcal}$$

45. True or False: A large equilibrium constant for a reaction means that, at equilibrium, the formation of reactants is favored over the formation of products.

46. Consider the carbonic acid/sodium carbonate buffer system in the blood.

$$H_2O + CO_2 \rightleftharpoons H_2CO_3 \rightleftharpoons H^{1+} + (HCO_3)^{1-}$$

When excess acid is added to the blood, this system accommodates to the "stress" in such a way as to maintain the pH of the blood. Explain how.

47. Write the equilibrium expression for the reaction $CaCO_3(s) \rightleftharpoons CaO(s) + CO_2(g)$.

48. Calculate the solubility product constant for $Pb(IO_3)_2$ if the concentration of the saturated solution of the salt is $4.0 \times 10^{-5}M$.

49. Calculate the solubility of Ag_2S in moles per liter, given that $K_{sp} = 6.0 \times 10^{-50}$.

50. Calculate the solubility of $PbSO_4$ in grams per liter, given that $K_{sp} = 6.0 \times 10^{-8}$.

51. Calculate the equilibrium concentrations of H^{1+}, $(C_2H_3O_2)^{1-}$, and $HC_2H_3O_2$ in a 0.75M acetic acid solution at 25°C, given that $K_a = 1.8 \times 10^{-5}$.

52. Write the K_{sp} expression for $Ba_3(PO_4)_2$.

53. Write the equilibrium expression for the reaction $HClO_2(aq) \rightleftharpoons H^{1+}(aq) + ClO_2^{1-}(aq)$.

54. Using Table 16.2, determine which compound has the greater molar solubility, AgCl, AgBr, or AgI.

55. Predict the direction in which the equilibrium will shift when heat is added to the reaction

$$4NH_3(g) + 3O_2(g) \rightleftharpoons 2N_2(g) + 6H_2O(g) + 366 \text{ kcal}$$

17

Special Topics:
Nuclear Chemistry, Electrochemistry, and Environmental Chemistry

Learning Goals

After you have studied this chapter, you should be able to:

1. Explain what is meant by radioactivity.
2. List the three major types of nuclear radiation and the characteristics of each.
3. Write balanced nuclear equations.
4. Define the term *half-life*.
5. Perform calculations involving the half-life of a substance.
6. Define the terms *nuclear fission* and *nuclear fusion*.
7. Discuss the emergence of nuclear power plants.
8. Explain how radiation affects human health.
9. Cite several methods of detecting radiation.
10. Explain how radiation is measured.
11. Give an example of a radioactive isotope that is used in medicine, and discuss how it is used.
12. Explain what we mean by a voltaic cell and an electrolytic cell.
13. Discuss the operation of a fuel cell and a lead storage battery.
14. List the normal ingredients of air and the common pollutants it may contain.
15. Describe some common reactions that produce air pollutants.
16. Discuss the health problems of pollution presented by the automobile engine and other pollution sources.
17. Describe some common ways of reducing air pollution.
18. Discuss the problem of hazardous wastes.

Introduction

It is August 6, 1945. A lone aircraft flies over Hiroshima, Japan, and releases a single bomb. Seconds later, the bomb explodes. A huge mushroom-shaped cloud forms, which grows until it fills the sky. Within hours, approximately 75,000 people are dead. During the next 30 years, an additional 800,000 people die of various diseases caused by the radiation from the bomb. The world has been catapulted into the atomic age (Figure 17.1).

It is the year 1800. In his laboratory, the Italian physicist Alessandro Volta is experimenting with solutions and metals that are able to conduct electricity. He finds that two metals and such a solution can be arranged to *produce* an electric current. Volta has invented the first *electrochemical battery*—a device that produces electrical energy from a chemical reaction.

It is late in October 1948. The place is Donora, Pennsylvania, a small industrial town located at the bottom of a river valley. An unusually thick smog has settled over the town. People look up at the leaden sky and sniff uneasily. The smog is so thick that it makes people's eyes sting and their throats become raw. Sulfur dioxide, dust, and waste products from the nearby zinc, iron, and steel mills all contribute to the smog. Local weather conditions have created an *atmospheric inversion*, which keeps the polluted air near the ground (Figure 17.2). By the time the smog lifts some 5 days later, 17 people have died as a direct result of it, and almost half the pop-

Figure 17.2

Downtown Los Angeles on a clear day (left), and on a day with ground-level smog (right)
(Courtesy South Coast Air Quality Management District, El Monte, Calif.)

Henri Antoine Becquerel
(Reprinted with permission from Torchbearers of Chemistry, by Henry Monmouth Smith, © Academic Press, Inc., 1949)

ulation of Donora has been affected to some degree. The nightmare of environmental pollution has made its first widely recognized appearance in the United States.

These three events, which took place at different times in history, all involve chemical processes. They represent, however, parts of three different areas of chemistry: nuclear chemistry, electrochemistry, and environmental chemistry. In this chapter we shall briefly survey these three important areas of study.

17.1 Nuclear Chemistry

Although the explosion of the atomic bomb over Hiroshima in 1945 is recognized as the beginning of the nuclear age, the study of nuclear chemistry actually began much earlier, in 1895, when the German physicist Wilhelm Konrad Röntgen accidentally discovered x rays. He found that x rays have a very great penetrating power—so great that they could easily pass through wood and plaster walls.

In 1896 the French physicist Antoine Becquerel placed a salt containing uranium on a photographic plate that was wrapped in black paper.

When he developed the photographic plate, he found the image of the pile of salt. The uranium salt had taken a picture of itself! Becquerel concluded that the uranium was giving off some type of penetrating rays and that these rays must be very strong to be able to pass through the black paper and expose the photographic plate. When Becquerel placed a thick barrier of lead between the salt and the photographic plate, the lead absorbed the rays and no image was formed. Becquerel thought that these penetrating rays were probably Röntgen's x rays.

At the same time, in Paris, a young Polish chemist named Marie Curie was working with her husband Pierre in the laboratories at the Sorbonne. The Curies became interested in Becquerel's discovery. In fact, it was Marie Curie who defined **radioactivity** as the ability of a substance to emit penetrating rays from its nucleus. The Curies found that a radioactive substance seems to keep on emitting these powerful penetrating rays, year after year. In 1898 they discovered that the element thorium is radioactive. So is polonium. And so, most of all, is radium.

Although early experimenters first thought that radioactive materials produced only x rays, they soon discovered that the situation was more complex. For example, when they allowed the radiation produced by uranium to pass through a magnetic field, they found that the radiation was deflected in three different directions (Figure 17.3). They had detected *three* types of radiation rather than only one. The English physicist Lord Rutherford named these three types of radiation *alpha particles*, *beta particles*, and *gamma rays*, from the first three letters of the Greek alphabet: α, β, and γ.

Gamma Rays (γ Rays)

Gamma rays *are made up of energy. They have no mass and carry no electrical charge. Gamma rays* travel straight through a magnetic field—they are not deflected. For this reason the early experimenters assumed that they car-

---

Learning Goal 1
Radioactivity and atomic energy

Learning Goal 2
Types of nuclear radiation

Marie Curie
(Reprinted with permission from Torchbearers of Chemistry, by Henry Monmouth Smith, © Academic Press, Inc., 1949)

Figure 17.3
Analyzing the radiation from uranium

ried no charge. Gamma rays seemed to have properties like those of x rays, but they were more energetic. In fact, gamma rays have a very high energy content and strong penetrating ability. Today we know that, because of this penetrating ability, a high concentration of gamma rays can harm human genes. The nucleus of an atom gives off gamma rays only under certain conditions, when the nucleus loses energy. And the atom, of course, has to be of a radioactive element.

Beta Particles (β Particles)

Beta particles *are identical to an electron in mass and size. These particles have a negative electrical charge.* When Becquerel passed beta particles through a magnetic field, he found that they behaved like cathode rays—that is, like the rays that emanate from a negative electrode. Therefore Becquerel concluded that these rays were composed of speeding particles, which he assumed to be electrons. However, beta particles, like gamma rays and alpha particles, originate in the *nucleus* of the atom, and electrons do not exist in the nucleus.

How, then, are beta particles formed? Scientists believe they are formed when neutrons decompose during a nuclear reaction:

$$\text{Neutron} \longrightarrow \text{proton} + \beta \text{ particle}$$

Like an electron, a beta particle has a zero mass number and a charge of -1. We can represent it in the following way:

$$_{-1}^{0}e = \beta \text{ particle}$$

Beta particles have less penetrating power than gamma rays. They can be stopped by a thin sheet (a few centimeters thick) of almost any metal.

Alpha Particles (α Particles)

Alpha particles *are helium nuclei. These particles carry a double positive charge.* Scientists of the early 1900s found that, in a magnetic field, a stream of alpha particles was deflected in the opposite direction from beta particles. They therefore assumed that alpha radiation was made up of positively charged particles. Experiments showed that these particles have a mass number of 4 and a charge of $+2$. Werner Heisenberg, a German physicist, suggested that we picture an alpha particle as a helium-4 atom with its electrons removed. In other words, we can imagine it as composed of two protons and two neutrons. We can thus represent an alpha particle in this way:

$$_{2}^{4}He = \alpha \text{ particle}$$

Alpha particles have little penetrating ability. They can be stopped by a thin sheet of paper.

Nuclear Transformations

. . . .
Learning Goal 3
Writing balanced nuclear
equations

Table 17.1 summarizes the basic properties of the three types of radioactivity. It also indicates the changes that take place in atoms that emit alpha or beta particles. Thus we can use the table to write the equations for nuclear reactions, and we shall do so shortly.

An element that is emitting radiation is said to be undergoing **radioactive decay** or **nuclear decay.** Some elements do so spontaneously—naturally—without help from scientists or laboratory equipment. An example is radium-226, $^{226}_{88}Ra$. When it undergoes nuclear decay, a radium atom emits an alpha particle and actually becomes a different element! We call this a **nuclear transformation** or **transmutation,** and we can represent it with the nuclear equation

$$^{226}_{88}Ra \longrightarrow \underset{\substack{\alpha \\ \text{particle}}}{^{4}_{2}He} + \, ?$$

But what element is formed in this transformation? To find out, we note that, in a balanced nuclear equation, the sum of the superscripts (mass numbers) and the sum of the subscripts (protons) must be the same on both sides of the equation. Subtraction then tells us that the element formed has the atomic number 86 and the mass number 222. In the periodic table (inside the front cover), we find that element 86 is radon, Rn. So we write the nuclear equation as

$$^{226}_{88}Ra \longrightarrow \, ^{4}_{2}He + \, ^{222}_{86}Rn$$

Here's another example: The isotope thorium-234 produces radiation by emitting beta particles. We can represent this nuclear reaction as

$$^{234}_{90}Th \longrightarrow \underset{\substack{\beta \\ \text{particle}}}{^{0}_{-1}e} + \, ?$$

What is the element that is formed? Table 17.1 tells us that the emission of a beta particle increases an element's atomic number by 1. Therefore

Table 17.1
Summary of Nuclear Particles

Type of emission	Mass number of particle emitted	Charge of particle emitted	Changes caused by emission: Mass number	Atomic number
Alpha	4	+2	Decreases by 4	Decreases by 2
Beta	0	−1	No change	Increases by 1
Gamma	0	0	No change*	No change*

*Gamma rays may accompany some radioactive decay reactions.

the unknown element must be the element whose atomic number is 91. That element is protactinium, so we can write our equation as

$$^{234}_{90}\text{Th} \longrightarrow \, ^{0}_{-1}\text{e} + \, ^{234}_{91}\text{Pa}$$

Again the sum of the superscripts and the sum of the subscripts are the same on both sides of the equation.

The Half-Life of a Radioactive Element

Learning Goal 4
Half-life

No one really knows why some elements are radioactive—that is, why they undergo nuclear decay. However, many scientists feel that certain combinations of protons and neutrons make the atomic nucleus unstable. These nuclei may become more stable by emitting alpha or beta particles.

And no one can tell when a particular atom of a radioactive element will decay. But, for a large number of atoms of any radioactive element, scientists can predict how many atoms will decay in a given time. This decay rate is usually expressed as the time required for half of the atoms to decay. It is called the half-life of the element. The **half-life** of a radioactive element is *the time required for the decay of half of the atoms in a sample of that element.* Half-lives of the radioactive elements vary in length. For example, the half-life of $^{238}_{92}\text{U}$ is 4.5 billion years, whereas the half-life of $^{257}_{103}\text{Lr}$ is eight seconds.

Example 17.1

The half-life of strontium-90 is 28 years. Suppose we have 100 g of strontium-90 today. How many grams of strontium-90 will be left in the sample in 84 years?

Solution

Because the half-life of strontium-90 is 28 years, half of the strontium-90 will decay every 28 years. After 84 years, the sample will contain 12.5 g of strontium-90.

Learning Goal 5
Calculation using half-life

Time in years	Grams of strontium-90 remaining
Now	100 g
28	50 g
56	25 g
84	12.5 g

Practice Exercise 17.1   The half-life of lawrencium-257 is 8 seconds. If we have 100 g of lawrencium-257 right now, how many grams of this radioactive element will be left in the sample in 64 seconds?

While discussing strontium-90, we should note that it has a chemical similarity to calcium (Ca and Sr are in the same chemical family). It will replace the calcium in bone and stay there, undergoing its radioactive decay. When nations test atomic bombs in the atmosphere, strontium-90 gets into the atmosphere and begins to float around the world. Gradually it falls to earth—on fields, for example. When cows eat the grass that grows in the fields, this strontium-90 gets into milk and other dairy products; it also gets into vegetables planted in the fields. People consume these farm products, and the radioactive isotope, strontium-90, accumulates in our bones, side by side with calcium. It then bombards nearby tissues and organs with beta particles. That 28-year half-life means that it stays around for a long time. And if a person is exposed to a large dose, it can cause bone cancer.

Artificially Produced Radiation—The Alchemists' Dreams Come True

In 1919, the British physicist Lord Rutherford bombarded nitrogen-14 with a stream of alpha particles. The results were astounding. Rutherford found that he had produced oxygen-17, a nonradioactive isotope of oxygen.

$$\underset{\substack{\alpha \\ \text{particle}}}{^{4}_{2}\text{He}} + {^{14}_{7}\text{N}} \longrightarrow {^{17}_{8}\text{O}} + \underset{\substack{\text{A} \\ \text{proton}}}{^{1}_{1}\text{H}}$$

The dreams of the alchemists had come true! By triggering the first nuclear reaction produced by humans, Rutherford had at last transmuted an element (Figure 17.4).

Fifteen years later, the French physicists Irène Joliot-Curie and Frédéric Joliot (daughter and son-in-law of Marie and Pierre Curie) bombarded boron-10 with alpha particles. They obtained nitrogen-13, the first artificially produced radioactive isotope.

$$\underset{\substack{\alpha \\ \text{particle}}}{^{10}_{5}\text{B}} + {^{4}_{2}\text{He}} \longrightarrow {^{13}_{7}\text{N}} + \underset{\substack{\text{A} \\ \text{neutron}}}{^{1}_{0}\text{n}}$$

By the 1930s, then, scientists were unraveling the atom's deepest secrets and were about to discover the power of the atom. They were, in fact, on the verge of producing a nuclear chain reaction, which could turn very small quantities of mass into huge quantities of energy.

Figure 17.4

Nitrogen-14, bombarded with an alpha particle, takes on an entirely new identity and becomes oxygen-17.

$${^{4}_{2}\text{He}} + {^{14}_{7}\text{N}} \longrightarrow {^{17}_{8}\text{O}} + {^{1}_{1}\text{H}}$$

$$\alpha + {^{14}\text{N}} \longrightarrow {^{17}\text{O}} + p$$

α particle   Nitrogen-14   Oxygen-17   A proton

The Chain Reaction and Nuclear Fission

Learning Goal 6
Nuclear fission

In 1934 the Italian physicist Enrico Fermi (who won the Nobel prize in 1938) bombarded uranium (element 92) with neutrons in an attempt to produce element 93 (neptunium). But instead, Fermi found himself with an isotope of barium (element 56), a mystifying outcome.

In 1938 the German physical chemist Otto Hahn, who was also working on atomic power, proposed an explanation. He said that uranium atoms split into different atoms when they are bombarded by neutrons. This splitting process is called **nuclear fission.** Furthermore, when these uranium atoms split, they produce *more* neutrons, which in turn are able to split still more uranium atoms in what is called a **chain reaction.** It can be likened to placing dominos in a triangular pattern. When you tip the first domino, it causes two others to tip over, and they can tip over several more, and so on down the line.

In September 1939, Europe exploded into World War II and all nations looked for new weapons. Many scientists who were studying nuclear reactions knew that nuclear fission could be turned into a decisive weapon of war. It was only a matter of time before someone perfected the technique. Would the Germans be first? Were they racing toward that goal? Everyone knew that their scientists were second to none.

The United States, though not yet at war, recognized the danger and launched a research program to produce an atomic (fission) bomb. Under the leadership of Enrico Fermi, the so-called Manhattan Project (a code name) was started. It operated out of a windy, makeshift laboratory beneath the bleachers of Stagg Field at the University of Chicago. On December 2, 1942, Enrico Fermi and his team achieved the first sustained nuclear chain reaction. By July of 1945, through hard work and perseverance, the scientists had managed to scrape together enough uranium-235 to make a fission bomb (Figure 17.5). To test it, the United States had to

Figure 17.5

An atomic bomb of the World War II variety
(Courtesy National Air and Space Museum, Smithsonian Institution)

fire it; the first explosion of an A-bomb took place in a deserted area near Alamogordo, New Mexico. It created sickening devastation. But, with feelings of deep foreboding, the scientists set about making more bombs.

By August of 1945 they had two more bombs in hand. Just two! Yet such was the crisis (Americans believed that they would lose a million soldiers if they tried to storm the Japanese mainland) that the United States went ahead and exploded them: the first at Hiroshima and the second, three days later, at Nagasaki, Japan. The Japanese surrendered shortly thereafter, which ended World War II.

The Emergence of Nuclear Power Plants

Learning Goal 7
The emergence of nuclear power plants

Although the first energy generated by nuclear fission was used in making war, scientists quickly learned to apply this energy to peaceful purposes. In 1946 Congress established the Atomic Energy Commission (AEC) to control the development of nuclear energy in the United States. Private industry soon became involved in the development of commercial nuclear power, and the decade of the 1950s saw the emergence of full-scale nuclear power plants in the United States and around the world. This trend escalated in the decades that followed.

Nuclear power plants generate electricity in much the same way as conventional fossil-fueled power plants. In a fossil-fueled power plant, the heat obtained from the burning of coal or oil is used to change water into steam, which is then used to turn a turbine and produce electricity. In a nuclear power plant, the heat obtained from the naturally occurring fission (splitting) of uranium-235 atoms is used to change water into steam, which is then used to turn a turbine and produce electricity (Figure 17.6). Control rods are used in the nuclear reactor to absorb neutrons

Figure 17.6
A nuclear reactor

and regulate the rate of the fission reaction. Boron is commonly used in control rods, because it absorbs neutrons well and does not undergo fission. The fuel rods that compose the reactor's core are submerged in water. This not only cools the core but also helps moderate the chain reaction (Figure 17.7).

The products of nuclear fission include free neutrons and fission fragments that produce alpha particles, beta particles, and gamma rays. Plutonium, one of the by-products of atomic fission, has a half-life of 24,000 years. Some argue that even though plutonium remains radioactive for so long, its most harmful high-level radiation dissipates after a few hundred years. Nevertheless, such radioactive waste products must be handled very carefully for a very long time.

Several problems are associated with the use of nuclear fission to produce energy. Many of the issues involved are universally acknowledged as problems that must be solved; the significance of others is widely debated. Most involve the impact of radiation on health, safety, and the environment. The problems no one denies concern the safe disposal of radioactive waste, the matter of security (uranium and plutonium must be kept out of the hands of unauthorized people), and the high cost of operating nuclear power plants. More controversial are the questions of whether functioning nuclear power plants routinely add significant amounts of radiation to the environment and whether a colossal nuclear accident could take place that would destroy life and property.

Figure 17.7
Artist's conception of the Salem Generating Station. Both units are now in operation.
(Photograph courtesy Public Service Electric & Gas Company, Newark, N.J.)

On both sides of the nuclear power question we find well-informed citizens. Those who argue for it cite the need for nuclear energy to maintain our standard of living and point to the relatively good safety record of the nuclear power industry. Those who oppose atomic energy mention the unknown effects of low levels of radiation on the incidence of such diseases as cancer. They argue that no safe method for disposing of nuclear waste exists and that no one can guarantee that nuclear materials will never fall into the hands of irresponsible people, terrorists, or criminals.

In addition, several recent accidents at nuclear power plants have tarnished the overall safety record of the industry. On March 28, 1979, a faulty water pump at the Three Mile Island Nuclear Power Plant in Harrisburg, Pennsylvania, failed. This was followed by a chain of mechanical and human errors. As a result, radioactive water was released from the emergency cooling system into the Susquehanna River. Radioactive steam was vented into the environment, and water was converted into hydrogen gas, making an explosion possible and raising the spectre of a reactor meltdown. Although experts believe that the low levels of radiation released could never threaten human health, others fear that long-term effects of the radiation will emerge in years to come.

On April 26, 1986, the worst accident in the history of nuclear power took place at the Chernobyl plant about 110 km north of Kiev in the Soviet Union. At the Chernobyl plant, the uranium was contained in fuel rods surrounded by graphite bricks that served to moderate the nuclear reaction. It is believed that the supply of cooling water was accidentally cut off. The water in the reactor turned to steam, and the steam reacted with the nuclear fuel and the graphite bricks. An explosive mixture of gases was formed and ignited. The reactor was destroyed, the roof was blown off of the building, and the graphite bricks caught fire. Soviet officials claimed the fire was extinguished on April 29. According to Soviet reports, several hundred people were hospitalized and the acknowledged death count stood at twenty. More ominously, the radioactive particles produced by the explosion have been dispersed all over the planet by the natural circulation of air. It will be years before all the effects of the Chernobyl disaster can be assessed.

Nuclear Fusion

Learning Goal 6
Nuclear fusion

Nuclear fusion is *the fusing or combining of small atoms into larger atoms.* The sun and other stars get their energy through a process of nuclear fusion in which hydrogen atoms are combined to form a helium atom. This process creates huge amounts of heat and high temperatures—on the order of tens of millions of degrees Celsius—and no waste products are left over. However, acute problems have so far prevented the use of this process to generate electricity. How do you contain such fantastic heat? What vessel or tools can handle it? And, to duplicate the sun's fusion process on

The explosion at the Chernobyl nuclear power station in the USSR caused damage to the building, destroyed the reactor, killed several people, and raised radiation levels throughout Europe.
(Wide World)

earth, you would need a great amount of energy to start with. So far, the only device that scientists have developed from their knowledge of nuclear fusion is the *hydrogen bomb*. This bomb is now possessed by the United States, Russia, Great Britain, France, China, and India.

Because it uses atomic fusion, rather than fission, the hydrogen bomb is much more powerful than the "old-fashioned" atomic bomb (in fact, in the hydrogen bomb, an atomic bomb acts as the initiator or detonator). It is clear that atomic fusion, used unwisely, could lead to our doom. However, once we learn how to control the process and tame it for peaceful uses, fusion may provide an unlimited power supply that burdens us with no radioactive waste products to pollute our environment.

Physicists at Princeton University's Plasma Physics Laboratory are conducting a series of major experiments as part of a national effort in that direction. One experiment utilizes a *tokamak* device, in which a very hot ionized gas called *plasma* is confined in a large stainless steel vacuum vessel shaped like a doughnut. The plasma must reach a temperature of about 100,000,000°C for 1 second with a density of about one one-hundred-thousandth that of air at sea level. Under these special conditions a sizable number of atomic nuclei within the plasma collide and fuse, producing helium nuclei and high-speed neutrons. The energy of motion of these neutrons can be converted to electricity. Experiments such as these will, it is hoped, provide a safe way of producing energy for all human needs even thousands of years after conventional fuels have been used up.

The fusion reactions being studied are

$$\underset{\text{Tritium}}{{}^3_1\text{H}} \ + \ \underset{\text{Deuterium}}{{}^2_1\text{D}} \ \longrightarrow \ {}^4_2\text{He} + {}^1_0\text{n} + \text{Energy}$$

$$\underset{\text{Tritium}}{{}^3_1\text{H}} \ + \ \underset{\text{Protium}}{{}^1_1\text{H}} \ \longrightarrow \ {}^4_2\text{He} + \text{Energy}$$

Radiation's Effects on Health

Learning Goal 8
How radiation affects human health

When an atom or molecule is hit by an alpha particle, beta particle, or gamma ray, there may be some transfer of energy. As a result, the particle or ray may cause the target to ionize when an electron is knocked out of its orbital, producing a cation.

Radioactive particles and rays are commonly called ionizing radiation (visible light does not commonly cause atoms or molecules to ionize). Sometimes ionizing radiation strikes a molecule in such a way that a bond is broken. If that bond lies inside a gene within a chromosome, the cell may undergo mutation, changing the genetic message it carries. The cell may even die. If either a sperm cell or egg cell undergoes mutation via ionizing radiation, the organism formed by that sperm and egg may contain a genetic defect.

Substances that emit radioactive particles are also dangerous if inhaled or ingested. The radioactive particles often settle in the body and damage a certain body part. This is how workers who used paint containing the element radium were harmed in the early 1900s. The paint was used to produce watch dials that glowed in the dark. To create a fine point at the end of the paintbrush, workers used their lips and mouths, thereby ingesting the radioactive paint. A large proportion of those workers fell victim to bone cancer, leukemia, and other bone marrow diseases. Because radium is chemically similar to calcium, the radium ions were metabolized as though they were calcium ions, and the ionizing radiation they gave off caused illness and death.

Detection of Radiation

Learning Goal 9
Methods of detecting
radiation

Several devices are commonly used to detect ionizing radiation. The best-known device is the Geiger–Müller counter, which consists of a detecting tube and a counter. The detecting tube is made of a pair of electrodes surrounded by a gas that can be ionized. As radiation enters the tube, the gas is ionized and the ions move toward an area of high voltage that surrounds the electrodes. Pulses of electric current caused by the presence of ions at the electrodes are detected and recorded by the counter. Because the Geiger–Müller tube is most sensitive to beta radiation, it is not useful for detecting gamma rays (which pass right through the tube undetected) or alpha radiation (which generally cannot pass through the window of the tube).

The Geiger–Müller tube records counts per minute entering the window, but it gives no indication of the energy of the radiation. A device that records the amount *and* the energy of radiation produced is a scintillation counter. This device operates when the radiation reacts with a crystal containing sodium iodide and thallium iodide, producing a series of flashes of varying intensity. The intensity of the flashes is proportional to the energy of the radiation.

People working in areas where radiation exists commonly wear badges that contain photographic film. Each month the film is developed to indicate the degree of that individual's exposure to radiation that month.

Measurement of Radiation

Learning Goal 10
Measurement of radiation

Several different kinds of measuring units are available to express the quantity of radiation emitted by a source or received by a living being. Some of the units express the activity of the radiation source; others relate the effect of a specific type of radiation to its impact on living tissue.

The simplest unit is the curie (symbol Ci), which is equal to 3.7×10^{10} disintegrations per second. This unit reveals little about the impact of radiation on human tissue because, in order to determine that, one must know the quantity of radiation received by the tissues.

Other units for the measurement of radiation include the roentgen, abbreviated *r*, and the rad, which stands for "radiation absorbed dose." The roentgen measures the ionizing ability of x rays and gamma rays. The amount of radioactivity that produces 2×10^9 ion pairs in 1 cubic centimeter of air is equivalent to 1 roentgen. A measurement taken in rads reveals the amount of radiation absorbed by living tissue, independent of the type of radiation.

Lethal dose is *a term used to express the toxicity of radiation.* **A lethal dose** is *the amount of radiation that will kill 50% of all exposed organisms within 30 days.* (This is written LD_{50}^{30}.) The most useful unit, however, is the rem, which stands for "roentgen equivalent man." The rem indicates the biological effect of radiation regardless of the source. Hence it includes the effect of medical x rays. The rem is calculated by taking the number of rads measured and multiplying it by a weighting factor. Each type of radiation has its own weighting factor.

In general, the recommended dose limit for the general public is 0.5 rem/year. Background radiation from cosmic rays, natural uranium and radon, and other isotopes ranges from 0.08 to 0.20 rem/year. A chest x ray is equivalent to 0.20 rem, whereas a series of x rays of the gastrointestinal tract, using fluoroscopy, produces about 22 rems. Nuclear power plants are estimated to add 0.001 rem/year to the general population.

17.2 Radioisotopes and Medicine

Learning Goal 11
Radioisotopes and medicine

Medical practitioners take advantage of the behavior of radioisotopes to diagnose and treat millions of patients each year. Specific radioactive isotopes, once introduced into a system, find their way to specific parts of the body. This behavior allows practitioners to follow an element with precision in order to diagnose a disease or target cancer cells for destruction. In medical diagnosis, the technique of radioactive tracing is routinely applied. Commonly used isotopes are chromium-51, iron-59, iodine-131, phosphorus-32, and technetium-99m.

A patient with an overactive thyroid can be administered iodine-131, a radioactive isotope of iodine. The isotope localizes in the thyroid gland, and its radiation destroys some of the thyroid cells, thereby decreasing the size of the thyroid gland and reducing its activity.

Because the body treats radioisotopes, also called tagged isotopes, the same way it treats nonradioactive elements, the destinations of element and isotope are the same. Iodine, for example, radioactive or not, travels to the thyroid gland where it is incorporated into the amino acid thyroxine. This is why iodine-131 is perfect for monitoring the function of the thyroid gland, as well as for reducing its activity. A scanning device is used to determine the rate at which the isotope is taken into the thyroid gland after it is administered.

The isotope technetium-99m emits only gamma rays, which can be detected by a scintillation counter. This isotope is used to create "images" of the brain, heart, kidneys, liver, spleen, lungs, and bones. Brain scans and bone scans both utilize technetium-99m, along with other elements.

Chromium, in the form of sodium chromate, attaches strongly to the hemoglobin of red blood cells. This makes radioactive chromium-51 an excellent isotope for determining the flow of blood through the heart. This isotope is also useful for determining the lifetime of red blood cells, which can be of great importance in the diagnosis of anemias.

Radioactive cobalt (cobalt-59 or cobalt-60) is used to study defects in vitamin B_{12} absorption. Cobalt is the metallic atom at the center of the B_{12} molecule. By injecting a patient with vitamin B_{12} that is labeled with (or contains) radioactive cobalt, a physician can study the path of the vitamin through the body and discover any irregularities.

Radiation can also be used to destroy cancer cells. The radiation can be delivered to the malignant area in three ways. In *teletherapy*, a high-energy beam of radiation is aimed at the cancerous tissue. In *brachytherapy*, a radioactive isotope is placed in the area to be treated. The radioactive source is usually enclosed in a seed (it could be a glass bead containing the isotope) that is implanted under the skin. In *radiopharmaceutical therapy*, the isotope is administered either orally or intravenously. Then the isotope travels along normal body pathways to seek its target. This method is used to get iodine-131 to the thyroid gland.

CAT Scans

Medical practitioners have long used x rays to diagnose problems with bones and some organs in the body. These photographs compress three-dimensional information into two dimensions and thereby lose a great deal of detail.

In 1979 the Nobel prize in physiology and medicine was given to Godfrey N. Hounsfield and Allan M. Corwin for developing a three-dimensional x-ray scanning technique that uses a computer along with a source of x rays. The CAT scan (CAT stands for "computerized axial tomography") rotates a source of x rays around a patient and uses a computer to come up with a three-dimensional picture of the patient's internal organs.

PET Scan

Whereas CAT scans show the different densities of human tissues, the PET scan (PET stands for "positron emission tomography") uses computers to analyze radiation coming from radioisotopes. This process informs medical practitioners of exactly when a radioisotope enters an organ and enables them to trace its mode of action.

PET scans have taught researchers a great deal about brain chemistry. They have revealed how the brain of a normal individual functions differently from the brain of an individual with schizophrenia. And degenerative brain diseases such as Alzheimer's disease are also being studied with this technique.

17.3 Electrochemistry

The Voltaic Cell

Learning Goal 12
Voltaic and electrolytic cells

As we noted in the introduction to this chapter, Volta devised the first electrochemical battery, or **voltaic cell,** in 1800. In his battery, Volta used the two metals silver and zinc, separated from each other by a piece

Voltmeter

Porous
partition

Ag cathode

Zn anode

e^- e^-

Ag^{1+} SO_4^{2-}

$NO_3^{1-} \rightarrow$ $\leftarrow Zn^{2+}$

AgNO$_3$ solution   ZnSO$_4$ solution

$Ag^{1+} + e^- \rightarrow Ag$ $Zn \rightarrow Zn^{2+} + 2e^-$

Figure 17.8
An example of a voltaic cell.
Although this cell has a more
advanced design, it is very
much like Volta's original cell.

of paper soaked in a salt-and-water solution. Wires were attached to the silver and zinc strips. When these wires were connected in a circuit, an electric current ran through them (Figure 17.8). Many scientists immediately tried to see how other combinations of metals would react when they were set up similarly.

The Electrolytic Cell

A few weeks after Volta's discovery was made known, two English chemists, William Nicholson and Anthony Carlisle, performed the reverse of Volta's experiment. Volta had used a chemical reaction to produce electricity; Nicholson and Carlisle used electricity to produce a chemical reaction. They ran electric current through water, using an experimental setup similar to the one shown in Figure 17.9. The current caused the water to decompose slowly into hydrogen gas and oxygen gas. What Nicholson and Carlisle had developed is called an **electrolytic cell.** It uses electricity to produce a chemical reaction.

The Lead Storage Battery

Today many batteries incorporate the principles of both the voltaic cell (when they are discharging) and the electrolytic cell (when they are being charged). An example is the *lead storage battery,* which supplies energy for automobiles, trucks, and other vehicles. The lead storage battery is actually a device that changes chemical energy into electrical energy.

Figure 17.9
An electrolytic cell

. . . .
Learning Goal 13
Fuel cells and the lead
storage battery

All batteries operate through the transfer of electrons from one substance to another substance. (An electric current is simply a flow of electrons.) For example, when a piece of zinc metal is dropped into a solution
of copper(II) sulfate, a single-replacement reaction occurs. The products
are copper metal and zinc sulfate.

$$Zn + CuSO_4 \longrightarrow Cu + ZnSO_4$$

In this reaction, electrons move from the zinc to copper ions. We can
show this by writing the reaction in the following way:

$$Zn + Cu^{2+} + SO_4^{2-} \longrightarrow Cu + Zn^{2+} + SO_4^{2-}$$

The Zn gives up two electrons and the Cu^{2+} picks them up. If we arrange
these substances so that they can't make direct contact with each other,
but are indirectly connected, we have a voltaic cell (Figure 17.10), which
can produce electricity.

The lead storage battery in an automobile uses lead(IV) oxide and lead,
separated by a sulfuric acid solution (Figure 17.11). Electricity flows from
the battery when the following reaction takes place.

$$Pb + PbO_2 + 2H_2SO_4 \longrightarrow 2PbSO_4 + 2H_2O \qquad \text{(Battery discharging)}$$

Figure 17.10

A zinc–copper voltaic cell

CuSO₄ solution   ZnSO₄ solution

$Cu^{2+} + 2e^- \rightarrow Cu$ $Zn \rightarrow Zn^{2+} + 2e^-$

In this reaction, lead metal reacts with sulfate ions to produce lead(II) sulfate, with the release of two electrons.

$$Pb + SO_4^{2-} \longrightarrow PbSO_4 + 2e^{-1}$$

While this is happening, lead(IV) oxide is reacting with sulfuric acid to produce lead(II) sulfate, with the gain of two electrons.

$$PbO_2 + 4H^{1+} + SO_4^{2-} + 2e^{-1} \longrightarrow PbSO_4 + 2H_2O$$

Electrons are thus transferred from the lead to the lead(IV) oxide. This produces electricity.

Figure 17.11

One cell of a lead storage battery

Each time a car is started, current is drained from its battery. The battery discharges as lead(II) sulfate is formed. Why, then, does the battery last so long? Because the lead storage battery has another characteristic. As the engine runs, the battery is recharged by the automobile's alternator (or generator). A current flows through the battery, moving in the opposite direction from the discharging current. The battery now becomes an electrolytic cell. The reaction that takes place during the charging of the battery is just the reverse of the discharging reaction:

$$2PbSO_4 + 2H_2O \longrightarrow Pb + PbO_2 + 2H_2SO_4 \qquad \text{(Battery charging)}$$

Lead and lead(IV) oxide are formed, and the battery is again ready to produce electricity.

Fuel Cells

When the United States put its space program into high gear, it needed a battery that would offer high efficiency, have few moving parts, and be light in weight. The **fuel cell** was the answer. This type of battery works like the voltaic cell, except that in a fuel cell the reactants are introduced continuously and the products are removed continuously. One kind of fuel cell used in space vehicles depends on the reaction between hydrogen and oxygen (Figure 17.12). The nickel in one electrode, and a combi-

Figure 17.12

The hydrogen–oxygen fuel cell

nation of nickel and nickel oxide in the other, act as *catalysts* for the reaction. The overall chemical reaction is

$$2H_2 + O_2 \longrightarrow 2H_2O$$

We can think of this reaction as occurring in two steps:

$$2H_2 \longrightarrow 4H^{1+} + 4e^{-1} \qquad \text{(The hydrogen releases electrons.)}$$

$$O_2 + 4e^{-1} + 4H^{1+} \longrightarrow 2H_2O \qquad \text{(The oxygen gains electrons.)}$$

An advantage of this fuel cell is that is produces a consumable end product—water! More important, it does not pollute the air. For this reason, research scientists are looking at fuel cells as a possible means of powering automobiles.

17.4 Environmental Chemistry and Air Pollution

Never before has our world been faced with so many environmental problems, because never before has the population been so large. The air we breathe, the water we drink, the oceans and lakes, and the land we live on are rapidly being polluted. We ask ourselves, "How can we stop this pollution?" To stop it we must first understand how it is caused. In this section we'll briefly look at the mechanisms of air pollution from the chemist's point of view.

We can define **air pollution** as *the presence of a contaminant in the outdoor atmosphere in a concentration large enough to injure human, plant, or animal life or to interfere with the enjoyment of life or property.*

The Air We Breathe

Learning Goal 14
Composition of the air and air
pollutants

Air consists, by volume, of 78% nitrogen, 21% oxygen, and 1% other gases (argon, helium, hydrogen, carbon dioxide, water vapor, and other inert gases). These percentages are fairly accurate, but unfortunately they don't tell the whole story. A number of substances that we often find in air are missing from this list. And although their concentrations may be extremely small, they have a devastating effect on human health. These unnamed ingredients include **gaseous air pollutants,** such as carbon monoxide, hydrocarbons, sulfur dioxide, sulfuric acid, nitrogen oxide, and nitrogen dioxide, and **particulate matter,** or tiny particles of smoke, dust, fumes, and aerosols in the atmosphere.

Reactions in the Air

Air pollution has been known since ancient times, but until quite recently the causes were natural. Active volcanoes spewed out huge amounts of ash, and swamps emitted foul-smelling gases. This natural pollution, since it was isolated and infrequent, wasn't much of a problem for human beings or the atmosphere. However, as the world population increased and people advanced technologically, a problem arose.

In England, for example, coal was the main source of fuel. Coal is basically carbon, along with some sulfur and other impurities. When the English burned coal to protect themselves from the cold and damp air, the smoke from their millions of fires threw out huge amounts of particulates and gases, and some of these gases were poisonous. It was only in 1956 (after a disastrous four-day smog had killed 4,000 Londoners in 1952) that Britain passed the Clean Air Act and made it illegal to burn coal in the cities. Britain even switched its railroads from coal to electricity and diesel fuel.

Here are some of the reactions that occur when coal is burned:

Learning Goal 15
Reactions that produce air pollution

$$C + O_2 \longrightarrow CO_2 \qquad\qquad S + O_2 \longrightarrow SO_2$$
$$\text{Carbon} \qquad\qquad\qquad\qquad\qquad \text{Sulfur}$$
$$\text{dioxide} \qquad\qquad\qquad\qquad\qquad \text{dioxide}$$

$$2SO_2 + O_2 \longrightarrow 2SO_3 \qquad SO_3 + H_2O \longrightarrow H_2SO_4$$
$$\text{Sulfur} \qquad\qquad\qquad\qquad \text{Sulfuric}$$
$$\text{trioxide} \qquad\qquad\qquad\qquad \text{acid}$$

The sulfur in the coal is oxidized to sulfur dioxide, which is eventually oxidized to sulfuric acid. The sulfuric acid then remains as a mist in the atmosphere. When it is inhaled, this acid mist can irritate the nasal passages. It can penetrate deep into the lungs, destroying parts of the lung tissue. Sulfuric acid can also corrode many metals and building materials. (This is happening, for example, to the beautiful old buildings and statues in Venice.) **Acid rain,** which is rain containing sulfurous and sulfuric acids, can ruin trees and farmland and destroy the aquatic life of ponds and lakes.

Coal isn't the only offender. Oil, the main source of fuel in the United States today, contains sulfur impurities also. Industry uses immense amounts of oil, and some power companies use it to produce electricity.

The Internal Combustion Engine

Learning Goal 16
Health problems caused by pollution

The automobile is another major source of air pollution. In this case, the problem is caused by flaws in the internal combustion engine. There are two major problems.

1. Not all the gasoline that enters the engine is burned completely. This leads to the emission of carbon monoxide and hydrocarbons, which are chemical compounds that contain the elements carbon and hydrogen. (Gasoline is a mixture of hydrocarbons.)

2. The nitrogen and oxygen in air react at the high operating temperatures of auto engines. This leads to the emission of oxides of nitrogen.

Incompletely burned gasoline travels out the exhaust pipe and into the air. Once in the air, the hydrocarbons can react with nitrogen oxides to produce photochemical smog. Smog is a mixture of smoke and fog. **Photochemical smog** is smog that results from atmospheric reactions in which sunlight is a catalyst. Both can be deadly to people with respiratory problems. And CO is poisonous to everything that breathes oxygen.

The other major by-products of the combustion of gasoline—oxides of nitrogen—are produced in a series of steps. First, nitrogen and oxygen produce nitric oxide within the engine.

$$N_2 + O_2 \longrightarrow \underset{\substack{\text{Nitric} \\ \text{oxide}}}{2NO}$$

The nitric oxide gets out into the atmosphere, where some of it reacts with oxygen in the air to produce nitrogen dioxide.

$$2NO + O_2 \longrightarrow \underset{\substack{\text{Nitrogen} \\ \text{dioxide}}}{2NO_2}$$

Nitrogen dioxide, which is unstable in the presence of sunlight, can decompose into nitric oxide and atomic oxygen.

$$NO_2 + \text{sunlight} \longrightarrow NO + \underset{\substack{\text{Atomic} \\ \text{oxygen}}}{O}$$

In the presence of sunlight, the atomic oxygen can react with unburned hydrocarbons to produce other harmful chemicals, but more commonly the atomic oxygen (O) reacts with molecular oxygen (O_2) to produce ozone (O_3). High concentrations of ozone near ground level pose a severe health problem, because ozone irritates the eyes and the respiratory system. It can also react with many different substances. For example, ozone can destroy rubber by causing it to harden and crack.

How Can Air Pollution Be Controlled?

Learning Goal 17
Ways of reducing pollution

The major drawbacks of both coal and oil as fuels are emissions of particulate matter and sulfur oxides. Emission of sulfur oxides can be reduced by using coal or oil that has a low sulfur content. We can also reduce emissions by passing exhaust gases through a *scrubber*, a device that absorbs gaseous pollutants. A scrubber for removing sulfur dioxide in a smokestack usually consists of a fine spray of water. Gas rising through the stack passes through the scrubber, where the water absorbs the sulfur dioxide (Figure 17.13). This process can even produce a useful by-product, sulfurous or sulfuric acid.

Figure 17.13
A scrubber

Particulate emissions can be reduced by using mechanical devices. One such device is the *electrostatic precipitator* (Figure 17.14), in which dust particles that have been electrically charged are collected between highly charged electrical plates. This eliminates the dust particles from the gases leaving the smokestack.

Because cars and trucks contribute half of all air pollution, eliminating their emissions is a top-priority problem for automobile manufacturers. One solution would be to switch from the internal combustion engine to

Figure 17.14
An electrostatic precipitator

electrically powered cars. However, for many technical reasons, this is not practical at the present time. Moreover, in a sense, it would merely shift the blame for pollution to power companies. So, at least for the near future, improvements must be made to the internal combustion engine.

A number of pollution-control devices are used in today's motor vehicles. Some are designed to reduce emissions of hydrocarbons by reducing the evaporation of gasoline from fuel tanks. Others aim at more nearly complete burning of gasoline; complete combustion would produce only carbon dioxide and water. For example, here is the equation for the complete combustion of octane, one ingredient of gasoline:

$$2C_8H_{18} + 25O_2 \longrightarrow 16CO_2 + 18H_2O$$
Octane

Catalytic converters were developed to reduce the emission of nitrogen oxides. Exhaust gases are passed through the converter, where the proper catalyst enhances the conversion of these oxides into less troublesome substances. Some converters also enhance the oxidation of partially burned hydrocarbons, to produce carbon dioxide and water.

The Problem of Hazardous Waste

As technology and industry have developed new processes, the problem of handling chemical waste has become crucial. Countless barrels of hazardous waste materials have been produced, threatening the integrity of our land, air, and waterways.

Hazardous waste is any waste that may cause death or irreversible or incapacitating illness and that threatens human health and the environment because of its composition. Many hazardous wastes do not decompose into harmless substances that cycle back into our ecological system. They will remain a threat to future generations unless we develop ways to dispose of them safely.

Most plastics, for example, don't decompose. A toy composed of polyethylene or a bottle made of PVC (polyvinyl chloride) may last for thousands of years. And when PVC is burned, it produces hydrogen chloride gas, which is toxic. Some plastics contain additives called plasticizers to make them more pliable. Among these are the polychlorinated biphenyls (PCBs) and phthalates. PCBs can leach out of plastics and, once free in the environment, can enter the food chain.

Besides plastics, there are pesticides, medicines, paints, petroleum products, metals, leathers, and textiles that all visit hazardous wastes upon us as a by-product of their manufacture. It's no surprise that with all the toxic waste that has been generated, hazardous-waste management is one of the most important applications of chemistry today. Like time bombs waiting to explode, thousands of barrels of these dangerous materials lie buried in illegal dumpsites or carelessly piled up in illegal

Learning Goal 18
Hazardous wastes

Figure 17.15
Chemical Control Corporation in Elizabeth, N.J., the former site of 40,000 drums of hazardous chemical wastes, (a) before the fire and explosion in April 1980, and (b) after the fire. (c) The U.S. Environmental Protection Agency's Mobile Incineration System, developed for the purpose of destroying hazardous wastes on-site, before such accidents take place.
(Photographs courtesy Environmental Protection Agency)

holding areas (Figure 17.15a,b). Researchers are seeking ways to reduce the amount of hazardous waste that has accumulated. Landfills to contain these materials are being developed, and injection wells are being considered in geologically stable regions. Other researchers are devising ways to treat the waste products through incineration, thermal destruction, high-temperature decomposition, and chemical stabilization (Figure 17.15c). The objective of these scientists is to rid the earth of these deadly materials. The quality of life available to our descendants may depend on their success.

Summary

Radioactivity is the ability of an atom spontaneously to emit penetrating rays from its nucleus. There are three types of radiation: (1) Alpha particles, which have a mass number of 4, an electrical charge of +2, and little penetrating ability. (2) Beta particles, which have a mass number of zero, a charge of −1, and only slight penetrating power. (3) Gamma rays, which have neither mass nor charge but have very high energy content and strong penetrating ability.

Some elements produce radiation naturally. When an atom of such an element emits an alpha or beta particle, it undergoes a nuclear transformation and actually becomes a different element. The half-life of a radioactive element is the time required for half the atoms in a sample of the element to decay in this manner.

Radioactive isotopes can be produced artificially, by bombarding certain elements with particles such as neutrons. It is also possible to "split" atoms by bombarding them, in a process called nuclear fission. Such a process, which results in a chain reaction leading to the splitting of more and more atoms, is the basis of the atomic bomb. Controlled nuclear fission is used in power plants to produce electricity.

The sun and stars get their energy through nuclear fusion, a process in which nuclei are combined, or fused together. This process is the basis of the hydrogen (fusion) bomb. Nuclear fusion may one day supply us with almost limitless amounts of energy for conversion to electricity.

Radioactive isotopes are used in medicine to diagnose and treat a variety of diseases. Radiation is also commonly used to destroy cancer cells.

Electrochemistry involves the production of electrical energy from chemical reactions. In the voltaic cell, electrons move from one metallic substance to another, to produce electricity. In the electrolytic cell, electricity is used to cause a chemical reaction. The common lead storage battery acts as a voltaic cell when it is discharging and as an electrolytic cell when it is being charged. The fuel cell is a battery that works like a voltaic cell, except that the reactants are introduced continously and the products are removed continously.

By volume, pure air consists of 78% nitrogen, 21% oxygen, and 1% other gases. Polluted air also contains such chemicals as carbon monoxide, hydrocarbons, sulfur dioxide, sulfuric acid, nitrogen oxide, nitrogen dioxide, and particulate matter. Some of these chemicals can react in the presence of sunlight to form dangerous photochemical smog. All these pollutants can cause health problems for people and animals and can damage plant life. However, several devices can be used to reduce emissions of harmful chemicals. They include industrial devices, such as the scrubber and the electrostatic precipitator, and emission-control devices on automobiles.

Hazardous wastes pollute the land in many areas. Attempts to solve this pollution problem have begun only recently.

Career Sketch

Nuclear Materials Handling Technicians

Working with nuclear scientists and engineers, nuclear materials handling technicians manage nuclear materials in many different stages. They fabricate, label, package, and transport nuclear fuel elements and deal with radioactive waste. They can work as industrial waste inspectors, atomic fuel assemblers, or radioactive waste dispatchers and decontaminators, as well as in several other areas.

From the mining of uranium ore to the disposing of radioactive waste, nuclear materials handling specialists take part in the milling, conversion, separation, and fabrication of nuclear materials and fuels. They ship and store spent fuel, dissolve cut-up fuel, and convert fuel into solid pellets for use in nuclear power plants.

A knowledge of nuclear chemistry is essential for specialists in this field, and they must be familiar with the techniques of solvent extraction, precipitation, centrifuging, diffusion, and enrichment of fuels. Remote-control devices are commonly used for safety, so that materials can be manipulated without being handled by humans.

Nuclear materials handling technicians at the Department of Energy's Hanford site wear protective clothing while producing weapons grade plutonium for use in the national defense program.
(Department of Energy)

A high school diploma is needed, as are college courses in nuclear technology, radiation physics, technical mathematics, electricity and electronics, technical communications, radiation detection and measurement, blueprint reading, meteorology, metallurgy, technical writing, and nuclear chemistry.

Career Sketch

Occupational Safety and Health Technicians

Occupational safety and health technicians work with safety engineers, industrial hygienists, and managers to analyze conditions in the workplace. They study engineering principles and safety-related legislation and must be familiar with techniques used to evaluate the environment. They assess physical hazards, air quality, water quality, and hazardous chemicals in the workplace, and they recommend methods of preventing unsafe working conditions in factories, offices, and places of business. They

Occupational safety and health workers ensure that conditions in the workplace are free of hazards.
(OSHA)

also sketch and lay out safety devices. Their aim is to reduce accidents and the incidence of work-related illnesses among industrial workers.

After graduating from high school, those planning careers as occupational safety and health technicians study fundamentals of fire protection, safety and health regulations, first aid, human relations, power sources and hazards, traffic safety, noise control, hazards deriving from industrial chemicals, environmental health, radiation safety, disaster preparedness, and chemical instrumentation and analysis.

· ·

Self-Test Exercises

Learning Goal 1: Radioactivity and Atomic Energy

1. List some advantages and some disadvantages of atomic energy for the human race.

2. What would be your concern about living near a nuclear power plant?

3. If you were a scientist working on the Manhattan Project, what questions would you ask yourself about the morality of your work?

4. What problems might be encountered in disposing of waste from conventional nuclear power plants?

5. What are the advantages and the disadvantages of having nuclear power plants supply all our electrical needs?

6. What is meant by the term *radioactivity?*

Learning Goal 2: Types of Nuclear Radiation

7. Describe the following: (*a*) alpha particles, (*b*) beta particles, (*c*) gamma rays.

8. (*a*) Which particle has the greater penetrating ability, a beta particle or a gamma ray?
(*b*) What kind of shield would you need to stop alpha particles?

9. What are the charge and mass of (*a*) an alpha particle, (*b*) a beta particle?

10. Why does the atomic number of an atom *increase* by 1 when the atom *emits* a beta particle?

Learning Goal 3: Writing Balanced Nuclear Equations

11. Complete each of the following nuclear equations.
(*a*) $^{197}_{78}\text{Pt} \longrightarrow \, ^{0}_{-1}\text{e} + ?$
(*b*) $^{212}_{84}\text{Po} \longrightarrow \, ^{4}_{2}\text{He} + ?$
(*c*) $^{10}_{5}\text{B} + \, ^{4}_{2}\text{He} \longrightarrow \, ^{1}_{0}\text{n} + ?$
(*d*) $^{238}_{92}\text{U} + \, ^{1}_{0}\text{n} \longrightarrow ? \longrightarrow \, ^{94}_{36}\text{Kr} + ?$

12. Complete each of the following nuclear equations.
(*a*) $^{226}_{88}\text{Ra} \longrightarrow \, ^{4}_{2}\text{He} + ?$
(*b*) $^{234}_{90}\text{Th} \longrightarrow \, ^{0}_{-1}\text{e} + ?$
(*c*) $^{234}_{91}\text{Pa} \longrightarrow \, ^{0}_{-1}\text{e} + ?$
(*d*) $^{210}_{82}\text{Pb} \longrightarrow \, ^{0}_{-1}\text{e} + ?$

13. Write the equation for the formation of a beta particle from a neutron.

14. What isotope is formed when $^{214}_{82}Pb$ emits a beta particle from its nucleus?

15. When an isotope emits an alpha particle, what happens to (a) the isotope's mass number, and (b) its atomic number?

16. What isotope is formed when $^{194}_{78}Pt$ emits an alpha particle from its nucleus?

17. When an isotope emits a beta particle, what happens to (a) the isotope's mass number, and (b) its atomic number?

18. What isotope is formed when $^{218}_{85}At$ emits an alpha particle from its nucleus?

19. Complete each of the following nuclear equations.

(a) $^{14}_{6}C \longrightarrow {}^{0}_{-1}e + ?$
(b) $? \longrightarrow {}^{0}_{-1}e + {}^{24}_{12}Mg$
(c) $? \longrightarrow {}^{4}_{2}He + {}^{234}_{90}Th$
(d) $^{84}_{36}Kr \longrightarrow {}^{1}_{0}n + ?$

20. Complete and balance each of the following nuclear equations.

(a) $^{137}_{55}Cs \longrightarrow {}^{0}_{-1}e + ?$
(b) $^{27}_{13}Al + {}^{4}_{2}He \longrightarrow {}^{30}_{15}P + ?$
(c) $^{214}_{83}Bi \longrightarrow {}^{4}_{2}He + ?$
(d) $^{0}_{-1}e + ? \longrightarrow {}^{7}_{3}Li$

Learning Goal 4: Half-life

21. Define the term *half-life* and explain how scientists use it.

22. Explain why the decay rates of radioactive elements are given in terms of half-life.

Learning Goal 5: Calculation Using Half-life

23. The half-life of carbon-14 is 5,770 years. Suppose you have a piece of wood that originally contained 4 g of carbon-14. How old is the piece of wood today if the amount of carbon-14 remaining is only 0.25 g?

24. The half-life of cobalt-60 is 5.3 years. Suppose a sample of cobalt-60 originally weighed 100.0 g. How old is the sample today if the amount of cobalt-60 remaining is only 12.5 g?

25. The isotope iodine-131 has a half-life of 8 days. Suppose a person is injected with $\overline{100}$ mg of iodine-131 today. How much iodine-131 will remain in the body of this individual after 24 days (assuming that none is lost through normal body channels)?

26. A patient is injected with the radioisotope ^{99}Tc in order to diagnose a possible malfunction of the liver or gall bladder. The radioisotope ^{99}Tc is a gamma ray emitter. The path of the isotope is traced with a gamma ray camera and projected on a monitor. The half-life of ^{99}Tc is 6 hours. How many hours must pass for only one-fourth of the original dosage to remain in the body?

Learning Goal 6: Nuclear Fission and Fusion

27. Define and give an example of (a) nuclear fission, (b) nuclear fusion.

28. Complete the following statement: The splitting of an atom is an example of nuclear _____ .

29. Complete the following statements. (a) The atomic bomb is a nuclear _____ (fusion/fission) device. (b) The hydrogen bomb is a nuclear _____ (fusion/fission) device.

30. Describe the research being done at the Princeton Plasma Physics Laboratory.

Learning Goal 7: Emergence of Nuclear Power Plants

31. What fuel is used to power conventional nuclear power plants?

32. How is that heat produced by nuclear fission used to generate electricity?

33. Discuss the problems associated with the disposal of radioactive waste.

34. Why are nuclear waste products dangerous?

35. Describe the accident that took place at Three Mile Island in Harrisburg, Pennsylvania.

36. What caused the accident at the Chernobyl nuclear power plant in the Ukraine to be so dangerous?

Learning Goal 8: How Radiation Affects Human Health

37. How might ionizing radiation induce a gene within a chromosome to undergo mutation?

38. Explain how the similarity between the elements radium and calcium led to the development of bone cancer in many workers who painted radium on watch dials in the early 1900s.

Learning Goal 9: Methods of Detecting Radiation

39. What causes the pulses of electricity that are detected in the tube of a Geiger–Müller counter?

40. To which type of radiation is the Geiger–Müller tube most sensitive?

41. A device that records the amount *and* the energy of radiation is the _____ counter.

42. Explain how a film badge works.

Learning Goal 10: Measurement of Radiation

43. Distinguish between the curie, the rad, and the roentgen.

44. Explain the meaning of the term LD_{50}^{30}.

45. What is the yearly recommended dose limit for the general public, measured in rems?

46. Is one exposed to more radiation from a chest x ray or a fluoroscope of the intestinal tract?

Learning Goal 11: Radioisotopes and Medicine

47. Name a radioactive isotope used in medicine, and discuss how it is used.

48. One treatment for malignancies known to be responsive to gamma radiation is to irradiate the area with the radioisotope cobalt-60. Is this an example of teletherapy or brachytherapy?

49. How is iodine-131 used to monitor the thyroid gland?

50. The isotope _____ can be used in performing both brain scans and bone scans.

Learning Goal 12: Voltaic and Electrolytic Cells

51. Define and give an example of (*a*) a voltaic cell, and (*b*) an electrolytic cell.

52. A battery is connected to a cell with two electrodes, both of which are immersed in a solution of hydrochloric acid. The battery supplies electrons to the cathode, which is the negative pole of the cell. When a switch is closed, an electric circuit is completed and the following chemical reaction takes place:

$$2HCl(aq) \longrightarrow H_2(g) + Cl_2(g)$$

Is this an electrolytic cell or a voltaic cell?

53. Balance the following equation for the discharging of the lead storage battery:

$$Pb + PbO_2 + H_2SO_4 \longrightarrow PbSO_4 + H_2O$$

54. The zinc–copper cell was used to provide energy for the first transcontinental telegraph lines. It consists of a piece of zinc immersed in a zinc sulfate solution and a piece of copper immersed in a copper(II) sulfate solution. The copper and zinc are connected by a wire, and a salt bridge provides contact between the two solutions. Is this an electrolytic cell or a voltaic cell?

Learning Goal 13: Fuel Cells and the Lead Storage Battery

55. (*a*) What is a fuel cell?
(*b*) Discuss the advantages and disadvantages of substituting a fuel cell for the internal combustion engine as a source of power for the automobile.

56. The reactions that occur during the discharging of a lead storage battery are

$$Pb + SO_4{}^{2-} \longrightarrow PbSO_4$$
$$PbO_2 + 4H^{1+} + SO_4{}^{2-} \longrightarrow PbSO_4 + 2H_2O$$

(*a*) Rewrite both half-reactions, supplying the proper number of electrons in each.
(*b*) Which reaction represents oxidation, and which represents reduction? (You may want to check Section 8.6 before answering this question.)

57. How does a fuel cell differ from a voltaic cell?

58. The fuel cell shown in Figure 17.12 depends on the reaction between hydrogen gas and oxygen gas. The two half-reactions for this fuel cell are

$$2H_2 \longrightarrow 4H^{1+}$$
$$O_2 + 4H^{1+} \longrightarrow 2H_2O$$

(*a*) Rewrite both half-reactions, supplying the proper number of electrons in each.
(*b*) Which reaction represents oxidation, and which represents reduction?

Learning Goal 14: Composition of the Air and Air Pollutants

59. Define the terms *gaseous air pollutants* and *particulate matter*.

60. Do you think that carbon dioxide in excessively high concentration would be classified as an air pollutant? Explain your answer.

61. List all the gaseous air pollutants you know of.

62. While you are standing outdoors, a small piece of ash enters your eye. What type of air pollutant has affected you?

63. List all the particulate air pollutants you know of.

64. When Mt. Saint Helens erupted on May 18, 1980, large amounts of ash were emitted into the atmosphere. What were some of the effects of the volcano on the people living in the area?

65. What two substances account for most of the composition of air?

66. Name four gaseous pollutants that are produced when coal is burned.

Learning Goal 15: Reactions That Produce Air Pollution

67. Would the introduction of electrically powered cars solve the problem of pollution of the air by cars?

68. Why is photochemical smog more dangerous than ordinary smog?

69. Complete and balance each of the following equations.
(a) $C + O_2 \longrightarrow$
(b) $S + O_2 \longrightarrow$
(c) $SO_2 + O_2 \longrightarrow$
(d) $? + ? \longrightarrow NO$

70. Write the equations for the formation of acid rain.

71. What is photochemical smog?

72. In the presence of sunlight, nitrogen dioxide decomposes into nitric oxide and atomic oxygen. Write an equation for this reaction.

73. How are oxides of nitrogen formed in the internal combustion engine?

74. Write the equation for the formation of ozone in the atmosphere.

Learning Goal 16: Health Problems Caused by Pollution

75. Of the gaseous and particulate air pollutants that we have discussed, which do you think would be most harmful to people suffering from respiratory and lung diseases?

76. Newspapers print a daily air-quality report that is issued by environmental experts. When the air quality is poor, people with heart problems or respiratory disease are advised to stay indoors. Why?

Learning Goal 17: Ways of Reducing Pollution

77. What type of control device would you place on a smokestack of a city incinerator that is spewing out black smoke?

78. What, if anything, can be done to eliminate pollution caused by emissions from a home fireplace?

79. What control devices would you place on the smokestack of a power company that is using oil as a fuel for generating electricity?

80. How have catalytic converters cut down on air pollution?

81. "Dilution is the solution to pollution." Discuss the logic of this statement.

82. What type of device can be used to control pollution that is due to an incinerator burning hazardous wastes that have a high chloride content?

83. To stop the pollution of our air, what steps must be taken by (*a*) government, (*b*) industry, (*c*) the average citizen?

84. What is the benefit of a power company burning low-sulfur-content fuel to generate electricity? Are there any disadvantages?

Learning Goal 18: Hazardous Wastes

85. Define the term *hazardous waste* and give an example of a hazardous waste.

86. How do PCBs affect people?

87. How can hazardous wastes enter the food chain?

88. Would you rather see hazardous wastes incinerated and risk air pollution, or see these wastes deposited in a landfill and risk ground-water pollution? List some advantages and disadvantages of each waste-disposal method.

Extra Exercises

89. The potassium–argon (^{40}K-^{40}Ar) method of dating rocks has been used to date the oldest known rocks on earth. These rocks, found in Greenland, have been analyzed as follows:

^{40}K remaining in crystal = 3 atoms

^{40}K originally present in crystal = 24 atoms

If ^{40}K has a half-life of 1.3 billion years, how old are the rocks?

90. Name three types of chemical compounds that are considered air pollutants. State how each originates and what its effects are.

91. Discuss the technological problems associated with the use of fusion as a source of energy.

92. In light of the near disaster at the Three Mile Island Atomic Power Plant near Harrisburg, Pennsylvania in March of 1979, and the Chernobyl disaster of 1986, what do you think should be the policy regarding our present-day nuclear power plants?

93. Discuss which kind of pollution—air or water pollution—is a bigger problem in your area.

94. A certain material is producing nuclear radiation. Assuming that you have the proper equipment, explain how you could determine what type of nuclear radiation the material is producing.

95. Contrast the beta particle, the electron, and the alpha particle.

96. The isotope ^{138}Cs has a half-life of 32.2 minutes. Is ^{138}Cs more or less stable than an isotope with a half-life of 3 days? Explain.

97. Complete the following equation by supplying the missing particle:

$$^{16}_{8}O \longrightarrow \ ^{12}_{6}C + ?$$

98. Suggest a way to prepare $^{13}_{7}$N from $^{10}_{5}$B.

99. The half-life of $^{45}_{19}$K is $3\overline{0}$ minutes. Suppose you have a sample of $3\overline{0}$ mg. How much time would elapse before the amount was reduced to 7.5 mg?

100. A nuclear power plant uses water from a nearby river as cooling water. The water returns to the river at an elevated temperature. What problems can this cause for fish?

18 *Organic Chemistry, Part 1:*
Hydrocarbons

Learning Goals

After you have studied this chapter, you should be able to:

1. Distinguish between organic compounds and inorganic compounds.
2. Write the Lewis electron-dot structure for the carbon atom.
3. Describe the geometry of the carbon bond.
4. Draw structural formulas when you are given the molecular formulas or the names of alkanes, alkenes, and alkynes.
5. Define and illustrate the terms *saturated hydrocarbon, unsaturated hydrocarbon,* and *cyclic hydrocarbon.*
6. Define and illustrate the terms *isomer* and *homologous series.*
7. Name alkanes, alkenes, and alkynes using the IUPAC system.
8. Define the term *aromatic hydrocarbon* and give a common example.

Introduction

During the 1700s and early 1800s most chemists believed that there were two distinct classes of chemical compounds: **organic** and **inorganic.** They said that organic compounds were derived from living or once-living organisms and that inorganic compounds were part of the nonliving world. Organic compounds were substances like sugar, fats, oils, and wood. Inorganic compounds were substances like salt or iron.

Many chemists of those days noted that organic substances could be converted into inorganic substances by heating but that inorganic substances could not be converted into organic substances. These observations gave rise to the theory of **vitalism.** According to this theory, (1) life and substances associated with life were not bound by the laws of science, and (2) these life-connected substances had a "vital force" that human beings could not control. Many scientists of this period thought that chemists, even though they could do many marvelous things in their laboratories, could not duplicate this "vital force" and would never be able to synthesize organic compounds (that is, make them artificially).

But in 1828 the German chemist Friedrich Wöhler made a startling discovery. In one of his experiments, Wöhler, who was working with cyanide compounds, heated the compound ammonium cyanate. At this time, chemists considered ammonium cyanate inorganic, so Wöhler wasn't expecting it to show any association with life. But after he had heated the ammonium cyanate, Wöhler found that this inorganic compound had undergone a chemical change and had formed the compound urea, an organic compound found in urine. Thus Wöhler became the first person to produce an organic compound artificially. He wrote a letter to his friend Berzelius, the Swedish chemist: "I must tell you that I can make urea, but without kidneys or even an animal being necessary, be it man or dog." This development put the theory of vitalism on shaky ground.

Now, some 160 years after Wöhler's discovery, chemists are able to synthesize more than 6 million organic compounds. In this chapter and Chapter 19, we shall look at a small number of these compounds.

18.1 Organic Chemistry and Carbon

Organic chemistry is primarily *the study of compounds that contain carbon.* The old definition of organic substances does not apply any more. Synthetic fibers, plastics, and pharmaceuticals (drugs, medicines) are all organic compounds because they all have numerous carbon atoms in their molecules. However, not *all* carbon-containing compounds are organic. For example, carbonate compounds (such as $CaCO_3$ and Na_2CO_3) and the oxides of carbon are considered *in*organic carbon compounds.

Learning Goal 1
Organic and inorganic
compounds

It's hard to establish an absolute definition of what an organic compound is. In most cases, **organic compounds** *are those that contain the element carbon,* along with elements such as hydrogen, oxygen, nitrogen, sulfur, and the Group VIIA elements. (Once we have examined the different types of organic compounds, a precise definition will no longer be necessary.)

Why Are There So Many Organic Compounds?

There are more than 6 million organic compounds, and there are only about 100,000 inorganic compounds, so the organic compounds far outnumber the inorganic. How is this possible? After all, organic compounds are made up primarily of carbon and a few other elements, whereas inorganic compounds are made up of combinations of all the 100 or so remaining elements.

The answer lies in *the ability of the carbon atom to combine with itself.* This unusual ability enables carbon atoms to form all kinds of chain-like and ring-shaped molecules, making the number of potential organic compounds almost infinite (Figure 18.1).

Carbon: Its Electronic Structure

Learning Goal 2
Lewis electron-dot structure of carbon $\cdot \overset{\cdot}{C} \cdot$

Look at the periodic table (inside the front cover) and you'll see that carbon is in Group IVA. This means that a carbon atom has four electrons in its outermost energy level—and that is an open invitation to bond: *A carbon atom can use all four of these electrons to bond with other atoms.* In organic compounds, carbon bonds covalently. In other words, it shares its four electrons with other atoms to form bonds that are very strong.

H—C—C—H   Ethane

H—C—C—C—C—H   Butane

H—C—C—C—C—C—C—C—C—C—C—H   Decane

Figure 18.1

Carbon atoms can form either long chains or ring-like compounds. Practically no other atoms can.

Cyclopentane

Cyclohexane

Carbon: Its Tetrahedral Bond

Learning Goal 3
Geometry of carbon bonds

When a carbon atom shares its four outer electrons with four other atoms, the result is a compound such as CH_4, the compound *methane*. In this compound, the central carbon atom is bonded to each of the four hydrogen atoms. We can write the electron-dot structure in the usual way:

$$H:\overset{\cdot\cdot}{\underset{\cdot\cdot}{C}}:H \quad \text{or} \quad H—\overset{\displaystyle H}{\underset{\displaystyle H}{C}}—H$$

These diagrams give the impression that the carbon-hydrogen bonds are at 90-degree angles to each other:

$$H—\overset{\displaystyle H}{\underset{\displaystyle H}{C}}—H \quad 90°$$

Figure 18.2

The tetrahedral structure of the CH_4 molecule

However, this is not the way the molecule actually exists. Molecules are *three*-dimensional, not two-dimensional. And the four outer electrons in the carbon move as far apart from each other as possible, because electrons repel each other. (Remember that particles with similar charges always repel one another.) The electrons can get as far apart as possible if the carbon-hydrogen bonds form a *tetrahedron* (Figure 18.2). A tetrahedron (*tetra* is from the Greek word for "four") is a pyramid-shaped figure with three upper sides plus a bottom side. You can make a tetrahedron this way: Take six pencils. Place three of them on a table in the form of a

Figure 18.3
How to make a tetrahedron using six pencils

triangle. Hold the other three pencils upright, placing one at each corner of the triangle. Now tilt the three upright pencils inward until they touch. You have formed a tetrahedron (Figure 18.3). This tetrahedral bonding is common to many organic compounds; methane is only one example. With the carbon inside the tetrahedron and the hydrogens (in methane) at the corners, the bonds are 109.5 degrees apart.

Carbon–Carbon Bonds

When two carbon atoms bond with each other, they can do so in three ways:

1. They can form a single bond: $\cdot\dot{C}-\dot{C}\cdot$

2. They can form a double bond: $\cdot\dot{C}=\dot{C}\cdot$

3. They can form a triple bond: $\cdot C\equiv C\cdot$

In the case of the single bond, we can visualize the two carbon atoms joining in such a way that the corners of two tetrahedrons are touching (Figure 18.4). The double and triple bonds are difficult to visualize or draw on a flat page. However, for the double bond, we can visualize two tetrahedrons with a common edge. For the triple bond, we can visualize two tetrahedrons with a common face.

Elemental Forms of Carbon

As an element, carbon exists in several forms. The two crystalline forms of carbon are diamond and graphite (Figure 18.5). Although graphite and diamond are both composed of pure carbon atoms, their properties are very different. Diamond is one of the hardest materials known. Graphite, on the other hand, is one of the softest. It is found in pencils (lead pencils

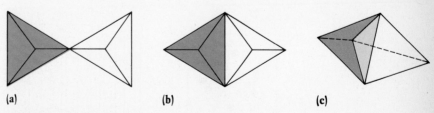

(a)   (b)   (c)

Figure 18.4
Visualization of two carbon atoms joined together by (a) a single bond, (b) a double bond, and (c) a triple bond.

Figure 18.5

The two crystalline forms of carbon.

[(left) Fundamental Photographs, New York. (right) Courtesy Carbon Products Division, Union Carbide Corp.]

don't really contain lead, but graphite) and in lubricants. Obviously the crystalline structures of diamond and graphite must differ vastly. Figure 18.6 shows how.

In diamond, the carbon atoms exhibit a tetrahedral arrangement wherein each carbon atom is bonded to four other carbon atoms. It is this structure that gives diamond its strength and hardness. In graphite, the carbon atoms are in parallel planes or sheets. Each sheet contains carbon atoms attached to one another, in six-sided structures. Two kinds of bonding forces hold the graphite structure together. These are the bonding between carbon atoms within each sheet and the bonding between the sheets of carbon atoms. The second bonding force is much weaker than the first. In other words, the forces holding the sheets together are very weak, so the sheets can slide over each other. That is why graphite is used to lubricate sticky locks and zippers.

Figure 18.6

The crystalline structures of diamond and graphite

          Graphite

18.2 Writing Formulas for Organic Compounds

· · · · · · · · · · · · ·
Learning Goal 4
Structural formulas of organic compounds

In earlier chapters, **molecular formulas** were written for various compounds. For example, water is written as H_2O. The molecular formula tells us the actual number of atoms of each element in a molecule of the compound.

For the organic chemist, however, the molecular formula is not enough. As we have seen, carbon atoms can bond together in different arrangements. For this reason, there can be many different compounds with the same molecular formula. For example, the molecular formula C_4H_{10} represents two distinct compounds, and the formula C_8H_{18} represents 18 distinct compounds! This is why, when organic chemists write the formula for an organic compound, they use a **structural formula.** *The structural formula indicates how the carbon atoms are bonded to each other, as well as the number of atoms of each element.* We can write the structural formulas for the two compounds represented by C_4H_{10} in one plane like this:

$$
\begin{array}{ccccc}
& H & H & H & H \\
& | & | & | & | \\
H- & C- & C- & C- & C-H \\
& | & | & | & | \\
& H & H & H & H
\end{array}
\quad \text{and} \quad
\begin{array}{ccc}
 & H\;H\;H & \\
 & \diagdown|\diagup & \\
 & C & \\
 & | & \\
H- C- & C- & C-H \\
 | & | & | \\
 H & H & H
\end{array}
$$

Structural formulas are sometimes written in a more convenient form called the **condensed form.** For example, the two formulas above can also be written in this way:

$$
CH_3-CH_2-CH_2-CH_3 \quad \text{and} \quad
\begin{array}{c}
CH_3 \\
| \\
CH_3-CH-CH_3
\end{array}
$$

In the first method, all bonds are shown. In the condensed form, only bonds between carbon atoms are shown. The other bonds are understood to be present, as in a molecular formula.

Example 18.1

Write each of the following structural formulas in the condensed form.

(a)
$$
\begin{array}{cccccc}
H & H & H & H & H & H \\
| & | & | & | & | & | \\
H-C- & C- & C- & C- & C- & C-H \\
| & | & | & | & | & | \\
H & H & H & H & H & H
\end{array}
$$

(b)
$$
\begin{array}{ccc}
H\;H\;H & \\
\diagdown|\diagup & \\
C & \\
| & \\
H-C- C- C-H \\
| \quad | \quad | \\
H \quad C \quad H \\
\diagup|\diagdown \\
H\;H\;H
\end{array}
$$

$$\begin{array}{c}
\qquad\qquad\ \ \ \overset{\displaystyle H\ \ H\ \ H}{\diagdown\ |\ \diagup} \\
\overset{\displaystyle H\ \ H}{|\ \ |}\ \ \overset{\displaystyle C}{|}\ \ \overset{\displaystyle H\ \ H}{|\ \ |} \\
\text{(c)}\quad H-\overset{|}{\underset{|}{C}}-\overset{|}{\underset{|}{C}}-\overset{|}{\underset{|}{C}}-\overset{|}{\underset{|}{C}}-\overset{|}{\underset{|}{C}}-H \\
\overset{\displaystyle H\ \ H}{}\ \ \overset{\displaystyle C}{\diagup\ |\ \diagdown}\ \ \overset{\displaystyle H\ \ H}{} \\
\qquad\qquad H\ \ H\ \ H
\end{array}$$

Solution

(a) $CH_3 - CH_2 - CH_2 - CH_2 - CH_2 - CH_3$

Always check to see that you have the same number of carbon atoms and hydrogen atoms as in the original.

(b) $CH_3 - \overset{\displaystyle CH_3}{\underset{\displaystyle CH_3}{\overset{|}{\underset{|}{C}}}} - CH_3$

Both this formula and the more complex formula show that the four CH_3 groups are attached to a central carbon atom.

(c) $CH_3 - CH_2 - \overset{\displaystyle CH_3}{\underset{\displaystyle CH_3}{\overset{|}{\underset{|}{C}}}} - CH_2 - CH_3$

Here it is easiest to start with the central carbon atom and work outward.

Practice Exercise 18.1 Write each of the following structural formulas in the condensed form.

$$\text{(a)}\quad H-\overset{\displaystyle H}{\underset{\displaystyle H}{\overset{|}{\underset{|}{C}}}}-\overset{\displaystyle H}{\underset{\displaystyle H}{\overset{|}{\underset{|}{C}}}}-\overset{\displaystyle H}{\underset{\displaystyle H}{\overset{|}{\underset{|}{C}}}}-\overset{\displaystyle H}{\underset{\displaystyle H}{\overset{|}{\underset{|}{C}}}}-H$$

$$\text{(b)}\quad H-\overset{\displaystyle H}{\underset{\displaystyle H}{\overset{|}{\underset{|}{C}}}}-\overset{\displaystyle H}{\underset{\displaystyle H}{\overset{|}{\underset{|}{C}}}}-\overset{\displaystyle H}{\underset{\displaystyle C}{\overset{|}{\underset{}{C}}}}-\overset{\displaystyle H}{\underset{\displaystyle H}{\overset{|}{\underset{|}{C}}}}-\overset{\displaystyle H}{\underset{\displaystyle H}{\overset{|}{\underset{|}{C}}}}-H$$

18.3 The Classification of Organic Compounds

We have said that there are more than 6 million known organic compounds. Is each of these compounds unique, or are some similar to each other? In other words, can we classify this great number of compounds into groups?

Most organic compounds can be classified into a few groups, called **series,** in such a way that compounds that react alike are placed together. We shall look at some of these series in this chapter and at others in Chapter 19.

18.4 The Alkanes

Learning Goal 5
Saturated hydrocarbons

Let us begin the study of organic series with the **alkanes.** This group of compounds is one of the simplest in composition and structure, for the following reasons.

1. All alkanes are **hydrocarbons.** Hydrocarbons are compounds that contain only carbon and hydrogen.

2. The bonds between carbon atoms in alkanes are all single bonds. *When all the carbon-carbon bonds in an organic compound are single bonds, we say that the compound is* **saturated.**

The simplest member of this series of hydrocarbons is *methane,* CH_4, with which you are already familiar.

$$
\begin{array}{c}
\quad\ \ H \\
\quad\ \ | \\
H - C - H \\
\quad\ \ | \\
\quad\ \ H
\end{array}
$$

The next alkane has two carbon atoms and is called *ethane,* $CH_3—CH_3$, or

$$
\begin{array}{c}
\ \ H\ \ \ H \\
\ \ |\ \ \ \ | \\
H - C - C - H \\
\ \ |\ \ \ \ | \\
\ \ H\ \ \ H
\end{array}
$$

Then we come to the member of this series that has three carbon atoms. This molecule is called *propane,* $CH_3—CH_2—CH_3$, or

$$
\begin{array}{c}
\ \ H\ \ \ H\ \ \ H \\
\ \ |\ \ \ \ |\ \ \ \ | \\
H - C - C - C - H \\
\ \ |\ \ \ \ |\ \ \ \ | \\
\ \ H\ \ \ H\ \ \ H
\end{array}
$$

Note that each member of the alkane series differs from the member before it by having one more CH_2 group. You can see this more clearly by looking at the molecular formulas of these compounds:

$$
\begin{array}{ccc}
CH_4 & C_2H_6 & C_3H_8 \\
\text{Methane} & \text{Ethane} & \text{Propane}
\end{array}
$$

Table 18.1
The First Ten Members of the Alkane Series

n	Molecular formula	Structural formula	Name	Comments
1	CH_4	CH_4	Methane	Main ingredient of natural gas
2	C_2H_6	$CH_3—CH_3$	Ethane	Minor ingredient of natural gas
3	C_3H_8	$CH_3—CH_2—CH_3$	Propane	Minor ingredient of natural gas; usually separated from other ingredients and sold as bottled gas
4	C_4H_{10}	$CH_3—(CH_2)_2—CH_3$	Butane	Also found in natural gas; usually separated from other ingredients and used as a fuel—for example, in butane lighters
5	C_5H_{12}	$CH_3—(CH_2)_3—CH_3$	Pentane	⎫
6	C_6H_{14}	$CH_3—(CH_2)_4—CH_3$	Hexane	⎪
7	C_7H_{16}	$CH_3—(CH_2)_5—CH_3$	Heptane	⎬ Ingredients of gasoline
8	C_8H_{18}	$CH_3—(CH_2)_6—CH_3$	Octane	⎪
9	C_9H_{20}	$CH_3—(CH_2)_7—CH_3$	Nonane	⎭
10	$C_{10}H_{22}$	$CH_3—(CH_2)_8—CH_3$	Decane	An ingredient of kerosene

Learning Goal 6
Isomers and homologous series

When the compounds in a series differ by the same structural group (like CH_2), we say that these compounds are members of a **homologous series.** This means that all the compounds in this series are of the same chemical type and differ only by fixed increments (an increment is an increase in quantity). In the alkane series, the fixed increment is CH_2. The general formula for this series of compounds can be written as C_nH_{2n+2}, where n represents the number of carbon atoms. If you substitute the numbers 1 through 10 for n in this formula, you will get each of the molecular formulas listed in Table 18.1, in turn.

It is important that you learn the names of the alkanes we have just discussed, because they are used in naming many other organic compounds.

18.5 Isomers

In Section 18.2 we noted that the molecular formula for an organic compound could actually represent many different compounds. As an example we showed that the formula C_4H_{10} represented two compounds. This phenomenon is known as **structural isomerism.**

Isomers *are compounds with the same* **molecular** *formula but different* **structural** *formulas. In other words, the atoms of isomers are bonded together differently.* The title of Table 18.1 says that it shows the first ten members of the alkane series. However, it really shows the first ten *straight-chain* members of the alkane series. This is because there are isomers for all of the alkane series except the first three members (Table 18.2). Let us write the structures for the isomers of C_4H_{10} once again:

$$CH_3 — CH_2 — CH_2 — CH_3$$

$$CH_3 — \underset{\underset{H}{|}}{\overset{\overset{CH_3}{|}}{C}} — CH_3$$

Normal butane   Isobutane

Table 18.2
Numbers of Isomers of Some Compounds in the Alkane Series

Molecular formula	Number of isomers
C_4H_{10}	2
C_5H_{12}	3
C_6H_{14}	5
C_7H_{16}	9
C_8H_{18}	18
C_9H_{20}	35
$C_{10}H_{22}$	75

The structure on the left is the **straight-chain** isomer, which we call normal butane (or *n*-butane). (The word "normal" or the letter *n* before the word "butane" signifies the straight-chain isomer.) The second structure is the **branched-chain** isomer, which is commonly called *iso*butane. *Before you continue reading, be sure you understand the differences between these two structures.*

The three isomers of pentane (C_5H_{12}) are

$$CH_3 - CH_2 - CH_2 - CH_2 - CH_3$$
n-pentane

$$CH_3 - CH_2 - \underset{\underset{H}{|}}{\overset{\overset{CH_3}{|}}{C}} - CH_3 \qquad CH_3 - \underset{\underset{CH_3}{|}}{\overset{\overset{CH_3}{|}}{C}} - CH_3$$

Isopentane   Neopentane

Again, be sure that you see the differences among the three structures.

Now suppose we wrote the structure of C_5H_{12} as

$$H - \underset{\underset{CH_3}{|}}{\overset{\overset{CH_3}{|}}{C}} - CH_2 - CH_3$$

Is this structure different from that of isopentane? The answer is no! We have simply flipped the molecule of isopentane around. The *bonding sequence* is still the same.

Example 18.2

Write the structural formulas for the five isomers of C_6H_{14}.

Solution
Try to work out the five different bonding sequences. Remember that *each carbon atom can be bonded to only four other atoms.*

(a) $CH_3 - CH_2 - CH_2 - CH_2 - CH_2 - CH_3$

(b) $CH_3 - CH_2 - CH_2 - \underset{\underset{H}{|}}{\overset{\overset{CH_3}{|}}{C}} - CH_3$

(c) $CH_3 - CH_2 - \underset{\underset{H}{|}}{\overset{\overset{CH_3}{|}}{C}} - CH_2 - CH_3$

(d)

$$CH_3-CH_2-\underset{\underset{CH_3}{\displaystyle |}}{\overset{\overset{CH_3}{\displaystyle |}}{C}}-CH_3$$

(e)

$$CH_3-\underset{\underset{H}{\displaystyle |}}{\overset{\overset{CH_3}{\displaystyle |}}{C}}\!\!-\!\!-\underset{\underset{H}{\displaystyle |}}{\overset{\overset{CH_3}{\displaystyle |}}{C}}-CH_3$$

Practice Exercise 18.2   Write the structural formulas for the nine iso-mers of C_7H_{16}.

Each of the five isomers in Example 18.2 has a distinct name. The way in which names are assigned to such compounds is called *organic nomen-clature*. Before we can discuss this topic, however, we need to discuss *alkyl groups*, which are somewhat similar to groups of atoms such as SO_4^{2-} and OH^{1-} in inorganic chemistry.

18.6 Alkyl Groups

Suppose that an organic chemist begins to study a compound with the formula

$$CH_3-C\equiv C-CH_3$$

The chemist then extends the study to other compounds with the same type of structure:

$$CH_3-CH_2-C\equiv C-CH_2-CH_3$$

$$CH_3-CH_2-CH_2-C\equiv C-CH_2-CH_2-CH_3$$

$$CH_3-CH_2-CH_2-CH_2-C\equiv C-CH_2-CH_2-CH_3$$

The chemist notes that all these compounds behave essentially alike. Moreover, their similar behavior is due to the triple carbon bond, which the chemist calls the *functional* part of the molecule, or the *functional group*. We can write the general structure for all these compounds as

$$R-C\equiv C-R'$$

where R and R' (which is pronounced "R prime") may stand for $-CH_3$, $-CH_2-CH_3$, or any of the other groups of atoms attached to the triple-bonded carbons. These particular groups of carbon and hydrogen atoms are called **alkyl groups.** (The names of the alkyl groups come from the names of the individual alkanes, with the final *-ane* changed to *-yl* in each case. Table 18.3 gives the first few alkyl groups.)

Table 18.3
Names of Some Alkyl Groups

Alkane name	*Alkyl group name**
CH_4 Methane	CH_3- Methyl
CH_3-CH_3 Ethane	CH_3-CH_2- Ethyl
$CH_3-CH_2-CH_3$ Propane	$CH_3-CH_2-CH_2-$ *n*-propyl (or propyl)
	$CH_3-CH-CH_3$ Isopropyl
	$\quad\quad\quad\vert$
$CH_3-CH_2-CH_2-CH_3$ Butane	$CH_3-CH_2-CH_2-CH_2-$ *n*-butyl (or butyl)
	$\quad\quad CH_3$
	$\quad\quad\vert$
	$CH_3-CH-CH_2-$ Isobutyl
	$\quad\quad\quad\quad CH_3$
	$\quad\quad\quad\quad\vert$
	CH_3-CH_2-CH- *sec*-butyl (secondary butyl)
	$\quad\quad CH_3$
	$\quad\quad\vert$
	CH_3-C- *tert*-butyl (tertiary butyl)
	$\quad\quad\vert$
	$\quad\quad CH_3$

*The lowercase *n* in *n*-propyl, *n*-butyl, and so on means the hydrocarbon chains are normal. That is, the alkyl group stands for an unbranched hydrocarbon chain that has a hydrogen atom missing from carbon number 1.

We use the alkyl-group nomenclature in naming organic compounds. However, the R-group notation, in which R or R′ can stand for any structural group attached to a functional group, has much wider application. We shall use it to discuss classes of organic compounds in Chapter 19.

18.7 Naming Organic Compounds

Learning Goal 7
Naming organic compounds

In the late 1800s, when many organic compounds were being discovered and put to use, organic chemists realized that they had to have some kind of logical system for naming them. Such a system—called the Geneva System—was developed in 1892 at an international meeting of chemists in Geneva, Switzerland. In 1930 this system was revised and extended at another meeting of chemists in Liège, Belgium. By this time chemists had organized, so the meeting was a convention of the International Union of Pure and Applied Chemistry (IUPAC, pronounced "I-you-pack"). The revised system is thus called the *IUPAC system*.

The chief feature of the IUPAC system is that it assigns numbers to carbon atoms to show where various R groups are located along the main carbon chain. In this section we introduce some simple rules for naming alkanes by the IUPAC method.

Rule 1 ▶ *Look for the longest carbon chain, and use this as the **base name** of the compound.*

The base names for alkanes are those given in Table 18.1. For example, a compound whose longest chain contains 10 carbon atoms has the base name *decane*.

Example 18.3

Find the longest carbon chain in each of the following, and name it. (To simplify the writing of structures, we'll show only the carbon skeletons, but keep in mind that there are always hydrogen atoms attached to the carbon atoms.)

(a)
```
C—C—C—C—C—C—C
        |
        C
        |
        C
```

(b)
```
C—C—C—C—C—C
          |
          C
          |
          C
```

(c)
```
      C
      |
      C
      |
C—C—C
|
C
```

Solution

We'll use color to show the longest chain. Don't be misled by the way the structure is written; just look for the longest continuous chain of carbon atoms.

(a)
```
C—C—C—C—C—C—C
        |
        C
        |
        C
```
The longest chain has seven carbon atoms, so the base name of this compound is *heptane*.

(b)
```
C—C—C—C—C—C
          |
          C
          |
          C
```
The longest chain in this compound has seven carbon atoms.

This compound would be better written as

$$C-C-C-C-C-C-C$$
$$|$$
$$C$$

The base name of this compound is also heptane.

(c)

$$
\begin{array}{c}
C \\
| \\
C \\
| \\
C-C-C \\
| \\
C
\end{array}
$$

The longest chain in this compound has five carbon atoms.

This compound would be better written as

$$C-C-C-C-C$$
$$|$$
$$C$$

The base name of this compound is *pentane*.

Practice Exercise 18.3   Find the longest carbon chain in each of the following, and name it.

(a)
$$
\begin{array}{c}
C-C-C-C-C \\
| \\
C \\
| \\
C
\end{array}
$$

(b)
$$
\begin{array}{c}
C-C-C \\
| \\
C \\
| \\
C
\end{array}
$$

(c)
$$
\begin{array}{cc}
 & C-C \\
 & | \\
C-C & C \\
| & | \\
C & C-C
\end{array}
$$

Rule 2 ▶   *Next add (as a prefix) the name of the side chain.*

The side chains are alkyl groups. Their names are derived from the names of the alkanes, with *-ane* changed to *-yl* (Table 18.3).

Example 18.4

Using the compounds in Example 18.3, add the names of the side chains to the names of the compounds.

Solution

(a) C—C—C—C—C—C—C
     |
     C
     |
     C

   Ethylheptane

(b) C—C—C—C—C—C—C
        |
        C

      Methylheptane

(c) C—C—C—C—C
     |
     C

  Methylpentane

Practice Exercise 18.4 Using the compounds in Practice Exercise 18.3, add the names of the side chains to the names of the compounds.

Rule 3 ▶ *Specify the location of the side chain by numbering the carbon atoms along the **longest** chain. Always number the longest chain so that the side-chain group will have the lowest possible number.*

  Suppose we have

C—C—C—C—C—C—C
      |
      C

Do we number from the right or the left? The answer is that we number from the side that will give the methyl side-chain group the lowest possible number. So, in this compound, we number from the right:

7 6 5 4 3 2 1
C—C—C—C—C—C—C
      |
      C

The name of this compound is *3-methylheptane*.

Example 18.5

Using the compounds in Example 18.3, give the entire name of each compound.

Solution

(a)

$$\overset{1}{C}-\overset{2}{C}-\overset{3}{C}-\overset{4}{C}-\overset{5}{C}-\overset{6}{C}-\overset{7}{C}$$
$$|$$
$$C$$
$$|$$
$$C$$

The name of this compound is *4-ethylheptane*. In this compound, we can number from left *or* right, because the side chain (the ethyl group) is attached to the middle carbon atom.

(b)

$$\overset{7}{C}-\overset{6}{C}-\overset{5}{C}-\overset{4}{C}-\overset{3}{C}-\overset{2}{C}-\overset{1}{C}$$
$$|$$
$$C$$

The name of this compound is *3-methylheptane*. In this compound we must number the carbon atoms from the right so that the methyl side chain has the lowest possible number, 3.

(c)

$$\overset{5}{C}-\overset{4}{C}-\overset{3}{C}-\overset{2}{C}-\overset{1}{C}$$
$$|$$
$$C$$

The name of this compound is *3-methylpentane*. In this compound we may number from either end.

Practice Exercise 18.5   Using the compounds in Practice Exercise 18.3, give the entire name of each compound.

Rule 4 ▶   *If **more than one side chain** is present, use the following procedure to name the compound.*
(a) *Find the **longest chain**, as usual, to get the base name of the compound.*
(b) *Locate the **side chains**.*
(c) *Number the **longest chain** so that the side chains will have the lowest possible numbers.*
(d) *Add (as a prefix) the name and position of the **side chains**, using alphabetical order.*
(e) *In determining alphabetical order, ignore prefixes such as **di, tri, tetra,** and so on. The actual name of the **alkyl group** is used to determine alphabetical order.*

Example 18.6

Name the compound

$$C-C-C-C-C-C$$

with a C branch below the second carbon and a C-C branch below the fourth carbon

Solution

We'll name this compound in steps:

(a) The longest chain is hexane.

$$C-C-C-C-C-C$$

with a C branch below the second carbon and a C-C branch below the fourth carbon

(b) The side chains are a methyl group and an ethyl group.

$$C-C-C-C-C-C$$

with a C branch below the second carbon and a C-C branch below the fourth carbon

(c) We number the longest chain from the left so that the side chains have the lowest possible numbers: 2 and 4.

$$\overset{1}{C}-\overset{2}{C}-\overset{3}{C}-\overset{4}{C}-\overset{5}{C}-\overset{6}{C}$$

with a C branch below carbon 2 and a C-C branch below carbon 4

(If we had numbered from the right, the side-chain numbers would have been 3 and 5.)

(d) We can now write the full name of this compound, using the information we have obtained. [Remember, side chains are added on alphabetically, left to right.]

Longest chain	hexane
Side chain and its position	2-methyl
Side chain and its position	4-ethyl
Full name of compound	4-ethyl-2-methylhexane

Practice Exercise 18.6   Name the compound

```
C—C—C—C—C—C—C
    |       |
    C       C
            |
            C
```

This proves—doesn't it?—that when you take the words apart, the way they are constructed isn't at all mysterious. Now we come to the final rule.

Rule 5 ▶   *If two side chains in a compound are the same, use a Greek prefix instead of repeating the name of the side chains.*

See Chapter 7 to review the Greek prefixes.

Example 18.7

Name the compound

```
      C       C
      |       |
C—C—C—C—C—C—C
```

Solution
We'll name this compound in steps:
(a) The longest chain is heptane.

```
      C       C
      |       |
C—C—C—C—C—C—C
```

(b) The side chains are two methyl groups. We number from the left, in order to keep the numbers as low as possible.

```
      C       C
      |       |
C—C—C—C—C—C—C
1  2  3  4  5  6  7
```

(c) We can now write the full name of the compound, using this information. (Instead of 2-methyl-4-methyl, we shall write 2,4-dimethyl.)

Longest chain	heptane
Side chains	
and their positions	2,4-dimethyl
Full name of compound	2,4-dimethylheptane

Practice Exercise 18.7   Name the compound

```
C — C — C — C — C — C
        |       |
        C       C
```

Example 18.8

Name each of the following compounds.

(a)
```
            C
            |
C — C — C — C — C — C
            |
            C
```

(b)
```
                        C
                        |
C — C — C — C — C — C — C — C
            |       |
            C       C
                    |
                    C
```

(c)
```
    C       C
    |       |
C — C — C — C — C
```

(d)
```
        C
        |
C — C — C — C — C
        |
        C
```

(e)
```
        C                       C
        |                       |
C — C — C — C — C — C — C — C — C — C
            |       |       |
            C       C       C
            |       |
            C       C
            |
            C
```

Solution

(a)
```
                    C
  6   5   4   3 |   2   1
  C — C — C — C — C — C
                |
                C
```

Longest chain	hexane
Side chains	
and their positions	3,3-dimethyl
Full name of compound	3,3-dimethylhexane

(b)

$$
\begin{array}{cccccccc}
 & & & & & & \overset{\displaystyle C}{|} & \\
\overset{8}{C}-\overset{7}{C}-\overset{6}{C}-\overset{5}{C}-\overset{4}{C}-\overset{3}{C}-\overset{2}{C}-\overset{1}{C} \\
 & & \underset{\displaystyle C}{|} & & & \underset{\displaystyle C}{|} & & \\
 & & & & & \underset{\displaystyle C}{|} & &
\end{array}
$$

Longest chain	octane
Side chains and their positions	3,5-dimethyl
Side chain and its position	3-ethyl
Full name of compound	3-ethyl-3,5-dimethyloctane

(c)

$$
\begin{array}{ccccc}
 & \overset{\displaystyle C}{|} & & \overset{\displaystyle C}{|} & \\
C-C-C-C-C \\
\underset{1}{} & \underset{2}{} & \underset{3}{} & \underset{4}{} & \underset{5}{}
\end{array}
$$

Longest chain	pentane
Side chains and their positions	2,4-dimethyl
Full name of compound	2,4-dimethylpentane

(d)

$$
\begin{array}{ccccc}
 & & & \overset{\displaystyle C}{|} & \\
\overset{5}{C}-\overset{4}{C}-\overset{3}{C}-\overset{2}{C}-\overset{1}{C} \\
 & & & \underset{\displaystyle C}{|} &
\end{array}
$$

Longest chain	pentane
Side chains and their positions	2,2-dimethyl
Full name of compound	2,2-dimethylpentane

(e)

$$
\begin{array}{cccccccccc}
 & & \overset{\displaystyle C}{|} & & & & & & \overset{\displaystyle C}{|} & \\
\overset{10}{C}-\overset{9}{C}-\overset{8}{C}-\overset{7}{C}-\overset{6}{C}-\overset{5}{C}-\overset{4}{C}-\overset{3}{C}-\overset{2}{C}-\overset{1}{C} \\
 & & & & \underset{\displaystyle C}{|} & & \underset{\displaystyle C}{|} & & \underset{\displaystyle C}{|} & \\
 & & & & \underset{\displaystyle C}{|} & & \underset{\displaystyle C}{|} & & & \\
 & & & & \underset{\displaystyle C}{|} & & & & &
\end{array}
$$

Longest chain	decane
Side chain and its position	6-propyl
Side chains and their positions	2,2,8-trimethyl
Side chain and its position	4-ethyl
Full name of compound	4-ethyl-2,2,8-trimethyl-6-propyldecane

Practice Exercise 18.8   Name each of the following compounds.

(a) C—C—C—C—C—C—C—C—C
           | |
           C   C

(b) C—C—C—C—C—C—C—C—C
           | | | |
           C C C C
                    |
                    C

(c) C—C—C—C—C—C
           | | | |
           C C C C

(d) C—C—C—C—C—C—C
                | |
                C C
                | |
                C C

(e) C—C—C—C—C—C—C—C
           | | | | |
           C C C C C
                    |
                    C

18.8 The Alkenes and Alkynes

Figure 18.7

The alkanes, alkenes, and alkynes

Learning Goal 5
Unsaturated hydrocarbons

The alk*enes* and alk*ynes*, like the alkanes, are hydrocarbons. But the alk**enes** have one carbon-carbon double bond and the alk**ynes** have a carbon-carbon triple bond (Figure 18.7).

The simplest *alkene* has the formula C_2H_4 and the structure

$$\underset{H}{\overset{H}{>}}C=C\underset{H}{\overset{H}{<}}$$

(Note the double bond linking the carbons.) The common name for this compound is ethylene. It is the substance from which we make polyethylene, a plastic used for containers, electrical insulation, and packaging.

The simplest *alkyne* has the formula C_2H_2 and the structure

$$H—C\equiv C—H$$

(Note the triple bond linking the carbons.) The common name for this compound is acetylene; it is used in torches that can cut through metals.

Each of these compounds is the first compound in its homologous series (Table 18.4). Both these series of hydrocarbons are said to be **unsaturated.** This means that there are double or triple carbon-carbon bonds in their molecules.

Table 18.4
The Alkene and Alkyne Series

	Alkenes	Alkynes
Structural formula of the first member of the series	$H_2C{=}CH_2$ (shown as $C{=}C$ with H atoms)	$H{-}C{\equiv}C{-}H$
Molecular formula	C_2H_4	C_2H_2
Common name	Ethylene	Acetylene
IUPAC name	Eth*ene*	Eth*yne*
General formula for the series	C_nH_{2n}	C_nH_{2n-2}

The IUPAC name for an alkene or alkyne can be found from the name of the alkane that has the same number of carbon atoms in its longest chain. We simply change the alkane ending to -*ene* for alkenes or to -*yne* for alkynes. For example, the IUPAC name for C_2H_4 is eth*ene*, and the IUPAC name for C_2H_2 is eth*yne*. Both names are derived from the name of the two-carbon alkane compound ethane; only the endings are changed.

Example 18.9

Name each of the following compounds. (Again, to simplify writing the structures, we'll show only the carbon skeletons; keep in mind that there are hydrogens attached to these carbon atoms.)

(a) $C{-}C{=}C$

(b) $C{-}C{-}C{=}C$

(c) $C{-}C{=}C{-}C$

(d) $C{-}\overset{\displaystyle C}{\underset{\displaystyle |}{C}}{-}C{=}C$

(e) $C{-}C{\equiv}C$

(f) $C{-}C{-}C{\equiv}C$

(g) $C{-}C{\equiv}C{-}C$

(h) $C{-}C{-}C{-}\overset{\displaystyle C}{\underset{\displaystyle |}{\underset{\displaystyle |}{\overset{\displaystyle |}{C}}}}{-}C{\equiv}C{-}C$ (with C, C branches shown below the fourth carbon)

Solution

(a) $C{-}C{=}C$ is called prop*ene*. It is a three-carbon compound that is the second member of the alkene series. It takes its name from the three-carbon alkane compound, except that it has an -*ene* ending.

(b) $\overset{4}{C}-\overset{3}{C}-\overset{2}{C}=\overset{1}{C}$ is called 1-but*ene*. This third member of the alkene series takes its name from the four-carbon alkane. But *there is one additional point:* The position of the double bond must be specified in the name. *To specify the position of the double bond, we number the carbon atoms, as we did before, keeping the number of the double bond as low as possible. Then, in the compound name, we give the number of the carbon atom from which the double bond* **starts** *(the smaller of the two numbers).*

(c) $\overset{4}{C}-\overset{3}{C}=\overset{2}{C}-\overset{1}{C}$ is called 2-but*ene*. This compound, similar to the compound in part (b), is also a butene. However, the position of its double bond is different.

(d)

$$\begin{array}{c} C \\ | \\ \underset{4}{C}-\underset{3}{C}-\underset{2}{C}=\underset{1}{C} \end{array}$$

When we number the carbon chain, we give preference to the double bond and not to the side chain. In other words, we want to keep the number of the double bond as low as possible.

Longest chain and position of double bond	1-butene
Side chain and its position	3-methyl-
Full name of compound	3-methyl-1-butene

(e) $C-C\equiv C$ is called prop*yne* and is the second member of the alkyne series. It takes its name from the three-carbon alkane compound, except that it has a *-yne* ending.

(f) $\overset{4}{C}-\overset{3}{C}-\overset{2}{C}\equiv\overset{1}{C}$ is called 1-but*yne*. It is the third member of the alkyne series and takes its name from the four-carbon alkane. Remember that we have to note the position of the triple bond when we name this compound, because there is another butyne compound.

(g) $\overset{4}{C}-\overset{3}{C}\equiv\overset{2}{C}-\overset{1}{C}$ is called 2-but*yne*. This compound is also a butyne, but it is different from the one in part (f), because the position of the triple bond is different.

(h)

$$\begin{array}{c} C \\ | \\ \underset{7}{C}-\underset{6}{C}-\underset{5}{C}-\underset{4}{C}-\underset{3}{C}\equiv\underset{2}{C}-\underset{1}{C} \\ | \\ C \\ | \\ C \end{array}$$

Longest chain and position of triple bond	2-heptyne
Side chain and its position	4-methyl
Side chain and its position	4-ethyl
Full name of compound	4-ethyl-4-methyl-2-heptyne

Practice Exercise 18.9   Name each of the following compounds.

(a) C—C—C—C=C

(b) C—C—C—C—C=C

(c) C—C—C—C=C—C—C

(d) C—C—C—C—C=C
              |
              C

(e) C—C—C—C—C≡C

(f) C—C—C≡C—C—C

(g) C—C—C—C—C≡C
        | |
        C   C

(h) C—C—C—C≡C
        | |
        C   C

18.9 Cyclic Hydrocarbons

Learning Goal 5
Cyclic hydrocarbons

Many hydrocarbons form **cyclic structures**—that is, structures that join to form a closed chain, like a snake biting its tail. An example is

or C_3H_6

The name of this compound—cyclopropane—is the name of the three-carbon alkane (propane) with the prefix *cyclo* added. Many cyclic hydrocarbons have useful properties. For example, cyclopropane is an anesthetic that is sometimes used in surgery. However, cyclopropane is very flammable, and great caution must be exercised in the operating room when it is used.

Another example of a cyclic hydrocarbon is

or C_5H_{10}

The name of this compound—cyclopentane—is the name of the five-carbon alkane (pentane) with the prefix *cyclo* added. To make the writing of cyclic hydrocarbons simpler, we generally show only the chain. For example, cyclopentane is shown as

The shape of the figure tells us the name of the compound. Each line represents a carbon-carbon bond, and each corner represents a CH_2 group.

Example 18.10

Name each of the following compounds.

(a)

(b)

(c) CH_3

(d) CH_2—CH_3

           CH_3

Solution

(a)   This is a cyclic hydrocarbon with six carbon atoms. It is called *cyclohexane*.

(b)   This is a cyclic hydrocarbon with seven carbon atoms. It is called *cycloheptane*.

(c) CH_3

This is a *substituted* cyclic hydrocarbon. The CH_3 side-chain group has replaced a hydrogen atom that was bonded to one of the carbon atoms. This compound is called *methylcyclohexane*.

(d) CH_2—CH_3

This is a *disubstituted* cyclic hydrocarbon, which means that *two* hydrogens on the ring have been replaced by side-chain groups.

           CH_3

When more than one side chain is attached to a cyclic hydrocarbon, it is necessary to number the C atoms in the ring. Remember to number them in such a way as to keep the numbers of the side chains as low as possible.

Name of cyclic structure	cyclohexane
Side chain	
and its position	3-methyl
Side chain	
and its position	1-ethyl
Full name of compound	1-ethyl-3-methylcyclohexane

Practice Exercise 18.10   Name each of the following compounds.

(a)   (b)

Some cyclic hydrocarbons have double and triple bonds..These hydrocarbons take their names from the corresponding alkene or alkyne, with the prefix *cyclo* added.

Example 18.11

Name each of the following compounds.

(a)   (b)

Solution

(a)   This is a six-carbon cyclic compound with a double bond. It takes its name from the corresponding alk*ene*. The name of this compound is *cyclohexene*. It can be used as a stabilizer for high-octane gasolines.

(b)   This is an eight-carbon cyclic compound with a triple bond. It takes its name from the corresponding alk*yne*. The name of this compound is *cyclooctyne*.

Practice Exercise 18.11   Name each of the following compounds.

(a)

(b)

(c)

(d)

18.10 Aromatic Hydrocarbons

Learning Goal 8
Aromatic hydrocarbons

One series of cyclic hydrocarbons has alternating single and double bonds. The members of this series are called **aromatic hydrocarbons,** because early organic chemists noticed that these compounds had pleasant odors. (Later organic chemists found that some of these compounds have unpleasant odors.) The parent compound of the aromatic hydrocarbons is benzene, C_6H_6, whose structure is

Although only three of the carbon-carbon bonds in benzene are written as double bonds, all the bonds seem to be equivalent. At one time chemists incorrectly thought that the double bonds shifted back and forth between carbon atoms. If that were true, either of these two diagrams would give a valid picture of the benzene molecule:

Today, chemists still don't agree on how to represent the structure of benzene. Any one of the following representations can be used.

The third diagram is meant to show that all the carbon-carbon bonds in benzene are equivalent, as they are known to be.

Figure 18.8

Some aromatic hydrocarbons

Naphthalene
$C_{10}H_8$

Anthracene
$C_{14}H_{10}$

Phenanthrene
$C_{14}H_{10}$

Benzene, which can be obtained from coal tar, is one of the most important organic compounds. It is the starting material for countless products that improve the human condition, including many synthetic fibers such as nylon, perfumes, synthetic dyes, drugs, and pesticides.

Other aromatic hydrocarbons are made up of benzene rings joined together (Figure 18.8). And many compounds consist of aromatic compounds with various side chains. An example is

$$CH_3$$

One name for this compound is methylbenzene, but organic chemists call it *toluene*. It is used in making explosives and dyes.

There are also rules for naming these types of compounds, but they will not be covered here.

18.11 Other Types of Organic Compounds

In this chapter we have discussed only hydrocarbons—organic compounds made up of carbon and hydrogen and nothing else. But there are many other kinds of organic compounds, and these contain other elements as well. In fact, the field of organic chemistry is enormous, and the products developed by organic chemists affect our lives in many ways every day.

In Chapter 19 we shall look briefly at several of these other kinds of organic compounds. For a "preview," you might want to look at Table 19.1.

Summary

Organic chemistry is essentially the study of compounds that contain carbon—along with such elements as hydrogen, oxygen, nitrogen, sulfur, and the Group VIIA elements. More than 6 million organic compounds are known.

The carbon atom has four electrons in its outermost energy level. It can use all four electrons to bond with other atoms, including other carbon atoms. Two carbon atoms can form a single, double, or triple bond. When a carbon atom bonds with four hydrogen atoms, the four carbon-hydrogen bonds define a tetrahedral shape, with the H atoms at the corners of the tetrahedron. This tetrahedral shape is common to many organic compounds.

Because carbon atoms can bond together in different ways, different organic compounds can have the same molecular formula. For this reason, structural formulas are used to specify organic compounds. The structural formula shows how the carbon atoms are bonded, as well as the elements that make up the compound.

Saturated compounds are compounds in which all bonds between carbon atoms are single bonds. The alkanes are a series of saturated hydrocarbons with the general formula C_nH_{2n+2}. The alkenes are hydrocarbons with double carbon-carbon bonds. The alkynes are hydrocarbons with triple carbon-carbon bonds. Organic compounds with double or triple bonds are said to be unsaturated. All three of these series—the alkanes, alkenes, and alkynes—are homologous series. That is, each compound in the series differs from the preceding compound by the same fixed increment.

Organic compounds are named according to the IUPAC system. In this system, the name of the compound indicates its structure. Thus isomers, or compounds with the same molecular formula but different structural formulas, are assigned different names in the IUPAC system.

Cyclic hydrocarbons are hydrocarbons in which the carbon atoms form a closed chain. An important series of cyclic hydrocarbons consists of the aromatic hydrocarbons, which include the compound benzene.

. . . .

Career Sketch

Technical writer

Technical writers put scientific and technical information into everyday language that the general public can readily understand. They write and edit technical materials. They may also produce publications or audiovisual materials. To ensure that their work is accurate, technical writers

Technical writers put scientific and technical information into simple language that can be understood by the general public.
(CL: Wang)

must know a good deal about the subject area in which they are writing—laser beams or pharmacology, for example. They often have to do considerable research before they can write on a given subject. Their writing must be clear and easy to follow. Technical writers must have a good command of the language, plus versatility of style, because they must be able to convey information in a way that is helpful to people who use it—technicians, executives, sales representatives, and others.

Some firms refer to technical writers by using job titles other than "technical writer." Depending on the employer, people in technical-writing jobs may be called publications engineers, communications specialists, industrial writers, medical writers, communicators, or developers of instructional materials.

Technical writers study English and writing. They often study chemistry, biology, physics, astronomy, or general science. They may major in English and minor in science, or they may major in science and take several courses in English and writing.

Career Sketch

Assayer

An assayer tests ores and minerals and analyzes the results of the tests to determine the value and properties of the materials being analyzed. The assayer must have available an up-to-date chemical laboratory with such equipment as spectroscopes, furnaces, beakers, graduates, pipettes, and crucibles. The assayer separates metals from the other components by mixing them in solutions, by the method of flotation, or by other processes involving liquids; or by using dry methods, such as applying heat to form slags of lead, borax, and other impurities. The assayer weighs residues on a balance scale to determine the proportion of pure gold, silver, platinum, or other metals present. An assayer who specializes in testing and analyzing precious metals is called a *gold-and-silver assayer*.

Self-Test Exercises

Learning Goal 1: Organic and Inorganic Compounds

1. Are carbon-carbon bonds in organic compounds ionic or covalent?

2. How is a diamond different from the graphite found in a "lead" pencil, from a structural point of view?

3. Name two crystalline forms of carbon. What accounts for their widely differing properties?

4. What are inorganic compounds?

5. List some examples of compounds that contain carbon but are *not* organic.

6. Approximately how many organic compounds are known?

7. What accounts for the large number of organic compounds?

8. What elements besides carbon and hydrogen are commonly found in organic compounds?

Learning Goal 2: Lewis Electron-Dot Structure of Carbon

9. Write the Lewis electron-dot structure for carbon.

10. How many electrons does carbon possess that are available for bonding?

Learning Goal 3: Geometry of Carbon Bonds

11. The bonding pattern of the carbon atom in many organic compounds is _____ .

12. What is meant by "tetrahedral carbon bonding"?

13. Tell whether each of the following descriptions indicates a carbon-carbon single bond, a double bond, or a triple bond.
(*a*) Two tetrahedral carbon atoms with a common edge
(*b*) Two tetrahedral carbon atoms with a common face
(*c*) Two tetrahedral carbon atoms with their corners touching

14. How many degrees apart are the carbon-hydrogen bonds in methane?

Learning Goal 4: Structural Formulas of Organic Compounds

15. Write structural formulas for the following compounds: (*a*) eth*ane*, (*b*) eth*ene*, (*c*) eth*yne*.

16. Why is a structural formula more important in organic chemistry than a molecular formula?

17. What is the difference between a structural formula and a molecular formula?

18. How many compounds are represented by the molecular formula C_4H_{10}?

19. Write the formulas and names for the first five straight-chain members of the alk*ene* series.

20. What are the two ways in which structural formulas can be written?

21. Write the formulas and names for the first five straight-chain members of the alk*yne* series.

22. Write the structural formula for *n*-heptane.

23. Write the following structural formula in the condensed form.

$$
\begin{array}{c}
\text{H} \quad \text{H} \quad \text{H} \quad \text{H} \quad \text{H} \\
| \quad\; | \quad\; | \quad\; | \quad\; | \\
\text{H—C—C—C—C—C—H} \\
| \quad\; | \quad\; | \quad\; | \quad\; | \\
\text{H} \quad \text{C} \quad \text{H} \quad \text{C} \quad \text{H} \\
\end{array}
$$

24. Write the structural formula for 1-heptene.

25. Rewrite each of the following structures by adding hydrogen atoms to the carbon skeleton.

(*a*) C—C—C—C—C—C
      | |
      C   C

(*b*) C—C—C=C—C—C
               |
               C

(*c*) C—C—C=C—C—C≡C
      |
      C

(*d*)

(*e*)

26. Write the structural formula for 1-heptyne.

27. Write the structure for each of the following hydrocarbons.
(*a*) 2-methylbutane
(*b*) 2,3,4-trimethylheptane
(*c*) 2,2-dimethylpropane
(*d*) 1-hexene
(*e*) 2-hexene
(*f*) 2-methyl-3-ethyl-1-pentene
(*g*) 3-methyl-1-butyne

28. Write the following structural formula in the condensed form.

29. Write the structure for each of the following cyclic hydrocarbons.
(*a*) cyclopropane
(*b*) cycloheptane
(*c*) 1-methyl-3-ethylcyclohexane
(*d*) cyclohexyne
(*e*) cyclohexene
(*f*) 1,2-dimethylbenzene
(*g*) 1,3-dimethylbenzene

30. Write the structure for each of the following hydrocarbons.
(*a*) 2,2-dimethylbutane
(*b*) 2,3-dimethylpentane
(*c*) 3-hexene
(*d*) 3-methyl-2-butene
(*e*) 2,3,5-trimethylheptane
(*f*) 3-methyl-3-ethyl-1-hexene

31. Write the structure for each of the nine isomers of C_7H_{16}.

32. Write the structure for each of the following cyclic hydrocarbons.
(*a*) 1-methylcyclopropane
(*b*) 3-methyl-1-ethylcycloheptane
(*c*) cyclopentene
(*d*) cyclopentyne
(*e*) 3-methyl-1-ethylbenzene
(*f*) 1-methylcyclohexene

Learning Goals 5 and 6: Saturated, Unsaturated, and Cyclic Hydrocarbons; Isomers and Homologous Series

33. Define the following terms, and give an example of each: saturated hydrocarbon, unsaturated hydrocarbon, isomer, and functional group.

34. Distinguish between a saturated and an unsaturated hydrocarbon.

35. Define the term *homologous series*, and give an example of such a series by listing some of its members.

36. What is a hydrocarbon?

37. Write the formulas and names of the first ten straight-chain members of the alkane series. (Don't look back in the chapter.)

38. How is a cyclic hydrocarbon different from an aromatic hydrocarbon?

39. The formula of an organic compound is C_5H_{10}. Write the structural formulas of all the possible compounds.

40. Name the first nine straight-chain members of the alkene series.

41. How many different straight-chain heptenes are there? Name each one.

42. How many different straight-chain hexenes are there? Explain.

Learning Goal 7: Naming Organic Compounds

43. Name each of the isomers in Self-Test Exercise 31.

44. Name each of the following alkanes by the IUPAC system.

(a) C—C—C—C—C   (b) C—C—C
              | |
              C   C

(c) C—C—C—C—C
        | |
        C   C

45. Name each of the following alkanes by the IUPAC system.

(a) C—C—C—C
        |
        C

(b)   C
         |
    C—C—C—C
         |
         C

(c) C—C—C—C—C—C
        | | |
        C   C   C

46. Name each of the following compounds by the IUPAC system.

(a) C—C—C=C
(b) C—C=C—C
(c) C—C—C—C—C—C=C
(d) C—C—C—C=C
        | |
        C   C
            |
            C

47. Name each of the following alkenes by the IUPAC system.

(a) C—C—C—C—C=C
(b) C—C—C—C=C—C
(c) C—C—C=C—C—C
(d) C—C—C—C=C
        | |
        C   C
        |
        C

48. Name each of the following alkynes by the IUPAC system.

(a) C—C≡C
(b) C—C—C—C≡C
        |
        C
        |
        C
        |
        C

(c) C—C—C—C≡C
(d) C—C—C—C—C≡C—C
        | |
        C   C
                |
                C

49. Name each of the following alkynes by the IUPAC system.

(a) C—C—C—C≡C   (b) C—C—C≡C—C
(c) C—C—C≡C
        |
        C

(d) C—C—C—C≡C—C—C
        | |
        C   C
        |
        C

50. Name each of the following cyclic hydrocarbons.

(a)

(b)

(c)

(d)

(e)

(f)

51. Name each of the following cyclic hydrocarbons.

(a)

(b)

(c)

(d)

(e)

(f)

(g)

52. Name each of the following compounds by the IUPAC system.

(a)

(b) $C-C\equiv C$

(c) $C=C-C-C-C-C-C$

(d) $C-C-C-C$
    with C branching to $C \quad C$

Learning Goal 8: Aromatic Hydrocarbons

53. Define and give an example of an aromatic hydrocarbon.

54. Write the structural formula for benzene.

Extra Exercises

55. Label the R group in each of the following compounds.
(a) $CH_3-CH_2-CH_2-O-CH_2-CH_2-CH_3$
(b) $CH_3-CH_2-CH_2-OH$
(c) CH_3-CH_2-C-OH with $\|$ O

56. Most organic compounds have different boiling points. Could a distillation apparatus as described in Chapter 13 be used to separate some of these compounds? Explain.

*****57.** Write a general formula for a dialkene, a carbon compound containing two double bonds.

58. What is the relationship between the compounds $CH_3-CH=CH-CH_3$ and $CH_2=CH-CH_2-CH_3$?

59. Name the compound

$CH_3CH_2CHCH=CHCHCH_3$
with CH_3 and CH_3 branches

60. What category of hydrocarbon does C_8H_{14} belong to?

61. Write a structural formula for 2-methylpropene.

62. Write a structural formula for methylcyclopentane.

63. Write a structural formula for 2-ethyl-1-butene.

64. Name the compound

$$
\begin{array}{c}
\ \ \ \ \ \ \ \ \ \ \ \ \text{C} \\
\ \ \ \ \ \ \ \ \ \ \ \ | \\
\text{C}-\text{C}-\text{C}-\text{C}-\text{C}=\text{C} \\
\ \ \ \ \ \ \ \ \ \ \ \ | \\
\ \ \ \ \ \ \ \ \ \ \ \ \text{C} \\
\ \ \ \ \ \ \ \ \ \ \ \ | \\
\ \ \ \ \ \ \ \ \ \ \ \ \text{C}
\end{array}
$$

65. Name all the possible straight-chain oct*enes*.

66. Rewrite the following structure of naphthalene, replacing the ring structure with the proper numbers of carbon atoms and hydrogen atoms.

19 Organic Chemistry, Part 2: Classes of Organic Compounds

Learning Goals

After you have studied this chapter, you should be able to:

1. Explain what a functional group is.
2. Discuss or state the general formula, names, structure, and uses of alcohols.
3. Discuss or state the general formula, common names, structure, and uses of ethers.
4. Discuss or state the general formula, names, structure, and uses of aldehydes.
5. Discuss or state the general formula, names, structure, and uses of ketones.
6. Discuss or state the general formula, names, structure, and uses of carboxylic acids.
7. Discuss or state the general formula, names, structure, and uses of esters.
8. Write the products of an esterification reaction.
9. Discuss or state the general formulas, names, structure, and uses of amines.

Introduction

In Chapter 18 we discussed only hydrocarbons—organic compounds made up of carbon and hydrogen. However, we also noted that carbon can form compounds with other elements besides hydrogen. And, just as hydrocarbons are grouped in several series, the non-hydrocarbons are grouped in *classes* of compounds. In this chapter we briefly survey seven of these classes. The basis for the classification of organic compounds is the functional group, which was introduced in Section 18.6.

19.1 Functional Groups

Learning Goal 1
Functional groups

A **functional group** is *an atom or group of atoms that determines the specific chemical properties of a class of organic compounds.* This definition actually contains three statements. First, all the members of each class of organic compounds have some basic properties in common. Second, all the compounds in a specific class contain the same functional group. And finally, the functional group determines the basic properties of the class.

Other atoms are attached to the functional group. It is these attached atoms (the R groups of Section 18.6) that give a specific compound its individual properties. In other words, we can write the general formula for the compounds in a class as

R—functional group

The functional group is *the same* for every compound in the class. The R group (or groups) is *different* for each compound in the class. The R groups are usually alkyl groups.

Table 19.1 shows the classes of compounds that we shall discuss in this chapter (those from alcohol down) and their general formulas. For each class, the functional group is the part that is attached to the R or R's.

Figure 19.1
If the R group of an alcohol consists of only a few carbon atoms, the alcohol is polar.

19.2 Alcohols

Learning Goal 2
Alcohols

The general formula for an **alcohol** *is* R—OH. The —OH is called the **hydroxyl functional group.** The R group is usually an alkyl group. If the R group consists of only a few carbon atoms, the alcohol tends to be polar because of the polarity of the —OH group (Figure 19.1). As a result, the alcohol dissolves in polar solvents such as water. For example, the alcohol CH_3CH_2OH (ethanol) dissolves in water.

However, if the R group consists of a long chain of nonpolar carbon atoms, the polarity of the —OH group is less obvious. In this case, the alcohol appears to be nonpolar (Figure 19.2) and dissolves in nonpolar solvents like benzene and carbon tetrachloride. For example, the alcohol $CH_3CH_2CH_2CH_2CH_2CH_2OH$ (hexanol) dissolves well in benzene.

Figure 19.2
If the R group of an alcohol consists of a long chain of carbon atoms, the alcohol tends to be nonpolar (because it is mostly an alkane chain) and behaves like a nonpolar alkane.

Table 19.1
Some Classes of Organic Compounds

Class	General formula	Example	Common name	IUPAC name
Alkane	C_nH_{2n+2}	CH_4	Methane (swamp gas)	Methane
Alkene	C_nH_{2n} (n = 2 or more)	$CH_2\!=\!CH_2$ $CH_3\!-\!CH\!=\!CH_2$	Ethylene Propylene	Ethene Propene
Alkyne	C_nH_{2n-2} (n = 2 or more)	$H\!-\!C\!\equiv\!C\!-\!H$ $CH_3\!-\!C\!\equiv\!C\!-\!H$	Acetylene Methylacetylene	Ethyne Propyne
Alcohol	$R\!-\!OH$	$CH_3\!-\!OH$	Methyl alcohol (wood alcohol)	Methanol
		$CH_3\!-\!CH_2\!-\!OH$	Ethyl alcohol (grain alcohol)	Ethanol
Ether	$R\!-\!O\!-\!R$	$CH_3\!-\!O\!-\!CH_3$ $CH_3\!-\!CH_2\!-\!O\!-\!CH_2\!-\!CH_3$	Dimethyl ether Diethyl ether	Methoxymethane Ethoxyethane
Aldehyde	$R\!-\!\underset{\underset{H}{\|}}{C}\!=\!O$	$H\!-\!\underset{\underset{H}{\|}}{C}\!=\!O$	Formaldehyde	Methanal
		$CH_3\!-\!CH_2\!-\!\underset{\underset{H}{\|}}{C}\!=\!O$	Propionaldehyde	Propanal
Ketone	$R\!-\!\underset{\overset{\|}{O}}{C}\!-\!R$	$CH_3\!-\!\underset{\overset{\|}{O}}{C}\!-\!CH_3$	Dimethyl ketone (acetone)	Propanone
		$CH_3\!-\!CH_2\!-\!\underset{\overset{\|}{O}}{C}\!-\!CH_2\!-\!CH_3$	Diethyl ketone	3-pentanone
Carboxylic acid	$R\!-\!\underset{\overset{\|}{O}}{C}\!-\!OH$	$H\!-\!\underset{\overset{\|}{O}}{C}\!-\!OH$	Formic acid	Methanoic acid
		$CH_3\!-\!\underset{\overset{\|}{O}}{C}\!-\!OH$	Acetic acid	Ethanoic acid
Ester	$R\!-\!\underset{\overset{\|}{O}}{C}\!-\!O\!-\!R$	$H\!-\!\underset{\overset{\|}{O}}{C}\!-\!O\!-\!CH_3$	Methyl formate	Methylmethanoate
		$CH_3\!-\!\underset{\overset{\|}{O}}{C}\!-\!O\!-\!CH_2\!-\!CH_3$	Ethyl acetate	Ethylethanoate
Amine	$R\!-\!NH_2$	$CH_3\!-\!NH_2$ $CH_3\!-\!CH_2\!-\!NH_2$	Methylamine Ethylamine	Aminomethane Aminoethane

Note: Remember that n = number of carbon atoms.

Example 19.1

Determine whether each of the following alcohols is polar or non-polar.

(a) CH_3OH **(b)** $CH_3—(CH_2)_9—OH$

     Methanol   Decanol

Solution

(a) Because methanol has a short R group, it is a polar alcohol.

(b) Decanol has a long R group, so it is nonpolar.

Practice Exercise 19.1   Determine whether each of the following alcohols is polar or nonpolar.

(a) $CH_3—CH_2—OH$ **(b)** $CH_3—(CH_2)_{10}—OH$

We can name alcohols in two ways. First, we can use their common names, which we find by adding the word *alcohol* to the alkyl group. For example, CH_3CH_2OH is called *ethyl alcohol*. Second, we can use their IUPAC names, which we find by changing the alkane ending *-ane* to the alcohol ending *-anol*. For example, CH_3CH_2OH is also called *ethanol*.

Example 19.2

Give both the common and the IUPAC name for each of the following alcohols.

(a) $CH_3CH_2CH_2CH_2OH$ **(b)**   OH

Solution

(a) $CH_3CH_2CH_2CH_2OH$ is a four-carbon alcohol. Therefore its common name is n-*butyl alcohol* and its IUPAC name is *butanol*.

(b)   OH   This six-carbon cyclic alcohol has

                        the common name *cyclohexyl alcohol*.

                        Its IUPAC name is *cyclohexanol*.

Practice Exercise 19.2   Give both the common and the IUPAC name for each of the following alcohols.

(a) $CH_3CH_2CH_2OH$ **(b)**

19.3 Some Important Alcohols

Methanol, CH_3OH, is the simplest alcohol. This very poisonous compound is also known as *wood alcohol* because it can be obtained from the heating of wood in the absence of air (a process called *destructive distillation*). A word of warning about methanol or wood alcohol: Death can result from drinking just 30 mL of it, and permanent blindness can result from drinking lesser amounts. During the years of Prohibition in the United States, when alcoholic drinks were prohibited by law, many people drank beverages containing methanol, not realizing the dangers involved.

Ethanol, CH_3CH_2OH, is the alcohol present in wine, liquor, or beer. The ethanol found in these beverages is produced through the process of fermentation, which involves the breakdown of the sugar *glucose*.

$$C_6H_{12}O_6 \xrightarrow{\text{Yeast}} 2CH_3CH_2OH + 2CO_2 + \text{Energy}$$
Glucose

A solution that is 70% ethanol by volume acts as an antiseptic by coagulating the proteins of bacteria.

Table 19.2 lists some additional alcohols and their uses.

Table 19.2
Some Alcohols and Their Uses*

Name	Structure	Use
Isopropyl alcohol	CH_3CHCH_3 | OH	Rubbing alcohol. Used on the skin when one is massaging sore muscles, and for sponge baths to reduce high fevers.
Glycerol	CH_2CHCH_2 | | | OH $OH$$OH$	Acts as a lubricant in hand and skin creams. Used in the preparation of pharmaceutical products.
Menthol	OH ... $CHCH_3$ ... CH_3 ... CH_3	Used in cough drops and nasal inhalers. This substance causes the mucous membranes of the throat to increase their secretions, thus soothing inflamed tissues.
BHT (butylated hydroxytoluene)	OH ... $(CH_3)_3C$ ... $C(CH_3)_3$ ... CH_3	Used in commercial food products to prevent spoilage.

*Reproduced in part from *The Merck Index, Tenth Edition*. Copyright under the Universal Copyright Convention and the International Copyright Convention 1983 by Merck & Co., Inc., Rahway, New Jersey, U.S.A.

19.4 Ethers

· · · · · · · · · · · · · · ·
Learning Goal 3
Ethers

The general formula for an **ether** *is* R—O—R'. If the R and R' groups are the same, we say the ether is **symmetrical.** If the R and R' groups are different, we say the ether is **unsymmetrical.**

Usually we refer to ethers by using their common names, which are obtained by naming the two alkyl groups in alphabetical order and adding the name *ether*. For example, $CH_3CH_2OCH_3$ is called *ethyl methyl ether.*

Example 19.3

Give the common name for each of the following ethers.
(a) $CH_3CH_2OCH_2CH_3$ **(b)** $CH_3CH_2CH_2OCH_3$

Solution

(a) $CH_3CH_2OCH_2CH_3$ has an ethyl group on both sides of the oxygen. Therefore the name of this ether is *diethyl ether,* or simply *ethyl ether* (it is not imperative to use the prefix *di-* on symmetrical ethers).
(b) $CH_3CH_2CH_2OCH_3$ has a methyl group on one side of the oxygen and a propyl group on the other. We list these groups alphabetically and add the word *ether*. Therefore the name of this compound is *methyl propyl ether.*

Practice Exercise 19.3   Give the common name for each of the following ethers.
(a) $CH_3—CH_2—CH_2—O—CH_2—CH_2—CH_3$
(b) $CH_3—CH_2—O—CH_3$

19.5 Some Important Ethers

Ethers tend to be very flammable, and many of them have some type of anesthetic properties. They also tend to vaporize easily.

Ethyl ether, $CH_3CH_2OCH_2CH_3$, is sometimes just called *ether*. It is probably the best-known member of this class of compounds. It was, and still is, used in hospitals as an anesthetic for various surgical procedures (Figure 19.3). This anesthetic is extremely safe because there is a significant

Figure 19.3
Ethyl ether was one of the earliest anesthetics used in hospitals. It is still used in certain procedures today.
(Courtesy St. Luke's—Roosevelt Hospital Center, New York)

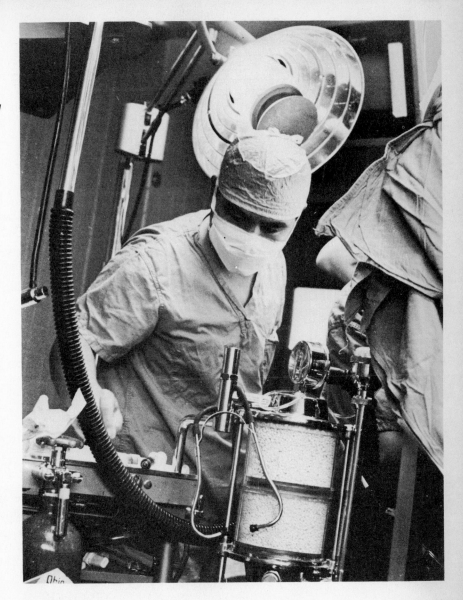

difference between the dosage needed to anesthetize and the dosage that will kill.

Ethylene oxide, $\overset{\displaystyle O}{\overset{\displaystyle /\backslash}{CH_2-CH_2}}$

is a gas at room temperature. It is a cyclic ether that can be used as a *fumigant* (disinfectant) for foodstuffs and as a fungicide in agriculture.

Table 19.3 lists some additional ethers and their uses.

Table 19.3
Some Important Ethers and Their Uses*

Name	Structure	Use
Eugenol	CH₂CH=CH₂ (benzene ring with OCH₃ and OH substituents)	Found in oil of cloves, which is used in toothache preparations for pain relief
Divinyl ether	$CH_2=CH-O-CH=CH_2$	A fast-acting anesthetic
Dimethyl ether	CH_3-O-CH_3	A gas used in refrigeration

*Reproduced in part from *The Merck Index, Tenth Edition.* Copyright under the Universal Copyright Convention and the International Copyright Convention 1983 by Merck & Co. Inc., Rahway, New Jersey, U.S.A.

19.6 Aldehydes

Learning Goal 4
Aldehydes

The general formula for an **aldehyde** *is*

$$R-\underset{\underset{O}{\|}}{C}-H$$

The functional group

$$-\underset{\underset{O}{\|}}{C}-H$$

is the **aldehyde** group. Many aldehydes have attractive odors and are used in perfumes. Some aldehydes have pleasant tastes and are used as flavoring agents.

We can name aldehydes in two ways. First, we can use the common name, which in many cases is simply the name of the alkyl group plus the word *aldehyde*. For example,

$$CH_3CH_2\underset{\underset{O}{\|}}{CH}$$

is called *propionaldehyde.*

Second, we can use the IUPAC name, which we derive by changing the alkane ending *-ane* to the aldehyde ending *-al*. For example,

$$CH_3CH_2\underset{\underset{O}{\|}}{CH}$$

is called *propanal.*

Example 19.4

Give both the IUPAC name and the common name for each of the following aldehydes.

(a) $CH_3CH_2CH_2CH$
$\parallel$
O

(b) $CH_3CH_2CH_2CH_2CH_2CH$
$\parallel$
O

Solution

(a) $CH_3CH_2CH_2CH$
$\parallel$
O

is a four-carbon aldehyde. Therefore the common name of this compound is *butyraldehyde;* its IUPAC name is *butanal.*

(b) $CH_3CH_2CH_2CH_2CH_2CH$
$\parallel$
O

is a six-carbon aldehyde that has the common name *hexaldehyde;* its IUPAC name is *hexanal.* (This compound also has another common name, *caproaldehyde,* because it is the aldehyde of caproic acid, which is found in goat fat. *Caper* is the Latin word for "goat.")

Practice Exercise 19.4 Give both the IUPAC name and the common name for each of the following aldehydes.

(a) $CH_3 - CH_2 - C - H$
$\parallel$
O

(b) $CH_3 - CH_2 - CH_2 - CH_2 - CH_2 - CH_2 - C - H$
$\parallel$
O

19.7 Some Important Aldehydes

Methanal is the simplest and best-known aldehyde.

$$H - C - H$$
$$\parallel$$
$$O$$

You are more likely to recognize it as *formaldehyde.* Although it is a gas at room temperature, it is usually dissolved in water and sold in a 40%-by-weight–volume solution. In this form it is used as a germicide, disinfectant, and preservative whose common name is *formalin.* Biologists use formalin solution to preserve tissues of various organisms. Formaldehyde

Table 19.4
Some Important Aldehydes and Their Uses

Name	Structure	Use
Benzaldehyde	(structure)	Almond flavoring
Cinnamaldehyde	(structure)	Cinnamon flavoring
Propionaldehyde	CH_3CH_2CH, ‖ O	Disinfectant

can be very irritating to your eyes, nose, and throat, and it has a pungent odor.

Vanillin is used both commercially and medically as a flavoring agent.

It has a very pleasant odor and flavor that can cover up the harsh tastes of other substances. Vanillin gives us the characteristic taste of vanilla ice cream.

Table 19.4 lists some additional aldehydes and their uses.

19.8 Ketones

The general formula for a **ketone** *is*

$$R-\underset{\underset{O}{\|}}{C}-R'$$

The functional group

$$-\underset{\underset{O}{\parallel}}{C}-$$

is called the **carbonyl** or **keto** group. Ketones are very similar in structure to aldehydes. However, ketones have two R groups attached to the carbonyl group, whereas aldehydes have only one R group and a hydrogen atom:

$$\underset{\text{Ketone}}{R-\underset{\underset{O}{\parallel}}{C}-R'} \qquad \underset{\text{Aldehyde}}{R-\underset{\underset{O}{\parallel}}{C}-H}$$

We can name ketones in two ways. First, we can use the common name, which we determine simply by naming each of the alkyl groups in alphabetical order and then adding the word *ketone*. For example,

$$CH_3CH_2-\underset{\underset{O}{\parallel}}{C}-CH_3$$

is called *ethyl methyl ketone*.

Second, we can use the IUPAC name, which we form by counting the number of carbon atoms in the compound (including those in the carbonyl group) and changing the alkane ending *-ane* to the ketone ending *-one*. For example,

$$CH_3CH_2-\underset{\underset{O}{\parallel}}{C}-CH_3$$

is called *butanone*. [*Note:* Sometimes we have to mention the carbon atom that is attached to the oxygen, if isomers are possible. Example 19.5 shows what we mean.]

Example 19.5

Give the IUPAC and the common name for each of the following ketones.

(a) $CH_3-\underset{\underset{O}{\parallel}}{C}-CH_3$ **(b)** $CH_3CH_2CH_2-\underset{\underset{O}{\parallel}}{C}-CH_2CH_3$

(c) $CH_3CH_2CH_2CH_2-\underset{\underset{O}{\parallel}}{C}-CH_3$

Solution

(a)
$$CH_3-\underset{\underset{O}{\|}}{C}-CH_3$$

is a compound whose IUPAC name is *propanone*. Its common name is *dimethyl ketone*. However, this compound is probably best known as *acetone*.

(b)
$$CH_3CH_2CH_2-\underset{\underset{O}{\|}}{C}-CH_2CH_3$$

is a six-carbon ketone; therefore its IUPAC name would be *hexanone*. However, other hexanones are possible [see part (c) of this example]. Thus we must number the carbon chain in such a way that the carbon atom attached to the oxygen atom will have the lowest possible number:

$$\overset{6}{C}H_3\overset{5}{C}H_2\overset{4}{C}H_2-\overset{3}{\underset{\underset{O}{\|}}{C}}-\overset{2}{C}H_2\overset{1}{C}H_3$$

In this case, the correct IUPAC name is *3-hexanone*. The common name of this compound is *ethyl propyl ketone*, which we derive from the alkyl groups attached to the carbonyl group.

(c)
$$\overset{6}{C}H_3\overset{5}{C}H_2\overset{4}{C}H_2\overset{3}{C}H_2-\overset{2}{\underset{\underset{O}{\|}}{C}}-\overset{1}{C}H_3$$

is a compound that is also a six-carbon ketone. The IUPAC name is *2-hexanone*. However, the common name of this compound is *butyl methyl ketone*.

Practice Exercise 19.5   Give the IUPAC and the common name for each of the following ketones.

(a) $CH_3-CH_2-\underset{\underset{O}{\|}}{C}-CH_2-CH_3$

(b) $CH_3-CH_2-CH_2-CH_2-\underset{\underset{O}{\|}}{C}-CH_2-CH_3$

(c) $CH_3-CH_2-\underset{\underset{O}{\|}}{C}-CH_3$

19.9 Some Important Ketones

Propanone (acetone) is the simplest and best-known ketone. It is an excellent solvent for many organic substances. It is found in paint removers and nail-polish removers. Persons with uncontrolled diabetes mellitus produce acetone in their blood and give off an acetone smell when they exhale. A physician can easily detect this *acetone breath* in an examination of such persons. Acetone has a unique odor and sweetish taste.

Methyl phenyl ketone, also known as *acetophenone*,

has an orange-blossom-like fragrance and therefore is used in the making of some perfumes. It has also been used in medicine as a hypnotic agent.

Table 19.5 lists some additional ketones and their uses.

Table 19.5
Some Important Ketones and Their Uses

Name	Structure	Use
Progesterone		One of the major female hormones. Used in oral contraceptives.
Testosterone		Male sex hormone
Camphor		Anesthetic and mild antiseptic

19.10 Carboxylic Acids

· · · · · · · · · · · · · ·
Learning Goal 6
Carboxylic acids

The general formula for a **carboxylic acid** *is*

$$R-\overset{\underset{\displaystyle \parallel}{}}{C}-OH$$
$$O$$

The functional group

$$-\overset{\underset{\displaystyle \parallel}{}}{C}-OH$$
$$O$$

is called the **carboxyl** group. This group acts like an acid because it re-leases hydrogen ions in aqueous solutions:

$$R-\overset{\underset{\displaystyle \parallel}{}}{C}-OH + H_2O \rightleftharpoons R-\overset{\underset{\displaystyle \parallel}{}}{C}-O^- + H_3O^+$$
$$O \qquad\qquad\qquad O$$

This ionization occurs only to a slight degree, so carboxylic acids are weak acids.

There are two ways of naming carboxylic acids. First, we can use the common name, which (in most cases) is the common root plus the ending -*ic*, followed by the word *acid*. For example,

$$CH_3CH_2-\overset{\underset{\displaystyle \parallel}{}}{C}-OH$$
$$O$$

is called *propionic acid*. Second, we can use the IUPAC name, which is the alkane name plus the ending -*oic*, followed by the word *acid*. For example,

$$CH_3CH_2-\overset{\underset{\displaystyle \parallel}{}}{C}-OH$$
$$O$$

is called *propanoic acid*.

Example 19.6

Give the IUPAC and the common name for each of the following car-boxylic acids.

(a) $CH_3CH_2CH_2-\overset{\underset{\displaystyle \parallel}{}}{C}-OH$ **(b)** $CH_3-\overset{\underset{\displaystyle \parallel}{}}{C}-OH$
$$\quad\quad\quad\quad\quad\quad\quad O \qquad\qquad\qquad\qquad\qquad O$$

Solution

(a)
$$CH_3CH_2CH_2 - \underset{\underset{O}{\parallel}}{C} - OH$$

is a four-carbon carboxylic acid; therefore its IUPAC name is *butanoic acid*. The common name of this acid is *butyric acid*.

(b)
$$CH_3 - \underset{\underset{O}{\parallel}}{C} - OH$$

The IUPAC name of this compound, which is a two-carbon carboxylic acid, is *ethanoic acid*. However, its common name is *acetic acid*.

Practice Exercise 19.6   Give the IUPAC and the common name for each of the following carboxylic acids.

(a) $CH_3 - CH_2 - \underset{\underset{O}{\parallel}}{C} - OH$ **(b)** $CH_3 - CH_2 - CH_2 - CH_2 - \underset{\underset{O}{\parallel}}{C} - OH$

19.11 Some Important Carboxylic Acids

Methanoic acid is the simplest carboxylic acid:

$$H - \underset{\underset{O}{\parallel}}{C} - OH$$

Its common name is *formic acid*. Ants and bees produce this acid and inject it into their victims when they bite or sting. Once a victim is stung by a bee, it is the formic acid that causes the skin to become inflamed.

Ethanoic acid, also known as *acetic acid*,

$$CH_3 - \underset{\underset{O}{\parallel}}{C} - OH$$

is the substance found in vinegar. We discussed some of the properties of this acid in Chapter 15.

Butanoic acid, also known as *butyric acid*,

$$CH_3CH_2CH_3 - \underset{\underset{O}{\parallel}}{C} - OH$$

is produced when butter becomes rancid. In its pure form, it has a very pungent odor. It is also one of the substances that causes body odor.

Table 19.6 lists some additional carboxylic acids and their uses.

Table 19.6
Some Important Carboxylic Acids and Their Uses*

Name	Structure	Use
Lactic acid	$CH_3CH-C-OH$ with OH and O below	Produced in muscles when you exercise.
Benzoic acid	$O=C-OH$ attached to benzene ring	Used as an antiseptic, antifungal agent, and preservative. Also used as a starting material for numerous other organic compounds.
Citric acid	$HO-C-CH_2-C-CH_2-C-OH$ with O, OH, O below and $O=C-OH$ above	This is what gives citrus fruits their taste. Used for treatment of rickets.
Propionic acid	CH_3CH_2-C-OH with O below	Along with its sodium and calcium salts, this compound is used as a food additive to inhibit the growth of molds.

*Reproduced in part from *The Merck Index, Tenth Edition.* Copyright under the Universal Copyright Convention and the International Copyright Convention 1983 by Merck & Co., Inc., Rahway, New Jersey, U.S.A.

19.12 Esters

Learning Goal 7
Esters

The general formula for an **ester** *is*

$$R-C-OR'$$
with O below C

Some of the most common flavors and fragrances are due to the properties of esters (Table 19.7).

Esters can be formed by the reaction of an alcohol with a carboxylic acid. For example,

$$CH_3CH_2-C-\boxed{OH}+\boxed{H}O-CH_3 \xrightarrow{H^+} CH_3CH_2-C-OCH_3 + H_2O$$

Propanoic acid   Methanol   An ester

Note how the acid and alcohol parts combine to produce the ester:

Acid   Alcohol
part   part

Learning Goal 8
Esterification reactions
The reaction between an alcohol and a carboxylic acid to produce an ester is called an **esterification reaction.**

Table 19.7
Some Esters Used for Their Flavoring or Fragrance

Name	Formula	Flavor or fragrance
*Amyl acetate	$CH_3-(CH_2)_4-O-\overset{\overset{\displaystyle O}{\|}}{C}-CH_3$	Banana
Amyl butyrate	$CH_3-(CH_2)_4-O-\overset{\overset{\displaystyle O}{\|}}{C}-CH_2CH_2CH_3$	Apricot
Ethyl butyrate	$CH_3CH_2-O-\overset{\overset{\displaystyle O}{\|}}{C}-CH_2CH_2CH_3$	Pineapple
Ethyl formate	$CH_3CH_2-O-\overset{\overset{\displaystyle O}{\|}}{C}-H$	Rum
Octyl acetate	$CH_3-(CH_2)_7-O-\overset{\overset{\displaystyle O}{\|}}{C}-CH_3$	Orange

*For a chain of five carbon atoms, the word *amyl* is often used as the root.

Example 19.7

Write the products of the following esterification reaction.

$$CH_3CH_2CH_2-\overset{\overset{\displaystyle O}{\|}}{C}-OH + HO-CH_2CH_3 \xrightarrow{H^+}$$

Solution
We'll "remove" the OH and H, as above, and combine the rest.

$$CH_3CH_2CH_2-\overset{\overset{\displaystyle O}{\|}}{C}\boxed{-OH} + \boxed{H}OCH_2CH_3 \xrightarrow{H^+}$$

$$CH_3CH_2CH_2-\overset{\overset{\displaystyle O}{\|}}{C}-OCH_2CH_3 + H_2O$$

Practice Exercise 19.7 Write the products of the following esterification reaction.

$$CH_3-CH_2-CH_2-CH_2-\overset{\overset{\displaystyle O}{\|}}{C}-OH + OH-CH_3 \xrightarrow{H^+}$$

There are two ways of naming esters. First, we can use the common name, which we find by adding the name of the alkane part of the alcohol group to the common name of the acid, and changing the *-ic* ending to *-ate*. For example,

$$CH_3CH_2-\underset{\underset{O}{\|}}{C}-OCH_3$$

is called *methyl propionate* (written as two words). Second, we can use the IUPAC name, which we get by adding the name of the alkane part of the alcohol group to the IUPAC name of the acid, and changing the *-ic* ending to *-ate*. For example,

$$CH_3CH_2-\underset{\underset{O}{\|}}{C}-OCH_3$$

is called *methylpropanoate* (written as one word).

Example 19.8

Give the IUPAC and the common name for each of the following esters.

(a) $CH_3CH_2-\underset{\underset{O}{\|}}{C}-OCH_2CH_3$ (b) $CH_3-\underset{\underset{O}{\|}}{C}-OCH_2CH_2CH_3$

Solution

(a) $CH_3CH_2-\underset{\underset{O}{\|}}{C}-OCH_2CH_3$

is the ester of ethyl alcohol and propionic acid. Therefore its common name is *ethyl propionate*. Its IUPAC name is *ethylpropanoate*.

(b) $CH_3-\underset{\underset{O}{\|}}{C}-OCH_2CH_2CH_3$

is the ester of propyl alcohol and acetic acid. Therefore its common name is *propyl acetate*. Its IUPAC name is *propylethanoate*. (Remember, the IUPAC name of acetic acid is ethanoic acid.)

Practice Exercise 19.8   Give the IUPAC and the common name for each of the following esters.

(a) $CH_3-CH_2-\underset{\underset{O}{\|}}{C}-OCH_3$

(b) $CH_3-CH_2-CH_2-\underset{\underset{O}{\|}}{C}-O-CH_2-CH_2-CH_3$

19.13 Some Important Esters

Ethyl acetate is a compound that is used in artificial fruit essences:

$$CH_3 - \overset{\overset{\displaystyle}{\underset{\displaystyle O}{\|}}}{C} - OCH_2CH_3$$

It is also an excellent solvent for varnishes, lacquers, and airplane lubricants. In addition, this compound is used in manufacturing smokeless powder, artificial leather, and perfumes and in cleaning textiles.

Methyl salicylate, commonly known as *oil of wintergreen*,

is the methyl ester of salicylic acid,

Salicylic acid

It has a pleasant spearmint-like odor. For this reason, it is used (in very low concentrations) in candies and perfumes. Because it also creates a sensation of warmth in the skin, which makes sore muscles feel better, methyl salicylate is found in many liniments. In higher concentrations it is toxic.

Table 19.8 lists some additional esters and their uses.

Table 19.8
Some Important Esters and Their Uses*

Name	Structure	Use
Aspirin	$CH_3 - \overset{\overset{O}{\|}}{C} - O$ (attached to benzene ring bearing $\overset{\overset{O}{\|}}{C} - OH$)	Pain reliever
Methyl acetate	$CH_3 - \overset{\overset{O}{\|}}{C} - OCH_3$	Used as a solvent for many oils and resins
Butyl acetate	$CH_3 - \overset{\overset{O}{\|}}{C} - OCH_2CH_2CH_2CH_3$	Used to make lacquer, artificial leather, plastics, and safety glass

*Reproduced in part from *The Merck Index, Tenth Edition.* Copyright under the Universal Copyright Convention and the International Copyright Convention 1983 by Merck & Co., Inc., Rahway, New Jersey, U.S.A.

Figure 19.4
The relationship of primary, secondary, and tertiary amines to ammonia

$$
\begin{array}{cccc}
\text{H—N—H} & \text{R—N—H} & \text{R—N—H} & \text{R—N—R''} \\
\quad| & \quad| & \quad| & \quad| \\
\text{H} & \text{H} & \text{R'} & \text{R'} \\
\text{Ammonia} & \text{Primary amine} & \text{Secondary amine} & \text{Tertiary amine}
\end{array}
$$

19.14 Amines

Learning Goal 9
Amines

There are three general formulas for **amines:**

1. $R—NH_2$ (primary amine)

2. $R—NH$ (secondary amine)
$\quad\quad|$
$\quad\quad R'$

3. $R—N—R''$ (tertiary amine)
$\quad\quad|$
$\quad\quad R'$

Amines are, essentially, ammonia (NH_3) with the hydrogen atoms replaced by R groups. If one R group has replaced a hydrogen atom, the compound is a **primary amine.** If two R groups have replaced two hydrogen atoms, the compound is a **secondary amine.** And if three R groups have replaced three hydrogen atoms, the compound is a **tertiary amine** (Figure 19.4).

Example 19.9

Determine whether each of the following compounds is a primary, secondary, or tertiary amine.

(a) $CH_3—N—CH_3$
$\quad\quad\quad\quad|$
$\quad\quad\quad\quad CH_3$

(b) $CH_3CH_2—NH$
$\quad\quad\quad\quad\quad|$
$\quad\quad\quad\quad\quad CH_2CH_3$

(c) [benzene ring with NH_2 substituent]

Solution

(a) $CH_3—N—CH_3$
$\quad\quad\quad\quad|$
$\quad\quad\quad\quad CH_3$

This compound has three R groups attached to the nitrogen atom, so it is a tertiary amine.

(b) $CH_3CH_2—NH$
$\quad\quad\quad\quad\quad|$
$\quad\quad\quad\quad\quad CH_2CH_3$

This compound has two R groups attached to the nitrogen atom, so it is a secondary amine.

(c) [benzene ring with NH_2 substituent]

This compound has one R group attached to the nitrogen atom, so it is a primary amine.

Practice Exercise 19.9   Determine whether each of the following compounds is a primary, secondary, or tertiary amine.

(a) $CH_3-\underset{\underset{CH_3}{|}}{N}-CH_3$ **(b)** $CH_3-\underset{\underset{CH_3}{|}}{N}-H$ **(c)** CH_3-NH_2

We can name amines in two ways. First, we can use the common name, which is the alkane name with *-yl* substituted for the *-ane*, plus the ending *amine*. For example, CH_3NH_2 is called *methylamine*. If two R groups are attached to the nitrogen atom, we name the groups in alphabetical order. For example,

$CH_3-\underset{\underset{CH_3}{|}}{NH}$ is called *dimethylamine*, and $CH_3CH_2-\underset{\underset{CH_3}{|}}{NH}$

is called *ethylmethylamine*. Second, we can use the IUPAC name, which for primary amines involves naming the $-NH_2$ group as an amino group along the alkane chain and numbering the carbon atoms to locate its position. For example,

$$CH_3-\underset{\underset{NH_2}{|}}{CH}-CH_3$$

is called *2-aminopropane*. For secondary and tertiary amines, the IUPAC rules become more complex and will not be covered here. Moreover, most people use the common names for amines.

Example 19.10

Give the common name for each of the following amines. If possible, give its IUPAC name also.

(a) $CH_3CH_2-NH_2$ **(b)** $CH_3CH_2CH_2-\underset{\underset{CH_3}{|}}{N}-CH_3$

(c) $CH_3CH_2CH_2CH_2-\underset{\underset{NH_2}{|}}{CH}-CH_2CH_3$

Solution

(a) $CH_3CH_2-NH_2$ is a two-carbon amine. Therefore the IUPAC name of this compound is *aminoethane*. Its common name is *ethylamine*.

(b)

$$CH_3CH_2CH_2-\underset{\underset{CH_3}{|}}{N}-CH_3$$

is a tertiary amine that consists of two methyl groups and a propyl group. Therefore the common name of this compound is *dimethyl-propylamine*.

$$
\overset{7}{C}H_3\overset{6}{C}H_2\overset{5}{C}H_2\overset{4}{C}H_2 - \overset{3}{C}H - \overset{2}{C}H_2\overset{1}{C}H_3
$$

(c) NH_2

is a primary amine that consists of seven carbon atoms, and the amino group is on the third carbon atom. Therefore the IUPAC name of this compound is *3-aminoheptane*.

Practice Exercise 19.10   Give the common name for each of the following amines. If possible, give its IUPAC name also.
(a) $CH_3CH_2 - NH_2$ **(b)** $CH_3 - N - CH_3$ **(c)** $CH_3CH_2 - N - H$
                                          CH_3 CH_3

19.15 Some Important Amines

The compound *1,5-diaminopentane*,

$$
H_2N - CH_2CH_2CH_2CH_2CH_2 - NH_2
$$

is also called *1,5-pentanediamine*, but it is better known as *cadaverine*. This compound is produced by decaying organisms. Such amines are called *ptomaines* (pronounced toe-mains) and, in fact, can also be formed by the action of bacteria on meat and fish (thus the name *ptomaine poisoning*). Cadaverine has a very pungent odor; it also acts as a base.

The compound 1-phenyl-2-aminopropane,

$$
CH_2 - CH - CH_3
$$
                          NH_2

also called *(phenylisopropyl)amine* but better known as *amphetamine* (or *benzedrine*), acts as a stimulant to the central nervous system. This compound and others related to it are called "uppers" because of their ability to keep one active and awake. Unfortunately, they can cause drug dependence. In other words, after a person has taken them for a while, he or she must continue to take them in order to stay alert. Amphetamine has also been used in nasal inhalers to relieve nasal congestion in people who have colds.

Table 19.9 lists some additional amines and their uses.

Table 19.9
Some Important Amines and Their Uses*

Name	Structure	Use
PABA (para-aminobenzoic acid)		Used in suntan lotions as a sunscreen (in other words, to prevent ultraviolet light from reaching the skin). Also found as part of the B-complex vitamins.
Methadone		Used as a substitute for heroin in the treatment of heroin addiction.
Methylamine	CH_3NH_2	Used in tanning leather.
Ephedrine		Used to constrict blood vessels.

*Reproduced in part from *The Merck Index, Tenth Edition.* Copyright under the Universal Copyright Convention and the International Copyright Convention 1983 by Merck & Co., Inc., Rahway, New Jersey, U.S.A.

Summary

The members of a class of organic compounds are composed of a functional group (which is the same for each compound in the class) and one or more R groups (which differ among the members of each class). The functional group consists of one or more atoms and gives the class of compounds its specific chemical properties. The R groups are usually alkyl groups, and they give the individual compounds their specific properties.

The *alcohols* all contain the —OH hydroxyl group. Their general formula is R—OH. Alcohols with short R groups, consisting of only a few carbon and hydrogen atoms, are polar. Those with long R groups are nonpolar.

Ethers have the general formula R—O—R'. They tend to be extremely flammable, and many have anesthetic properties.

Aldehydes have the general formula

$$R-\overset{\displaystyle \underset{\|}{} }{C}-H$$
$$\underset{O}{\|}$$

Some are used in perfumes and as flavoring agents. Formaldehyde (which is not used in perfumes) is probably the best-known member of this class. It is used as a germicide, disinfectant, and tissue preservative.

Ketones have the general formula

$$R-\overset{}{C}-R' \text{ in which the functional group } -\overset{}{C}-$$
$$\underset{O}{\|} \qquad\qquad\qquad\qquad \underset{O}{\|}$$

is called the carbonyl or keto group. The male hormone testosterone and the female hormone progesterone are ketones.

The general formula for the *carboxylic acids* is

$$R-\overset{}{C}-OH$$
$$\underset{O}{\|}$$

in which the functional group is called the carboxyl group. These compounds ionize to a slight degree in water.

Esters have the general formula

$$R-\overset{}{C}-OR'$$
$$\underset{O}{\|}$$

They are formed in the reaction of an alcohol and a carboxylic acid, which is called an esterification reaction. Aspirin is an ester.

Amines are, essentially, ammonia in which R groups have replaced one, two, or all three hydrogen atoms. In a primary amine, with the general formula $R-NH_2$, one hydrogen atom has been replaced by an R group. In secondary and tertiary amines, two and three hydrogen atoms, respectively, have been replaced.

The compounds in all seven of these classes have both common names and IUPAC names. The IUPAC name of a compound is derived from the name of the hydrocarbon that the compound resembles in structure.

. . . .

Career Sketch

Perfumer

A perfumer evaluates the odors of aromatic chemicals, sets standards for producing perfumes, and sees to it that workers engaged in compounding perfumes work accurately. The perfumer is responsible for the quality of the finished product and thus must visit the compounding and distillation areas to determine, by smelling, the quality of the materials being pre-

A perfumer evaluates odors of aromatic chemicals, sets standards for producing perfumes, and sees to it that workers engaged in compounding perfumes work accurately.
(International Flavors and Fragrances, Inc.)

pared and to approve batches of perfume for finishing (or bottling). It is necessary for the perfumer to have a keen enough sense of smell to distinguish characteristics of various odors and blends of odors and to determine whether the sample being smelled conforms to the customer's (or laboratory's) standards. The perfumer directs those who prepare work orders for the compounding department and checks batch sheets prepared for the distillation department to be sure that the formulas devised in the perfume laboratory are adhered to. The perfumer resmells samples in an air-filtered room, using strips of blotter dipped in aromatic chemicals, to evaluate fragrances for specific characteristics, such as odor, body, harmony, strength, and permanence. The perfumer compares the odor and color of the current sample with samples of previous batches, rejects any batches that do not meet company criteria, and asks the distillation department to redistill the product if it does not live up to standards. A so-called *creative perfumer* may devise formulas by which new perfumes are to be made.

Self-Test Exercises:

Learning Goal 1: Functional Groups

1. Define the term *functional group*.

2. What is the difference between a functional group and an R group?

3. Circle the functional group in each of the following compounds, and state what it is called.

(a) CH_3OH

(b) $CH_3CH_2OCH_3$

(c) CH_3CH
   ‖
   O

(d) $CH_3CH_2 — \overset{\overset{\displaystyle O}{\|}}{C} — OH$

4. Circle the functional group in each of the following compounds, and state what it is called.

(a) CH_3CH_2OH

(b) CH_3OCH_3

(c) CH_3CH_2CH
   ‖
   O

(d) $CH_3 — \overset{}{\underset{\underset{\displaystyle CH_3}{|}}{N}} — CH_3$

Learning Goal 2: Alcohols

5. Write the general formula for an alcohol.

6. Complete this sentence: The —OH group is called the _____ functional group.

7. Is each of the following alcohols polar or non-polar?

(a) $CH_3CH_2CH_2CH_2CH_2CH_2CH_2OH$ (b) CH_3CH_2OH

8. Is each of the following alcohols polar or non-polar?

(a) $CH_3 — (CH_2)_{12} — CH_2OH$ (b) $CH_3 — OH$

9. Give the IUPAC or the common name for each of the following alcohols.

(a) $CH_3CH_2CH_2OH$ (b) $CH_3CH_2CH_2CH_2CH_2CH_2OH$

10. Give the IUPAC or the common name for each of the alcohols listed in Self-Test Exercise 8.

11. Give an example of an important alcohol, and state its uses.

12. Why is it dangerous to drink methanol even once, but not so dangerous to drink small amounts of ethanol occasionally?

13. Use the names of the following alcohols to write their structures.

(a) 3-hexanol

(b) Butyl alcohol

(c) 2-octanol

(d) Methyl alcohol

14. Use the name of each of the following alcohols to write its structure.

(a) 2-hexanol

(b) Ethyl alcohol

(c) 2-pentanol

(d) Propyl alcohol

Learning Goal 3: Ethers

15. Give the general formula for an ether.

16. Distinguish between a symmetrical and an unsymmetrical ether.

17. What is the common name for each of the following ethers?

(a) $CH_3CH_2 — O — CH_2CH_2CH_2CH_3$

(b) $CH_3CH_2 — O — CH_3$

18. What is the common name of each of the following ethers?

(a) $CH_3 — O — CH_2CH_2CH_2CH_3$

(b) $CH_3CH_2 — O — CH_2CH_3$

19. Use the common name for each of the following ethers to write its structure.

(a) Methyl butyl ether   (b) Dimethyl ether

20. Use the common name of each of the following ethers to write its structure.

(a) Ethyl methyl ether   (b) Dibutyl ether

21. Give an example of an important ether, and state its uses.

22. Name the ether that is commonly used in hospitals as an anesthetic for various surgical procedures.

Learning Goal 4: Aldehydes

23. Write the general formula for an aldehyde.

24. What is the general formula of an aldehyde?

25. Give the common or the IUPAC name for each of the following aldehydes.

(a) $CH_3 — (CH_2)_4 — \overset{\overset{\displaystyle O}{\|}}{CH}$ (b) $CH_3CH_2 — \overset{\overset{\displaystyle O}{\|}}{CH}$

26. Give the common or the IUPAC name for each of the following aldehydes.

(a) $CH_3 — (CH_2)_3 — \overset{\overset{\displaystyle O}{\|}}{C} — H$ (b) $CH_3 — \overset{\overset{\displaystyle O}{\|}}{C} — H$

27. Write the structure of each of the following aldehydes from its name.
(*a*) Pentanal   (*b*) Acetaldehyde

29. Give an example of an important aldehyde, and state its uses.

28. Write the structure of each of the following aldehydes from its name.
(*a*) Hexanal   (*b*) Butanal

30. Name the aldehyde that gives vanilla ice cream its flavor.

Learning Goal 5: Ketones

31. What is the general formula for a ketone?

33. Give the IUPAC or the common name for each of the following ketones.
(*a*) $CH_3CH_2\text{—}\underset{\underset{O}{\|}}{C}\text{—}CH_2CH_3$

(*b*) $CH_3CH_2CH_2CH_2\text{—}\underset{\underset{O}{\|}}{C}\text{—}CH_3$

(*c*)

32. Write the general structure of a carbonyl group.

34. Give the IUPAC or the common name for each of the following ketones.
(*a*) $CH_3\text{—}\underset{\underset{O}{\|}}{C}\text{—}CH_3$ (*b*) $CH_3CH_2\text{—}\underset{\underset{O}{\|}}{C}\text{—}CH_3$

(*c*) O

35. Write the structure of each of the following ketones from its name.
(*a*) 4-octanone   (*b*) Methyl propyl ketone

37. Give an example of an important ketone, and state its uses.

36. Write the structure of each of the following ketones from its name.
(*a*) 3-octanone   (*b*) Methyl ethyl ketone

38. Describe some of the uses of acetone.

Learning Goal 6: Carboxylic Acids

39. What is the general formula for a carboxylic acid?

41. Give the IUPAC or the common name for each of the following carboxylic acids.
(*a*) $CH_3\text{—}(CH_2)_5\text{—}\underset{\underset{O}{\|}}{C}\text{—}OH$ (*b*) $H\text{—}\underset{\underset{O}{\|}}{C}\text{—}OH$

(*c*) $CH_3CH_2CH_2\text{—}\underset{\underset{O}{\|}}{C}\text{—}OH$

43. Write the structure of each of the following carboxylic acids from its name.
(*a*) Heptanoic acid   (*b*) Propionic acid

45. Give an example of an important carboxylic acid, and state its uses.

40. What is the formula for the carboxyl group?

42. Give the IUPAC or the common name for each of the following carboxylic acids.
(*a*) $CH_3\text{—}(CH_2)_3\text{—}\underset{\underset{O}{\|}}{C}\text{—}OH$ (*b*) $CH_3\text{—}\underset{\underset{O}{\|}}{C}\text{—}OH$

(*c*) $CH_3\text{—}CH_2\text{—}\underset{\underset{O}{\|}}{C}\text{—}OH$

44. Write the structure of each of the following carboxylic acids from its name.
(*a*) Butanoic acid   (*b*) Ethanoic acid

46. Name the carboxylic acid that ants and bees produce and inject into their victims when they bite or sting.

Learning Goal 7: Esters

47. Write the general formula for an ester.

48. What two classes of organic compounds react to form an ester?

49. Give the IUPAC or the common name for each of the following esters.

(a) $CH_3-(CH_2)_4-\overset{\overset{\displaystyle O}{\|}}{C}-OCH_2CH_3$

(b) $CH_3-\overset{\overset{\displaystyle O}{\|}}{C}-OCH_3$

50. Give the IUPAC or the common name for each of the following esters.

(a) $CH_3-(CH_2)_3-\overset{\overset{\displaystyle O}{\|}}{C}OCH_3$

(b) $CH_3CH_2-\overset{\overset{\displaystyle O}{\|}}{C}OCH_2CH_3$

51. Write the structure of each of the following esters from its name.

(a) Ethyl butyrate   (b) Butylethanoate

52. Write the structure of each of the following esters from its name.

(a) Ethyl acetate   (b) Propylethanoate

53. Give an example of an important ester, and state its uses.

54. Which ester smells like a banana?

Learning Goal 8: Esterification Reactions

55. Write the products of each of the following esterification reactions.

(a) $CH_3CH_2-\overset{\overset{\displaystyle O}{\|}}{C}-OH + HO-CH_2CH_2CH_3 \overset{H^+}{\longrightarrow}$

(b) $CH_3-\overset{\overset{\displaystyle O}{\|}}{C}-OH + HO-CH_3 \overset{H^+}{\longrightarrow}$

56. Write the products of each of the following esterification reactions.

(a) $CH_3CH_2CH_2-\overset{\overset{\displaystyle O}{\|}}{C}-OH + OH-CH_3 \overset{H^{+1}}{\longrightarrow}$

(b) $CH_3CH_2-\overset{\overset{\displaystyle O}{\|}}{C}-OH + OH-CH_2CH_3 \overset{H^{+1}}{\longrightarrow}$

57. Given the products of the following esterification reaction, provide the reactants.

_____ + _____ $\overset{H^+}{\longrightarrow}$

$CH_3CH_2CH_2CH_2-\overset{\overset{\displaystyle O}{\|}}{C}-OCH_2CH_3 + H_2O$

58. Given the products of the following esterification reaction, provide the reactants.

_____ + _____ $\overset{H^+}{\longrightarrow}$

$CH_3CH_2-\overset{\overset{\displaystyle O}{\|}}{C}-OCH_3 + H_2O$

Learning Goal 9: Amines

59. What are the three general formulas for amines?

60. How are amines related to the compound ammonia?

61. Is each of the following amines primary, secondary, or tertiary?

(a) $CH_3-\overset{\overset{\displaystyle }{\underset{\underset{\displaystyle CH_3}{|}}{N}}}-CH_3$

(b) $CH_3CH_2-\overset{\underset{\underset{\displaystyle CH_3}{|}}{N}H}{}$

62. Is each of the following amines primary, secondary, or tertiary?

(a) $CH_3-\overset{\underset{\underset{\displaystyle CH_2CH_3}{|}}{N}}{}-CH_2CH_3$

(b) $CH_3-\overset{\underset{\underset{\displaystyle H}{|}}{N}}{}-CH_2CH_3$

(c)

(d)

(c)

(d)

63. Give the IUPAC or the common name for each of the following amines.

(a) CH₃—N—CH₃ with CH₃ below N

(b) NH₂ (on cyclohexane)

(c) CH₃CH₂—NH (on cyclopentane)

64. Give the IUPAC or the common name for each of the following amines.

(a) CH₃—N—CH₃ with H below N

(b) (tricyclohexylamine, N with three cyclohexyl groups)

(c) NH₂ (on cyclopentane)

65. Write the structure of each of the following amines from its name.

(a) Diethylamine
(b) Ethylpropylamine
(c) 2-aminopentane
(d) 3-aminopentane

66. Write the structure of each of the following amines from its name.

(a) Dimethylamine
(b) Butylethylamine
(c) 3-aminoheptane
(d) 1,5-diaminopentane

67. Give an example of an important amine, and state its uses.

68. What is amphetamine?

Extra Exercises

69. Explain why ethanol (C_2H_5OH) is soluble in water but ethane (C_2H_6) is not.

70. Why is it important to know whether an alcohol product to be used for human consumption contains any methanol?

71. Distinguish between a ketone and an aldehyde.

72. Match each class on the left with the correct general formula on the right.

(a) Amine
(b) Acid
(c) Alcohol
(d) Ester
(e) Alkane
(f) Aromatic
(g) Ether

1. ROH
2. R—C—OH with O double bonded to C
3. R—C—OR′ with O double bonded to C
4. R—O—R′
5. R—NH₂
6. R—CH₃
7. R (with benzene ring)

73. Isomers are compounds that have the same percentage composition and molecular weight, different structures, and the same molecular formula. Draw all possible isomers for trichlorobenzene. (This is a benzene molecule in which three hydrogens have been replaced by three Cl atoms.)

74. Circle the functional group in each of the following compounds, and state what it is called.

(a) CH₃—NH₂
(b) CH₃—O—CH₃
(c) CH₃—C—OH with O double bonded to C
(d) CH₃—C—O—CH₃ with O double bonded to C

75. Write the structure for 2-butanol.

76. Write the structure for 3-aminohexane.

77. Write the structure for dibutyl ether.

78. Name the compound

$$CH_3—(CH_2)_5—C—OH$$ with O double bonded to C

79. Write the formula for propionaldehyde.

80. Write the formula for 2-pentanone.

81. Write the formula for butanoic acid.

82. Name each of the following compounds.

(a) $H{-}\overset{\displaystyle\|}{\underset{\displaystyle O}{C}}{-}OCH_2CH_3$ (b) $H{-}\overset{\displaystyle\|}{\underset{\displaystyle O}{C}}{-}OCH_3$

83. What class of organic compounds do you think is the most important? Explain your answer.

84. Can a compound contain more than one functional group? Can you give an example of such a compound?

20 *Biochemistry: The Chemistry of Life*

Learning Goals

After you have studied this chapter, you should be able to:

1. Describe the processes of metabolism, anabolism, and catabolism.
2. Define and classify carbohydrates, and give examples of the various classes.
3. Explain what lipids are, and describe the two main types of lipids.
4. Define and give the general formulas for fatty acids and fats.
5. Define and give examples of a wax, a steroid, and a terpene.
6. Explain the various functions of proteins.
7. Describe the structure of amino acids.
8. Explain what is meant by the terms *essential amino acid, complete protein,* and *incomplete protein.*
9. Explain how proteins are formed, and describe their structure.
10. Define and give examples of vitamins, explain their sources, and identify the diseases that arise from vitamin deficiencies.
11. Define and give examples of minerals.

Introduction

Like living laboratories, our bodies are factories in which countless chemical reactions take place continually. The reactions start with the foods we eat, and the ultimate products range from water to some of the most complex molecules known. These reactions provide the energy and materials needed to make growth and development possible and to keep our bodies working properly.

This chapter briefly surveys some of the chemistry of the human body and serves as a simple introduction to a very complex topic—biochemistry, or the chemistry of living organisms.

20.1 Metabolism

. . . .
Learning Goal 1
Metabolism, anabolism, and catabolism

When we swallow food, it first enters the digestive system. There, the large food molecules are broken down into smaller molecules in the process known as *digestion*. The smaller molecules are then transported to cells throughout the body. In the cells, the molecules undergo a variety of chemical reactions that together are called **metabolism.**

Metabolic reactions are of two general types. In one type, the digested food molecules serve as the raw materials for the synthesis (or build-up) of more complex substances. This process, called **anabolism,** results in the formation of body tissues.

In the second type of metabolic reaction, the digested food molecules are broken down into even smaller molecules. This process is called **catabolism.** It results in the release of energy within the cells, where the energy can be used to power the body.

The foods we eat are metabolized and provide energy to power the body.

Thus our bodies make use of the food we eat in two steps. In the first step, digestion, the food is broken down into molecules of "convenient" size. In the second step, metabolism, these molecules are either built up into tissue or broken down further in chemical reactions that release energy to the cells. Of course, these processes must begin with the proper foods, which contain the reactants needed for digestion, anabolism, and catabolism. Let us look at the basic chemical components of foods and at some of the products of digestion.

20.2 Carbohydrates

Learning Goal 2
Carbohydrates

Carbohydrates make up a class of organic compounds that are composed of the elements carbon, hydrogen, and oxygen. Carbohydrate molecules supply us with energy: Our bodies produce four kilocalories of energy for every gram of carbohydrate that is taken in and metabolized.

Carbohydrate molecules have many functions. In plants, carbohydrates make up the structures that support the plant and those that store the plant's energy supply. The carbohydrate molecule called *cellulose*, the major component of wood, is a structural carbohydrate. A plant stores its energy in the form of *starch*, which is a carbohydrate located in the plant's seeds, leaves, and roots.

Our blood and tissues also contain a type of carbohydrate called *glucose*, or *blood sugar*. *Heparin* is an important carbohydrate that serves as a blood anticoagulant; it inhibits the formation of blood clots.

The foods we eat contain varying amounts of carbohydrates. Examples of carbohydrate foods are sugars and starches, which are found in high concentrations in cookies, cakes, breads, whole-grain cereals, wheat, rice, corn, and fruits.

20.3 Classification of Carbohydrates

We classify carbohydrates according to the size of the molecule. The three major categories of carbohydrates are based on the number of sugar groups found in the molecule. These categories are *monosaccharides, disaccharides,* and *polysaccharides.*

Monosaccharides are simple sugars. They have the general formula $(CH_2O)_x$, where x is 3, 4, 5, or 6. Monosaccharides cannot be changed into simpler sugars upon reaction with water.

Ribose and *deoxyribose* are five-carbon monosaccharides that compose part of the genetic molecules *ribonucleic acid* (*RNA*) and *deoxyribonucleic acid* (*DNA*) (Figure 20.1). The most important monosaccharides are *glucose, fructose,* and *galactose* (Figure 20.2). Although they all have the same molecular formula, $C_6H_{12}O_6$, their structural formulas are different. That

$$
\begin{array}{cc}
\overset{\displaystyle O}{\underset{\displaystyle \|}{}} & \overset{\displaystyle O}{\underset{\displaystyle \|}{}} \\
C-H & C-H \\
| & | \\
H-C-OH & H-C-H \\
| & | \\
H-C-OH & H-C-OH \\
| & | \\
H-C-OH & H-C-OH \\
| & | \\
CH_2OH & CH_2OH \\
\text{Ribose} & \text{Deoxyribose (contains one} \\
 & \text{less oxygen than ribose)}
\end{array}
$$

Figure 20.1

Ribose and deoxyribose are five-carbon monosaccharides that compose part of the genetic molecules, ribonucleic acid and deoxyribonucleic acid.

is, they are isomers (for a refresher course on isomers, look back at Chapter 18).

Glucose—also called *dextrose* or *blood sugar*—is found in the blood stream and doesn't have to be digested to be used by the body. People who can't take food orally are given glucose intravenously (directly into the veins).

The monosaccharide *fructose*—also called *levulose* or *fruit sugar*—is the sugar that is found in honey. *Galactose*, unlike glucose and fructose, isn't found free in nature. It is found in compounds such as *lactose* (milk sugar), *agar*, and *pectin*. *Agar* is a complex sugar prepared from seaweed; it is often used by biologists as a medium in which to grow bacteria. *Pectin* is used along with sugar to thicken fruit juices in the production of jelly.

Disaccharides are compounds formed from two monosaccharide molecules that are bonded together by an oxygen atom. The three most important disaccharide molecules are *maltose, lactose,* and *sucrose.* They all have the molecular formula $C_{12}H_{22}O_{11}$. Their structural formulas differ, however, so they are isomers (Figure 20.3).

Figure 20.2

Glucose, galactose, and fructose—the three most important monosaccharides

$$
\begin{array}{ccc}
H & H & \\
| & | & \\
C=O & C=O & CH_2OH \\
| & | & | \\
H-C-OH & H-C-OH & C=O \\
| & | & | \\
HO-C-H & HO-C-H & HO-C-H \\
| & | & | \\
H-C-OH & HO-C-H & H-C-OH \\
| & | & | \\
H-C-OH & H-C-OH & H-C-OH \\
| & | & | \\
CH_2OH & CH_2OH & CH_2OH \\
\text{Glucose} & \text{Galactose} & \text{Fructose}
\end{array}
$$

Carbohydrates are an important part of a balanced diet.

Sucrose, found in sugar cane and sugar beets, is a disaccharide that is also called *table sugar.* The two monosaccharides that form sucrose are fructose and glucose. *Lactose* is sometimes referred to as *milk sugar.* It is a component of mother's milk and cow's milk. When certain bacteria act on lactose, they cause it to become lactic acid. It is this reaction that causes milk to become sour. (You will recall from Chapter 15 that acids have a sour taste.) The two monosaccharides that produce lactose are glucose and galactose. *Maltose*—also called *malt sugar*—is not found free in nature but is produced when certain more complex carbohydrates (such as starch, dextrin, and glycogen) break down. The two monosaccharides that produce maltose are two molecules of glucose.

Polysaccharides are molecules made up of three or more monosaccharides bonded together. These molecules are very large. Figure 20.4 shows the structures of some important polysaccharides.

Starch is a polysaccharide. Foods such as potatoes (and some other vegetables), wheat, corn, and rye contain large amounts of starch. The polysaccharide *cellulose* is important in the diet. We can't digest cellulose, although it serves as bulk to help move our waste products through the intestinal tract.

Carbohydrates are stored in the body in the form of *glycogen.* Glycogen is found in the liver and muscles. When our bodies need sugar to burn, glycogen breaks down to form glucose.

Glucose + Glucose ⟶ Maltose + H₂O

Maltose

Glucose + Fructose ⟶ Sucrose + H₂O

Sucrose

Glucose + Galactose ⟶ Lactose + H₂O

Lactose

Figure 20.3
Maltose, sucrose, and
lactose—the three most
important disaccharides

Example 20.1

Name the three most important monosaccharides.

Solution

The three most important monosaccharides are glucose, galactose, and fructose.

Practice Exercise 20.1   Name some sources of the three most important monosaccharides.

Amylose—a straight-chain glucose polymer consisting of from 250 to 500 alpha glucose molecules linked together

Amylopectin—a highly branched glucose polymer consisting of about 1,000 alpha-glucose molecules

Figure 20.4

Some important poly-saccharide molecules are amylose, amylopectin, and cellulose. Starch is a mixture of amylose and amylopectin.

Cellulose—consists of from 900 to 6,000 glucose units linked together in a straight chain. This polymer consists of beta-glucose units.

Example 20.2

Name the three most important disaccharides, and state which monosaccharides compose them.

Solution

The three most important disaccharides are *maltose,* composed of two glucose units; *lactose,* composed of glucose and galactose; and *sucrose,* composed of fructose and glucose.

Practice Exercise 20.2   Name some sources of the three most important disaccharides.

20.4 Lipids

Learning Goal 3
Lipids

Lipids constitute another important group of biochemical molecules. This group includes fats and other compounds whose physical properties are similar to those of fats. Lipids are found in both plants and animals. In plants, the seeds and fruits contain most of the fats. Vegetable fats, which are called *oils,* are found in the seeds. Waxes are also lipids; the leaves of plants are often coated with waxes for protection.

Prostaglandins are a type of lipids found in animal tissues. They affect such important systems as digestion, reproduction, and respiration in humans. Body fat is also a type of lipid. It is stored in special cells called *adipose cells,* which keep a reserve of energy for the body in the form of fat. For every gram of fat that the body metabolizes, 9 kcal of energy are produced. (Carbohydrates produce only 4 kcal/g.) One of the most important functions of lipids is in the construction of the membrane that surrounds each of our body cells and determines what enters and leaves the cell.

20.5 Classification of Lipids

Learning Goal 4
Fatty acids and fats

There are two main types of lipids: those *with* fatty acids in their molecules and those *without* fatty acids in their molecules. A **fatty acid** is an organic acid composed of a straight chain of carbon atoms connected to a carboxyl group,

$$-\overset{\displaystyle}{\underset{\displaystyle \overset{\|}{O}}{C}}-OH$$

which is found at the end of the molecule.

$$
\begin{array}{c}
CH_2OH \\
| \\
CHOH \\
| \\
CH_2OH
\end{array}
\quad + \quad 3C_{15}H_{31}-\overset{\overset{\displaystyle }{}}{\underset{\underset{\displaystyle O}{\|}}{C}}-OH \longrightarrow
$$

Glycerol
(an alcohol)

Palmitic acid
(a fatty acid)

$$
\begin{array}{c}
CH_2-O-\overset{\overset{\displaystyle }{}}{\underset{\underset{\displaystyle O}{\|}}{C}}-C_{15}H_{31} \\
| \\
CH-O-\overset{\overset{\displaystyle }{}}{\underset{\underset{\displaystyle O}{\|}}{C}}-C_{15}H_{31} \\
| \\
CH_2-O-\overset{\overset{\displaystyle }{}}{\underset{\underset{\displaystyle O}{\|}}{C}}-C_{15}H_{31}
\end{array}
$$

Tripalmitin
(a fat)

Figure 20.5

Fats are esters of fatty acids and glycerol.

$$
\begin{array}{c}
CH_2OH \\
| \\
CHOH \\
| \\
CH_2OH
\end{array}
\quad or \quad
\begin{array}{c}
H \\
| \\
H-C-OH \\
| \\
H-C-OH \\
| \\
H-C-OH \\
| \\
H
\end{array}
$$

Figure 20.6

The structure of the alcohol glycerol

· · · · · · · · · · · · · ·

Learning Goal 5

Waxes, steroids, and terpenes

Fats are esters of a fatty acid and the alcohol *glycerol* (Figure 20.5). The glycerol molecule contains three —OH groups (Figure 20.6), and one, two, or all three of these hydroxyl groups may undergo esterification. Thus a fat may contain one, two, or three fatty-acid chains bonded to a molecule of glycerol. Fats are called *monoglycerides*, *diglycerides*, or *triglycerides*, depending on whether they contain one, two, or three fatty-acid chains, respectively. Figure 20.7 shows the general structural formula for a triglyceride.

Waxes are formed when fatty acids and high-molecular-weight alcohols other than glycerol undergo esterification. They have a variety of uses. *Carnauba wax* is applied to cars as a protective coating. *Lanolin* is another wax frequently found as an ingredient in many cosmetic preparations. It is used to coat the skin and prevent loss of water by evaporation. It occurs naturally in wool.

Steroids are lipids that *do not* contain fatty acids in their molecules. Instead, they are composed of an interlocking structure made of four rings. Three are cyclohexane rings, and the fourth is a cyclopentane ring (Figure 20.8). These compounds are extremely important to humans. The male

Figure 20.7

The general formula for a fat (a triglyceride)

(R_1, R_2, and R_3 are fatty acid chains)

Figure 20.8
The general structure of a steroid

hormone *testosterone* and the female hormones *estrogen* and *progesterone* are steroids. Testosterone is the hormone that stimulates the development of such male characteristics as muscle structure and facial hair. Similarly, estrogen leads to the development of the female structural characteristics, and progesterone plays an important role during pregnancy. Another important steroid is vitamin D, which we'll cover in more detail later.

Terpenes are also lipids that *do not* contain fatty acids in their molecules. They are made up of molecules of *isoprene*, which link together to form large molecules (Figure 20.9). Vitamins A, E, and K are terpenes. These vitamins, along with vitamin D, are called the *fat-soluble vitamins* because they can be stored and transported in the body.

$$H-\underset{\underset{H}{|}}{C}=\underset{\underset{}{\overset{\overset{CH_3}{|}}{C}}}{}-\underset{\underset{H}{|}}{C}=\underset{\underset{H}{|}}{C}-H$$

or

$$CH_2=\underset{\overset{\overset{CH_3}{|}}{}}{C}-CH=CH_2$$

Figure 20.9
A molecule of isoprene

Example 20.3

What are the two main types of lipids?

Solution

The two main types of lipids are those *with* fatty acids in the molecule and those *without* fatty acids in the molecule.

Practice Exercise 20.3   Give an example of a lipid that contains fatty acids and an example of one that does not contain fatty acids.

Example 20.4

Give two examples of each of the following classes of lipids:
(*a*) fats,   (*b*) waxes,   (*c*) steroids,   (*d*) terpenes.

Solution

(**a**) Fats: monoglycerides, diglycerides, triglycerides
(**b**) Waxes: carnauba wax, lanolin
(**c**) Steroids: testosterone, estrogen, progesterone, vitamin D
(**d**) Terpenes: vitamins A, E, and K

Practice Exercise 20.4   Draw the structure of a molecule of isoprene.

20.6 Proteins

Learning Goal 6
Proteins

Proteins perform a great variety of functions in our bodies. They protect us against disease, help our blood clot, and form the membranes, muscles, and connective tissues that build our bodies.

Proteins are very large molecules made up of carbon, oxygen, hydrogen, nitrogen, and sometimes sulfur, iodine, iron, or phosphorus. They

Figure 20.10

The general structure of an amino acid

are formed when molecules called *α-amino acids* link together in special arrangements (peptide linkages).

An **α-amino acid** is an organic acid that contains an amino group, $-NH_2$, on the carbon adjacent to the carboxyl group,

$$-\underset{\underset{O}{\|}}{C}-OH$$

Learning Goal 7
Amino acids

Learning Goal 8
Essential amino acids

which can occur only at an end of the molecule (Figure 20.10). Each different amino acid contains a unique R group. There are 20 amino acids that can join to form proteins (Table 20.1). The body can synthesize some amino acids but not others, so we must include those that it cannot synthesize in the foods we eat. The eight amino acids that our bodies cannot synthesize are called the **essential amino acids** (Table 20.2). Foods with all the essential amino acids are said to contain *complete proteins;* these include eggs, meat, milk and fish. Foods that do not contain all the essential amino acids are said to be *incomplete;* such foods include nuts and cereals.

The bonds that hold amino acids together to form proteins are called **peptide bonds.** Because an amino acid molecule has a carboxyl group at

Table 20.1
The 20 Amino Acids Commonly Found in Proteins

Those containing nonpolar R groups		Those containing polar R groups	
Alanine (Ala)	$H_2N-CH-\underset{\underset{O}{\|}}{\overset{\overset{CH_3}{\|}}{C}}-OH$	Glycine (Gly)	$H_2N-CH-\underset{\underset{O}{\|}}{\overset{\overset{H}{\|}}{C}}-OH$
Valine (Val)	$H_2N-CH-\underset{\underset{O}{\|}}{\overset{\overset{\overset{CH_3 \diagdown \diagup CH_3}{CH}}{\|}}{C}}-OH$	Serine (Ser)	$H_2N-CH-\underset{\underset{O}{\|}}{\overset{\overset{CH_2OH}{\|}}{C}}-OH$

Table 20.1
The 20 Amino Acids Commonly Found in Proteins (*continued*)

Those containing nonpolar R groups		Those containing polar R groups	
Leucine (Leu)	CH$_2$—CH(CH$_3$)$_2$ / H$_2$N—CH—C—OH ‖ O	Threonine (Thr)	CH$_3$ / CH—OH / H$_2$N—CH—C—OH ‖ O
Isoleucine (Ile)	CH$_3$ / CH$_2$ / CH$_3$—CH / H$_2$N—CH—C—OH ‖ O	Tyrosine (Tyr)	OH—(ring)—CH$_2$ / H$_2$N—CH—C—OH ‖ O
Methionine (Met)	CH$_2$—CH$_2$—S—CH$_3$ / H$_2$N—CH—C—OH ‖ O	Cysteine (Cys)	CH$_2$SH / H$_2$N—CH—C—OH ‖ O
Phenylalanine (Phe)	(ring)—CH$_2$ / H$_2$N—CH—C—OH ‖ O	Asparagine (Asn)	O ‖ CH$_2$—C—NH$_2$ / H$_2$N—CH—C—OH ‖ O
Proline (Pro)	CH$_2$ / CH$_2$ CH$_2$ / HN—CH—C—OH ‖ O	Glutamine (Gln)	O ‖ CH$_2$—CH$_2$—C—NH$_2$ / H$_2$N—CH—C—OH ‖ O
Tryptophan (Trp)	(indole ring)—CH$_2$ / H$_2$N—CH—C—OH ‖ O		

Table 20.1
The 20 Amino Acids Commonly Found in Proteins (*continued*)

*Those containing
ionic R groups*

Aspartic acid (Asp)

$$CH_2-\overset{\overset{\displaystyle O}{\|}}{C}-O^-$$

$$H_2N-CH-\underset{\underset{\displaystyle O}{\|}}{C}-OH$$

Arginine (Arg)

$$CH_2-CH_2-CH_2-NH-\overset{\overset{\displaystyle NH_2}{|}}{C}=NH_2{}^+$$

$$H_2N-CH-\underset{\underset{\displaystyle O}{\|}}{C}-OH$$

Glutamic acid (Glu)

$$CH_2-CH_2-\overset{\overset{\displaystyle O}{\|}}{C}-O^-$$

$$H_2N-CH-\underset{\underset{\displaystyle O}{\|}}{C}-OH$$

Lysine (Lys)

$$CH_2-CH_2-CH_2-CH_2-NH_3{}^+$$

$$H_2N-CH-\underset{\underset{\displaystyle O}{\|}}{C}-OH$$

Histidine (His)

$$H_2N-CH-\underset{\underset{\displaystyle O}{\|}}{C}-OH$$

Table 20.2
Essential Amino Acids

Essential amino acid	Daily requirement, grams/day (for a 150-pound adult)
Isoleucine	0.84
Leucine	1.12
Lysine	0.84
Methionine[a]	0.70
Phenylalanine[b]	1.12
Threonine	0.56
Tryptophan	0.21
Valine	0.98

[a] Includes cystine.

[b] Includes tyrosine.

SOURCE: The National Research Council—National Academy of Sciences, 1974.

$$H_2N-\underset{\underset{H}{|}}{\overset{\overset{R_1}{|}}{C}}-\underset{\underset{O}{\parallel}}{C}-\boxed{OH} + \underset{\boxed{H}}{\overset{\overset{H}{|}}{N}}-\underset{\underset{H}{|}}{\overset{\overset{R_2}{|}}{C}}-\underset{\underset{O}{\parallel}}{C}-OH$$

$$\longrightarrow H_2N-\underset{\underset{H}{|}}{\overset{\overset{R_1}{|}}{C}}-\underset{\underset{O}{\parallel}}{C}-\underset{\underset{H}{|}}{N}-\underset{\underset{H}{|}}{\overset{\overset{R_2}{|}}{C}}-\underset{\underset{O}{\parallel}}{C}-OH + H_2O$$

Figure 20.11
When two amino acids combine to form a peptide bond, one molecule of water is eliminated in the reaction.

one end and an amino group at the other end, the carboxyl group of each amino acid bonds to the amino group of the next (Figure 20.11). A molecule of water is eliminated in each bonding reaction.

Long chains of amino acids can be connected in this way and form extremely large protein molecules. And each different combination of amino acids constitutes a different protein.

We refer to the sequence, or order, in which the amino acids make up a protein as the protein's **primary structure.** Each protein also has a special geometrical arrangement in space known as its **secondary structure.** One type of secondary structure may take the shape of a coil called a *helix* (Figure 20.12), although other arrangements are possible. A protein also has a **tertiary structure,** which is the way the secondary structure folds back on itself. For example, imagine that a protein molecule has the shape of a coil, like a Slinky® toy, as its secondary structure. Now imagine the Slinky as folding back over on itself and sort of curling up, and you will have an idea of what a tertiary structure is like (Figure 20.13). All three structures are important to the functioning of each protein. In other words, if the amino acids are not in the proper order, if the shape of a protein molecule is altered, or if the manner in which it is folded is changed, the protein will not function as it should.

Learning Goal 9
Structure of proteins

Figure 20.12
The secondary structure of a protein can take the form of a type of coil called a helix.

COOH

H_2N

Figure 20.13
The folding pattern of the secondary structure is the protein's tertiary structure.

Proteins are the main constituents of all the cells in the body. Protein molecules are needed to build new cells, to maintain existing cells, and to replace old, worn-out cells. The chemicals that take part in body reactions and those that help control the rates at which body reactions occur are called enzymes and hormones and are made of protein molecules. In addition, the all-important molecules that carry our genetic information are made of proteins.

Example 20.5

What are proteins composed of?

Solution

Proteins are composed of amino acids linked together by peptide bonds.

Practice Exercise 20.5   How many different amino acids have been found to compose proteins?

Example 20.6

What makes one amino acid different from another amino acid?

Solution

Each amino acid has a different R group attached to it, and this is what makes one amino acid different from another.

Practice Exercise 20.6   How many essential amino acids are there? Name them.

Example 20.7

What is a complete protein food? What is an incomplete protein food?

Solution

A complete protein food is one in which all eight essential amino acids are present. An incomplete protein food is one that lacks one or more of the eight essential amino acids.

Practice Exercise 20.7   Name two foods that are sources of incomplete protein.

Example 20.8

Distinguish between the primary, secondary, and tertiary structures of a protein.

Solution

The *primary* structure is the sequence or arrangement of amino acids in the protein chain. The way in which the primary structure is oriented in space constitutes the *secondary* structure. The *tertiary* structure is the way in which the secondary structure folds back on itself.

Practice Exercise 20.8   What are some of the functions of proteins?

20.7 Vitamins and Minerals

Learning Goal 10
Vitamins

To sustain life, we must include generous amounts of carbohydrates, lipids, and proteins in our diets. Very small amounts of substances called **vitamins** are also necessary to health. Vitamins must be obtained from the foods we eat because they are not synthesized by the body. Most vitamins serve as catalysts in the body. They take part in chemical reactions, but they are not changed chemically when the reactions are completed.

Vitamins are either **water-soluble** or **fat-soluble.** The water-soluble vitamins are, in general, polar molecules. The fat-soluble vitamins are soluble in lipids and are generally absorbed from the intestines along with lipids that have been digested by the body. Some of the fat-soluble vitamins can be stored in the body for various amounts of time. When a vitamin is missing from the diet, or when sufficient quantitites of a vitamin are not taken in, a *deficiency disease* results. Table 20.3 lists the sources, daily requirements, and functions of the various vitamins and the deficiency diseases that result when each is lacking in the diet.

Learning Goal 11
Minerals

Minerals are part of the system that helps control the rates of the chemical reactions that sustain life. Inorganic substances such as sodium, calcium, chlorine, copper, magnesium, iodine, and zinc are minerals that the body needs. The first three—sodium, calcium, and chlorine—are minerals we need in large amounts. They are called *macronutrients*. Copper, magnesium, iodine, and zinc are minerals we need in small amounts; they are called *micronutrients*.

A lack of minerals in the diet can lead to certain diseases. For example, insufficient iron intake causes *anemia*, a blood disorder. A lack of sufficient iodine causes *goiter*, which is a disease of the thyroid gland.

Table 20.3
Some Common Vitamins

Name	Minimum daily adult requirement	Function of vitamin	Deficiency disease or symptoms	Source of vitamin in diet
Fat-soluble				
Vitamin A (retinol and de-hydroretinol)	5000 I.U.* (1.5 mg)	Helps to form visual pigments and maintains the normal epithelial structure	Night blindness	Fish-liver oils, liver, egg yolk, green leafy or yellow vegetables
Vitamin D (ergocalciforal and cholecalciforal)	400 I.U.	Needed for formation of good bones and teeth	Rickets, which results in defective bone formation	Fish-liver oil, butter, egg yolk, along with ultraviolet light (that is, sunlight)
Vitamin E (tocopherol)	30 mg as synthetic *dl-α*-tocopherol acetate	Stabilizes biological membranes and acts as an antioxidant	Red blood cells with a greater tendency for hemolysis	Vegetable oil, wheat germ, leafy vegetables
Vitamin K (K_2-phylloquinone)	About 0.03 mg per kg of body weight	Needed for formation of prothrombin and for normal coagulation of blood	Hemorrhage	Leafy vegetables, vegetable oils; is also produced by intestinal flora after the fourth day of life
Water-soluble				
Vitamin C (ascorbic acid)	60 mg	Needed for maintenance of connective tissues, vascular function, tissue respiration, and wound healing	Scurvy; loose teeth and bleeding gums	Citrus fruits, tomatoes, potatoes
Folic acid (folacin)	0.1 to 0.5 mg, depending on source	Needed for synthesis of purines and pyrimidines and for maturation of red blood cells	Anemia	Fresh leafy green vegetables, fruits, organ meats, liver, dried yeast
Niacin (nicotinic acid)	15–20 mg	Acts as a coenzyme in hydrogen transport, also used in metabolism of carbohydrates and tryptophan	Pellagra	Dried yeast, liver, meat, fish, whole grains

*I.U. stands for international unit. One international unit equals 0.3 μg of retinol.

Table 20.3
Some Common Vitamins (*continued*)

Name	Minimum daily adult requirement	Function of vitamin	Deficiency disease or symptoms	Source of vitamin in diet
Vitamin B_1 (thiamine)	1–1.5 mg	Needed for metabolism of carbohydrates and for function of nerve cells	Beriberi (acute cardiac symptoms and heart failure, also weight loss and nerve inflammation)	Whole grains, pork, liver, nuts, enriched cereal products
Vitamin B_2 (riboflavin)	1.0–1.7 mg	Coenzyme in energy transport; also involved in protein metabolism	Visual problems; skin fissures	Milk, cheese, liver, meat, eggs
Vitamin B_6 group (pyridoxine, pyridoxal, pyridoxamine)	2 mg	Needed for cellular function and for metabolism of certain amino and fatty acids	Convulsions (in infants); skin disorders (in adults)	Dried yeast, liver, organ meats, fish, whole-grain cereals
Vitamin B_{12} (cyanocobalamine)	5 μg	Needed for DNA synthesis related to folate coenzymes and for maturation of red blood cells	Pernicious anemia and certain psychiatric syndromes	Liver, beef, pork, eggs, milk and milk products
Vitamin H (biotin)	0.15–30 mg	Needed for protein synthesis, transamination, and CO_2 fixation	Skin disorders	Liver, dried peas and lima beans, egg white; is also produced by bacteria in alimentary canal
Pantothenic acid	10 mg	Forms part of coenzyme A	Fatigue, malaise, neuromotor and digestive disorders	Dried yeast, liver, eggs

Example 20.9

What are the two types of vitamins?

Solution
The two types of vitamins are water-soluble vitamins and fat-soluble vitamins.

Practice Exercise 20.9   Which vitamins are fat-soluble?

Example 20.10

What is a deficiency disease?

Solution

A deficiency disease is a disease caused by the lack of a particular vitamin in the diet.

Practice Exercise 20.10   Which vitamins are water-soluble?

Example 20.11

What are minerals, in the context of biochemistry?

Solution

Minerals are inorganic elements that the body needs because they form part of the system that helps control the rates of life-sustaining reactions.

Practice Exercise 20.11   Distinguish between macronutrients and micronutrients.

Summary

Large food molecules are broken down into smaller molecules via digestion. These smaller molecules are then transported to body cells, where they become the reactants in a variety of chemical processes. In some of the reactions, the smaller molecules are anabolized into larger molecules for the formation of body tissues. In other reactions, they are catabolized, or broken down, to supply energy to the body. Both types of reactions are part of the process called metabolism.

Carbohydrate molecules are organic compounds that are composed of the elements carbon, hydrogen, and oxygen. The three major categories of carbohydrates, on the basis of the number of sugar groups found in the molecule, are monosaccharides, disaccharides, and polysaccharides. A disaccharide is composed of two monosaccharide molecules, and a polysaccharide is composed of three or more monosaccharide molecules.

Lipids include fats and other compounds with physical properties similar to those of fats. For example, oils (vegetable fats), waxes, steroids, and terpenes are lipids. Lipids are classified in two groups: those with fatty acids in their molecules and those without fatty acids in their molecules. A fatty acid is an organic acid composed of a straight chain of carbon atoms with a carboxyl group at one end. And a fat is an ester of a fatty acid and the alcohol glycerol. Body fat, when it is metabolized, provides more energy per gram than carbohydrate.

Proteins are very large molecules made up of carbon, oxygen, hydrogen, nitrogen and sometimes sulfur, iodine, iron, or phosphorus. They are formed when molecules called amino acids are linked together by peptide bonds. An amino acid is an organic acid that contains an amino group at one end of its molecule and a carboxyl group at the other end. A peptide bond forms when an amino group in one amino acid molecule combines with a carboxyl group in another amino acid molecule; long chains of amino acids can develop in this way. Each protein has a specific sequence of amino acids and a definite shape.

There are eight essential acids that the body cannot synthesize. Foods containing all these essential amino acids are called complete protein foods, and foods lacking one or more essential amino acids are called incomplete protein foods.

The body requires very small amounts of vitamins and minerals. Vitamins serve as catalysts. They take part in chemical reactions in the body, but they are not changed chemically when the reactions are completed. There are two types of vitamins: those that are water-soluble and those that are fat-soluble. Minerals are inorganic substances that help control reaction rates within the body. Lack of any vitamin or mineral can lead to disease.

. . . .

Career Sketch

Physician Assistant

A physician assistant provides patient services under the direct supervision of a doctor of medicine or osteopathy who is responsible for the quality of the services. Some of the duties of the physician assistant are to elicit detailed patient histories and do complete physical examinations, reach tentative diagnoses and order appropriate laboratory tests, counsel patient and family on health care and disease, suture minor wounds, assist in surgery, apply and remove casts and traction apparatus, and perform other therapeutic procedures under the supervision of a doctor of medicine or osteopathy. The physician assistant may have special training in a particular medical specialty and may be designated a child health associate, gynecologist assistant, orthopedic physician assistant, surgeon's assistant, or urologic physician assistant.

. . . .

Self-Test Exercises:

Learning Goal 1: Metabolism, Anabolism, and Catabolism

1. What is meant by the terms *metabolism, anabolism,* and *catabolism?*

2. How do anabolism and catabolism differ?

Learning Goal 2: Carbohydrates

3. What is a carbohydrate?

4. What are the main sources of carbohydrates?

5. List three sources of carbohydrates.

6. What is a starch?

7. How are carbohydrates classified?

8. What is meant by the term *saccharide?*

9. Distinguish between monosaccharides, disaccharides, and polysaccharides.

10. Give examples of monosaccharides, disaccharides, and polysaccharides.

11. Write the structural formulas for glucose, fructose, and galactose.

12. Where is glucose found in the body?

13. Write structures for sucrose, lactose, and maltose.

14. Name the three carbohydrates that have the formula $C_{12}H_{22}O_{11}$.

15. Name two polysaccharides, and state their uses.

16. What is the name given to the carbohydrate that is stored in the liver and muscles?

Learning Goal 3: Lipids

17. What is a lipid? Give an example of a lipid.

18. Where are the lipids commonly stored in plants?

19. Give three examples of lipids, and state the use of each.

20. What are prostaglandins?

21. How are lipids classified?

22. Using the lipid classification scheme, classify fats, waxes, steroids, and terpenes.

Learning Goal 4: Fatty Acids and Fats

23. Define, and give the general formula for, a fatty acid.

24. Write the structural formula for a fatty acid.

25. Define, and give the general formula for, a fat.

26. How are fats formed?

27. Distinguish among monoglycerides, diglycerides, and triglycerides.

28. What do we call the compound that is formed when all three of the glycerol —OH groups are esterified?

Learning Goal 5: Waxes, Steroids, and Terpenes

29. What is a wax? Give two examples of waxes.

30. What is carnauba wax?

31. Draw the steroid ring structure.

*32. Find out what role progesterone plays in pregnancy.

33. List three important steroids and discuss the roles they play in the human body.

34. What are the functions of vitamin D in the body?

35. Write the structure for an isoprene unit.

36. Do terpenes contain fatty acids in their molecules?

37. Name the vitamins that are terpenes.

38. Are all the fat-soluble vitamins terpenes?

Learning Goal 6: Proteins

39. Define the term *protein*, and give an example of a protein molecule.

40. What are the basic building blocks of proteins?

41. List four important functions of proteins.

42. What elements compose proteins?

Learning Goal 7: Amino Acids

43. Write the general formula for an amino acid.

44. What two functional groups are present in an amino acid?

45. Give five examples of amino acids, and write their structures.

46. What makes one amino acid different from another?

Learning Goal 8: Essential Amino Acids

47. What is meant by essential amino acids?

48. How many essential amino acids are there?

49. Explain the difference between a complete protein and an incomplete protein.

50. True or False: Essential amino acids are synthesized by the body.

Learning Goal 9: Structure of Proteins

51. Explain how a peptide bond is formed.

52. Complete this sentence: The bonds that hold amino acids together to form proteins are called _____ bonds.

53. Distinguish between the primary, secondary, and tertiary structures of a protein.

54. Complete this sentence: The sequence of amino acids that make up the protein is called the protein's _____ structure.

Learning Goal 10: Vitamins

55. Define the term *vitamin*, and give an example of a vitamin.

56. What is a deficiency disease?

57. List four vitamins, along with their sources. State what diseases result when these vitamins are lacking in the diet.

58. Read the labels on a variety of foods, and find the list of vitamins that they contain. What conclusion can you come to about the vitamin content of these foods?

59. Explain what is meant by water-soluble vitamins and fat-soluble vitamins.

60. Name the fat-soluble vitamins.

Learning Goal 11: Minerals

61. Define the term *mineral,* and give an example of a mineral.

62. Name the minerals that are macronutrients.

63. What is the importance of minerals to the body?

64. What is iron-deficiency anemia?

Extra Exercises

65. For each of the following four symptoms, suggest a vitamin deficiency that might be responsible and a diet that would correct the problem: (*a*) night blindness; (*b*) anemia; (*c*) internal bleeding, bleeding of gums, defective teeth; (*d*) dermatitis.

66. Carbohydrates and proteins are both high-molecular-weight polymers. What are the basic structural differences between them?

67. What are the functions of proteins?

68. Name four micronutrients important for proper functioning of the human body.

69. Hydrolysis is a reaction with water. In each of the following cases, list the products of the indicated hydrolysis reaction.
(*a*)Proteins + water $\longrightarrow$
(*b*)Carbohydrates + water $\longrightarrow$
(*c*)Triglycerides + water $\longrightarrow$

70. What is the origin of the word *carbohydrate?*

71. Name the three classifications of foodstuffs, and give an example of each.

*72. Sweetness is a physical property of all sugars. Explain why a teaspoon of sugar (sucrose) in cold water is not as sweet-tasting as the same teaspoon of sugar (sucrose) in hot water.

*73. Why does a strong alkali (strong base) aid in unstopping a sink drain clogged by grease?

74. Distinguish between the primary structure and the secondary structure of a protein.

Cumulative Review/Chapters 17–20

1. Distinguish between alpha, beta, and gamma radiation.

2. Complete the following nuclear equation.

$$^{10}_{5}B + ^{4}_{2}He \longrightarrow ^{1}_{0}n + ?$$

3. What isotope will be formed when $^{214}_{83}Bi$ emits beta radiation from its nucleus?

4. An isotope of bismuth called bismuth-210 has a half-life of 5.00 days and emits beta particles.
(a) Write the nuclear equation for this reaction.
(b) How many grams of bismuth-210 remain after a 25.0-g sample of bismuth-210 decays over a period of 20 days?

5. If 100 mg of radium-226, which has a half-life of 1,620 years, begins to decay today, how many milligrams of the isotope will be left after 3,240 years?

6. At the University of California at Berkeley, Albert Ghiorso and his research team bombarded californium-246, ^{246}Cf, with oxygen-18, ^{18}O, to form element 106, whose mass number is 263. Four neutrons were also formed. Write the nuclear equation for this reaction.

7. Element 106 has a half-life of 0.9 seconds and decays to form $^{259}_{104}Unq$ and an alpha particle. Write the nuclear equation for this reaction.

8. Element 109 is formed when bismuth-209 is bombarded with iron-58. A neutron is released during the reaction. Write the nuclear equation for this reaction.

9. Element 107 is formed when element-109 decays, releasing an alpha particle. Write the nuclear equation for this reaction.

10. True or False: The process by which electrical energy is used to bring about a chemical change is called electrolysis.

11. When a lead storage battery discharges, the following reactions take place:

$$Pb + SO_4^{2-} \longrightarrow PbSO_4 \quad \text{(not balanced)}$$
$$PbO_2 + SO_4^{2-} + H^{1+} \longrightarrow PbSO_4 + H_2O$$
$$\text{(not balanced)}$$

Complete and balance the half-cell equations for these reactions, indicating where electrons are supplied.

12. How is an electrolytic cell different from a voltaic cell?

13. True or False: In a redox reaction, the oxidizing agent is oxidized and the reducing agent is reduced.

14. How is a gaseous air pollutant different from a particulate?

15. How is acid rain formed in the environment?

16. Why is PVC potentially hazardous?

17. Name three ways to safely dispose of hazardous waste.

18. How are PCB's dangerous to the environment?

19. Distinguish between smog and photochemical smog.

20. True or False: All gasoline that enters an automobile engine is burned completely.

21. Write the Lewis electron-dot notation for carbon, indicating how many electrons are available for bonding.

22. True or False: Two tetrahedral carbon atoms sharing a common side form a double bond.

23. Write the names and formulas for the first five straight-chain members of the alkane, alkene, and alkyne series.

24. Write the structure for each of the following.
(a) 2-hexene
(b) 4,5,5-trimethyl-2-hexene

25. Complete this sentence: _____ are compounds that have the same molecular formula but different structural formulas.

26. Write the names and draw the structural formulas for straight-chain alkyl groups that have the formula C_nH_{2n+1} and contain from six to ten carbon atoms.

27. How is a cyclic hydrocarbon different from an aromatic hydrocarbon?

28. Define the term *homologous series*. Choose one series and list three of its members.

29. Name the cyclic hydrocarbon represented as

30. How many electron pairs are associated with a covalently bonded carbon atom?

31. Write the structural formulas for all of the isomers that have the molecular formula C_4H_7I.

32. What is a functional group?

33. Write the structures for each of the following.
(*a*) 2-hexanol
(*b*) 3-octanol

34. Which of the following alcohols is nonpolar?

$$CH_3CH_2OH \quad \text{or} \quad CH_3-(CH_2)_{10}-CH_2OH$$

35. Give the common name for

$$CH_3CH_2-O-(CH_2)_2-CH_3$$

36. Give the IUPAC or the common name for

37. Write the structure of 2-octanone.

38. Decide whether each of the following is a primary, a secondary, or a tertiary amine.
(*a*) $CH_3-CH_2-N-CH_2-CH_3$
              |
              H

(*b*)

(*c*)

39. Given the products of the following esterification reaction, name the reactants.

? + ? $\longrightarrow$

$$CH_3-CH_2-\overset{\displaystyle O}{\underset{\displaystyle \|}{C}}-O-CH_2-CH_2-CH_2-CH_3$$

$+\ H_2O$

40. Write the structure of the ester ethylethanoate.

41. Write the structure of the aldehyde butanal.

42. Write the structure of methyl propyl ketone.

43. Distinguish among the terms *metabolism*, *anabolism*, and *catabolism*.

44. How are monosaccharides, disaccharides, and polysaccharides different from each other?

45. Explain the classification scheme for lipids.

46. Explain how fats are formed.

47. How are waxes, steroids, and terpenes different from each other?

48. Write the general formula for an amino acid.

49. What is an essential amino acid?

50. What is the difference between a fat-soluble vitamin and a water-soluble vitamin?

Supplement A
Basic Mathematics for Chemistry

Learning Goals

After you have studied this supplement, you should be able to:

1. Add and subtract algebraically.
2. Multiply and divide fractions.
3. Multiply and divide exponential numbers.
4. Use the factor-unit method in solving problems.
5. Add and subtract decimals.
6. Multiply and divide decimals.
7. Solve simple algebraic equations.
8. Set up and solve proportions.
9. Solve simple density problems.
10. Solve problems involving percentages.
11. Perform mathematical operations using a calculator.

Introduction

Full appreciation of many ideas in chemistry requires some basic mathematics. (There are also concepts in chemistry that require absolutely *no* mathematics.) This supplement is here so that you can review and master the mathematical skills you need to be successful in chemistry. Whenever in doubt, as you work your way through the book, turn to this supplement for a quick solution to your mathematical problem.

You'll find that chemical mathematics is not very difficult, as long as you can perform some basic operations. Work your way through this supplement carefully, and try your hand at all the sample problems. Then do the exercises at the end. Remember that the knowledge you gain here will greatly increase your success in chemistry.

A.1 Adding and Subtracting Algebraically

Learning Goal 1
Adding and subtracting algebraically

In chemistry we deal with both positive and negative numbers, so it is important to know how to add and subtract both. The process is called algebraic addition and subtraction. Here are the rules to follow.

Addition

1. *When the signs are alike*   When the signs of two numbers are the same, add the numbers and keep the same sign.

$$+4 + 2 = +6$$
$$-4 - 2 = -6$$

2. *When the signs are different*   When the signs of two numbers are not the same, subtract the numbers and keep the sign of the larger number.

$$-4 + 2 = -2$$
$$+4 - 2 = +2$$

Subtraction

Change the sign of the *number being subtracted* and follow the rules for addition. [*Note:* Plus signs before positive numbers are usually left out, but minus signs before negative numbers are always written.]

$4 - (-2)$	becomes	$4 + 2 =$ 6
$4 - (+2)$	becomes	$4 - 2 =$ 2
$-4 - (+2)$	becomes	$-4 - 2 = -6$
$-4 - (-2)$	becomes	$-4 + 2 = -2$

A.2 Fractions

A ratio of two numbers is called a **fraction.** Some examples of fractions are $\frac{1}{4}, \frac{1}{2}, \frac{2}{5}, \frac{1}{11}$, and $\frac{2}{1}$. A fraction has two parts: the **numerator,** which is the top number, and the **denominator,** which is the bottom number.

. . . .
Learning Goal 2
Multiplying and dividing
fractions

Here are some basic rules to follow in dealing with fractions.

Multiplication

Multiply numerators; then multiply denominators.

$$\frac{1}{2} \times \frac{3}{4} = \frac{3}{8} \qquad \frac{5}{9} \times \frac{9}{5} = \frac{45}{45} = 1$$

Division

Invert (turn upside down) the fraction to the right of the division sign, and follow the rules for multiplication.

$$\frac{1}{2} \div \frac{1}{4} \qquad \text{becomes} \qquad \frac{1}{2} \times \frac{4}{1} = \frac{4}{2} = 2$$

$$\frac{3}{4} \div \frac{5}{9} \qquad \text{becomes} \qquad \frac{3}{4} \times \frac{9}{5} = \frac{27}{20}$$

A.3 Exponents

Exponents are numbers written to the right of and above another number (called the **base number**). A positive exponent tells you to multiply the base number by itself. The exponent tells you how many times the base number is taken. For example, $(4)^2$ means take the number 4 twice: 4×4. Here 4 is the base number and 2 is the exponent.

. . . .
Learning Goal 3
Multiplying and dividing
exponential numbers

$$(4)^2 \longleftarrow \text{Exponent}$$
$$\uparrow \longrightarrow \text{Base number}$$

Sometimes exponents are negative. For example, we may have

$$(4)^{-2} \qquad \text{which means} \qquad \frac{1}{(4)^2}$$

$$\frac{1}{(4)^{-2}} \qquad \text{which means} \qquad \frac{(4)^2}{1}$$

The following problems will show how exponents are used.

Example A.1

Perform the indicated operation.

(a) $(3)^2$ **(b)** $(4)^3$ **(c)** $(3)^1$ **(d)** $(3)^{-2}$ **(e)** $\dfrac{1}{(2)^{-3}}$

Solution

(a) $(3)^2 = 3 \times 3 = 9$ **(b)** $(4)^3 = 4 \times 4 \times 4 = 64$

(c) $(3)^1 = 3$ **(d)** $(3)^{-2} = \dfrac{1}{(3)^2} = \dfrac{1}{3 \times 3} = \dfrac{1}{9}$

(e) $\dfrac{1}{(2)^{-3}} = \dfrac{(2)^3}{1} = 2 \times 2 \times 2 = 8$

Sometimes it is necessary to *multiply* exponential numbers that have the *same base*. This is done as shown in Example A.2.

Example A.2

Perform the indicated operation.
(a) $(4)^8(4)^2$ **(b)** $(3)^2(3)^{-1}$ **(c)** $(8)^2(8)^1$ **(d)** $(2)^{-2}(2)^{-3}$
(e) $(3)^{-2}(3)^2$

Solution

Keep the same base, and *add the exponents algebraically*.
(a) $(4)^8(4)^2 = (4)^{10}$ **(b)** $(3)^2(3)^{-1} = (3)^1 = 3$
(c) $(8)^2(8)^1 = (8)^3$ **(d)** $(2)^{-2}(2)^{-3} = (2)^{-5}$
(e) $(3)^{-2}(3)^2 = (3)^0 = 1$ (Any number to the zero power is equal to 1.)

Sometimes it is necessary to *divide* exponential numbers that have the *same base*. This is done as shown in Example A.3.

Example A.3

Perform the indicated operation.
(a) $\dfrac{(2)^5}{(2)^2}$ **(b)** $\dfrac{(2)^5}{(2)^{-2}}$ **(c)** $\dfrac{(2)^{-5}}{(2)^2}$ **(d)** $\dfrac{(2)^{-5}}{(2)^{-2}}$

Solution

Keep the same base, and subtract the exponent in the denominator from the exponent in the numerator *algebraically.*

(a) $\dfrac{(2)^5}{(2)^2} = (2)^{5-2} = (2)^3$

(b) $\dfrac{(2)^5}{(2)^{-2}} = (2)^{5-(-2)} = (2)^{5+2} = (2)^7$

(c) $\dfrac{(2)^{-5}}{(2)^2} = (2)^{-5-(+2)} = (2)^{-5-2} = (2)^{-7}$

(d) $\dfrac{(2)^{-5}}{(2)^{-2}} = (2)^{-5-(-2)} = (2)^{-5+2} = (2)^{-3}$

In these examples, only two numbers have been multiplied or divided. However, sometimes there are more than two numbers in the calculation.

Example A.4

Perform the indicated operation.

(a) $(5)^3(5)^3(5)^2$ **(b)** $(a)^5(a)^3(a)^{-4}$ **(c)** $\dfrac{(b)^3(b)^6}{(b)^2}$

(d) $\dfrac{(2)^5(2)^{-6}}{(2)^{-9}}$ **(e)** $\dfrac{(a)^{-3}(a)^4(a)^6}{(a)^{-2}(a)^7}$

Solution

(a) $(5)^3(5)^3(5)^2 = (5)^8$ The exponents were added algebraically.

(b) $(a)^5(a)^3(a)^{-4} = (a)^4$ The exponents were added algebraically.

(c) $\dfrac{(b)^3(b)^6}{(b)^2} = \dfrac{(b)^9}{(b)^2} = (b)^{9-2} = (b)^7$ When you work out a problem one step at a time, it becomes easy.

(d) $\dfrac{(2)^5(2)^{-6}}{(2)^{-9}} = \dfrac{(2)^{-1}}{(2)^{-9}} = (2)^8$ **(e)** $\dfrac{(a)^{-3}(a)^4(a)^6}{(a)^{-2}(a)^7} = \dfrac{(a)^7}{(a)^5} = (a)^2$

A.4 Working with Units

Almost all numbers, in chemistry or anywhere else, are accompanied by units. It is important to be able to work with these units, as well as with the numbers. Study the following examples carefully.

Example A.5

How many feet are there in 36 inches?

Solution

We are going to use what is known as the **factor-unit method** of analysis. Here's how it works. Because there are 12 inches in 1 foot, we can express this relationship mathematically as

$$\frac{12 \text{ inches}}{1 \text{ foot}} \qquad \text{(reads "12 inches per 1 foot")}$$

or as

$$\frac{1 \text{ foot}}{12 \text{ inches}} \qquad \text{(reads "1 foot per 12 inches")}$$

These are called factor units. We set the problem up so that, when we multiply the quantity in question (36 inches) by the appropriate factor unit, we end up with what we want (feet). Here, we have

$$? \text{ feet} = (36 \text{ inches}) \left(\frac{1 \text{ foot}}{12 \text{ inches}} \right) = 3 \text{ feet}$$

Notice how the unit *inches* cancels out in Example A.5 when the numbers are multiplied. You may wonder how you'd know that you must use the factor unit

$$\left(\frac{1 \text{ foot}}{12 \text{ inches}} \right) \qquad \text{and not the factor unit} \qquad \left(\frac{12 \text{ inches}}{1 \text{ foot}} \right)$$

That's easy. If you used the wrong factor unit, the term *inches* wouldn't cancel out. You would arrive at

$$? \text{ feet} = (36 \text{ inches}) \left(\frac{12 \text{ inches}}{1 \text{ foot}} \right) = \frac{432 \text{ (inches)}^2}{1 \text{ foot}}$$

which has no meaning to you. Try some more problems using the factor-unit method.

Example A.6

How many inches are there in 5 feet?

Solution

Set the problem up this way:

$$? \text{ inches} = (5 \text{ feet})\left(\frac{12 \text{ inches}}{1 \text{ foot}}\right) = 60 \text{ inches}$$

Note how the unit *feet* cancels out when the numbers are multiplied with their units.

Example A.7

How many minutes are there in 4 hours? How many seconds are there in 4 hours?

Solution

There are 60 minutes in 1 hour. So the factor unit is

$$\frac{60 \text{ minutes}}{1 \text{ hour}}$$

The problem is solved this way:

$$? \text{ minutes} = (4 \text{ hours})\left(\frac{60 \text{ minutes}}{1 \text{ hour}}\right) = 240 \text{ minutes}$$

There are 60 seconds in 1 minute. Our factor unit for this part of the problem is

$$\frac{60 \text{ seconds}}{1 \text{ minute}}$$

So the problem is solved as follows:

$$? \text{ seconds} = (240 \text{ minutes})\left(\frac{60 \text{ seconds}}{1 \text{ minute}}\right) = 14{,}400 \text{ seconds}$$

Example A.8

How many gallons are there in 60 quarts?

Solution

There are 4 quarts in 1 gallon. Therefore

$$? \text{ gallons} = (60 \text{ quarts})\left(\frac{1 \text{ gallon}}{4 \text{ quarts}}\right) = 15 \text{ gallons}$$

Example A.9

Change 60 miles per hour to feet per second.

Solution

Here you can really see the value of the factor-unit method. However, we have to know that 1 mile equals 5280 feet and that 1 hour equals 3600 seconds. Now, 60 miles per hour means

$$60 \, \frac{\text{miles}}{\text{hour}} \qquad \text{which is the same as} \qquad \frac{60 \text{ miles}}{1 \text{ hour}}$$

So the complete calculation reads

$$? \, \frac{\text{feet}}{\text{seconds}} = \left(\frac{60 \text{ miles}}{1 \text{ hour}} \right) \left(\frac{5,280 \text{ feet}}{1 \text{ mile}} \right) \left(\frac{1 \text{ hour}}{3600 \text{ seconds}} \right)$$

$$= 88 \, \frac{\text{feet}}{\text{second}}$$

A.5 Decimals

• • • • • • • • • • • • • •
Learning Goal 5
Adding and subtracting decimals

Chemistry is such an exact science that the measurements we take usually consist of a string of numbers with a decimal point—for example, 2.7135 grams. When we are working with decimal numbers, we have to keep certain rules in mind.

Addition and Subtraction

In adding or subtracting numbers, it is important to line up the columns correctly. Keep the decimal points stacked up one over the other.

Example A.10

Perform the indicated operation.
(a) 25.8 + 107.09 + 88.004 + 0.011   **(b)** 342.78 − 14.99

Solution
Line up the numbers properly.

```
(a)     25.8        (b)     342.78
       107.09              − 14.99
        88.004             ───────
         0.011              327.79
       ───────
       220.905
```

Multiplication

When we multiply 2.4×1.6, where does the decimal point go? First we write the numbers and perform the multiplication, ignoring the decimal point.

$$
\begin{array}{r}
2.4 \\
\times\ 1.6 \\
\hline
144 \\
24\ \ \\
\hline
384 \\
\end{array}
$$

To find out where the decimal point goes, we count the number of digits to the right of the decimal point in each term. (There is one digit to the right of the decimal point in each term.) Then we add these numbers of digits (one plus one equals two) and count off this new number from the right-hand side of the answer. In our example, we count off two digits from the right, and we place the decimal point between the 3 and the 8. This gives us an answer of 3.84.

Example A.11

Perform the indicated operation.
(a) $(4.25)(5)$ **(b)** $(7.12)(3.64)$ **(c)** $(5.222)(4.11)$

Solution

$$
\textbf{(a)}\quad
\begin{array}{r}
4.25 \\
\times\ 5 \\
\hline
21.25 \\
\end{array}
\qquad
\textbf{(b)}\quad
\begin{array}{r}
7.12 \\
\times\ 3.64 \\
\hline
2848 \\
4272\ \ \\
2136\ \ \ \\
\hline
25.9168 \\
\end{array}
\qquad
\textbf{(c)}\quad
\begin{array}{r}
5.222 \\
\times\ 4.11 \\
\hline
5222 \\
5222\ \ \\
20888\ \ \ \\
\hline
21.46242 \\
\end{array}
$$

Division

Where does the decimal point go when we divide 11 by 0.56? First we write the problem in a more familiar way.

$$0.56\overline{)11}$$

(Here 0.56 is called the **divisor,** and 11 is called the **dividend.**) Our next step is to move the decimal point in the divisor so that all digits are to the left of the decimal point: 0.56 becomes 56. We must also move the decimal point in the dividend the same number of places: 11 becomes 1100.

$$0.56 \overline{)11.00} \quad \text{becomes} \quad 56 \overline{)1100.00}$$

We now perform the division, locating the decimal point in the answer directly above the decimal point in the dividend.

$$
\begin{array}{r}
19.64 \\
56 \overline{)1100.00} \\
\underline{56\text{x xx}} \\
540 \\
\underline{504} \\
360 \\
\underline{336} \\
240 \\
\underline{224} \\
16
\end{array}
$$

Example A.12

Perform the indicated operation.

(a) $\dfrac{9.25}{2.5}$ **(b)** $\dfrac{24.138}{7.45}$

Solution

(a) $2.5 \overline{)9.2\,5}$ **(b)** $7.45 \overline{)24.13\,8}$

$$
\begin{array}{r}
3.7 \\
25 \overline{)92.5} \\
\underline{75\,\text{x}} \\
175 \\
\underline{175} \\
0
\end{array}
$$

$$
\begin{array}{r}
3.24 \\
745 \overline{)2413.80} \\
\underline{2235\,\text{xx}} \\
1788 \\
\underline{1490} \\
2980 \\
\underline{2980} \\
0
\end{array}
$$

A.6 Solving Algebraic Equations

Learning Goal 7
Solving simple algebraic equations

Very often in chemistry, we need to solve algebraic equations ("find the unknown"). So study the following examples carefully.

Example A.13

Solve each of the following equations for the unknowns a, y, and z.
(a) $3a = 9$ **(b)** $4y + 5 = 37$ **(c)** $9z - 3 = 22$

Solution

To solve an algebraic equation, we try to isolate the unknown.

(a) In $3a = 9$, we can isolate the a by dividing both sides of the equation by 3. This operation removes the 3 from the left-hand side of the equation. Remember, what we do to one side of the equation we *must* do to the other side, in order to maintain the equality.

$$\frac{\cancel{3}a}{\cancel{3}} = \frac{9}{3}$$

After simplifying both sides of the equation, we get the equality

$$a = 3$$

(b) In $4y + 5 = 37$, we want to isolate the y. Our first step is to remove the 5 from the left-hand side of the equation. We do this by subtracting 5 from both sides of the equation.

$$4y + 5 - 5 = 37 - 5 \qquad \text{so} \qquad 4y = 32$$

We now divide both sides of the equation by 4.

$$\frac{\cancel{4}y}{\cancel{4}} = \frac{32}{4}$$

After simplifying both sides of the equation, we get the equality

$$y = 8$$

(c) In $9z - 3 = 22$, the first step is to remove the 3 from the left-hand side of the equation. We do this by adding 3 to both sides of the equation.

$$9z - 3 + 3 = 22 + 3 \qquad \text{so} \qquad 9z = 25$$

Now divide both sides of the equation by 9.

$$\frac{\cancel{9}z}{\cancel{9}} = \frac{25}{9}$$

After simplifying both sides of the equation, we have the equality

$$z = \frac{25}{9}$$

Example A.14

Solve each of the following equations for the unknown.

(a) $\dfrac{2a}{3} = \dfrac{12}{9}$ **(b)** $5y + 3 = 2y - 42$

Solution

(a) In the equation

$$\frac{2a}{3} = \frac{12}{9}$$

we can remove the 3 on the left-hand side by multiplying both sides of the equation by 3.

$$(3)\left(\frac{2a}{3}\right) = \left(\frac{12}{9}\right)(3) \qquad 2a = \frac{36}{9} \qquad 2a = 4$$

Now divide both sides of the equation by 2.

$$\frac{2a}{2} = \frac{4}{2}$$

After simplifying both sides of the equation, we have the equality

$$a = 2$$

(b) In the equation

$$5y + 3 = 2y - 42$$

we first gather all the terms containing the unknown on one side. Place all y terms on the left-hand side of the equation by subtracting $2y$ from each side of the equation.

$$5y + 3 - 2y = 2y - 42 - 2y \qquad \text{so} \qquad 3y + 3 = -42$$

Then subtract 3 from each side of the equation.

$$3y + 3 - 3 = -42 - 3 \qquad \text{so} \qquad 3y = -45$$

Now divide each side of the equation by 3.

$$\frac{3y}{3} = \frac{-45}{3}$$

After simplifying both sides of the equation, we have the equality

$$y = -15$$

A.7 Ratios and Proportions

Learning Goal 8
Solving proportions

There are four wheels on a car. This relationship can be expressed mathematically in the following ways:

$$4 \text{ wheels} : 1 \text{ car} \qquad \text{or} \qquad \frac{4 \text{ wheels}}{1 \text{ car}}$$

Both statements express the fact that there are four wheels on one car. Such statements are known as **ratios** because they reveal the numerical relationship between different things. A ratio is just like a factor unit, and it can be used as an alternative to the factor-unit method to solve many types of problems. But to use ratios in problem solving, you must understand the concept of proportions. A **proportion** is *an equality between two ratios*, as in

$$\frac{1}{2} = \frac{5}{10} \qquad \text{or} \qquad 1:2 = 5:10$$

When the second expression is used, the symbol : is often read "is to" and the symbol = is read "as.") In a problem involving a proportion, you are usually given three of the four numbers and are asked to determine the missing number. Example A.15 shows how this works.

Example A.15

If there are 4 wheels on 1 car, how many wheels are there on 5 cars?

Solution

The proportion is set up in the following way:

$$\frac{4 \text{ wheels}}{1 \text{ car}} = \frac{y \text{ wheels}}{5 \text{ cars}}$$

To solve for y, multiply each side of the equation by 5 cars.

$$(5 \text{ cars})\left(\frac{4 \text{ wheels}}{1 \text{ car}}\right) = \left(\frac{y \text{ wheels}}{5 \text{ cars}}\right)(5 \text{ cars})$$

$$20 \text{ wheels} = y \text{ wheels}$$

In other words, $y = 20$ wheels, so there are 20 wheels on 5 cars.

A.8 Solving Word Equations

Learning Goal 9
Solving simple density
problems

Many of the mathematical equations we use in chemistry are word equations, such as

$$\text{Density} = \frac{\text{mass}}{\text{volume}} \qquad \text{sometimes abbreviated as} \qquad D = \frac{m}{V}$$

As written, the equation is set up so that we can solve for the density of a substance if we are given its mass and volume. In some problems, however, we are given the density and volume of a substance and are asked to solve for its mass. We do so as shown in Example A.16.

Example A.16

Given the following information, solve for the mass m.

$$D = 0.8 \; \frac{g}{mL} \qquad V = 30 \text{ mL}$$

Solution

First write the formula.

$$D = \frac{m}{V}$$

Then isolate the m by multiplying both sides of the equation by V.

$$(D)(V) = \left(\frac{m}{\cancel{V}}\right)(\cancel{V}) \qquad \text{so} \qquad DV = m$$

Now substitute the numbers given in the problem, and solve for the mass in grams.

$$\left(0.8 \; \frac{g}{\cancel{mL}}\right)(30 \; \cancel{mL}) = m \qquad \text{so} \qquad 24 \text{ g} = m$$

In some problems we are given the density and mass of a substance and asked to solve for its volume.

Example A.17

Given the following information, solve for the volume V.

$$D = 2 \; \frac{g}{mL} \qquad m = 20 \text{ g}$$

Solution

Write the density formula in its original form.

$$D = \frac{m}{V}$$

We must get the V into the numerator by itself. This can be done in two steps:

1. Multiply both sides of the equation by V.

$$(D)(V) = \left(\frac{m}{\cancel{V}}\right)(\cancel{V}) \qquad \text{so} \qquad DV = m$$

2. Divide both sides of the equation by D.

$$\frac{\cancel{D}V}{\cancel{D}} = \frac{m}{D} \qquad \text{so} \qquad V = \frac{m}{D}$$

Now substitute the numbers given in the problem, and solve for the volume in milliliters.

$$V = \frac{m}{D} \quad \text{so} \quad V = \frac{20 \text{ g}}{2 \text{ g/mL}} = 10 \text{ mL}$$

Do you understand how the unit *gram* cancels out to give milliliters in the answer? We actually have

$$V = 20 \text{ g} \div \frac{2 \text{ g}}{1 \text{ mL}}$$

When we apply the *rules for division of fractions* discussed in Section A.2, this expression becomes

$$20 \text{ g} \times \frac{1 \text{ mL}}{2 \text{ g}} = 10 \text{ mL}$$

and the unit *gram* cancels out.

A.9 Calculating and Using Percentages

Learning Goal 10
Solving problems involving percentages

A **percentage** is *the number of parts of something out of 100 parts*. For example, if a chemistry class has 100 students and 40 of the students are women, the percentage of women is 40 percent (also written as 40%).

$$40 \text{ women per } 100 \text{ students} = 40\%$$

We can state this mathematically (using the factor-unit method) as follows:

$$\text{Percentage women} = \left(\frac{40 \text{ women}}{100 \text{ students}} \right) (100 \text{ students})$$

$$= 40 \text{ percent women}$$

However, we usually set the formula up as follows:

$$\textbf{Percentage} = \frac{\text{number of items of interest}}{\text{total number of items}} \times 100$$

Suppose that we have a chemistry class with 60 students and that 48 of them are woman. What is the percentage of women in the class? By the formula,

$$\text{Percentage} = \frac{\text{number of items of interest}}{\text{total number of items}} \times 100$$

$$= \frac{48 \text{ women}}{60 \text{ students}} \times 100 = 80\% \text{ women}$$

Example A.18

A box of mixed vegetables has 25 carrots, 30 tomatoes, and 95 heads of lettuce. What is the percentage of each vegetable in the box?

Solution

First obtain the total number of vegetables in the box.

$$25 \text{ carrots} + 30 \text{ tomatoes} + 95 \text{ lettuces} = 150 \text{ vegetables}$$

Now calculate the percentage of each vegetable, using our formula.

$$\% \text{ carrots} = \frac{25 \text{ carrots}}{150 \text{ vegetables}} \times 100 = 16.7\% \text{ carrots}$$

$$\% \text{ tomatoes} = \frac{30 \text{ tomatoes}}{150 \text{ vegetables}} \times 100 = 20\% \text{ tomatoes}$$

$$\% \text{ lettuces} = \frac{95 \text{ lettuces}}{150 \text{ vegetables}} \times 100 = 63.3\% \text{ lettuces}$$

Note that the total of the percentages equals 100%. In other words, the whole is the sum of the parts.

Sometimes we know the percentage of something and we want to solve for the particular number of items. For example, say that 30% of the people in a particular community own a car. If this community has 800 people, how many people own cars? We can solve this problem using the factor-unit method if we remember that 30% car ownership means

$$\frac{30 \text{ people own cars}}{100 \text{ people in town}}$$

Therefore,

$$? \text{ people owning cars} = (800 \text{ people}) \left(\frac{30 \text{ people owning cars}}{100 \text{ people}} \right)$$

$$= 240 \text{ people own cars}$$

However, a simpler method of solving this problem is to move the decimal point on the percent number two places to the left and multiply it by the number of people in town. In this method, 30% becomes 0.30. Therefore

$$800 \text{ people} \times 0.30 = 240 \text{ people own cars}$$

Example A.19

A box of fruit contains apples, oranges, and bananas. There are 500 pieces of fruit in the box; 20% are apples, 30% are oranges, and 50% are bananas. How many pieces of each kind of fruit are in the box?

Solution
Turn each percent into its decimal equivalent and multiply it by the total number of fruit in the box.

20% becomes 0.20 (for apples)

30% becomes 0.30 (for oranges)

50% becomes 0.50 (for bananas)

Therefore,

500 fruit $\times$ 0.20 = 100 apples

500 fruit $\times$ 0.30 = 150 oranges

500 fruit $\times$ 0.50 = 250 bananas

Note that the total number of pieces of fruit is 500.

A.10 Using the Calculator

Learning Goal 11
Performing mathematical operations with the use of a calculator

If they are used correctly, electronic calculators speed up the process of making mathematical computations. One common error is the simple mistake of pushing the buttons incorrectly (in the wrong order, for instance). Many such errors can be detected by first estimating the answer so you can tell whether the calculated answer makes sense.

You should follow the specific directions given for your particular calculator, of course, but the following rules apply to most calculators.

1. Enter the operations in the order in which they are written.

$$\boxed{4}\ \boxed{\times}\ \boxed{5}\ \boxed{=}\ \boxed{20}$$

First press the "4" key, then the "$\times$" key, then the "5" key, followed by the "=" key. The answer "20" will appear in the electronic display position.

Example A.20

Perform the following operations.

(a) $\boxed{10}\ \boxed{\times}\ \boxed{5}\ \boxed{=}$

(b) $\boxed{10}\ \boxed{+}\ \boxed{5}\ \boxed{=}$

(c) $\boxed{10}\ \boxed{\div}\ \boxed{5}\ \boxed{=}$

(d) $\boxed{10}\ \boxed{-}\ \boxed{5}\ \boxed{=}$

Solution

(a) 50

(b) 15

(c) 2

(d) 5

2. Once the calculator is told to add, subtract, multiply, or divide two numbers, the results of the mathematical operation are stored in its memory. Another operation key $(+, -, \times, \text{ or } \div)$ can then be pressed and another calculation carried out.

Example A.21

Perform the following operations.

(a) $\boxed{10}\ \boxed{\times}\ \boxed{5}\ \boxed{=}\ \boxed{+}\ \boxed{4}\ \boxed{=}$

(b) $\boxed{20}\ \boxed{\div}\ \boxed{5}\ \boxed{-}\ \boxed{4}\ \boxed{=}$

(c) $\boxed{6}\ \boxed{-}\ \boxed{4}\ \boxed{\times}\ \boxed{20}\ \boxed{=}$

(d) $\boxed{8}\ \boxed{+}\ \boxed{6}\ \boxed{\times}\ \boxed{2}\ \boxed{=}$

Solution

(a) 54

(b) 0

(c) 40

(d) 28

3. Your calculator probably has a floating decimal point. Once you reach the proper place for the decimal point, press the $\boxed{\cdot}$ key to position the decimal point.

4. Final zeroes to the right of the decimal point are not shown in the electronic display after a calculation is complete.

Self-Test Exercises

Learning Goal 1: Adding and Subtracting Algebraically

1. *Add* the following numbers algebraically.
(*a*) 5 + 12   (*b*) −5 − 10   (*c*) 5 + 10   (*d*) −5 + 10

2. *Subtract* the following numbers algebraically.
(*a*) 3 − (+2)   (*b*) 3 − (−2)
(*c*) −3 − (+2)   (*d*) −3 − (−2)

3. *Add* the following numbers algebraically.
(*a*) 8 + 25   (*b*) 8 − 25   (*c*) −8 + 25   (*d*) −8 − 25

4. *Subtract* the following numbers algebraically.
(*a*) 55 − (+44)   (*b*) 55 − (−44)   (*c*) −55 − (+44)
(*d*) −55 − (−44)

Learning Goal 2: Multiplying and Dividing Fractions

5. Perform the indicated operation.

(a) $\frac{1}{8} \times \frac{1}{2}$ (b) $\frac{1}{8} \div \frac{1}{2}$ (c) 8^3 (d) $(2)^{-5}$

6. Perform the indicated operation.

(a) $\frac{3}{5} \times \frac{2}{7}$ (b) $\frac{20}{50} \times \frac{9}{5}$ (c) $\frac{5}{6} \div \frac{10}{3}$ (d) $\frac{25}{35} \div \frac{7}{5}$

Learning Goal 3: Multiplying and Dividing Exponential Numbers

7. Perform the indicated operation.

(a) $(4)^{-2}(4)^{-5}$ (b) $(4)^2(4)^3$ (c) $(6)^4(6)^{-2}$
(d) $(5)^{-8}(5)^9$

8. Perform the indicated operation.

(a) $\frac{(4)^5}{(4)^2}$ (b) $\frac{(4)^5}{(4)^{-3}}$ (c) $\frac{(a)^3(a)^2}{(a)^4}$ (d) $\frac{(b)^6(b)^4}{(b)^3(b)^5}$

9. Perform the indicated operation.

(a) $(25)^2$ (b) $(10)^3$ (c) $(99)^1$ (d) $(8)^{-3}$

(e) $\frac{1}{(4)^{-2}}$

10. Perform the indicated operation.

(a) $(5)^2(5)^3$ (b) $(12)^3(12)^9$ (c) $(10)^2(10)^{-2}$
(d) $(4)^{-8}(4)^{-2}$

11. Perform the indicated operation.

(a) $\frac{(4)^5}{(4)^3}$ (b) $\frac{(12)^3}{(12)^{-2}}$ (c) $\frac{(20)^{-4}}{(20)^3}$ (d) $\frac{(20)^{-4}}{(20)^{-3}}$

12. Perform the indicated operation.

(a) $(4)^4(4)^3(4)^2$ (b) $(x)^2(x)^3(x)^{-4}$

(c) $\frac{(y)^2(y)^8}{(y)^3}$ (d) $\frac{(p)^2(p)^{-9}}{(p)^{-8}}$

(e) $\frac{(10)^{-3}(10)^5(10)^7}{(10)^{-8}(10)^9}$

13. Perform the indicated operation.

(a) $(2)^6$ (b) $(3)^{-4}$ (c) $\frac{1}{(4)^3}$ (d) $\frac{1}{(4)^{-3}}$

14. Perform the indicated operation.

(a) $(10)^5(10)^{-2}$ (b) $(2)^{-3}(2)^{-2}$ (c) $(3)^{-5}(3)^2$
(d) $(6)^{-8}(6)^{10}$

15. Perform the indicated operation.

(a) $\frac{(3)^7}{(3)^5}$ (b) $\frac{(2)^4(2)^{-6}}{(2)^7(2)^{-12}}$ (c) $\frac{(a)^{12}(a)^{-10}}{(a)^{-7}(a)^{-6}}$

(d) $\frac{(b)^{-20}(b)^{-10}}{(b)^{-9}(b)^{-8}}$

Learning Goal 4: Solving Problems Using the Factor Unit Method

16. $(4 \text{ feet})\left(\frac{12 \text{ inches}}{1 \text{ foot}}\right) = ?$

17. $A = l \times w$; therefore $w = ?$ Solve the equation for w in terms of A and l.

18. How many yards are there in 18 feet?

19. How many eggs are there in 4 dozen eggs?

20. (a) How many hours are there in 360 minutes?
(b) How many hours are there in 7200 seconds?

21. How many pints are there in 8 quarts?

22. Change 1 mile per hour to feet per second.

23. $(72 \text{ inches})\left(\frac{1 \text{ foot}}{12 \text{ inches}}\right) = ?$

24. $(15 \text{ feet})\left(\frac{12 \text{ inches}}{1 \text{ foot}}\right) = ?$

25. How many feet are there in 100 yards?

26. How many inches are there in 5 yards?

27. How many dozen eggs do you have if you have 288 eggs?

28. How many minutes are there in 24 hours? How many seconds are there in 24 hours?

29. How many gallons are there in 40 quarts?

30. A rocket is traveling at 22,000 feet per second. How many miles per hour is it traveling?

Learning Goal 5: Adding and Subtracting Decimals

31. Perform the indicated operation.
(a) $15.4 + 117.33 + 16.909 + 0.044$
(b) $171.82 - 30.41$

32. Perform the indicated operation.
(a) $25.431 + 0.761 + 0.325 + 0.008$
(b) $123.25 - 19.54$
(c) $98.77 - 38.25 + 45.62$
(d) $254.37 - 68.26 - 38.33$

Learning Goal 6: Multiplying and Dividing Decimals

33. Perform the indicated operation.
(a) $(7.33)(4)$ (b) $(9.01)(4.28)$ (c) $(3.111)(8.7)$

34. Perform the indicated operation.
(a) $\dfrac{6.324}{3.1}$ (b) $\dfrac{25.30}{5.06}$

35. Perform the indicated operation.

(a) $(9.54)(3.27)$ (b) $(6.5)(7.1)$ (c) $\dfrac{80.4}{2.02}$

(d) $\dfrac{640.75}{8.5}$

Learning Goal 7: Solving Simple Algebraic Equations

36. Solve each of the following equations for the unknown.
(a) $25a = 100$ (b) $40y + 13 = 26$
(c) $12z - 12 = 24$

37. Solve each of the following equations for the unknown.
(a) $\dfrac{6x}{3} = \dfrac{36}{12}$ (b) $12y + 6 = 6y + 12$

38. Solve the equation $PV = nRT$ for: (a) P, (b) n, (c) T, (d) n/V.

39. Solve each of the following equations for the unknown.
(a) $4t = 64$ (b) $6y + 3 = 45$
(c) $2x - 7 = 33 - 3x$ (d) $40k + 8 = 96 + 51k$

40. Solve the equation $A = bcd$ for: (a) b, (b) c, (c) A/b.

Learning Goal 8: Solving Proportions

41. Solve each of the following equations for the unknown.
(a) $\dfrac{x}{25} = \dfrac{4}{5}$ (b) $\dfrac{3}{y} = \dfrac{7}{63}$ (c) $\dfrac{4t}{12} = \dfrac{2}{3}$ (d) $\dfrac{2}{5k} = \dfrac{5}{50}$

42. Solve each of the following proportions.
(a) $\dfrac{k}{5} = \dfrac{9}{45}$ (b) $\dfrac{2}{7} = \dfrac{h}{28}$ (c) $\dfrac{72}{f} = \dfrac{9}{2}$ (d) $\dfrac{18}{6} = \dfrac{12}{y}$

43. Solve each example for the missing value or values.
(a) $2:5:9 = x:20:y$ (b) $7:9 = h:54$
(c) $2:3:5:7 = 1:a:b:c$ (d) $2:5 = 5:d$

Learning Goal 9: Solving Simple Density Problems

44. Given that $D = \dfrac{m}{V}$
solve this formula for (a) m and (b) V.

45. A substance has a mass of 25 g and a volume of 5 mL. What is its density?

Learning Goal 10: Solving Problems Involving Percentages

46. A university has the following enrollments: (a) liberal arts, 200 students; (b) laboratory technology, 100 students; and (c) health science technology, 300 students. This accounts for all the students at the university. What percentage of the students is enrolled in each curriculum?

47. A survey of 2,000 people shows that 40% want to travel to Europe, 30% to Asia, 10% to Alaska, and 20% to Miami. How many people want to travel to each place?

Learning Goal 11: Performing Mathematical Operations with the Use of a Calculator

48. Perform the following operations using a calculator.
(a) $11 \times 6 = ?$
(b) $11 + 6 = ?$
(c) $30 \div 5 = ?$
(d) $30 - 5 = ?$

49. Perform the following operations.
(a) $(12 \times 5) + 5 = ?$
(b) $(20 \div 4) - 2 = ?$
(c) $(10 + 20) \div (5 \times 6) = ?$
(d) $(100 \times 50) \div (10 \times 40) = ?$

50. Perform the following operations.
(a) $(50 \times 6) \div 12 = ?$
(b) $(100 + 50) \div (15 \times 0.01) = ?$
(c) $(0.05 \times 0.30) \div 4 = ?$
(d) $(10,000 \times 5) \div 4000 = ?$

Supplement B
Important Chemical Tables

Supplement B

Table B.1
Solubilities

	Acetate	Arsenate	Bromide	Carbonate	Chlorate	Chloride	Chromate	Hydroxide	Iodide	Nitrate	Oxide	Phosphate	Sulfate	Sulfide
Aluminum	W	a	W	—	W	W	—	A	W	W	a	A	W	d
Ammonium	W	W	W	W	W	W	W	W	W	W	—	W	W	W
Barium	W	w	W	w	W	W	A	W	W	W	W	A	a	d
Cadmium	W	A	W	A	W	W	A	A	W	W	A	A	W	A
Calcium	W	w	W	w	W	W	W	W	W	W	w	w	w	w
Chromium	W	—	W*	W	—	I	—	A	W	W	a	w	W†	d
Cobalt	W	A	W	A	W	W	A	A	W	W	A	A	W	A
Copper(II)	W	A	W	—	W	W	—	A	a	W	A	A	W	A
Hydrogen	W	W	W	—	W	W	—	—	W	W	W	W	W	W
Iron(II)	W	A	W	w	W	W	—	A	W	W	A	A	W	A
Iron(III)	W	A	W	—	W	W	A	A	W	W	A	w	w	d
Lead(II)	W	A	W	A	W	W	A	w	w	W	w	A	w	A
Magnesium	W	A	W	w	W	W	W	A	W	W	A	w	W	d
Mercury(I)	w	A	A	A	W	a	w	—	A	W	A	A	w	I
Mercury(II)	W	w	W	—	W	W	w	A	w	W	w	A	d	I
Nickel	W	A	W	w	W	W	A	w	W	W	A	A	W	A
Potassium	W	W	W	W	W	W	W	W	W	W	W	W	W	W
Silver	w	A	a	A	W	a	w	—	I	W	w	A	w	A
Sodium	W	W	W	W	W	W	W	W	W	W	d	W	W	W
Strontium	W	w	W	w	W	W	w	W	W	W	W	A	w	W
Tin(II)	d	—	W	—	W	W	A	A	W	d	A	A	W	A
Tin(IV)	W	—	W	—	—	W	W	w	d	—	A	—	W	A
Zinc	W	A	W	w	W	W	w	A	W	W	w	A	W	A

Reprinted with permission from *CRC Handbook of Chemistry and Physics*, 66th ed. Copyright © The Chemical Rubber Co., CRC Press, Inc.

Abbreviations: W = soluble in water; A = insoluble in water, but soluble in acids; w = only slightly soluble in water, but soluble in acids; a = insoluble in water, and only slightly soluble in acids; I = insoluble in both water and acids; d = decomposes in water.

*CrBr$_3$ †Cr$_2$(SO$_4$)$_3$

Table B.2
Charges of Ions Frequently Used in Chemistry

1+		2+		3+	
Hydrogen	H^{1+}	Calcium	Ca^{2+}	Iron(III)	Fe^{3+}
Lithium	Li^{1+}	Magnesium	Mg^{2+}	Aluminum	Al^{3+}
Sodium	Na^{1+}	Barium	Ba^{2+}		
Potassium	K^{1+}	Zinc	Zn^{2+}		
Mercury(I)*	Hg_2^{2+}*	Mercury(II)	Hg^{2+}		
Copper(I)	Cu^{1+}	Tin(II)	Sn^{2+}		
Silver	Ag^{1+}	Iron(II)	Fe^{2+}		
Ammonium	$(NH_4)^{1+}$	Lead(II)	Pb^{2+}		
		Copper(II)	Cu^{2+}		

1−		2−		3−	
Fluoride	F^{1-}	Oxide	O^{2-}	Nitride	N^{3-}
Chloride	Cl^{1-}	Sulfide	S^{2-}	Phosphide	P^{3-}
Hydroxide	$(OH)^{1-}$	Sulfite	$(SO_3)^{2-}$	Phosphate	$(PO_4)^{3-}$
Nitrite	$(NO_2)^{1-}$	Sulfate	$(SO_4)^{2-}$	Arsenate	$(AsO_4)^{3-}$
Nitrate	$(NO_3)^{1-}$	Carbonate	$(CO_3)^{2-}$	Borate	$(BO_3)^{3-}$
Acetate	$(C_2H_3O_2)^{1-}$	Chromate	$(CrO_4)^{2-}$		
Chlorate	$(ClO_3)^{1-}$	Dichromate	$(Cr_2O_7)^{2-}$		
Permanganate	$(MnO_4)^{1-}$	Oxalate	$(C_2O_4)^{2-}$		
Hydrogen carbonate (bicarbonate)	$(HCO_3)^{1-}$				
Hydrogen sulfate (bisulfate)	$(HSO_4)^{1-}$				
Cyanide	$(CN)^{1-}$				

*Note that the mercury(I) ion is a diatomic ion. In other words, you never find an Hg^{1+} ion alone, but always as Hg^{1+}—Hg^{1+}. The two Hg^{1+} ions are bonded to each other.

Table B.3
Prefixes and Abbreviations

nano-	= 0.000000001	= 10^{-9}	nanometer = nm	decimeter	= dm
micro-	= 0.000001	= 10^{-6}	micrometer = μm (Greek letter mu)	decigram	= dg
milli-	= 0.001	= 10^{-3}	millimeter = mm	kilometer	= km
centi-	= 0.01	= 10^{-2}	milliliter = mL	kilogram	= kg
deci-	= 0.1	= 10^{-1}	milligram = mg		
deka-	= 10	= 10^{1}	centimeter = cm		
kilo-	= 1000	= 10^{3}	centigram = cg		

Table B.4
The Metric System

Length

1 millimeter = 0.001 meter = $\frac{1}{1000}$ meter	or	1 meter = 1000 millimeters
1 centimeter = 0.01 meter = $\frac{1}{100}$ meter	or	1 meter = 100 centimeters
1 decimeter = 0.1 meter = $\frac{1}{10}$ meter	or	1 meter = 10 decimeters
1 kilometer = 1000 meters	or	1 meter = 0.001 kilometer

Mass

1 microgram = 0.000001 gram	1 gram = 1,000,000 micrograms
1 milligram = 0.001 gram	1 gram = 1000 milligrams
1 centigram = 0.01 gram	1 gram = 100 centigrams
1 decigram = 0.1 gram	1 gram = 10 decigrams
1 kilogram = 1000 grams	1 gram = 0.001 kilogram

Volume

1 milliliter = 0.001 liter

1 milliliter = 1 cubic centimeter
(Cubic centimeter is abbreviated cm^3.)

Table B.5
Conversion of Units (English–Metric)

To convert	into	multiply by
Length		
inches	centimeters	2.540 cm/in
centimeters	inches	0.3937 in/cm
feet	meters	0.30 m/ft
meters	feet	3.28 ft/m
Weight (mass)		
ounces	grams	28.35 g/oz
grams	ounces	0.035 oz/g
pounds	grams	454 g/lb
grams	pounds	0.0022 lb/g
Volume		
liters	quarts	1.057 qt/L
quarts	liters	0.9463 L/qt

Centimeters / Inches

Table B.6
Naturally Occurring Isotopes of the First 15 Elements

Name	Symbol	Atomic number	Mass number	Percentage natural abundance
Hydrogen-1	$^{1}_{1}H$	1	1	99.985
Hydrogen-2	$^{2}_{1}H$	1	2	0.015
Hydrogen-3	$^{3}_{1}H$	1	3	Negligible
Helium-3	$^{3}_{2}He$	2	3	0.00013
Helium-4	$^{4}_{2}He$	2	4	99.99987
Lithium-6	$^{6}_{3}Li$	3	6	7.42
Lithium-7	$^{7}_{3}Li$	3	7	92.58
Beryllium-9	$^{9}_{4}Be$	4	9	100
Boron-10	$^{10}_{5}B$	5	10	19.6
Boron-11	$^{11}_{5}B$	5	11	80.4
Carbon-12	$^{12}_{6}C$	6	12	98.89
Carbon-13	$^{13}_{6}C$	6	13	1.11
Nitrogen-14	$^{14}_{7}N$	7	14	99.63
Nitrogen-15	$^{15}_{7}N$	7	15	0.37
Oxygen-16	$^{16}_{8}O$	8	16	99.759
Oxygen-17	$^{17}_{8}O$	8	17	0.037
Oxygen-18	$^{18}_{8}O$	8	18	0.204
Fluorine-19	$^{19}_{9}F$	9	19	100
Neon-20	$^{20}_{10}Ne$	10	20	90.92
Neon-21	$^{21}_{10}Ne$	10	21	0.257
Neon-22	$^{22}_{10}Ne$	10	22	8.82
Sodium-23	$^{23}_{11}Na$	11	23	100
Magnesium-24	$^{24}_{12}Mg$	12	24	78.70
Magnesium-25	$^{25}_{12}Mg$	12	25	10.13
Magnesium-26	$^{26}_{12}Mg$	12	26	11.17
Aluminum-27	$^{27}_{13}Al$	13	27	100
Silicon-28	$^{28}_{14}Si$	14	28	92.21
Silicon-29	$^{29}_{14}Si$	14	29	4.70
Silicon-30	$^{30}_{14}Si$	14	30	3.09
Phosphorus-31	$^{31}_{15}P$	15	31	100

Table B.7
Electron Configurations of the Elements

Atomic number	Element	1	2		3			4				5				6				7
		s	*s*	*p*	*s*	*p*	*d*	*s*	*p*	*d*	*f*	*s*	*p*	*d*	*f*	*s*	*p*	*d*	*f*	*s*
1	H	1																		
2	He	2																		
3	Li	2	1																	
4	Be	2	2																	
5	B	2	2	1																
6	C	2	2	2																
7	N	2	2	3																
8	O	2	2	4																
9	F	2	2	5																
10	Ne	2	2	6																
11	Na	2	2	6	1															
12	Mg	2	2	6	2															
13	Al	2	2	6	2	1														
14	Si	2	2	6	2	2														
15	P	2	2	6	2	3														
16	S	2	2	6	2	4														
17	Cl	2	2	6	2	5														
18	Ar	2	2	6	2	6														
19	K	2	2	6	2	6		1												
20	Ca	2	2	6	2	6		2												
21	Sc	2	2	6	2	6	1	2												
22	Ti	2	2	6	2	6	2	2												
23	V	2	2	6	2	6	3	2												
24	Cr	2	2	6	2	6	5	1												
25	Mn	2	2	6	2	6	5	2												
26	Fe	2	2	6	2	6	6	2												
27	Co	2	2	6	2	6	7	2												
28	Ni	2	2	6	2	6	8	2												
29	Cu	2	2	6	2	6	10	1												
30	Zn	2	2	6	2	6	10	2												
31	Ga	2	2	6	2	6	10	2	1											
32	Ge	2	2	6	2	6	10	2	2											
33	As	2	2	6	2	6	10	2	3											
34	Se	2	2	6	2	6	10	2	4											
35	Br	2	2	6	2	6	10	2	5											
36	Kr	2	2	6	2	6	10	2	6											

Table B.7
Electron Configurations of the Elements (Continued)

Atomic number	Element	1	2		3			4				5				6				7
		s	s	p	s	p	d	s	p	d	f	s	p	d	f	s	p	d	f	s
37	Rb	2	2	6	2	6	10	2	6			1								
38	Sr	2	2	6	2	6	10	2	6			2								
39	Y	2	2	6	2	6	10	2	6	1		2								
40	Zr	2	2	6	2	6	10	2	6	2		2								
41	Nb	2	2	6	2	6	10	2	6	4		1								
42	Mo	2	2	6	2	6	10	2	6	5		1								
43	Tc	2	2	6	2	6	10	2	6	6		1								
44	Ru	2	2	6	2	6	10	2	6	7		1								
45	Rh	2	2	6	2	6	10	2	6	8		1								
46	Pd	2	2	6	2	6	10	2	6	10										
47	Ag	2	2	6	2	6	10	2	6	10		1								
48	Cd	2	2	6	2	6	10	2	6	10		2								
49	In	2	2	6	2	6	10	2	6	10		2	1							
50	Sn	2	2	6	2	6	10	2	6	10		2	2							
51	Sb	2	2	6	2	6	10	2	6	10		2	3							
52	Te	2	2	6	2	6	10	2	6	10		2	4							
53	I	2	2	6	2	6	10	2	6	10		2	5							
54	Xe	2	2	6	2	6	10	2	6	10		2	6							
55	Cs	2	2	6	2	6	10	2	6	10		2	6			1				
56	Ba	2	2	6	2	6	10	2	6	10		2	6			2				
57	La	2	2	6	2	6	10	2	6	10		2	6	1		2				
58	Ce	2	2	6	2	6	10	2	6	10	2	2	6			2				
59	Pr	2	2	6	2	6	10	2	6	10	3	2	6			2				
60	Nd	2	2	6	2	6	10	2	6	10	4	2	6			2				
61	Pm	2	2	6	2	6	10	2	6	10	5	2	6			2				
62	Sm	2	2	6	2	6	10	2	6	10	6	2	6			2				
63	Eu	2	2	6	2	6	10	2	6	10	7	2	6			2				
64	Gd	2	2	6	2	6	10	2	6	10	7	2	6	1		2				
65	Tb	2	2	6	2	6	10	2	6	10	9	2	6			2				
66	Dy	2	2	6	2	6	10	2	6	10	10	2	6			2				
67	Ho	2	2	6	2	6	10	2	6	10	11	2	6			2				
68	Er	2	2	6	2	6	10	2	6	10	12	2	6			2				
69	Tm	2	2	6	2	6	10	2	6	10	13	2	6			2				
70	Yb	2	2	6	2	6	10	2	6	10	14	2	6			2				
71	Lu	2	2	6	2	6	10	2	6	10	14	2	6	1		2				
72	Hf	2	2	6	2	6	10	2	6	10	14	2	6	2		2				

Table B.7
Electron Configurations of the Elements (Continued)

Atomic number	Element	1	2		3			4				5				6				7
		s	s	p	s	p	d	s	p	d	f	s	p	d	f	s	p	d	f	s
73	Ta	2	2	6	2	6	10	2	6	10	14	2	6	3		2				
74	W	2	2	6	2	6	10	2	6	10	14	2	6	4		2				
75	Re	2	2	6	2	6	10	2	6	10	14	2	6	5		2				
76	Os	2	2	6	2	6	10	2	6	10	14	2	6	6		2				
77	Ir	2	2	6	2	6	10	2	6	10	14	2	6	7		2				
78	Pt	2	2	6	2	6	10	2	6	10	14	2	6	9		1				
79	Au	2	2	6	2	6	10	2	6	10	14	2	6	10		1				
80	Hg	2	2	6	2	6	10	2	6	10	14	2	6	10		2				
81	Tl	2	2	6	2	6	10	2	6	10	14	2	6	10		2	1			
82	Pb	2	2	6	2	6	10	2	6	10	14	2	6	10		2	2			
83	Bi	2	2	6	2	6	10	2	6	10	14	2	6	10		2	3			
84	Po	2	2	6	2	6	10	2	6	10	14	2	6	10		2	4			
85	At	2	2	6	2	6	10	2	6	10	14	2	6	10		2	5			
86	Rn	2	2	6	2	6	10	2	6	10	14	2	6	10		2	6			
87	Fr	2	2	6	2	6	10	2	6	10	14	2	6	10		2	6			1
88	Ra	2	2	6	2	6	10	2	6	10	14	2	6	10		2	6			2
89	Ac	2	2	6	2	6	10	2	6	10	14	2	6	10		2	6	1		2
90	Th	2	2	6	2	6	10	2	6	10	14	2	6	10		2	6	2		2
91	Pa	2	2	6	2	6	10	2	6	10	14	2	6	10	2	2	6	1		2
92	U	2	2	6	2	6	10	2	6	10	14	2	6	10	3	2	6	1		2
93	Np	2	2	6	2	6	10	2	6	10	14	2	6	10	4	2	6	1		2
94	Pu	2	2	6	2	6	10	2	6	10	14	2	6	10	6	2	6			2
95	Am	2	2	6	2	6	10	2	6	10	14	2	6	10	7	2	6			2
96	Cm	2	2	6	2	6	10	2	6	10	14	2	6	10	7	2	6	1		2
97	Bk	2	2	6	2	6	10	2	6	10	14	2	6	10	8	2	6	1		2
98	Cf	2	2	6	2	6	10	2	6	10	14	2	6	10	10	2	6			2
99	Es	2	2	6	2	6	10	2	6	10	14	2	6	10	11	2	6			2
100	Fm	2	2	6	2	6	10	2	6	10	14	2	6	10	12	2	6			2
101	Md	2	2	6	2	6	10	2	6	10	14	2	6	10	13	2	6			2
102	No	2	2	6	2	6	10	2	6	10	14	2	6	10	14	2	6			2
103	Lr	2	2	6	2	6	10	2	6	10	14	2	6	10	14	2	6	1		2
104	Unq	2	2	6	2	6	10	2	6	10	14	2	6	10	14	2	6	2		2
105	Unp	2	2	6	2	6	10	2	6	10	14	2	6	10	14	2	6	3		2
106	Unh	2	2	6	2	6	10	2	6	10	14	2	6	10	14	2	6	4		2
107	Uns	2	2	6	2	6	10	2	6	10	14	2	6	10	14	2	6	5		2
108	Uno	2	2	6	2	6	10	2	6	10	14	2	6	10	14	2	6	6		2
109	Une	2	2	6	2	6	10	2	6	10	14	2	6	10	14	2	6	7		2

Table B.8
Heats of Formation at 25°C and 1 Atmosphere Pressure

Substance	ΔH_f^0 (kcal/mole)	Substance	ΔH_f^0 (kcal/mole)
$BaCl_2(s)$	− 205.56	$H_2O(g)$	− 57.8
$BaO_2(s)$	− 150.5	$H_2O(l)$	− 68.3
$CH_4(g)$	− 17.9	$H_2O_2(l)$	− 44.8
$C_2H_2(g)$	+ 54.2	$H_2S(g)$	− 4.8
$C_2H_6(g)$	− 20.2	$H_2SO_4(aq)$	− 217
$C_8H_{18}(l)$	− 59.97	$H_3PO_4(aq)$	− 308.2
$C_6H_{12}O_6(s)$	− 304.4	$SO_2(g)$	− 70.96
$CCl_4(l)$	− 33.4	$SO_3(g)$	− 94.45
$CO(g)$	− 26.4	$NH_3(g)$	− 11.04
$CO_2(g)$	− 94.1	$NO(g)$	+ 21.5
$CaO(s)$	− 151.9	$NO_2(g)$	+ 8.1
$CaCO_3(s)$	− 288.5	$NaCl(aq)$	− 97.3
$HF(g)$	− 64.2	$NaOH(aq)$	− 112
$HCl(g)$	− 22.1	$Fe_2O_3(s)$	− 197
$HBr(g)$	− 8.7	$KOH(aq)$	− 115
$HI(g)$	+ 6.2	$K_2O(s)$	− 86.4
$HCl(aq)$	− 39.6		

Table B.9
Pressure of Water Vapor, P_{H_2O}, at Various Temperatures

Temperature (°C)	Pressure (torr)	Temperature (°C)	Pressure (torr)
0	4.58	32	35.66
5	6.54	33	37.73
10	9.21	34	39.90
15	12.79	35	42.18
16	13.63	36	44.56
17	14.53	37	47.07
18	15.48	38	49.69
19	16.48	39	52.44
20	17.54	40	55.32
21	18.65	45	71.88
22	19.83	50	92.51
23	21.07	55	118.04
24	22.38	60	149.38
25	23.76	65	187.54
26	25.21	70	233.7
27	26.74	75	289.1
28	28.35	80	355.1
29	30.04	85	433.6
30	31.82	90	525.8
31	33.70	95	633.9
		100	760.0

Supplement C
Glossary

In each entry, the number in parentheses is the number of the chapter in which the subject is first discussed.

absolute zero (11) The coldest possible temperature that matter can reach. It is equal to 0 K ($-273°$C).

accuracy (2) How close a measurement is to the true value

acid (15) A substance that releases hydrogen ions in solution, counteracts bases, and is a proton donor

activated-sludge process (12) A method used to treat sewage and waste water

activation energy (16) The energy required to begin a chemical reaction

activity series (8) The elements listed according to their reactivities

alcohols (19) A class of organic compounds with the general formula R—OH

aldehydes (19) A class of organic compounds with the general formula

$$R-\underset{\underset{O}{\|}}{C}-H$$

alkali (15) Any substance that dissolves to give a basic solution

alkanes (18) A series of organic hydrocarbons each of which contains only single carbon-carbon bonds

alkenes (18)   A series of organic hydrocarbons, each of which contains a carbon-carbon double bond in its structure

alkynes (18)   A series of organic hydrocarbons, each of which contains a carbon-carbon triple bond in its structure

alpha particles (4)   Helium atoms with their electrons removed

amines (19)   A class of organic compounds with the general Formula $R\!-\!NH_2$

amino acids (20)   Organic acids that contain a carboxyl group at one end of the molecule and an amino group at the other end of the molecule

amorphous solids (13)   Solids that do not have a well-defined crystalline structure

anabolism (20)   The process in which organic molecules such as sugars, amino acids, fatty acids, and glycerol—molecules that have been absorbed by the cells in the body—serve as the building blocks for the synthesis of more complex organic substances such as polysaccharides, lipids, nucleic acids, and proteins

anhydrous salt (12)   A salt molecule that has no water of hydration bonded to it

anode (4)   A positive electrode

aromatic hydrocarbons (18)   A series of cyclic hydrocarbons that contain alternating single and double bonds

atom (3)   The smallest part of an element that can enter into chemical combinations

atomic mass (or **weight**) (3)   The mass of an element in relation to the mass of an atom of carbon-12

atomic number (4)   The number of electrons or protons in a neutral atom

atomic radius (6)   The distance from the center of the nucleus to the outermost electron

Avogadro's number (3)   The number of atoms whose weight is the gram-atomic weight of any element. It is equal to 6.02×10^{23}.

Avogadro's principle (11)   Equal volumes of all gases, under the same conditions of temperature and pressure, contain the same number of molecules.

balanced equation (8)   A chemical equation that has the same number of atoms of each element on the reactant side as on the product side

base (15)   A substance that releases hydroxide ions in solution, counteracts acids, and accepts protons

beta particles (17)   Electron-like high-speed particles that come from the nuclei of radioactive atoms

boiling point (13)   The temperature at which the equilibrium vapor pressure of a liquid equals the atmospheric pressure

Boyle's law (11)   At constant temperature, the volume of a given mass of gas varies inversely with the pressure.

buffer solution (6)   A solution prepared by mixing a weak acid and its salt or a weak base and its salt. It tends to maintain its pH when an acid or base is added to it.

calorie (10)   A measure of heat energy. It is the amount of heat needed to raise the temperature of 1 gram of water 1 degree Celsius.

carbohydrates (20)   Polyhydroxy aldehydes or polyhydroxy ketones or compounds that upon hydrolysis yield polyhydroxy aldehydes or polyhydroxy ketones.

carbonyl group (19)   An organic functional group with the formula

$$-\!\underset{\underset{O}{\|}}{C}\!-$$

carboxyl group (19)   An organic functional group with the formula

$$-\!\underset{\underset{O}{\|}}{C}\!-\!OH$$

carboxylic acids (19)   A class of organic compounds with the general formula

$$R\!-\!\underset{\underset{O}{\|}}{C}\!-\!OH$$

catabolism (20)   The breaking down of digested food molecules into smaller molecules, usually accompanied by a release of energy

catalyst (17)   A substance that accelerates a chemical reaction, without itself being used up (or undergoing any permanent change)

cathode (4)   A negative electrode

Celsius (centigrade) scale (2)   A temperature scale

on which the freezing point of water is zero degrees and the normal boiling point of water is 100 degrees

chain reaction (17) A self-sustaining nuclear reaction

Charles's law (11) At constant pressure, the volume of a given amount of gas varies directly with the temperature in kelvins.

chemical equilibrium (16) A dynamic state in which two or more opposing processes take place simultaneously and at the same rate

chemical formula (3) The combination of the symbols of the particular elements that form a chemical compound, showing the number of atoms of each element.

chemical kinetics (16) The study of the rate at which chemical reactions occur

chemical properties (3) Properties that indicate how a particular substance changes in chemical reactions

coefficient (8) A number placed before an element or compound in a chemical equation to balance the equation

colligative properties (14) Collective properties of solutions; for example, boiling-point elevation and freezing-point depression

combination reactions (8) Reactions in which two or more substances combine to form a more complex substance

combustion (10) A chemical reaction that takes place so quickly that heat and light are produced. Usually refers to the reaction of a substance with oxygen.

coordinate covalent bond (7) A bond in which one atom donates all the electrons used to form the bond

covalent bond (7) A bond formed by the sharing of electrons between two atoms

crystal (13) A solid that has geometrically arranged plane surfaces and a symmetrical structure on the inside

crystal lattice (13) The symmetrical structure formed by the particles throughout a crystal

decay, nuclear (or radioactive) (17) Emission of radiation by a radioactive atom

decomposition reactions (8) Reactions that involve the breakdown of a complex substance into simpler substances

deficiency disease (20) A disease that results from the lack of a particular vitamin or mineral in the diet

deliquescence (12) The ability to absorb enough water from the atmosphere to form a solution

density (2) The mass per unit volume of a substance

desiccant (12) A hygroscopic substance that is used as a drying agent

diatomic elements (7) Elements that are found in molecules with two atoms apiece; for example, H_2, O_2, N_2

diglycerides (20) Lipids in which two glycerol —OH groups are combined with two fatty acids in ester linkages

disaccharide (20) A carbohydrate containing two sugar groups in the molecule

dissociation (14) The process by which an ionic substance separates into ions by the action of a solvent

double-replacement reaction (8) A reaction in which two compounds exchange ions with each other

dynamic equilibrium (14) The state that is achieved by two opposing processes that proceed at equal rates

efflorescence (12) The spontaneous loss of water of hydration by a hydrated salt when exposed to the atmosphere

electrochemical battery (voltaic cell) (17) A device that produces electrical energy from a chemical reaction

electrolysis (12) Decomposition of a compound into simpler substances when electricity is passed through it

electrolytic solution (14) A solution that conducts electric current

electromagnetic spectrum (5) The full range of electromagnetic radiation, from radio waves to x rays.

electron (4) A particle with a relative negative charge of one unit and a mass of 0.0005486 amu

electron affinity (6) The energy released when an additional electron is added to a neutral atom

electron configuration (5) The positions of the electrons in the various energy levels of an atom

electron-dot structure (Lewis electron-dot structure) (7) A notation that shows the symbol for an element and, by the use of dots, the number of outer electrons in an atom of the element

electronegativity (7) The attraction that an element has for shared electrons in a molecule

elements (3) The basic building blocks of matter, which cannot be broken down physically or by chemical means

empirical formula (3) A chemical formula showing the simplest whole-number ratio of the atoms in a molecule of a compound

endothermic reaction (10) A reaction that absorbs energy from the surroundings

energy (3) The ability to do work. Energy appears in many forms—for example, as heat and as chemical, electrical, mechanical, and radiant (light) energy.

energy levels (5) Various regions, outside the nucleus of an atom, in which the electrons move

energy sublevels (6) A breakdown of energy levels into more specific regions (outside the nucleus) in which the electrons move

enthalpy (10) The heat content of a chemical substance, represented by the symbol H

equilibrium vapor pressure (13) The pressure exerted by a vapor when it is in equilibrium with its liquid at any given temperature

equivalent (of an acid or base) (14) The mass of the acid or base that contains Avogadro's number of hydrogen ions or hydroxide ions

essential amino acids (20) Amino acids that can't be synthesized by the body and must be included in the diet

esters (19) A class of organic compounds with the general formula

$$R—C—O—R'$$
$$\|$$
$$O$$

ethers (19) A class of organic compounds with the general formula

$$R—O—R'$$

excited state (5) The state of an electron when it is at an energy level higher than its ground-state level; an electron is in an excited state after it has absorbed energy.

exothermic reactions (10) A reaction that releases energy to the surroundings

Fahrenheit scale (2) A temperature scale on which the freezing point of water is 32 degrees and the nor-

mal boiling point of water is 212 degrees

fats (20) Lipids composed of fatty acids and glycerol

fat-soluble vitamins (20) Vitamins A, E, K, and D. They are stored and carried by fats in the body.

fatty acids (20) Organic compounds composed of a long chain of carbon atoms connected to a carboxyl group

fission (17) The splitting of an atom into two or more different atoms when it is bombarded by neutrons

freezing point (2) The temperature at which the solid and liquid phases of a substance exist together

fuel cell (17) A high-efficiency battery in which reactants are introduced continuously and products are removed continuously

fumigant (19) A substance that has disinfecting properties

functional group (18) The part of an organic molecule that gives it its unique properties

fusion, nuclear (17) The combination of two atoms to form a new, heavier atom. The process releases huge amounts of energy.

gamma rays (17) A high-energy form of electromagnetic radiation given off by radioactive substances

gas (11) One of the states of matter and the least compact of the three physical states. (Gases have no definite shape or volume.)

Gay-Lussac's law of combining volumes (11) For gases under similar conditions of temperature and pressure, the coefficients of the balanced equation are the ratios for combination by volume.

gram-atomic weight (3) The atomic weight of an element expressed in grams

gram-molecular weight (3) The molecular weight of a substance expressed in grams

ground state (5) The most stable state of an atom. In this state the electrons are in the lowest possible energy levels.

group (of elements) (6) A vertical column of elements in the periodic table

Haber process (9) A method used to synthesize ammonia from its elements, nitrogen and hydrogen

half-life (17) The time required for the decay of half the atoms originally present in a sample of a radioactive substance

heat of formation (10) The heat released or absorbed when 1 mole of a compound is formed from its elements

heat of reaction (10) The change in heat content of a substance during a chemical reaction

heterogeneous matter (3) Matter that has parts with different properties; nonuniform matter

homogeneous matter (3) Matter that has similar properties throughout; uniform matter

Homologous series (18) A series of compounds of the same chemical type that differ only by fixed increments of the constituent elements

hydrates (12) Compounds that contain chemically combined water in definite proportions

hydrocarbons (18) Organic compounds composed only of carbon and hydrogen

hydrogen bond (as applied to water molecules) (12) A chemical bond formed between the hydrogen atom of one water molecule and the oxygen atom of another water molecule

hydrogenation (12) Addition of hydrogen to a compound by a chemical reaction

hydronium ion (15) The ion formed by the addition of a hydrogen ion to a water molecule. It can be written as H_3O^{1+}.

hygroscopic substance (12) A substance that can absorb water from the atmosphere

ideal gas (11) A gas that obeys the Ideal Gas Law, $PV = nRT$

indicators (15) Chemical substances that have the ability to change color depending on the pH of the solution they are placed in

ionic bond (7) A bond formed by the transfer of electrons from one atom to another. The atoms are always of different elements.

ionization (14) The process by which ions are formed from atoms or molecules by the transfer of electrons

ionization constant (16) The equilibrium constant for an acid or base. This constant expresses the degree of dissociation of the acid or base.

ionization potential (6) The energy needed to remove an electron from an isolated atom

ions (6) Atoms or groups of atoms that have gained or lost electrons and therefore have a positive or negative charge

isomers (18) Compounds with the same molecular formula but different structural formulas

isotopes (4) Atoms that have the same number of electrons and protons but different numbers of neutrons

Kelvin temperature scale (11) An absolute temperature scale in which kelvins = degrees Celsius + 273

ketones (19) A class of organic compounds with the general formula

$$R\!-\!\underset{\underset{O}{\|}}{C}\!-\!R'$$

kinetic theory of gases (11) An attempt to explain the properties of gases in terms of the forces between the gas molecules and the energy of these molecules

law of conservation of mass and energy (3) Matter and energy are neither created nor destroyed, but they can change from one form to another.

law of definite composition (or **definite proportions**) (3) A given compound always has the same elements in the same proportions by weight.

lipids (20) Biochemical substances found in plant and animal tissues. They include fats and other compounds that resemble fats in physical properties.

liquid (13) One of the states of matter in which units are held together but are free to move about. (Liquids have a definite volume but take the shape of their container.)

macronutrients (20) Minerals that the body needs in large amounts—for example, sodium, calcium, and chlorine

mass (2) A measure of the amount of matter that an object has

matter (3) Anything that occupies space and has mass

melting point (14) The temperature at which the solid and the liquid phase of a substance exist together

metabolism (20) The process by which the food we eat is transformed in the body, through a series of complex chemical reactions, to provide material for cellular function and energy to power the body

metalloids (3) Elements that have some of the properties of metals and some of the properties of nonmetals

metals (3)   Elements that conduct electricity and heat, have luster, and take on a positive oxidation number when they bond

metric system (2)   A system of measurement based on multiples of ten

micronutrients (20)   Minerals that the body needs in small amounts—for example, copper, magnesium, iodine, and zinc

minerals (20)   Inorganic substances that the body needs to control the rates of life-sustaining reactions. Examples include sodium, calcium, copper, iodine and zinc.

miscible (14)   Capable of forming a solution when mixed

mixture (3)   A material that consists of two or more substances, each substance retaining its own characteristic properties

molality (14)   A concentration unit for solutions: moles of solute per kilogram of solvent

molar volume of a gas (11)   The volume of 1 mole of a gas at STP, which is 22.4 liters for an ideal gas

molarity (14)   A concentration unit for solutions: moles of solute per liter of solution

mole (3) 6.02×10^{23} items

molecular formula (3)   A chemical formula showing the number of atoms of each element in a molecule of a compound

molecular weight (3)   The sum of the atomic weights of all of the atoms that make up a molecule

molecule (3)   The smallest part of a compound that can enter into chemical reactions

monoglycerides (20)   Lipids in which one glycerol —OH group is combined with one fatty acid in an ester linkage

monosaccharide (20)   A carbohydrate containing one sugar group

neutralization reactions (8)   Reactions in which an acid and a base react to form a salt and water

neutron (4)   A particle with a mass of 1.008665 amu and zero charge

nomenclature (7)   A system of names (for example, the system of names for chemical compounds)

nonelectrolytic solution (14)   A solution that does not conduct electric current

nonmetals (3)   Elements that are not good conductors of heat and electricity and that usually take on a negative oxidation number when they bond

normality (14)   A concentration unit for solutions: equivalents of solute per liter of solution

orbital (6)   A region in space, near the atomic nucleus, in which there is a 95% probability of finding an electron

oxidation (8)   Loss of electrons

oxidation number (7)   A number that expresses the apparent charge of an element or a polyatomic ion in a compound

oxidizing agent (8)   A substance that causes something else to be oxidized

particulate matter (17)   Particles of smoke, dust, fumes, and aerosols that are found in the atmosphere

peptide bond (20)   The bond that holds amino acids together in a protein molecule

period (6)   A horizontal row of elements in the periodic table

pH scale (15)   A scale that expresses the acidity of a solution

photochemical smog (17)   Smog that results from reactions of chemicals in the atmosphere; in these reactions sunlight is a catalyst.

physical properties (3)   Properties such as color, odor, taste, boiling point, and melting point that can be measured by nonchemical means

polar covalent bond (7)   A covalent bond in which there is unequal sharing of electrons

polyatomic ion (7)   A charged group of covalently bonded atoms

polysaccharides (20)   Carbohydrates containing many sugar groups per molecule

precision (2)   The closeness of repeated measurements to each other

pressure (11)   Force per unit area

primary protein structure (20)   The sequence of amino acids in a protein

proton (4)   A particle with a mass of 1.0072766 amu and a relative positive charge of one unit

R group (18)   Any number of carbon atoms attached

to the reactive part of the molecule (the functional group)

radioactivity (17)   Emissions from the nucleus of an atom as it undergoes nuclear decay, usually in the form of alpha or beta particles, gamma rays, or x rays

reactants (3)   The starting materials in a chemical reaction

reduction (8)   Gain of electrons

reducing agent (8)   A substance that causes something else to be reduced while it, itself, is oxidized

salt (15)   A compound composed of the positive ion of a base and the negative ion of an acid

saturated hydrocarbon (18)   An organic compound that contains only carbon and hydrogen, bonded only by single bonds

saturated solution (14)   A solution in which no more solute can be dissolved

secondary protein structure (20)   The way in which a protein molecule (amino acid chain) is arranged in space

significant figures (2)   In a number, the digits that are known to be correct

single-replacement reactions (8)   Reactions in which an uncombined or free element replaces another element in a compound

solid (13)   One of the states of matter in which the units are held together in a definite arrangement. (Solids have a definite shape and definite volume.)

solute (15)   In a solution, the substance that is being dissolved

solution (3)   A homogeneous mixture of two or more substances

solvent (14)   In a solution, the substance that is doing the dissolving

specific heat (10)   The number of calories needed to raise the temperature of 1 gram of a substance by 1 degree Celsius

standard-state conditions (10)   In thermochemistry, 25° C and 1 atmosphere pressure

standard temperature and pressure (11)   273 K (0° C) and 1 atmosphere pressure (760 torr)

stoichiometry (9)   Calculation of the quantities of chemical elements or compounds involved in chemical reactions

steroids (20)   Physiologically active lipids having the following basic structure:

structural formula (18)   A representation of the bonding of the carbon atoms in an organic compound

synthesis (9)   The formation of the whole from its constituents

temperature (2)   A measure of the intensity of heat

terpenes (20)   Polymers of isoprene—for example, vitamins A, E, and K

tertiary protein structure (20)   The folding pattern of a protein's secondary structure

thermochemical equation (10)   A chemical equation that includes information about the physical state of the reactants and products and the amount of heat released or absorbed during the chemical reaction

titration (15)   The process of determining the amount of a substance present in a solution by measuring the volume of a different solution of known strength that must be added to complete a chemical change

triglyceride (20)   A lipid, commonly called a fat, in which all three —OH groups of the alcohol glycerol have been esterified

unit cell (13)   A repetitive unit in a crystal lattice

unsaturated hydrocarbon (18)   An organic compound that contains only carbon and hydrogen and that has some carbon–carbon double bonds or carbon–carbon triple bonds

vacuum (4)   An enclosed space from which all matter has been removed

vitamins (20)   Nutrients that the body needs to maintain good health. Most vitamins serve as catalysts in the body.

volatile (13)   Easily vaporized

volume (2)   The capacity of a three-dimensional object

waxes (20)   A class of lipids formed from fatty acids and high-molecular-weight alcohols other than glycerol

weight (2)   The gravitational attraction of an object to the earth or any other body

Supplement D
Expanded Rules for Writing and Naming Chemical Compounds

D.1 Common Names

There are no rules governing the common names of compounds. Most common names have been derived from common usage or just handed down through chemical history. For example, table salt is sodium chloride, and lye is sodium hydroxide. Table D.1 lists some additional examples of common names.

Table D.1
Common Names of Some Chemical Compounds

Common name	Formula	Chemical name
Baking Soda	$NaHCO_3$	Sodium hydrogen carbonate
Borax	$Na_2B_4O_7 \cdot 10H_2O$	Sodium tetraborate decahydrate
Epsom salts	$MgSO_4 \cdot 7H_2O$	Magnesium sulfate heptahydrate
Gypsum	$CaSO_4 \cdot 2H_2O$	Calcium sulfate dihydrate
Laughing gas	N_2O	Nitrous oxide
Quicklime	CaO	Calcium oxide
Marble	$CaCO_3$	Calcium carbonate
Milk of magnesia	$Mg(OH)_2$	Magnesium hydroxide
Muriatic acid	HCl	Hydrochloric acid
Oil of vitriol	H_2SO_4	Sulfuric acid
Saltpeter	$NaNO_3$	Sodium nitrate

D.2 Names for Binary Compounds

Binary compounds are derived from the names of the ions or elements that compose them. Let's first look at the rules for positive ions and then at the rules for negative ions.

Positive Ions

The name of the element is the name of the ion. For example, Na^{1+} is the sodium ion, Mg^{2+} is the magnesium ion, and Al^{3+} is the aluminum ion. For those elements that can exist as either one of two (or more) positive ions, the Stock System, which is recommended by the IUPAC, is used to name the ion. In this system, a Roman numeral is used to designate the charge on the ion. For example, Cu^{1+} is called copper(I) and Cu^{2+} is called copper(II). The ion Fe^{2+} is called iron(II) and Fe^{3+} is called iron(III). (Note that the roman numeral refers to the *charge on the ion*.)

Although the Stock System is the system of choice, the old system of naming positive ions that can exist with more than one charge is still used today. Therefore it is important to review the rules of this system. The ion with the lower charge is given the ending *-ous*, and the ion with the higher charge is given the ending *-ic*. Also the Latin form of the element is used. For example, Fe^{2+} is called ferr*ous*, and Fe^{3+} is called ferr*ic*. Note the use of the suffixes *-ous* and *-ic* and the use of *ferr-*, not iron, as the main stem of the element. Another example is copper. The Cu^{1+} ion is called cupr*ous*, and Cu^{2+} is called cupr*ic*.

Table D.2 lists some additional examples of positive ions and their names, using the rules of both systems.

Negative Ions

The name of the negative ion is the name of the element with the ending *-ide*. For example, Cl^{1-} is called chlor*ide* ion, and O^{2-} is called ox*ide* ion. The ion S^{2-} is called sulf*ide*, and P^{3-} is called phosph*ide*.

If the negative ion is a polyatomic ion (one consisting of two or more elements), the name is *best learned* by memorizing it along with its formula. For example, $(PO_4)^{3-}$ is called the phosphate ion, and $(SO_4)^{2-}$ is called the sulfate ion. The polyatomic ion $(NO_2)^{1-}$ is the nitrite ion, and $(NO_3)^{1-}$ is the nitrate ion. Table D.3 gives a summary of negative ions, including polyatomic ions.

Writing the Names and Formulas of Binary Compounds

Using Tables D.2 and D.3, we can now write the formulas of binary compounds if we are given their names, and we can write the names of binary compounds if we are given their formulas.

Table D.2
Some Positive Ions

1+		2+	
Hydrogen	H^{1+}	Calcium	Ca^{2+}
Lithium	Li^{1+}	Magnesium	Mg^{2+}
Sodium	Na^{1+}	Barium	Ba^{2+}
Potassium	K^{1+}	Zinc	Zn^{2+}
Mercury(I)*	Hg^{1+} (also called mercurous)	Mercury(II)	Hg^{2+} (also called mercuric)
Copper(I)	Cu^{1+} (also called cuprous)	Tin(II)	Sn^{2+} (also called stannous)
Silver	Ag^{1+}	Iron(II)	Fe^{2+} (also called ferrous)
Ammonium	$(NH_4)^{1+}$	Lead(II)	Pb^{2+} (also called plumbous)
Rubidium	Rb^{1+}	Copper(II)	Cu^{2+} (also called cupric)
Cesium	Cs^{1+}	Strontium	Sr^{2+}
		Nickel(II)	Ni^{2+}
		Chromium(II)	Cr^{2+} (also called chromous)
		Cobalt(II)	Co^{2+} (also called cobaltous)
		Manganese(II)	Mn^{2+} (also called manganous)

3+		4+	
Aluminum	Al^{3+}	Tin(IV)	Sn^{4+} (also called stannic)
Iron(III)	Fe^{3+} (also called ferric)	Lead(IV)	Pb^{4+}
Bismuth(III)	Bi^{3+}	Manganese(IV)	Mn^{4+}
Chromium(III)	Cr^{3+} (also called chromic)		
Cobalt(III)	Co^{3+} (also called cobaltic)		

*Note that the mercury(I) ion is a diatomic ion. In other words, you never find Hg^{1+} ion alone, but always as Hg^{1+}—Hg^{1+}. The two Hg^{1+} ions are bonded to each other.

Here are suggestions to help you choose the right subscripts.

1. Write the symbol of each element of the compound, along with its oxidation number. For example, in the compound zinc chloride, write

$$Zn^{2+} Cl^{1-}$$

2. You want the positive side of the compound to balance the negative side. You accomplish this by crisscrossing the numbers.

$$Zn^{2+}\;Cl^{1-} \qquad \text{or} \qquad Zn_1{}^{2+}Cl_2{}^{1-} \qquad \text{or} \qquad ZnCl_2$$

3. Note that the subscript numbers are written without charge (that is, without a plus or a minus sign). For example, the formula for aluminum oxide is

$$Al^{3+}\;O^{2-} \qquad \text{or} \qquad Al_2{}^{3+}O_3{}^{2-} \qquad \text{or} \qquad Al_2O_3$$

Table D.3
Some Negative Ions

1^-		2^-		3^-	
Fluoride	F^{1-}	Oxide	O^{2-}	Nitride	N^{3-}
Chloride	Cl^{1-}	Sulfide	S^{2-}	Phosphide	P^{3-}
Hydroxide	$(OH)^{1-}$	Sulfite	$(SO_3)^{2-}$	Phosphate	$(PO_4)^{3-}$
Nitrite	$(NO_2)^{1-}$	Sulfate	$(SO_4)^{2-}$	Arsenate	$(AsO_4)^{3-}$
Nitrate	$(NO_3)^{1-}$	Carbonate	$(CO_3)^{2-}$	Borate	$(BO_3)^{3-}$
Acetate	$(C_2H_3O_2)^{1-}$	Chromate	$(CrO_4)^{2-}$		
Bromide	Br^{1-}	Dichromate	$(Cr_2O_7)^{2-}$		
Iodide	I^{1-}	Oxalate	$(C_2O_4)^{2-}$		
Hypochlorite	$(ClO)^{1-}$				
Chlorite	$(ClO_2)^{1-}$				
Chlorate	$(ClO_3)^{1-}$				
Perchlorate	$(ClO_4)^{1-}$				
Permanganate	$(MnO_4)^{1-}$				
Cyanide	$(CN)^{1-}$				
Hydrogen sulfite	$(HSO_3)^{1-}$				
Hydrogen sulfate	$(HSO_4)^{1-}$				
Hydrogen carbonate	$(HCO_3)^{1-}$				

4. Also note that the subscript numbers should be in least-common-denominator form. For example, the formula for barium oxide is

$$Ba^{2+}O^{2-} \quad \text{or} \quad Ba_2^{2+}O_2^{2-} \quad \text{or} \quad Ba_1^{2+}O_1^{2-} \quad \text{or} \quad BaO$$

Example D.1

Write the chemical formula for sodium chloride.

Solution

Because the sodium ion has a charge of $+1$ (Na^{1+}), and the chloride ion has a charge of -1 (Cl^{1-}), the formula for sodium chloride is

$$Na_1^{1+}Cl_1^{1-}$$

The subscripts show that we need one ion of each element to obtain a neutral compound. The subscript 1 is usually not written. Likewise, the charges $1+$ and $1-$ are left out, giving us the chemical formula NaCl.

Let's look at the formation of chemical compounds in more detail. We assume that any compound has a positive part and a negative part. The oxidation number tells us *how* positive or negative a specific ion is. It is our job to choose the right subscripts to balance the charges, so that the compound is electrically neutral.

Example D.2

Write the formula for aluminum sulfide

Solution

From Tables D.2 and D.3, we obtain

$$Al^{3+} S^{2-}$$

Now we choose the right subscripts to balance the charges.

$$Al_2^{3+} S_3^{2-} \quad \text{or} \quad Al_2S_3$$

You can use the same procedure when dealing with polyatomic ions. If two or more polyatomic ion groups are needed to balance the charges, remember to put parentheses around the polyatomic ion.

Example D.3

Write the chemical formula of aluminum sulfate.

$$Al^{3+} (SO_4)^{2-}$$

Solution

Now choose the right subscripts to balance the charges.

$$Al_2^{3+} (SO_4)_3^{2-} \quad \text{or} \quad Al_2(SO_4)_3$$

Example D.4

Write the chemical formula for copper chloride.

Solution

Which copper compound do we mean? We can have copper(I) chloride or copper(II) chloride. The roman numeral tells us the oxidation number of the copper compound that is formed.

If we mean copper(I) chloride, we find the formula in the following way. We start with

$$Cu^{1+}Cl^{1-}$$

Now we choose the right subscripts to balance the charges.

$$Cu_1^{1+}Cl_1^{1-} \quad \text{or} \quad CuCl$$

If we mean copper(II) chloride, we find the formula in the following way. We start with

$$Cu^{2+}Cl^{1-}$$

Now we choose the right subscripts to balance the charges.

$$Cu_1^{2+}Cl_2^{1-} \quad \text{or} \quad CuCl_2$$

Each copper chloride is a distinctly different compound, with its own properties.

We may write the names of chemical compounds from their formulas by following the rules we have just studied. Remember that, if the positive-ion part of the compound can exist as more than one positive ion, we must give the charge of the ion in that particular compound.

Example D.5

Name each of the following compounds,
(a) $CuCl_2$ **(b)** $HgSO_4$ **(c)** $Cu(NO_3)_2$ **(d)** $Fe_2(SO_3)_3$
(e) Ag_2CrO_4 **(f)** $Ba_3(PO_4)_2$ **(g)** $MnCl_2$ **(h)** MnO_2
(i) Cu_2S **(j)** SnO_2 **(k)** Cs_2CO_3 **(l)** UBr_4

Solution
(a) $CuCl_2$ is called copper(II) chloride or cup*ric* chloride.
(b) $HgSO_4$ is called mercury(II) sulfate or mercu*ric* sulfate.
(c) $Cu(NO_3)_3$ is called copper(II) nitrate or cup*ric* nitrate.
(d) $Fe_2(SO_3)_3$ is called iron(III) sulfite or fer*ric* sulfite.
(e) Ag_2CrO_4 is called silver chromate.
(f) $Ba_3(PO_4)_2$ is called barium phosphate.
(g) $MnCl_2$ is called manganese(II) chloride or manga*nous* chloride.
(h) MnO_2 is manganese(IV) oxide (commonly called manganese dioxide).
(i) Cu_2S is called copper(I) sulfide or cup*rous* sulfide.
(j) SnO_2 is called tin(IV) oxide or stan*nic* oxide (commonly called tin dioxide).
(k) Cs_2CO_3 is called cesium carbonate.
(l) UBr_4 is called uranium(IV) bromide (commonly called uranium tetrabromide).

D.3 Names for Ternary and Higher Compounds (Three or More Elements)

For compounds that are ternary or higher, we follow the rules for binary compounds. This is because the negative part of ternary compounds is usually composed of polyatomic ions. (That's what makes them ternary.) For example, Na_3PO_4 is sodium phosphate and $Fe(NO_3)_3$ is iron(III) nitrate.

Example D.6

Name each of the following ternary compounds.
(a) $Ca(NO_3)_2$ **(b)** $Zn_3(PO_4)_2$ **(c)** $Cu(C_2H_3O_2)_2$ **(d)** K_2SO_4

Solution

(a) Calcium nitrate   **(b)** Zinc phosphate   **(c)** Copper(II) acetate
(d) Potassium sulfate

D.4 Names for Inorganic Acids

According to Arrhenius, an acid is a substance that releases hydrogen ions, H^{1+}, in aqueous solutions. Two major classes of inorganic acids exists, and the naming rules differ for the two classes. First we'll examine the rules for naming the non-oxygen-containing acids. Then we'll examine the rules for naming the oxygen-containing acids.

Non-Oxygen-Containing Acids

These acids usually consist of hydrogen plus a nonmetal ion. For example, HCl, which is hydrogen chloride gas in its pure form, becomes hydrochloric acid when dissolved in water. The covalently bonded HCl molecules ionize in water to form hydrogen ions and chloride ions. We derive the names of acids such as HCl by adding the prefix *hydro-* to the name of the nonmetal, which is given an *-ic* suffix. For example,

hydro——— chlor ——— ic ———acid

Prefix   Name of   Suffix
         nonmetal

Example D.7

Name each of the following non-oxygen-containing acids.
(a) HF   **(b)** HBr   **(c)** H_2S **(d)** HCN

Solution

We'll give the name of each substance first as a covalent compound and then as an acid in aqueous solution.

Name as covalent compound	*Name as acid in aqueous solution*
(a) Hydrogen fluoride	Hydrofluoric acid
(b) Hydrogen bromide	Hydrobromic acid
(c) Hydrogen sulfide	Hydrosulfuric acid
(d) Hydrogen cyanide	Hydrocyanic acid

Oxygen-Containing Acids

Oxygen-containing acids formed from hydrogen and polyatomic ions (containing oxygen) are named as follows:

1. If the polyatomic ion has an *-ate* suffix, we name the acid by replacing the *-ate* suffix with an *-ic* suffix and adding the word *acid*. For example, HNO_3, composed of a hydrogen ion and nitr*ate* ion, is called nitr*ic* acid in aqueous solution.

Example D.8

Name each of the following oxygen-containing acids
(a) H_2SO_4 **(b)** $HC_2H_3O_2$ **(c)** $HBrO_3$

Solution

These three acids are composed of hydrogen plus a polyatomic ion that has the suffix *-ate*.
(a) H_2SO_4 is called sulfur*ic* acid. (SO_4^{2-} is the sulf*ate* ion.)
(b) $HC_2H_3O_2$ is called acet*ic* acid. ($C_2H_3O_2^{1-}$ is the acet*ate* ion.)
(c) $HBrO_3$ is called brom*ic* acid. (BrO_3^{1-} is the brom*ate* ion.)
(Note that H_2SO_4 is an exception to our rule, in that the whole name of the element is used rather than the name of the sulfate group.

2. If the polyatomic ion has an *-ite* suffix, we name the acid by replacing the *-ite* suffix with an *-ous* suffix and adding the word *acid*. For example, HNO_2, composed of a hydrogen ion and a nitr*ite* ion, is called nitr*ous* acid.

Example D.9

Name each of the following oxygen-containing acids
(a) H_2SO_3 **(b)** H_3PO_3 **(c)** $HBrO_2$

Solution

These three acids are composed of hydrogen plus a polyatomic ion that has the suffix *-ite*.

(a) H_2SO_3 is called sufur*ous* acid. (SO_3^{2-} is the sulf*ite* ion.)

(b) H_3PO_3 is called phosphor*ous*. (PO_3^{2-} is the phosph*ite* ion.)

(c) $HBrO_2$ is called brom*ous* acid. (BrO_2^{1-} is the brom*ite* ion.)

(Note that H_2SO_3 is an exception to our rule, in that the whole name of the element is used rather than the name of the sulfite group.

3. Some elements form more than two oxygen-containing acids. Most notable are the acids formed by the elements chlorine, bromine, and iodine. This occurs because these elements can form four polyatomic ions. For example, bromine forms the following ions:

BrO_4^{1-}, perbrom*ate* ion BrO_3^{1-}, brom*ate* ion

BrO_2^{1-}, brom*ite* ion BrO^{1-}, hypobrom*ite* ion

Note that the first two ions listed have the suffix *-ate* and that the last two ions listed have the suffix *-ite*. Therefore we use rules 1 and 2. Keep the name of the polyatomic ion, but change the *-ate* suffix to *-ic* or the *-ite* suffix to *-ous*.

Example D.10

Name each of the following oxygen-containing acids.
(a) $HBrO_4$ **(b)** $HBrO_3$ **(c)** $HBrO_2$ **(d)** $HBrO$

Solution

(a) $HBrO_4$ is called perbrom*ic* acid. (BrO_4^{1-} is the perbrom*ate* ion.)

(b) $HBrO_3$ is called brom*ic* acid. (BrO_3^{1-} is the brom*ate* ion.)

(c) $HBrO_2$ is called brom*ous* acid. (BrO_2^{1-} is the brom*ite* ion.)

(d) $HBrO$ is called hypobrom*ous* acid. (BrO^{1-} is the hypobrom*ite* ion.)

Supplement E
Study Skills

About Learning Chemistry

Chemistry is an exciting subject. It affects our lives each day in many ways. The foods we eat and the clothes we wear, the materials used to construct our homes, the medications that cure our illnesses, and the energy we depend on are all products of the science of chemistry coupled with technology.

As you learn chemistry, you will gain knowledge that will help you understand more about our technological society. You'll be able to apply your understanding of chemistry to help solve some of the problems you will face in the real world. You'll be able to approach familiar situations with new insight and to make decisions based on the knowledge you'll acquire in this course.

It will be easier for you to learn chemistry if you start out with a positive attitude and the knowledge of how to approach the subject most effectively. Although some of the material in the course requires memorization, a large part of it involves learning how to solve problems. *Basic Concepts of Chemistry* focuses on developing your problem-solving skills. Throughout the text are worked-out examples with solutions given step by step. Following each worked-out example is a practice exercise that offers the same type of problem whose solution has just been demonstrated. The practice exercises give you an opportunity to test your understanding of the skills presented in the chapter. At the end of each chapter, two additional sets of problems give you even more practice in problem solving.

The learning goals listed at the beginning of each chapter will help you organize the material you'll need to master. How can you be sure, as you go on with your studies, that your knowledge of the material is adequate? As you read each chapter you will notice that the learning goals are printed in color for easy location of important material. The examples and practice exercises described above are included wherever problem-solving skills are essential to your understanding of the topic. At the end of each chapter, self-test exercises are keyed to each of the learning goals. If you are able to correctly work through the examples, the practice exercises, and the self-test exercises, you can assume you have mastered the necessary problem-solving skills. For sections that require you to memorize reactions or descriptive material, self-test exercises are designed to help you master these skills, too.

Before an exam, it is important to speak with your instructor to find out what material will be covered on the test. Instructors emphasize important concepts in class, and some will help guide your studies by offering information on the material that deserves special attention.

You can successfully complete basic chemistry if you develop good study skills. The tools to help you learn the subject are in the classroom and in the textbook. If you use your time efficiently, and ask for help from your instructor when you encounter material that needs further explanation, you should be able to do well in the course.

The sections that follow discuss some common-sense methods of achieving success in college and offer additional advice you can apply to your day-to-day study of chemistry. As you become a well-prepared and successful student, you will approach examinations with confidence. The successful habits you form will stay with you as you complete your education and will benefit you throughout your life.

Time Management for College Students

You've got an exam on Friday and a paper due on Monday. You know you've got to get to work. Just then the phone rings. A group of friends from your dorm is going out for a snack. You know you shouldn't join them, but you've got to eat anyway, so you go along. When you return home you need a little time to unwind. You turn on the television. Two hours later you're relaxed, but you're also tired. You decide to call it a night. There's always tomorrow!

If this scenario sounds familiar, it should. It happens to all of us now and then. As a college student, however, it's very important to learn to manage your time effectively.

Setting Realistic Goals

The first step to managing your time effectively is to know where you're going. It helps to set goals for yourself. Three different types of goals that you should consider are long-range goals, medium-range goals, and short-range goals.

Long-range goals are usually personal wishes. They have to do with your career aims, your educational plans, and your social desires. Think about where you would like to be five or ten years from now. The education you are now receiving in college should be a steppingstone to your long-range goals. Depending on your career plans, the grades you earn in your courses will help determine whether or not you will be able to fulfill these goals.

Medium-range goals help you achieve your long-range goals. They can be set two or three times a year. For example, if you plan to enter medical school after graduation, you will need a considerable number of A's in your courses. A medium-range goal would be to get four or five A's in your courses. Another medium-range goal might be to join a club or to improve your skills in your favorite sport. Let's say that your grades last semester weren't the best. A medium-range goal for you, then, might be to improve them. If you're saving money to buy a car, watching your budget more carefully might be a reasonable goal to set.

Short-range goals can be set each week. These goals include taking care of your daily tasks and keeping up with your assignments. To read a chapter in a book, complete an assignment, or write a paper is an example of a short-range goal.

Planning to Reach Your Goals

Achieving your goals often requires the setting up of a plan. Although we'd like to think we have the willpower to do all we have to do, very often we become distracted. A well-planned, flexible schedule is a useful tool to use to get things done more effectively.

The first step to managing your time is to know yourself. When do you function the best? If you're a morning person, it's best to do as much of your work as you can in the morning when you are at your best. If you function better later in the day, try to schedule most of your work in the afternoon or evening.

Setting up a monthly plan is the next step in becoming organized. Obtain a calendar and write in, on the appropriate date, the important assignments and events you need to remember for each month of the semester (Figure E.1). Term papers, exams, sports meets, and social activities can all fit in if you see the overall picture and plan in advance. For example, if you know you have a paper due the same week of two scheduled exams, you'll know you have to keep your schedule free to allow for your school work.

After you've completed your monthly plan, devise a weekly plan. Map out a schedule for each day of the week. Think over how much time you need to complete an assignment, or to study for an exam, in order to fit all of your responsibilities into your schedule (Figure E.2).

Figure E.1
Monthly Planning Calendar

MONTH_____ YEAR_____

SUNDAY	MONDAY	TUESDAY	WEDNESDAY	THURSDAY	FRIDAY	SATURDAY

It's usually best to find a quiet place to work. Accomplish your tasks in priority order, breaking the large tasks into smaller ones. Don't try to do too much in one day. It's better to do a little less than to do too much. Complete one task at a time, and avoid working on two projects at once.

You may find that after a month or so, you can manage your time effectively without a written schedule. For some students, having a written schedule is a necessity. When you are designing your plan, don't forget that you need time for enjoyment, as well as for work. Giving yourself positive reinforcement is a good idea. Reward yourself when you've accomplished something worthwhile. It is important to enjoy what you're doing and to set enough time for rest and relaxation.

Study Skills for Learning Chemistry

Developing good study skills is an important goal for every student. They'll form the basis of your success in school by enabling you to keep pace with the material covered in your classes. One of the primary reasons many students are forced to drop out of school is that they lack good study skills. In this section we'll review some of the basics of effective studying, and then hone in on some special tips for learning chemistry.

Figure E.2
Weekly Plan

Your main goal this week: _____

	Monday	Tuesday	Wednesday	Thursday	Friday	Saturday	Sunday
6 A.M.							
7 A.M.							
8 A.M.							
9 A.M.							
10 A.M.							
11 A.M.							
12 A.M.							
1 P.M.							
2 P.M.							
3 P.M.							
4 P.M.							
5 P.M.							
6 P.M.							
7 P.M.							
8 P.M.							
9 P.M.							
10 P.M.							
11 P.M.							
12 A.M.							

Study Skill Basics

There are several skills that a good student must develop. The ability to read with comprehension, to memorize when necessary, to take thorough notes, and to solve problems and analyze the material at hand are the basics of studying effectively. But there are also some negative factors that can stand in the way of good skills. Some students spend a great deal of time worrying about school work, which wastes energy and time. And some students come into chemistry class convinced that they won't be able to learn the material, which is an attitude that only invites failure.

Students are also often overburdened with activities. There are only seven days in a week, and it's impossible to fit eight days worth of activities into seven days. Trying to do this often results in your doing things haphazardly and poorly. Effective time management is the key to overcoming the dilemma of having more to do than there are hours in a day.

In addition, it's important for you to care about your studies. If you lack interest in what you're learning and feel no enthusiasm for the subject, you'll be reluctant to put in the time required for studying chemistry. Sometimes lack of interest results from uncertainty about your long-range goals. If you know why you're going to study chemistry, your interest level will remain high.

Finally, lack of intellectual ability can be a reason for not achieving success in a subject. Your instructor is the best person to consult with if you come to believe you don't have the basic knowledge to succeed in chemistry. He or she can help you evaluate your basic skills and determine whether they are adequate.

Getting Yourself to Study

Many people find it difficult to sit down and study. They fall prey to countless distractions to avoid facing the task. If you have a problem getting started, convince yourself you'll be able to get through the material quickly by breaking it up into small, manageable units. After learning each small unit, stop and take a short break, and then continue.

Developing the ability to concentrate is an important study skill. Keep your mind on what you're doing. Don't allow yourself to be distracted. Remove the telephone, the television set, and try to eliminate interruptions. Find a quiet place where you won't be disturbed. Low-level noise such as instrumental music or a steady flow of traffic sometimes helps stimulate concentration. And don't try to study if you are hungry. That's a sure way to break your concentration.

The Importance of Attending Class

Nothing is more crucial to your study of chemistry than attending class. Regular attendance allows you to have all of the lecture notes you'll need to organize your studies, and the experiments you perform in the laboratory give you hands-on experience in chemical principles. Except for illnesses and emergencies, there is no valid reason for missing class. Punctuality is also important. Being late interrupts the class and causes you to miss some of the material being discussed.

How to Study Chemistry

The first step in your studies is to read the relevent chapter in the textbook before going to the lecture. Survey the material for the topics to be covered. Examine the types of problems to be solved. Know what the chapter is about.

Each chapter in *Basic Concepts of Chemistry* begins with a list of learning goals. Look these over and locate them in the chapter. These topics are likely to have special emphasis in class. Also look over the figures and tables to see how they relate to the material. Glance over the summary at the end of each chapter.

In class you will take notes. Review them the next time you study, and reread each section of the textbook slowly, for understanding. Combine what you learned in class with what you read in the textbook. Where problem solving takes place, go over the step-by-step solutions in the textbook. Then answer the practice exercises to be sure you understand how to find the solutions yourself. If you're stuck, ask the instructor for help. *Do not ignore material you do not understand.*

The next time you study, go back over your class notes and reread for a second time the sections you have learned in the textbook. *Basic Concepts of Chemistry* is known for its simple, conversational, highly readable style. You should not have a difficult time understanding the book. Spend some time being sure you can recall what you have learned. Sometimes it helps to write as you read, recalling what you have learned in your own words on paper. The last step is to reread the entire chapter and go over all of your notes.

Taking Good Notes

Note taking is one of the most important skills for a student to develop. Your notes must be accurate to be useful. You will have to develop your own shorthand in order to take notes quickly. There are many symbols and abbreviations you can use in devising a shorthand. Mathematical symbols such as $>$ (greater than) and $<$ (less than), the plus sign ($+$), and the equals sign ($=$) are useful. Spelling a word phonetically by leaving out the vowels also speeds things up.

As you work through the exercises in each chapter of the textbook, clearly indicate each type of problem you are solving. Write a short description of the type of problem and explain why you are solving this particular one. When there are several variables in a problem, be sure you can solve for each of the variables, depending on the information given.

Combine your lecture notes with your textbook notes. Be certain you understand your notes. To test your knowledge, turn to the end of the chapter and do the self-test exercises. If you can answer them, you're on your way.

Taking a Test

Being well prepared and confident is the way to approach taking a test. Cramming doesn't work in chemistry. You cannot open your book the night before the exam and learn a few week's worth of material in one sitting. Chemistry must be studied gradually. If you have trouble solving a particular type of problem, you may need to consult your instructor. That's difficult to do if you try to study at the last minute.

Try to study a little each day. You'll find this method more effective and less stressful. *Basic Concepts of Chemistry* contains an entire package of helpful materials. Besides the textbook, there is a study guide for students written by James R. Braun, and a solutions manual, containing step-by-step solutions to all exercises that have specific answers. These ancillary materials can give you the extra practice you may need to achieve success in the course.

Get a good night's sleep before an exam. Be as relaxed as possible so you can think easily. You'll know whether you're prepared or not. If you've studied properly, you should do well.

Avoiding School-Related Stress

Your college years are likely to be among the most demanding and enjoyable years of your life. During this time many activities will compete for your attention. Studying, developing relationships, and handling your financial affairs are among the most important challenges that will require your energy, creativity, and brain power. You'll be able to successfully manage the different facets of your life by developing good coping skills. These are skills that will remain with you throughout your life and that can be applied to just about any situation you'll encounter.

Maintaining a Good Balance

School-related stress usually results when you allow yourself to become overwhelmed and overloaded. Taking on too much responsibility often results in a high stress level, which can cause you a whole range of physical, emotional, and social problems.

When you register for your courses, be sure you're not taking on too much. Most colleges schedule a full load for you. This means that you sign up for a set number of credits that school personnel believe a full-time student can handle successfully. The full load is calculated so that a full-time student can finish the requirements for a two-year degree in two years, or a bachelor's degree in four years.

Students who work diligently and have good study skills and few other demands on their time can normally handle a full load. Problems often arise when a student signs up for more credits than are recommended for a full load, or when he or she has other time-consuming responsibilities. When you have too much responsibility to handle at once, stress can readily develop. Instead of completing your course work with ease, you may find yourself playing "catch-up" all semester.

Arranging a course load to fit your needs is one of the most important objectives as a college student. You can avoid overloading yourself by taking a night course during the summer at a local college, or by attending summer school. Spending an extra semester or two in college is a reasonable price to pay in order to achieve success and avoid stress while you're in school.

When Money Becomes a Problem

College students very often find that they are low on money. Instead of wasting time worrying about how you'll get by, work on figuring out how you can make ends meet.

Students with a full academic load can usually work a maximum of ten hours per week. If you work more than ten hours per week you should not carry a full load or you're likely to become overloaded. You might be able to keep up with such a schedule for a month or two, but you'll probably find yourself falling behind in your school work, and putting too much stress on your body.

If you have financial problems, find out if your school has a financial aid office where you can get information about scholarships and student loans. A financial aid counselor can often help you make up a budget. Your school may also have a job-placement office to help you locate part-time work.

Being on Your Own

If you are like most young adults, your college years represent the first time you are on your own and away from your family. Loneliness is a feeling commonly shared by many students at this time. It is important to learn to meet people and make new friends when you enter college. Although it's often difficult to strike up a conversation with people you don't know, it's a skill you'll need to develop to combat loneliness.

College campuses have many clubs and organizations that provide opportunities to meet people. The college newspaper, a sports team, a political party, or the school governing body are places to get started. Make appointments and schedule social meetings with your new acquaintances. It'll make being on your own in a new environment less threatening.

When You Socialize Too Much

Whereas some people find it difficult to socialize, others become overly involved in social activities. Attending college can be very exciting, and for some young adults the endless opportunities to work for a cause, so-cialize with friends, or attend frequent parties is irresistible. It is not un-common for some college students to use study time for socialization. But around the time of midterms, panic sets in for these students and lasts until final exams. Don't get caught in this trap. Be sure you are not be-coming involved in too many social activities in order to avoid your stud-ies. If you're serious about receiving a college education, your study time will be important to you.

Maintaining Your Individuality

Peer pressure is a powerful force at all ages, and it plays a large role dur-ing the college years. As humans, we all want approval from our friends and are sometimes afraid we'll lose their respect or companionship if we disagree with them or refuse to go along with the group.

The truth is, you must set your own goals for yourself. If doing well in school is very important to you, and your friends criticize you for working so hard, then they probably don't share your goals or values. This doesn't necessarily mean you should give up your studies or your friends. It means that it's time to assert yourself and do what you think is right. If you feel you must compromise too much to remain a member of your cir-cle of friends, then perhaps it's time to develop new relationships.

If College Isn't Right for You

Sometimes we find ourselves in situations that aren't right for us at a par-ticular time in our lives. Some students may have to work a full-time job, leaving little time for course work or studies. Problems with relationships, drugs, alcohol, or one's family may be other reasons that college isn't working out. A lack of good study skills or simply a general lack of inter-est in school may also suggest that it's not the right time for an individual to be in college. If you find yourself in one of these situations, consider taking a leave of absence from college, or even dropping out. You can re-turn when the time is right for you.

If a major problem causes you to consider leaving college, the first task at hand is to work on solving that problem. If you can't solve the problem yourself, don't hesitate to seek professional help. Your college counseling office may offer some useful solutions.

Sometimes leaving college works out for the best. It might give you the opportunity to develop a skill or talent that you truly enjoy. It might en-courage you to sit down and think about your goals for the future. Col-leges today enroll a large number of adults who return to begin or com-plete their studies. They tend to be serious students who are interested in learning and so are an asset to their classes.

Words of Encouragement

As you begin your study of chemistry, we hope you do so with an open mind and a positive attitude. You are probably enrolled in a chemistry course because it is part of the path you have chosen in order to reach your career goals. We hope you will enjoy the course and find the textbook a useful tool to help you learn the subject matter.

This supplement has been added to the text to provide guidance for those who will benefit from it. Our goal is not to offer advice, but rather to offer suggestions. Chemistry may be new to you, and we hope that some of these tips for studying more effectively will be useful to you. If you have taken a chemistry course before, or if you have a different method of studying that works for you, we invite you to write to us and share your suggestions.

As you enter the fascinating world of chemistry, we hope your experience is a good one and that you come away from the course with an understanding of why we remain so enthusiastic about this exciting subject.

Answers to Selected Exercises

This part of the text contains answers to the odd-numbered Self-Test Exercises and to the Extra Exercise problems at the end of each chapter. Only those exercises that have either numerical or quantitative answers are given. Answers to even-numbered Self-Test Exercises are *not* given. In addition, answers to all of the Cumulative Review Exercises are included.

CHAPTER 2

1. 18 m^2
3. 426 cm^2 or 66.1 in^2 (depending on how you measure the page)
5. $6\bar{0}0$ m^2
7. $45\bar{0}$ cm^3
9. 42,400 cm^3
11. $45\bar{0}0$ cm^3
13. $17\bar{0}0$ cm^3
15. 1728 in^3
17. (a) 150 g   (b) 1500 dg   (c) 150,000 mg
19. (a) 31 dm   (b) 310 cm   (c) 3100 mm
21. (a) 14.9 cm   (b) 0.149 m
    (c) 0.000149 km
23. (a) $785\bar{0}$ mm   (b) 785.0 cm
    (c) 78.50 dm   (d) 7.850 m
    (e) 0.007850 km
25. (a) 2500 m   (b) 25,000 dm
    (c) 250,000 cm
27. (a) 5678.0 cg   (b) 567.80 dg
    (c) 56.780 g   (d) 0.056780 kg
29. (a) 3.500 L   (b) 35.00 dL
31. 1.00×10^6 cm^3
33. (a) 82.0 ft   (b) 985 in.

35. (a) 0.32 gal (b) 1.3 qt
37. (a) 0.0110 lb (b) 0.176 oz
39. 35.40 ft^3/m^3
43. 10.8 ft^2/m^2
49. (a) 3.0 m (b) 3.0 × 10^2 cm
    (c) 3.0 × 10^3 mm
51. (a) 1.83 m (b) 183 cm
53. (a) 91.4 m (b) 9140 cm
55. (a) 18.9 L (c) 18,900 mL
59. 4.80 g/cm^3
61. 203 cm^3 or 2$\overline{0}$0 cm^3 for the proper sig. figs.
63. 53 g
65. 0.019 g
67. 6.83 g/cm^3
69. 1.25 g/cm^3
71. 110 g
73. 163 g
75. 9.32 g/cm^3
77. 28,950 g or 29,$\overline{0}$00 for right number of sig. figs.
79. 22.1 cm^3
81. 3.67 g/cm^3
83. 19,500 g
85. 50.0 cm^3
87. (a) Five (b) One (c) Three
    (d) Five (e) Seven (f) One
    (g) Four (h) Three (i) Three
    (j) Four
89. (a) Seven (b) Two (c) One
    (d) Three (e) Seven (f) Three
    (g) Six (h) Four (i) Six (j) Four
91. 12 cm; two significant figures
93. 1.7 cm^2; two significant figures

95. 20.60 cm
97. 214 cm^2
99. (a) Two (b) Three (c) Four
    (d) Four (e) One (f) Two
    (g) Three (h) Four (i) Four
    (j) Three
101. (a) 8 × 10^6 (b) 2 × 10^{-3} (c) 4 × 10^{-5}
     (d) 9.05 × 10^5 (e) 2.07 × 10^{-3}
     (f) 3.05 × 10^8
103. (a) 1.0581 × 10^4 (b) 2.05 × 10^{-3}
     (c) 1 × 10^6 (d) 8.02 × 10^2
105. (a) 8.5 × 10^8 (b) 6.07 × 10^{-6}
     (c) 6.308 × 10^6 (d) 6.005 × 10^{-2}
     (e) 5 × 10^2 (f) 5.0 × 10^2
     (g) 5.00 × 10^2 (h) 5.000 × 10^2
     (i) 2.300 × 10^7 (j) 9.30 × 10^{-8}
107. −2°C
109. −40°C; at this temperature degrees F equal degrees C
111. 68°F
113. (a) 10°C (b) −70°C (c) 215°C
     (d) −90.0°C
115. (a) 203°F (b) −112°F (c) 176°F
     (d) 410°F
118. No, the person would have to be about 13.1 feet tall and weigh 44$\overline{0}$ pounds.
119. 48 cm^3
120. 218°C
123. 820
124. −8.89°C
125. (a) 61°F (b) 392°F
126. (a) 149°C (b) −101°C
127. 22.5 miles/hour

CHAPTER 3

5. (1) e (2) a (3) c (4) d (5) b
7. (a) chemical (b) physical (c) chemical
   (d) chemical (e) physical (f) chemical
13. 6.023 × 10^{23} atoms
15. 2.11 moles of Os atoms
17. 2.53 × 10^8 dollars or 3 × 10^8 dollars (for one significant figure)
19. (a) 0.600 mole (b) 5.00 moles
    (c) 2.499 × 10^{-3} mole (d) 15.0 moles
21. 81.25 g
23. 64 g
25. (a) 270.1 g Ni (b) 0.240 g Br
    (c) 8020 g Ca (d) 1.282 g S
27. (a) 10.8 g water (b) 1280 g SO$_2$

(c) 0.022 g CO$_2$ (d) 2.05 g Al$_2$(SO$_4$)$_3$
29. 1.8 × 10^{24} atoms
31. 9 × 10^{-20} dollars/atom
33. 1 × 10^{10} miles
35. NaCl, sodium chloride (table salt)
37. C$_{27}$H$_{46}$O
39. Empirical formula: CH$_3$; molecular formula: C$_2$H$_6$
41. Empirical formula: C$_2$H$_4$O; molecular formula: C$_4$H$_8$O$_2$
43. (a) 85.5 (b) 52.0 (c) 238 (d) 79.0
    (e) 74.9
45. (a) 102 (b) 223 (c) 98.0 (d) 42.4
    (e) 74.0 (f) 400 (g) 68.1 (h) 130

47. (*a*) 60.1   (*b*) 82.1   (*c*) 122   (*d*) 104
     (*e*) 188   (*f*) 219   (*g*) 149   (*h*) 60.0
49. (*a*) 0.20 mole   (*b*) 1.2×10^{23} molecules
     (*c*) 0.40 mole C atoms
     (*d*) 1.2 moles H atoms
     (*e*) 0.20 mole O atoms
     (*f*) 2.4×10^{23} C atoms
     (*g*) 7.2×10^{23} H atoms
     (*h*) 1.2×10^{23} O atoms
51. (*a*) 0.400 mole H_2O (*b*) 3.000 moles MnO_2
     (*c*) 2.5×10^{-3} mole N_2
     (*d*) 15.0 moles $(NH_4)_2SO_4$
53. (*a*) 4.00 moles   (*b*) 2.41×10^{24} molecules
     (*c*) 8.00 moles C   (*d*) 20.00 moles H
     (*e*) 8.00 moles O   (*f*) 4.00 moles N
     (*g*) 96.0 g C   (*h*) 20.2 g H   (*i*) 128 g O
     (*j*) 56.7 g N   (*k*) They should be the same.
55. 2.71×10^{24} molecules
57. $C_6H_8O_6$
59. C_6H_6
61. $C_3H_6N_6$
65. H = 2.0%, S = 32.7%, O = 65.3%
67. (*a*) B = 16%, F = 84%

     (*b*) U = 67.6%, F = 32.4%
     (*c*) C = 39.1%, H = 8.7%, O = 52.2%
69. 34 g of sugar present
71. (*a*) S = $5\overline{0}$%, O = $5\overline{0}$%
     (*b*) C = 75%, H = 25%
     (*c*) C = 42.11%, H = 6.43%, O = 51.46%
     (*d*) $\overline{Ca}$ = $4\overline{0}$%, C = 12%, O = 48%
73. $128\overline{0}$ g of copper
75. 245 g of iron
77. $2\overline{0}0$ g of oxygen
79. 305 g sodium sulfide and 195 g iron(III) oxide
82. 1.9×10^{16} years
83. CH_2Cl
84. $C_2H_4Cl_2$
85. 363 g
88. 1.76×10^{22} atoms
89. Hematite, 70% Fe, and magnetite, 72.3% Fe
90. Cu_5FeS_4 and $CuFeS_2$
91. $5\overline{0}$%
92. 1.79×10^{-21} g
93. 4.00 moles
94. 124 g
95.- 1.7×10^{-24} g

CUMULATIVE REVIEW/CHAPTERS 1–3

1. True
2. False
3. False
4. True
5. False
6. True
7. False
8. True
9. True
10. True
11. 128 m^2
12. 4548 cm^2
13. 16,540 cm^2
14. 280 g; 2800 dg; 280,000 mg
15. 68 dg; 680 cm; 6,800 mm
16. 12.5 cm; 0.125 m; 0.000125 km
17. 25.595 L, 255.95 dL
18. 164 ft, 1970 in
19. 0.0562 lb, 0.899 oz
20. 16.4 cm^3/in^3
21. 0.363 lb
22. 45.3, 1540 fl oz
23. 0.289 g/cm^3
24. 0.177 g/mL. The density of air is 1.18×10^{-3}

g/mL, so this gas is heavier than air and will not float in air.
25. 6.00 g/mL
26. 4.00 cm
27. (*a*) 31.8   (*b*) 9.29   (*c*) 79.3   (*d*) 98.3
28. (*a*) 5×10^3 (*b*) 5×10^{-4}
     (*c*) 6.023×10^8 (*d*) 3.5000×10^7
29. (*a*) 5   (*b*) 4   (*c*) 5   (*d*) 4
30. (*a*) 7.22°C   (*b*) −23°C   (*c*) 230°C
     (*d*) −73.33°C
31. (*a*) $19\overline{0}$°F   (*b*) 9.50°F   (*c*) $15\overline{0}$°F
     (*d*) 752°F
32. Amorphous solids have no definite internal structure or form. Crystalline solids have a fixed, regularly repeating symmetrical inner structure.
33. (*a*) chemical   (*b*) physical   (*c*) chemical
     (*d*) physical
34. Heterogeneous matter is made up of different parts with different properties. Homogeneous matter has the same properties throughout.
35. Dozen = 12 items; couple = 2 items
36. 0.0100 mole
37. 0.40 mole

38. (a) 73.5 g H_2SO_4
(b) 90.0 g H_2O
(c) 0.434 g H_2CO_3
(d) 3.000×10^3 moles H_2CO_3

39. 1.2×10^{23} atoms

40. SnO_2

41. $C_{40}H_{64}O_{12}$

42. $C_6H_{12}O_6$

43. (a) 0.200 mole   (b) 1.20×10^{23} molecules
(c) 1.80 moles C atoms
(d) 1.60 moles H atoms
(e) 0.800 moles O atoms

44. N_2O_4

45. (a) 310   (b) 16   (c) 46   (d) 255

46. (a) 6% H, 94% S   (b) 39.3% Na, 60.7% Cl
(c) 24% Fe, 31% C, 4% H, 41% O
(d) 45% K, 18% S, 37% O

47. C_6H_{14}

48. 74.1% C, 8.6% H, 17.3% N

49. 2.41×10^{22} molecules

50. 2.5 g $CaCO_3$

CHAPTER 4

13. $^{23}_{11}Na$ has 11 p, 11 e, and 12 n; $^{223}_{87}Fr$ has 87 p, 87 e, and 136 n; $^{238}_{92}U$ has 92 p, 92 e, and 146 n

15. $^{16}_{8}O$ has 8 p, 8 e, and 8 n; $^{17}_{8}O$ has 8 p, 8 e, and 9 n; $^{18}_{8}O$ has 8 p, 8 e, and 10 n; $^{20}_{10}Ne$ has 10 p, 10 e, and 10 n; $^{21}_{10}Ne$ has 10 p, 10 e, and 11 n; $^{22}_{10}Ne$ has 10 p, 10 e, and 12 n

17. (b) is not an isotope of the others

19. $^{57}_{25}Mn$ has 25 p, 25 e, and 32 n; $^{80}_{36}Kr$ has 36 p, 36 e, and 44 n; $^{128}_{52}Te$ has 52 p, 52 e, and 76 n

21. 3 p, 3 e, and 8 n

27. Protons, electrons, neutrons

29. (a) $^{103}_{48}Cd$ (b) $^{123}_{55}Cs$ (c) $^{223}_{87}Fr$ (d) $^{32}_{17}Cl$

31. Atomic weight of Ga = 69.80. (This is not the same as the atomic weight in the periodic table. The reason is that we have not used the precise mass for each isotope.)

33. 35.49

35. 52.1

37. The percentage abundance of ^{79}Br is 55% and the percentage abundance of ^{81}Br is 45%.

39. $^{121}Sb = 62.5\%$, $^{123}Sb = 37.5\%$

41. 1500 atoms ^2H

43. 2500 g oxygen

45. 7.42×10^6 atoms ^6Li

47. 5.59×10^5 atoms ^{26}Mg

49. 222 atoms ^{13}C

51. 3.7×10^{12} atoms ^{15}N

53. 33,600 g of hydrogen needed

55. 7770 g ^{24}Mg

58. (a) C   (b) D   (c) 222   (d) D   (e) Hg

59. They are the same.

61. The Lenard experiment using ultraviolet light

62. (a) 6 p, 6 e, 7 n   (b) 5 p, 5 e, 5 n
(c) 7 p, 7 e, 7 n

63. Isotopes exist

64. 2570 Ne-21 atoms

65. 1836 electrons

CHAPTER 5

3. By examining the line spectrum of an extraterrestrial body with a spectroscope, scientists can determine which elements are present.

9. Energy levels in atoms

11. Electrons

13. Seven energy levels, K, L, M, N, O, P, Q or 1, 2, 3, 4, 5, 6, 7, respectively

17. Excited state

19. (a) K has 2, L has 8, M has 1
(b) K has 2, L has 8, M has 3
(c) K has 2, L has 6
(d) K has 2, L has 8, M has 6

(e) K has 2, L has 3   (f) K has 2, L has 1
The elements that are chemically similar are (a) and (f), (b) and (e), (c) and (d)

21. (a) One   (b) Four   (c) Five

23. (a) nitrogen   (b) magnesium
(c) chlorine   (d) argon   (e) calcium

27. The group VIIIA elements tend to be chemically unreactive.

29. Ne, K has 2 e, L has 8 e
Ar, K has 2 e, L has 8 e, M has 8 e

30. Be, K has 2 e, L has 2 e
Ca, K has 2 e, L has 8 e, M has 8 e, N has 2 e

32. (*a*) 8 e in the outermost energy level
(*b*) 2 e in the outermost energy level
(*c*) 7 e in the outermost energy level

38. It could leave the atom. The neutral atom would then become a positive ion.

39. 280 would be the atomic number.

CHAPTER 6

7. Period, group or family
9. The same number of electrons in the outermost energy level
11. The number of electrons in the outermost energy level
13. A-group elements, B-group elements
15. Rb, Sr, Ca, S, Cl
17. P, As, I, Sb, In
25. (*a*) decreases (*b*) increases (*c*) decreases
27. Cl, S, Ca, Sr, Rb
29. Sb, I, As, P
31. (*d*)
33. (*c*)
35. (*d*)
39. Copper, silver, and gold
41. Element 118 should have eight electrons in its outermost energy level and behave like the Group VIIIA elements.

45. (*a*) Ba (*b*) CA (*c*) Sr (*d*) Be
47. (*a*) 31 (*b*) 8 (*c*) 13 (*d*) 10 (*e*) 31
49. The elements fall into three groups: 6 and 14 go together, 3 and 19 go together, and 17 and 35 go together.
51. 98 electrons
57. (*a*) 60 (*b*) 12 (*c*) 24 (*d*) 20 (*e*) 4 (*f*) 60
61. Elements 15 and 33, 38 and 56, 50 and 14 go together.
65. Two electrons
67. Five *d* orbitals
78. Atomic number 118
79. (*a*) Group IA (*b*) Group VIIA
81. (*a*) K (*b*) K (*c*) K
84. (*a*) K (*b*) Se
85. (*a*) K (*b*) Se
86. Group IVA

CUMULATIVE REVIEW/CHAPTERS 4–6

1. Protons and neutrons in nucleus, electrons surrounding nucleus
2. Cathode rays came from the cathode.
3. A magnet would deflect particles, not waves. An electric field would easily deflect particles.
4. UV light is directed onto a metal and makes it emit electrons.
5. An atom is a sphere of positive electricity in which electrons are embedded. The positive particles balance the negative electrons.
6. Rutherford found that one alpha particle in 20,000 ricocheted. This led him to believe that an atom has a small but dense center of positive charge.
7. New evidence disproved Thomson's theory.
8. Neutrons
9. Isotopes are atoms of the same element with different atomic masses.
10. $^{59}_{27}Co$ 27 protons
$^{90}_{40}Zr$ 40 protons
$^{74}_{34}Se$ 34 protons

11. 20.17
12. 370 N atoms
13. (*a*) $^{35}_{17}Cl$ (*b*) $^{28}_{14}Si$
14. 75% ^{50}Z and 25% ^{52}Z
15. 32 p, 32 e, 41 n
16. 26 electrons
17. ^{16}O is most abundant
18. 1900 ^{7}Li atoms
19. 93,360 g
20. 1.07 g Li
21. Electrons
22. (*a*)
23. 6 electrons
24. $_{1}H$ *K* has 1
$_{3}Li$ *K* has 2, *L* has 1
$_{11}Na$ *K* has 2, *L* has 8, *M* has 1
$_{19}K$ *K* has 2, *L* has 8, *M* has 8, *N* has 1
25. 2 electrons
26. Sulfur
27. (*b*)
28. It has 5 electrons in its outermost energy level.

29. K has 1

K has 2, L has 1

K has 2, L has 2

K has 2, L has 8, M has 1

K has 2, L has 8, M has 8, N has 1

30. $_{60}$Nd K has 2, L has 8, M has 18, N has 32

31. Raindrops act as prisms

32. Line spectra are used to analyze the sun and other extraterrestrial bodies.

33. Line spectra of his hair indicated the presence of the element arsenic.

34. Our major source of radiant energy is the sun.

35. Electromagnetic waves travel through the vacuum of space.

36. (c)

37. By producing electromagnetic waves that were longer than visible light waves

38. Luminous watch dials absorb light energy and re-emit it when they glow.

39. A sample of an element is heated and it begins to glow.

40. Quanta

41. The periodic law states that the chemical properties are periodic functions of their atomic numbers. This means that elements with similar chemical properties recur at regular intervals and are placed accordingly on the periodic table.

42. Moseley was able to determine the nuclear charge or the atomic numbers of the atoms of known elements.

43. Element 107 is a VIIB element

$1s^2 2s^2 2p^6 3s^2 3p^6 3d^{10} 4s^2 4p^6 4d^{10} 4f^{14} 5s^2 5p^6 5d^{10}$
$5f^{14} 6s^2 6p^6 6d^5 7s^2$

44. Element 106 is a VIB element

$1s^2 2s^2 2p^6 3s^2 3p^6 3d^{10} 4s^2 4p^6 4d^{10} 4f^{14} 5s^2 5p^6 5d^{10}$
$5f^{14} 6s^2 6p^6 6d^4 7s^2$

45. #18–#19, #27–#28, #52–#53 and #92–#93

46. False

47. True

48. False

49. Fe, Ru, Os

50. $_{11}$Na

51. True

52. False

53. Fluorine

54. $1s^2 2s^2 2p^6 3s^2 3p^2$

55. True

56. (a) Group IIA

(b) Group VIA

57. (a) Na

(b) Ca

58. Mg

59. $_{104}$Unq, $_{105}$Unp, $_{106}$Unh, $_{107}$Uns

60. Cs

CHAPTER 7

3. (a) ·C̈l: (b) :C̈l:$^{1-}$

5. (a) Be·   (b) Be^{2+}

9. (a) Mg^{2+} (b) Na^{1+} (c) Al^{3+}

11. (a) :B̈r:$^{1-}$ (b) :S̈e:$^{2-}$ (c)   (d) :P̈:$^{3-}$

15. Double

19. (a) H—C̈—B̈r: (b) H—C≡C—H
             |
             :B̈r:

(c) H—P̈—H   (d) H—C≡N̈
        |
        H

21. (a) :F̈—C̈—F̈:   (b) H—C=C—H
             | | |
             :F̈:   H   H

23. :C̈l: C̈l: 10 electrons surround
        \   /   the P in this compound
          P
        /   \
    :Cl   :Cl: C̈l:

25. (a) :Ö=N—Ö—H   (b) H—Ö—S—Ö—H
                                      ↑
                                     :Ö:

(with :Ö: above N in (a), and :Ö: above S in (b))

29. (a) ionic   (b) covalent   (c) ionic
(d) covalent   (e) covalent

31. (a) 4% ionic, 96% covalent
(b) 0% ionic, 100% covalent
(c) 51% ionic, 49% covalent
(d) 63% ionic, 37% covalent

33. HI, HBr, HCl, HF
Most   Least
covalent   covalent

(c) H—Äs—H   (d) H—S̈:
        | |
        H   H

35. (a) 43% ionic, 57% covalent
(b) 0.5% ionic, 99.5% covalent
(c) 74% ionic, 26% covalent
(d) 47% ionic, 53% covalent

37. H_2O, H_2S, H_2Se, H_2Te
Least   Most
covalent   covalent

39. CsCl is the most ionic

41. (a) covalent   (b) covalent   (c) ionic
(d) covalent

43. (a) Cu_2SO_3 (b) $Hg_3(PO_4)_2$
(c) $Sn_3(AsO_4)_2$ (d) $Fe(C_2H_3O_2)_3$
(e) Al_2O_3 (f) $(NH_4)_2CO_3$
(g) UF_6 (h) SO_2 (i) $Zn(NO_2)_2$
(j) $LiOH$

49. (a) 1−   (b) 2.67+   (c) 4+   (d) 5+
(e) 1−

53. (a) silver carbonate   (b) osmium(VIII) oxide
(c) mercury(II) phosphate   (d) iron(III) sulfate   (e) sodium nitrate   (f) silicon dioxide   (g) cobalt(II) chloride   (h) copper(I) nitride   (i) copper(II) chromate   (j) zinc hydroxide

59. (a) bonds are polar, molecule is polar
(b) bonds are polar, molecule is polar
(c) bonds are polar, molecule is polar
(d) bond is nonpolar, molecule is nonpolar

61. (a) bonds are polar, molecule is nonpolar
(b) bonds are polar, molecule is polar

(c) bonds are polar, molecule is nonpolar
(d) bond is nonpolar, molecule is nonpolar

63. (a) 1−   (b) 3−   (c) 2+   (d) 2−
(e) 1+   (f) 3+

64. (a) A_2B (b) DC_2

65. (a) Na_2O (b) Al_2O_3

66. 1^{1-}, Xe^0, Cs^{1+}

68. (a) $\left[H-\overset{\overset{\displaystyle H}{\uparrow}}{\underset{\underset{\displaystyle H}{|}}{P}}-H \right]^{1+}$ (b) $O=C=O$

(c)

70. $O=O$ $N-O$ $C-O$
Nonpolar   Most polar bond
bond   of the three

71. $C-S$ bond

68. (a) $\left[H-\overset{\overset{\displaystyle H}{\uparrow}}{\underset{\underset{\displaystyle H}{|}}{N}}-H \right]^{1+}$ (b) $Cl-Br$

73. (a) MgO (b) Al_2O_3 (c) Cs_2S

74. (a) AlN (b) V_2O_5

75. (a) aluminum nitride
(b) vanadium(V) oxide or vanadium pentoxide

76. (a) SnS (b) Cu_3PO_4 (c) $Fe(NO_2)_3$
(d) $Mg(C_2H_3O_2)_2$

CHAPTER 8

1. (a) $2Li + 2H_2O \longrightarrow 2LiOH + H_2$
(b) $2HNO_3 + Ca(OH)_2 \longrightarrow Ca(NO_3)_2 + 2H_2O$
(c) $Mg + Zn(NO_3)_2 \longrightarrow Mg(NO_3)_2 + Zn$
(d) $K_2O + H_2O \longrightarrow 2KOH$
(e) $2ZnS + 3O_2 \longrightarrow 2ZnO + 2SO_2$
(f) $2Al + Fe_2O_3 \longrightarrow Al_2O_3 + 2Fe$

3. (a) $4Fe + 3O_2 \longrightarrow 2Fe_2O_3$
(b) $3H_2SO_4 + 2Al(OH)_3 \longrightarrow$
$$Al_2(SO_4)_3 + 6H_2O$$
(c) $2AgNO_3 + BaCl_2 \longrightarrow 2AgCl + Ba(NO_3)_2$
(d) $Cu_2S + O_2 \longrightarrow 2Cu + SO_2$

5. (a) decomposition   (b) double replacement
(c) single replacement   (d) combination
(e) combination   (f) double replacement

7. (a) decomposition   (b) decomposition
(c) decomposition   (d) combination
(e) combination   (f) single replacement
(g) no reaction   (h) double replacement

9. (a) combination   (b) double replacement

(c) double replacement
(d) single replacement

11. (a) $H_2 + Br_2 \longrightarrow 2HBr$
(b) $BaO + H_2O \longrightarrow Ba(OH)_2$
(c) $2Na + Cl_2 \longrightarrow 2NaCl$
(d) $N_2 + 2O_2 \longrightarrow 2NO_2$
(e) $CaCO_3 \longrightarrow CaO + CO_2$
(f) $2KOH \longrightarrow K_2O + H_2O$
(g) $Hg(ClO_3)_2 \longrightarrow HgCl_2 + 3O_2$
(h) $PbCl_2 \longrightarrow Pb + Cl_2$
(i) $2Li + 2H_2O \longrightarrow 2LiOH + H_2$
(j) $Zn + H_2SO_4 \longrightarrow ZnSO_4 + H_2$
(k) $Ni + Al(NO_3)_3 \longrightarrow$ no reaction
(Aluminum is above nickel in the activity series.)
(l) $2Al + 3Hg(C_2H_3O_2)_2 \longrightarrow$
$$2Al(C_2H_3O_2)_3 + 3Hg$$
(m) $H_2SO_4 + 2NH_4OH \longrightarrow$
$$(NH_4)_2SO_4 + 2H_2O$$

(n) $2AgNO_3(aq) + BaCl_2(aq) \longrightarrow$
$Ba(NO_3)_2(aq) + 2AgCl(s)$

(o) $3H_2SO_3 + 2Al(OH)_3 \longrightarrow$
$Al_2(SO_3)_3 + 6H_2O$

(p) $NaNO_3(aq) + KCl(aq) \longrightarrow$ no reaction

13. (a) $2N_2O \longrightarrow 2N_2 + O_2$
(b) $H_2CO_3 \longrightarrow H_2O + CO_2$
(c) $2NaNO_3 \longrightarrow 2NaNO_2 + O_2$
(d) $H_2 + F_2 \longrightarrow 2HF$
(e) $N_2 + 3H_2 \longrightarrow 2NH_3$
(f) $Ca + 2HCl \longrightarrow CaCl_2 + H_2$
(g) $Cu + NiCl_2 \longrightarrow$ no reaction
(h) $3AgNO_3(aq) + K_3AsO_4(aq) \longrightarrow$
$3KNO_3(aq) + Ag_3AsO_4(s)$

15. (a) $2K + Cl_2 \longrightarrow 2KCl$
(b) $K_2O + H_2O \longrightarrow 2KOH$
(c) $MgO + H_2O \longrightarrow Mg(OH)_2$
(d) $CaO + CO_2 \longrightarrow CaCO_3$
(e) $2NH_3 \longrightarrow N_2 + 3H_2$
(f) $SrCO_3 \longrightarrow SrO + CO_2$
(g) $2NaClO_3 \longrightarrow 2NaCl + 3O_2$
(h) $Mg(OH)_2 \longrightarrow MgO + H_2O$

17. (a) NaOH and HCl   (b) KOH and H_2SO_4
(c) $Al(OH)_3$ and HNO_3

19. (a) $Zn + 2HNO_2 \longrightarrow Zn(NO_2)_2 + H_2$
(b) $Ag + NiCl_2 \longrightarrow$ No reaction
(c) $Zn + 2AgNO_3 \longrightarrow Zn(NO_3)_2 + 2Ag$
(d) $2Cs + 2H_2O \longrightarrow 2CsOH + H_2$
(e) $3HCl + Al(OH)_3 \longrightarrow AlCl_3 + 3H_2O$
(f) $KNO_3 + ZnCl_2 \longrightarrow$ No reaction
(g) $Al(NO_3)_3 + 3NaOH \longrightarrow$
$Al(OH)_3 + 3NaNO_3$
(h) $K_2CrO_4 + Pb(NO_3)_2 \longrightarrow$
$PbCrO_4 + 2KNO_3$

23. (a) $2K + Cl_2 \longrightarrow 2KCl$ (the K is oxidized; the Cl is reduced)
(b) $2NH_3 \longrightarrow N_2 + 3H_2$ (the N is oxidized; the H is reduced)
(c) $CuO + H_2 \longrightarrow Cu + H_2O$ (the H is oxidized; the Cu is reduced)
(d) $Sn + 2Cl_2 \longrightarrow SnCl_4$ (Sn is oxidized; Cl is reduced)

25. (1) K is the reducing agent; Cl_2 is the oxidizing agent.
(2) The NH_3 is both oxidizing and reducing agent.
(3) H_2 is the reducing agent; CuO is the oxidizing agent.
(4) Sn is the reducing agent; Cl_2 is the oxidizing agent.

27. (a) Zn is the reducing agent; HNO_3 is the oxidizing agent.
(b) No reaction took place.
(c) Zn is the reducing agent; $AgNO_3$ the oxidizing agent.
(d) Cs is the reducing agent; H_2O the oxidizing agent.

29. (a) vanadium pentoxide   (b) zinc
(c) zinc   (d) vanadium

30. $MgCO_3 \longrightarrow MgO + CO_2$
$CaCO_3 \longrightarrow CaO + CO_2$

31. (a) $N_2 + 3H_2 \longrightarrow 2NH_3$
(b) $3Fe + 4H_2O \longrightarrow Fe_3O_4 + 4H_2$
(c) Balanced

33. (a) $2C + O_2 \longrightarrow 2CO$
(b) $H_2 + Cl_2 \longrightarrow 2HCl$
(c) $2Na + 2H_2O \longrightarrow 2NaOH + H_2$
(d) No reaction

34. (a) No reaction
(b) $Mg + Zn(NO_3)_2 \longrightarrow Mg(NO_3)_2 + Zn$
(c) No reaction
(d) $Zn + Cu(NO_3)_2 \longrightarrow Zn(NO_3)_2 + Cu$

36. (a) $Ca(OH)_2 + 2HCl \longrightarrow$
$CaCl_2 + 2H_2O$ (double replacement)
(b) $Zn + H_2SO_4 \longrightarrow$
$ZnSO_4 + H_2$ (single replacement)
(c) $2Ba + O_2 \longrightarrow 2BaO$ (combination)
(d) $Cs_2O + H_2O \longrightarrow 2CsOH$ (combination)

37. $4NH_3 + 7O_2 \longrightarrow 2N_2O_4 + 6H_2O$

38. (a) Fe^{2+} (b) chromium in $Cr_2O_7^{2-}$
(c) $Cr_2O_7^{2-}$ (d) Fe^{2+}

39. $\underset{\text{Sodium}}{12Na} + \underset{\text{Phosphorus}}{P_4} \longrightarrow \underset{\substack{\text{Sodium} \\ \text{phosphate}}}{4Na_3P}$

CHAPTER 9

1. (a) $H_2SO_4 + 2KCN \longrightarrow K_2SO_4 + 2HCN$
(b) 9.80 g sulfuric acid
(c) 17.4 g potassium sulfate, 5.40 g hydrogen cyanide

3. (a) $NH_4NO_3 \overset{\text{Heat}}{\longrightarrow} N_2O + 2H_2O$
(b) 4.0 g ammonium nitrate   (c) 1.8 g water

5. (a) $C_7H_8 + 3HNO_3 \longrightarrow C_7H_5N_3O_6 + 3H_2O$
(b) 405 g of toluene, 832 g nitric acid
(c) 238 g water

7. (a) $2NH_4Cl + H_2SO_4 \longrightarrow (NH_4)_2SO_4 + 2HCl$
   (b) 14.7 g sulfuric acid
   (c) 19.8 g ammonium sulfate, 10.8 g hydrogen chloride
9. Percent $KClO_3 = 51.2\%$
11. (a) $3BaO_2 + 2H_3PO_4 \longrightarrow 3H_2O_2 + Ba_3(PO_4)_2$
    (b) 68 g hydrogen peroxide
    (c) 130 g phosphoric acid
13. 12 g hydrogen
15. 1460 g air (or 1500 g air to two significant figures)
17. (a) $2C_4H_{10} + 13O_2 \longrightarrow 8CO_2 + 10H_2O$
    (b) 83.2 g oxygen
    (c) 70.4 g CO_2 and 36.0 g H_2O
19. 64.0 g ozone
21. (a) $2Al(OH)_3 + 3H_2SO_4 \longrightarrow$
    $$Al_2(SO_4)_3 + 6H_2O$$
    (b) 228 g $Al(OH)_3$ and 429 g H_2SO_4
    (c) 158 g H_2O
23. (a) $Fe_2O_3 + 3CO \longrightarrow 2Fe + 3CO_2$
    (b) 649.2 g Fe_2O_3
    (c) 341.3 g CO and 536.4 g CO_2
25. (a) $4NH_3 + 5O_2 \longrightarrow 4NO + 6H_2O$
    (b) $10\overline{0}0$ g O_2 (c) $75\overline{0}$ g NO and 675 g H_2O
27. 502 g $Zn(NO_3)_2$
29. (a) $CaC_2 + 2H_2O \longrightarrow Ca(OH)_2 + C_2H_2$
    (b) 148 g $Ca(OH)_2$ and 52.0 g C_2H_2

31. (a) $CO + 2H_2 \longrightarrow CH_3OH$
    (b) 228 g of methyl alcohol
33. (a) $2H_2 + O_2 \longrightarrow 2H_2O$ (b) 72 g water
35. (a) $Ca_3P_2 + 6H_2O \longrightarrow 2PH_3 + 3Ca(OH)_2$
    (b) 136 g PH_3 (c) 444 g $Ca(OH)_2$
37. $2\overline{0}$ moles $CaSiO_3$, 33 moles CO and 6.7 moles P_2
39. (a) $3Fe + 4H_2O \longrightarrow Fe_3O_4 + 4H_2$
    (b) 10.7 g H_2 (c) 310 g Fe_3O_4
41. (a) 0.128 mole   (b) 19.1 g
42. (a) $2Cu(NO_3)_2 \longrightarrow 2CuO + 4NO_2 + O_2$
    (b) 8.51 g   (c) $11\overline{0}$ g
43. (a) oxygen   (b) 206 g
44. (a) $2H_2 + O_2 \longrightarrow 2H_2O$
    (b) 0.70 g H_2 left and 72.0 g H_2O produced
45. 393 g
46. (a) $Zn + H_2SO_4 \longrightarrow ZnSO_4 + H_2$
    (b) 19,600 g (to three significant figures)
47. (a) The second experiment,
    $$Pb + Cu(NO_3)_2 \longrightarrow Pb(NO_3)_2 + Cu$$
    (b) 80.1 g $Pb(NO_3)_2$ and 15.4 g Cu
48. 6.66 g
49. 58.5 g Na Cl produced and 40.0 g NaOH left
50. 112 g

CUMULATIVE REVIEW/CHAPTERS 7-9

1. (a) $\dot{C}a\cdot$ (b) $[Ca]^{2+}$
2. (a) $\cdot\ddot{C}l\colon$ (c) $[\colon\!\ddot{C}l\colon]^{1-}$
3. An ionic bond is formed by the transfer of electrons from one atom to another. The atoms are always of different elements. A covalent bond is formed by the sharing of electrons between two atoms. A coordinate covalent bond is a bond in which one element donates the electrons to form the bond.
4. $H\overset{\text{H H H}}{\underset{\text{H H H}}{\colon C \cdot \cdot C \cdot \cdot C\colon H}}$
5. HCl = 0.9, $CaCl_2 = 2.0$, $BaCl_2 = 2.1$,
   $CaF_2 = 3.0$
6. CsF
7. (a) Cu_2SO_3 (b) OsO_4
   (c) $(NH_4)_2CO_3$ (d) N_2O
8. (a) W^{5+} (b) Sn^{4+} (c) V^{5+} (d) In^{3+}

9. (a) ammonium iodide   (b) calcium carbide
   (c) copper(I) chromate
10. X_2Y_3
11. False
12. True
13. True
14. False
15. True
16. MgO
17. C—H
18. Na_2SO_4
19. $2-$
20. (a) $\left[\begin{array}{c} \overset{xx}{O}\overset{x}{x}\colon N\colon\overset{xx}{x}\overset{}{O} \\ xx \quad \cdot\cdot \quad xx \\ \overset{xx}{x}O\overset{}{x} \end{array} \right]^{1-}$
    (b) $\left[\begin{array}{c} H \\ \cdot\cdot \\ H\colon N\colon H \\ \cdot x \\ H \end{array} \right]^{1+}$

21. $CaBr_2 + H_2SO_4 \longrightarrow 2HBr + CaSO_4$
22. $2Mg + O_2 \longrightarrow 2MgO$
23. $NH_3(g) + HCl(g) \longrightarrow NH_4Cl(s)$
24. (a) decomposition   (b) combination
    (c) single replacement   (d) double replacement
25. (a) $H_2 + Cl_2 \longrightarrow 2\ HCl$
    (b) $2NaCl + Pb(NO_3)_2 \longrightarrow 2NaNO_3 + PbCl_2$
    (c) $2Al + 3H_2SO_4 \longrightarrow Al_2(SO_4)_3 + 3H_2$
    (d) $SrCO_3 \longrightarrow SrO + CO_2$
26. $CaCO_3 \longrightarrow CaO + CO_2$
27. (a) $TiCl_4 + 2H_2O \longrightarrow TiO_2 + 4HCl$
    (b) $P_4O_{10} + 6H_2O \longrightarrow 4H_3PO_4$
28. (a) $Fe(OH)_3 + H_3PO_4 \longrightarrow FePO_4 + 3H_2O$
    (b) $Pb(OH)_2 + 2HNO_3 \longrightarrow Pb(NO_3)_2 + 2H_2O$
29. $SO_3 + H_2O \longrightarrow H_2SO_4$
30. $2C_8H_{18} + 25O_2 \longrightarrow 16CO_2 + 18H_2O$
31. $2C_3H_7OH + 9O_2 \longrightarrow 6CO_2 + 8H_2O$
32. $CaSO_4 + Na_2CO_3 \longrightarrow CaCO_3(s) + Na_2SO_4$
33. $C_{12}H_{22}O_{11} \longrightarrow 12C + 11H_2O$
34. True

35. False
36. True
37. True
38. True
39. True
40. $C + 2H_2SO_4 \longrightarrow CO_2 + 2SO_2 + 2H_2O$
41. 5 moles CuO
42. 0.20 mole NaOH
43. 88 grams
44. 730 grams HCl
45. 12.6 grams CO_2
46. 1850 grams H_3PO_4
47. 77.5 grams $Ca_3(PO_4)_2$
    55.5 grams $Ca(OH)_2$
48. 15.0 moles, $18\overline{0}0$ grams
49. 119 grams H_2O_2
50. 13.9 grams $PbCl_2$
51. 96.0 grams LiOH
52. 72.9 grams Mg
53. 171.3 grams $Ba(OH)_2$, 2.000 grams H_2, 36.00 grams excess H_2O

CHAPTER 10

7. 20.0 kcal
9. 0.345 kcal
11. 2.09 kcal
13. 30.0°C
15. $\Delta H_R = -68.5$ kcal/mole
21. -14.4 kcal
23. 3210 kcal
25. (a) $\Delta H_R = -621.4$ kcal   (b) -62140 kcal
27. (a) $\Delta H_R = -55.7$ kcal   (b) 1520 g
29. $\Delta H_f = -183.0$ kcal
31. ΔH_f (calculated) $= -94.5$ kcal
33. $\Delta H_f = -107$ kcal
35. -14.2 kcal
37. 2.5 kcal
39. -47 kcal
41. (a) -47.0 kcal   (b) 141 kcal
43. -21.36 kcal
45. (a) $PCl_5 + 4H_2O \longrightarrow H_3PO_4 + 5HCl$

    (b) -139.8 kcal   (c) -27.96 kcal
47. $\Delta H_f = -37.3$ kcal/mole
49. $\Delta H_R = -73.2$ kcal
51. $\Delta H_R = -288.5$ kcal
53. $\Delta H_R = -65.4$ kcal
55. 6.30 kcal
56. -288.5 kcal
57. (a) and (d), energy is absorbed; (b) and (c), energy is released
58. Acetylene releases more energy; $\Delta H = -300.2$ kcal/mole
59. Endothermic
60. $55\overline{0}0$ kcal
61. H_2 produces 29 kcal/g,
    CH_4 produces 13.3 kcal/g,
    C produces 7.8 kcal/g
63. 10,600 kcal
64. 10.5 kcal/mole

CHAPTER 11

1. (a) 0.329 atm   (b) 2.04 atm   (c) $\overline{1}0$ atm
3. (a) 2280 torr   (b) 228 torr   (c) 114 torr

5. (a) 0.10 atm   (b) 3.50 atm
   (c) 0.00500 atm

7. (*a*) 7.60 torr   (*b*) 9120 torr   (*c*) 4180 torr
9. V_f = 180 L
11. V_f = 3$\overline{0}$00 mL or 3.00 L
13. P_f = 95$\overline{0}$0 torr
15. t_f = 427°C
17. t_f = 527°C
19. V_f = 33.3 L
21. (*a*) 227°C   (*b*) −269°C
23. (*a*) 283 K   (*b*) 263 K   (*c*) 200 K
25. (*a*) −273°C   (*b*) −173°C   (*c*) −73°C
    (*d*) 27°C
27. (*a*) 293 K   (*b*) 273 K   (*c*) 173 K
    (*d*) 373 K
29. P_f = 2 atm
31. P_f = 1590 torr
33. V_f = 421 mL
35. V_f = 709 mL
37. V_f = 394 mL
39. It would increase to 2.7 times its original volume.
41. P_f = 59.8 atm
43. V_f = 819 mL
45. V_f = 276 mL
47. (1) increases   (2) increases   (3) increases
49. P_{total} = 700 torr
51. 608 torr or 610 torr for two sig. figs.
53. 24 L H_2O vapor
55. 89.6 L
57. 20.5 g Na

59. O_2 uncombined = 40 ML,
    $H_2O(g)$ formed = 80 mL
61. (*a*) $2SO_2 + O_2 \longrightarrow 2SO_3$
    (*b*) 75.0 L   (*c*) 15$\overline{0}$ L
63. (*a*) $2C_4H_{10} + 13O_2 \longrightarrow 8CO_2 + 10H_2O$
    (*b*) 97.5 L   (*c*) 60.0 L CO_2 and 75.0 L H_2O
65. (*a*) $N_2 + 2O_2 \longrightarrow 2NO_2$ (*b*) 37. 5 mL
    (*c*) 75.0 mL
67. 6.83 g N_2
69. R = 0.0821 $\dfrac{\text{liter-atm}}{\text{K-mole}}$
71. 11.2 L
73. 2.24 g
75. 1$\overline{0}$0 atm
77. 1.52 K or −271.5°C
79. (*a*) C_2H_5 (*b*) MW = 59   (*c*) C_4H_{10}
81. MW = 41
83. D_{O_2} = 1.43 g/L
85. MW = 24.0
87. (*a*) NO_2 (*b*) MW = 92.3   (*c*) N_2O_4
92. 50.0 L
93. 124 L
94. 1797 L or 18$\overline{0}$0 L for three sig. figs.
95. (*a*) 277 K   (*b*) 355 K   (*c*) 373 K
96. 22.4 L
97. 0.714 g/L
99. 429 mL
100. 31.5 g/mole; O_2 is a likely candidate

CHAPTER 12

5. (*a*) $2HgO \longrightarrow 2Hg + O_2$
    (*b*) $2KClO_3 \longrightarrow 2KCl + 3O_2$
    (*c*) $2H_2O \longrightarrow 2H_2 + O_2$
19. CO_2 and H_2O
21. (*a*) $2Zn + O_2 \longrightarrow 2ZnO$
    (*b*) $2C_2H_6 + 7O_2 \longrightarrow 4CO_2 + 6H_2O$
    (*c*) $4Al + 3O_2 \longrightarrow 2Al_2O_3$
23. 1650 kcal
33. (*a*) HI   (*b*) LiH
35. (*a*) $3H_2 + N_2 \longrightarrow 2NH_3$
    (*b*) $H_2 + Br_2 \longrightarrow 2HBr$
    (*c*) $2H_2 + O_2 \longrightarrow 2H_2O$
43. 59 g
47. 1.00 L (1$\overline{0}$00 mL)
49. (*a*) $Mg(OH)_2$ (*b*) H_2SO_4
    (*c*) $NaOH + H_2$

51. (*a*) $SrO + H_2O \longrightarrow Sr(OH)_2$
    (*b*) $SO_2 + H_2O \longrightarrow H_2SO_3$
    (*c*) $2Li + 2H_2O \longrightarrow 2LiOH + H_2$
53. 36.0%
57. (*a*) 51.2%   (*b*) 62.9%   (*c*) 45.6%
59. (*a*) $CaSO_4 \cdot 2H_2O \longrightarrow CaSO_4 + 2H_2O$
    (*b*) $KAl(SO_4)_2 \cdot 12H_2O \longrightarrow$
                        $KAl(SO_4)_2 + 12H_2O$
61. $CoSO_4 \cdot 6H_2O$
73. $MgSO_4 \cdot 7H_2O$
74. 14.7%
75. (*a*) $2K + 2H_2O \longrightarrow 2KOH + H_2$
    (*b*) $MgO + H_2O \longrightarrow Mg(OH)_2$
77. 22.4 L
79. Yes

CHAPTER 13

1. (a) The boiling point of a liquid increases with increasing atmospheric pressure. This is because a greater vapor pressure is needed for molecules to escape from the liquid. This can happen only at a higher temperature.
   (b) No effect. The melting phenomenon is not affected by changes in atmospheric pressure.
5. (a) liquid   (b) gas   (c) solid
9. Once the water is boiling, it doesn't get any hotter. Turning up the heat on the stove doesn't change the temperature of the water and doesn't speed up the cooking of the potatoes.
11. If the mountain is high enough, the water boils at a significantly lower temperature because of the decrease in atmospheric pressure. Because the temperature of the water is lower when it begins to boil, it takes longer to cook a hard-boiled egg on a mountain than at sea level.
15. It is the high-energy molecules of liquid that escape from the container. The low-energy ones get left behind. The liquid therefore has less energy—in other words, less heat. The temperature of the liquid drops.
37. Amorphous
39. Well-defined (lattice) structure
43. A molecular solid
45. An ionic solid
47. (a) metallic   (b) ionic
49. (a) atoms   (b) molecules   (c) ions
59. It is a repeating unit.
60. Yes, some substances are already gases at *normal* atmospheric pressure (1.00 atm), for example, CO_2. The *normal boiling point* is defined as the boiling point at one atmosphere pressure. Therefore many substances do not have normal boiling points.
63. Glass is an amorphous solid.
64. (a) solid and liquid   (b) solid   (c) gas
66. Higher

CUMULATIVE REVIEW/CHAPTERS 10–13

1. Endothermic: heat taken in from the surroundings; exothermic: heat given off to the environment
2. 10,000 calories
3. 138 calories
4. 291 calories
5. We use standard state conditions so we can measure heat given up or absorbed during thermochemical reactions, as opposed to measuring heat involved in temperature or pressure changes.
6. 25,000 calories
7. $H_2(g) + \frac{1}{2}O_2(g) \longrightarrow H_2O(l) + 68.3$ kcal
8. −54 calories
9. 284 calories
10. 7.05 calories
11. 24 mL
12. 18.8 mL
13. 616 torr
14. 2500 mL
15. 551°C
16. 5.9 L
17. 7.31 L
18. 0.361 L
19. 4.16 L
20. 16.0 moles
21. 336 L
22. 273 L
23. 0.0413 mole
24. MW = 285
25. 10.0 L
26. 3.13 g/L
27. Because $P_f < P_i$, therefore $V_f > V_i$
28. 2.63 L
29. MW = 93.6
30. Forces of attraction between gas molecules vary.
31. Combined oxygen
32. Electrolysis of water
33. (1) Free air of CO_2 and H_2O
    (2) Liquefy remaining air
    (3) Allow liquid air to evaporate
    (4) Pure O_2 is left behind
34. Marine organisms "breathe" using the dissolved oxygen.
35. It will decrease.

36. $CO_2 = 342.3$ torr, $H_2 = 326$ torr, $N_2 = 137.7$ torr, $CH_4 = 0.245$ torr

37. (a) $2Li(s) + H_2(g) \longrightarrow 2 LiH$
(b) $2Sr(s) + H_2(g) \longrightarrow 2 SrH$

38. ammonia, Haber

39. $CO(g) + 2H_2(g) \longrightarrow CH_3OH(l)$

40. 3a, 2b, 1c

41. $CaCO_3 \cdot 6H_2O$

42. 49.5% H_2O

43. DDT enters the bodies of marine organisms and people consume these organisms.

44. A body of water becomes enriched with nutrients, which causes overgrowth of algae. They remove oxygen from the water, causing the deterioration of the body of water.

45. 20.9%

46. No, safe water is no greater than 250 mg/L.

47. $BaCl_2 \cdot 2H_2O$

48. $Na_2CO_3 \cdot 10H_2O$

49. 28.12 grams

50. 3a, 1b, 2c

51. 2a, 3b, 1c

52. b

53. Sublimation occurred

54. Molecular

55. Ionic

56. Amorphous

57. Molecules gain energy, kinetic energy increases, attractive forces between molecules decrease

58. Benzene

59. In some crystals electrons are free to wander throughout the crystal.

60. Attraction of opposite charges makes the bonds so strong

CHAPTER 14

7. True

9. Point of saturation

11. (a) $2\overline{0}$% by weight   (b) 1.5% by weight
(c) 15% by weight

13. 20 g

15. (a) $2\overline{0}$% by volume   (b) 6% by volume

17. 12.5 mL

19. $2\overline{0}$ mL

21. (a) 3.3% by weight–volume
(b) 5.0% by weight–volume

23. 15 g

25. 0.05% by weight

27. (a) 12.5%   (b) 5.00%

29. $3\overline{0}$ g

31. (a) 5.00%   (b) 0.833%

33. 28 mL

35. $2\overline{00}$ mL

37. (a) 3.13%   (b) 5.00%

39. 12.5 g

41. (a) 0.30M (b) 0.5M

43. 2.74M

45. 8.4 g

47. 0.11M

49. 0.15M

51. (a) 0.167M (b) 0.200M

53. 0.526M

55. 16.4 g

57. (a) 0.10 equivalent   (b) 0.20 equivalent
(c) 0.900 equivalent

61. (a) 0.0300 equivalent   (b) 0.748 equivalent

63. (a) 1120 g   (b) 9.27 g

65. (a) 2.97N (b) 0.13N (c) 0.99M, 0.063M

67. 59 g

69. 250 mL

71. (a) 1.00N (b) 0.500N

73. 3.25 g

75. $6\overline{00}$ mL

77. Take 125 mL of $16N$ H_2SO_4; dilute to $5\overline{00}$ mL with water.

79. Take 50 mL of $6N$ NaOH; dilute to 3L with water.

81. Take 104 mL of $5.00M$ HCl; dilute to $25\overline{0}$ mL with water.

83. Take 42 mL of $3.00M$ H_2SO_4; dilute to 5.00 L with water.

93. 0.15 molal

95. (a) 0.80 molal   (b) 0.333 molal

101. $-33.1°C$

103. $109.26°C$

105. f.p. = $-2.79°C$, b.p. = $100.78°C$

108. $1 \times 10^{-2}M$ or 0.01M

109. 0.01N and 0.005M

110. $5\overline{00}$ g

111. 0.510M and 1.02N

114. Ethanol
115. 41.0 mL
116. $15\bar{0}$ g/mole
117. 2.00 *molal* solution
118. 3.71 g/mole
119. A 0.104°C increase in boiling point; therefore, if the boiling point of the pure water is 100.000°C, the boiling point of the solution is 100.104°C
120. (*a*) 2.00 equivalents (*b*) 2.00 equivalents
121. (*a*) 1.00 equivalent (*b*) 2.00 equivalents

CHAPTER 15

3. Acetic acid
7. $NH_3 + H_2O \rightleftharpoons NH_4^{1+} + OH^{1-}$
15. Organic
17. (*a*) blue, red   (*b*) red, blue
19. $HBr + H_2O \longrightarrow H_3O^{1+} + Br^{1-}$
23. (*a*) $K_2O + H_2O \longrightarrow 2KOH$
    (*b*) $MgO + H_2O \longrightarrow Mg(OH)_2$
25. (*a*) $SO_2 + H_2O \longrightarrow H_2SO_3$
    (*b*) $CO_2 + H_2O \longrightarrow H_2CO_3$
27. (*a*) $2H_3PO_4 + 3Mg(OH)_2 \longrightarrow$
$$Mg_3(PO_4)_2 + 6H_2O$$
    (*b*) $P_2O_5 + 3H_2O \longrightarrow 2H_3PO_4$
    (*c*) $K_2O + H_2O \longrightarrow 2KOH$
    (*d*) $Mg(OH)_2 + CO_2 \longrightarrow MgCO_3 + H_2O$
    (*e*) $3Zn + 2H_3PO_4 \longrightarrow Zn_3(PO_4)_2 + 3H_2$
29. Use a *weak* acid; for example, dilute acetic acid. Under no circumstances should you use a strong mineral acid (even if it is diluted).
35. (*a*) $Zn + H_2SO_4 \longrightarrow ZnSO_4 + H_2$
    (*b*) $HCl + LiOH \longrightarrow LiCl + H_2O$
    (*c*) $Ca + H_2SO_4 \longrightarrow CaSO_4 + H_2$
    (*d*) $3Ca(OH)_2 + 2H_3PO_4 \longrightarrow$
$$Ca_3(PO_4)_2 + 6H_2O$$
37. (*a*) $Zn + 2HCl \longrightarrow ZnCl_2 + H_2$
    (*b*) $2HCl + Ca(OH)_2 \longrightarrow CaCl_2 + 2H_2O$
    (*c*) $Mg(OH)_2 + CO_2 \longrightarrow MgCO_3 + H_2O$
41. (*a*) HNO_3 (*b*) HF   (*c*) HI
43. (*a*) sodium hydroxide
    (*b*) calcium hydroxide
    (*c*) iron(II) hydroxide
    (*d*) aluminum hydroxide
45. (*a*) nitric acid   (*b*) phosphoric acid
    (*c*) hydroiodic acid   (*d*) sulfuric acid
47. (*a*) LiOH   (*b*) $Ba(OH)_2$
    (*c*) $Fe(OH)_2$ (*d*) $Fe(OH)_3$
59. (*a*) 100 times more acidic
    (*b*) 100,000 times more acidic
63. (*a*) 3   (*b*) 4   (*c*) 2   (*d*) 11   (*e*) 10
    (*f*) 9
65. (*a*) $10^{-5}M$ (*b*) $10^{-9}M$ (*c*) $10^{-2}M$

(*d*) $10^{-12}M$ (All answers for [H⁺] concentration)
67. (*a*) $[H^{1+}] = 10^{-9}M$ (*b*) $[H^{1+}] = 10^{-6}M$
    (*c*) $[H^{1+}] = 10^{-4}M$ (*d*) $[H^{1+}] = 10^{-12}M$
69. (*a*) $10^{-11}M$ (*b*) $10^{-10}M$ (*c*) $10^{-6}M$
    (*d*) $10^{-5}M$ (*e*) $10^{-9}M$ (*f*) $10^{-12}M$
71. (*a*) sodium sulfide   (*b*) potassium bromide
    (*c*) sodium sulfate   (*d*) potassium nitrate
    (*e*) calcium carbonate   (*f*) barium iodide
73. (*a*) $AgNO_3$ (*b*) $CoCl_2$ (*c*) Cu_2SO_3
    (*d*) $Hg_3(PO_4)_2$
79. $2LiOH + H_2SO_4 \longrightarrow Li_2SO_4 + 2H_2O$
81. $Zn + 2HBr \longrightarrow ZnBr_2 + H_2$
83. (*a*) $Mg + 2HCl \longrightarrow MgCl_2 + H_2$
    (*b*) $Zn + H_2SO_4 \longrightarrow ZnSO_4 + H_2$
85. (*a*) $3KOH + H_3PO_4 \longrightarrow K_3PO_4 + 3H_2O$
    (*b*) $Ca(OH)_2 + 2HNO_3 \longrightarrow Ca(NO_3)_2 + 2H_2O$
87. (*a*) $2K + Cl_2 \longrightarrow 2KCl$
    (*b*) $Ca + Br_2 \longrightarrow CaBr_2$
89. $0.63N$
91. $0.2778N$
93. $0.180N$
95. $0.667N$
97. (*a*) acid   (*b*) base
98. (*a*) base   (*b*) acid
99. (*a*) base   (*b*) acid
100. pOH = 3, pH = 11
101. (*a*) $HCl + NaOH \longrightarrow NaCl + H_2O$
    (*b*) $BaO + 2HCl \longrightarrow BaCl_2 + H_2O$
    (*c*) $Zn + H_2SO_4 \longrightarrow ZnSO_4 + H_2$
102. (*a*) $OH^{1-} + H_2O \longrightarrow H-OH + OH^{1-}$
    (*b*) $NH_3 + H_2O \longrightarrow NH_4^{1+} + OH^{1-}$
    (*c*) $HSO_4^{1-} + H_2O \longrightarrow H_2SO_4 + OH^{1-}$
103. (*a*) $H_2O + H-OH \longrightarrow H_3O^{1+} + OH^{1-}$
    (*b*) $HCO_3^{1-} + H_2O \longrightarrow H_3O^{1+} + CO_3^{2-}$
    (*c*) $HNO_3 + H_2O \longrightarrow H_3O^{1+} + NO_3^{1-}$

105. $H^{1+} + :PCl_3 \longrightarrow \left[\begin{array}{c} Cl \\ | \\ H \leftarrow P-Cl \\ | \\ Cl \end{array} \right]^{1+}$

CHAPTER 16

1. The matching pairs are as follows:
   (a) 2   (b) 3   (c) 1
9. The catalyst lowers the activation energy of a reaction.
11. The four factors affecting the rate of a chemical reaction are: a change in the concentration of the reactants, a change in the temperature, a change in the pressure on gaseous products and reactants, the presence of a catalyst.
15. Reaction rate approximately doubles for each 10°C rise in the temperature.
17. 80 torr per minute
23. (a) False   (b) True   (c) False   (d) True
25. $HC_2H_3O_2 \xrightarrow{H_2O} H^{1+} + C_2H_3O_2{}^{1-}$ (forward)

   $H^{1+} + C_2H_3O_2{}^{1-} \xrightarrow{H_2O} HC_2H_3O_2$ (reverse)

27. $NaCl(s) \xrightarrow{H_2O} Na^{1+} + Cl^{1-}$ (forward)

   $Na^{1+} + Cl^{1-} \xrightarrow{H_2O} NaCl(s)$

29. (a) $K_{eq} = \dfrac{[C]^5[D]^4}{[A]^2[B]^3}$ (b) $K_{eq} = \dfrac{[HI]^2}{[H_2][I_2]}$

   (c) $K_{eq} = \dfrac{[HCl]^2}{[H_2][Cl_2]}$ (d) $K_{eq} = \dfrac{[N_2O_4]}{[NO_2]^2}$

31. (a) $K_{eq} = \dfrac{[NH_3]^2}{[N_2][H_2]^3}$ (b) $K_{eq} = \dfrac{[NO_2]^2}{[NO]^2[O_2]}$

   (c) $K_{eq} = \dfrac{[PCl_3][Cl_2]}{[PCl_5]}$ (d) $K_{eq} = \dfrac{[CO]^2[O_2]}{[CO_2]^2}$

33. $K_{eq} = 67.2$
35. $K_{eq} = 51.5$
37. $K_a = 6.4 \times 10^{-5}$
39. $K_a = 2.1 \times 10^{-4}$
41. $[H^{1+}] = [C_2H_3O_2{}^{1-}] = 1.34 \times 10^{-3}M$
43. $[H^{1+}] = [CN^{1-}] = 1.41 \times 10^{-5}M$
45. $[CuOH^{1+}] = [OH^{1-}] = 7.07 \times 10^{-5}M$

47. $[NH_4{}^{1+}] = [OH^{1-}] = 1.34 \times 10^{-3}M$
49. (a) $K_{sp} = [Ag^{1+}][I^{1-}]$
   (b) $K_{sp} = [Fe^{3+}][OH^{1-}]^3$
   (c) $K_{sp} = [Ag^{1+}]^2[CrO_4{}^{2-}]$
51. (a) $K_{sp} = [Ba^{2+}][CrO_4{}^{2-}]$
   (b) $K_{sp} = [Ca^{2+}][F^{1+}]^2$
   (c) $K_{sp} = [Cu^{2+}][S^{2-}]$
53. $K_{sp} = 1.5 \times 10^{-9}$
55. $K_{sp} = 1.90 \times 10^{-12}$
57. Solubility of CuS = 9.1×10^{-21} g/L
59. Solubility of Ag_2CrO_4 = 2.6×10^{-2} g/L
63. $H_2O + CO_2 \rightleftarrows H_2CO_3 \rightleftarrows H^{1+} + HCO_3{}^{1-}$
65. (a) Equilibrium shifts to the left
   (b) Equilibrium shifts to the right
   (c) Equilibrium shifts to the right
67. $HCO_3{}^{1-} + H^{1+} \longrightarrow H_2CO_3 \longrightarrow H_2O + CO_2$
69. (a) No shift
   (b) Equilibrium shifts to the right
   (c) Equilibrium shifts to the right
70. (a) Equilibrium shifts to the left
   (b) Equilibrium shifts to the right
   (c) Equilibrium shifts to the right
71. (a) $K_{eq} = \dfrac{[NO]^2}{[N_2][O_2]}$ (b) $K_{eq} = \dfrac{[CO]^2[O_2]}{[CO_2]^2}$

   (c) $K_{eq} = \dfrac{[SO_3]^2}{[SO_2]^2[O_2]}$

72. $K_{eq} = 3.5 \times 10^{-3}$
73. $K_a = 1.8 \times 10^{-4}$
74. $[H^{1+}] = [NO_2{}^{1-}] = 6.7 \times 10^{-3}M$
75. $K = 100$
76. (a) $K_{sp} = [Mn^{2+}][OH^{1-}]^2$
   (b) $K_{sp} = [Ag^{1+}][Cl^{1-}]$
   (c) $K_{sp} = [A^{3+}]^2[X^2]^3$
77. Solubility of CaF_2 = 1.7×10^{-2} g/L
78. Ag_2CrO_4

CUMULATIVE REVIEW/CHAPTERS 14–16

1. (a) oxygen = solute and nitrogen = solvent
   (b) sulfur dioxide = solute and water = solvent
   (c) lemon juice and sugar = solutes and water = solvent
2. (a) H—F polar, H_2O polar
   (b) benzene: nonpolar; *trans*-dichloroethane: nonpolar
3. (a) 66.7% by weight salt water
   (b) 11.7% by weight sugar water

4. (a) 5.49% by volume alcohol in water
   (b) 36.8% by volume benzene in CCl_4
5. 6.7% wt-volume benzene in CCl_4
6. 0.027M
7. 13.3 g NaOH
8. 236 mL
9. Normality is the number of equivalents of solute per liter of solution, and molarity is the number of moles of solute per liter of solution.

10. 65.4 g H_3PO_4

11. 1.00N

12. Dilute 0.203 L of 36.0 N with water to 5.50 L

13. 51.0 g $AgNO_3$

14. False

15. 176 g H_2SO_4

16. (a) 545 g of sugar in 2180 g solution
    (b) 0.795 M solution

17. 0.90 g solute

18. Dilute the solution to a total volume of 3.82 L

19. 101.7°C

20. 0.1 m

21. Arrhenius acid is a substance that releases hydrogen ions when dissolved in water. Brønsted-Lowry acid is a substance that is a proton donor.

22. False

23. (a) $2HC_2H_3O_2 + Ca(OH)_2 \longrightarrow$
$$Ca(C_2H_3O_2)_2 + 2H_2O$$
    (b) $N_2O_5 + H_2O \longrightarrow 2HNO_3$

24. $H_2O + SO_3 \longrightarrow H_2SO_4$ or
    $H_2O + SO_2 \longrightarrow H_2SO_3$

25. pH = 2

26. pH = 6

27. 22,500 mL

28. 1 M Na^{1+} and 1 M Cl^{1-}

29. 1.20 M HCl

30. $M_{NaOH} = 0.0450$

31. Ionization is the process by which ions are formed from atoms or molecules by the transfer of electrons. Dissociation is the process by which an ionic substance separates into ions by the action of a solvent.

32. 1 M sulfuric acid is more acidic than 1 N sulfuric acid

33. 1 M hydrochloric acid is more acidic than 1 M acetic acid

34. 0.040 M ions

35. If the Cl^{1-} concentration is 0.500 M, the Sr^{2+} concentration must be half of that (or 0.250 M).

36. 0.125 M

37. The term refers to ionization.

38. Twice as many Cl^{1-} as Sr^{2+}

39. pH = 4

40. $[OH^{1-}] = 10^{-6}$

41. Concentration, temperature, presence of catalyst and the reactants themselves affect the rate of a chemical reaction.

42. As equilibrium approaches, the rate of the forward reaction slows and the rate of the reverse reaction increases. At equilibrium the forward reaction and the reverse reactions occur at the same rate.

43. $H_2O(l) \rightleftharpoons H_2O(g)$

44. Exothermic means heat is released as a result of the reaction.

45. False

46. The equilibrium shifts to the left to remove H^{1+} ions.

47. $K_{eq} = [CO_2]$

48. $K_{sp} = 2.6 \times 10^{-13}$

49. Solubility is 2.5×10^{-17} moles/L

50. Solubility is 0.073 g/mole

51. 0.75 mole/Liter

54. AgCl has the greatest molar solubility.

55. It will shift to the left to relieve the stress.

CHAPTER 17

11. (a) $^{197}_{78}Pt \longrightarrow ^{0}_{-1}e + ^{197}_{79}Au$
    (b) $^{212}_{84}Po \longrightarrow ^{4}_{2}He + ^{208}_{82}Pb$
    (c) $^{10}_{5}B + ^{4}_{2}He \longrightarrow ^{1}_{0}n + ^{13}_{7}N$
    (d) $^{238}_{92}U + ^{1}_{0}n \longrightarrow ^{239}_{92}U \longrightarrow ^{94}_{36}Kr + ^{145}_{56}Ba$

19. (a) $^{14}_{6}C \longrightarrow ^{0}_{-1}e + ^{14}_{7}N$
    (b) $^{24}_{11}Na \longrightarrow ^{0}_{-1}e + ^{24}_{12}Mg$
    (c) $^{238}_{92}U \longrightarrow ^{4}_{2}He + ^{234}_{90}Th$
    (d) $^{84}_{36}Kr \longrightarrow ^{1}_{0}n + ^{83}_{36}Kr$

23. 23,080 years

25. 12.5 g

29. (a) fission   (b) fusion

31. ^{235}U

40. beta

41. scintillation counter

45. 0.5 rem/year

53. $Pb + PbO_2 + 2H_2SO_4 \longrightarrow 2PbSO_4 + 2H_2O$

65. Nitrogen and oxygen

69. (a) $C + O_2 \longrightarrow CO_2$ (b) $S + O_2 \longrightarrow SO_2$
    (c) $2SO_2 + O_2 \longrightarrow 2SO_3$
    (d) $N_2 + O_2 \longrightarrow 2NO$

77. An electrostatic precipitator to remove particulates

79. An electrostatic precipitator for particulates and a scrubber for sulfur oxides

89. 3.9 billion years

96. Less stable

97. $^{16}_{8}O \longrightarrow ^{12}_{6}C + ^{4}_{2}He$

98. $^{10}_{5}B + ^{4}_{2}He \longrightarrow ^{13}_{7}N + ^{1}_{0}n$

99. 60 minutes
100. The warmer the water, the less the amount of dissolved oxygen. Fish could die from lack of oxygen.

CHAPTER 18

1. Covalent
3. Diamond and graphite
9. ·C·
11. Tetrahedral
13. (a) double bond   (b) triple bond
    (c) single bond
15. (a) $CH_3—CH_3$ (b) $CH_2\!=\!CH_2$
    (c) $CH\!\equiv\!CH$
19. $CH_2\!=\!CH_2$ $CH_3CH\!=\!CH_2$ $CH_3CH_2CH\!=\!CH_2$
    Ethene   Propene   1-butene

    $CH_3CH_2CH_2CH\!=\!CH_2$
    1-pentene

    $CH_3CH_2CH_2CH_2CH\!=\!CH_2$
    1-hexene

23. $CH_3—CH—CH_2—CH—CH_3$
             $|$ $|$
            CH_3 CH_3
27. (a) $CH_3CH_2—CH—CH_3$
                     $|$
                    CH_3

    (b) $CH_3—CH—CH—CH—CH_2CH_2CH_3$
                $|$ $|$ $|$
               CH_3 CH_3 CH_3

    (c) $CH_3—\overset{\displaystyle CH_3}{\underset{\displaystyle CH_3}{\overset{|}{\underset{|}{C}}}}—CH_3$

    (d) $CH_3CH_2CH_2CH_2CH\!=\!CH_2$
    (e) $CH_3CH_2CH_2CH\!=\!CHCH_3$
    (f) $CH_2\!=\!C—CHCH_2CH_3$ (g) $CH\!\equiv\!CCHCH_3$
                $|$ $|$ $|$
             CH_3 CH_2CH_3 CH_3

31 and 43
$C—C—C—C—C—C—C$ *n*-heptane

$C—C—C—C—C—C$ 2-methylhexane
                $|$
                C

$C—C—C—C—C—C$ 3-methylhexane
             $|$
             C
             $|$
             C

$C—C—C—\overset{\displaystyle C}{\underset{\displaystyle C}{\overset{|}{\underset{|}{C}}}}—C$ 2,2-dimethylpentane

$C—C—\overset{\displaystyle C}{\overset{|}{C}}—C—C$ 3,3-dimethylpentane

$C—C—C—C—C$ 2,3-dimethylpentane
     $|$ $|$
     C C

$C—C—C—C—C$ 2,4-dimethylpentane
     $|$ $|$
     C C

$C—C—C—C—C$ 3-ethylpentane
             $|$
             C
             $|$
             C
             $|$
             C

$C—C—C—C$ 2,2,3-trimethylbutane
     $|$ $|$
     C C

41. Three straight-chain heptenes, 1-heptene,
    2-heptene, and 3-heptene.
45. (a) 2-methylbutane   (b) 2,2-dimethylpropane
    (c) 2,3,4-trimethylhexane
47. (a) 1-hexene   (b) 2-hexene   (c) 3-hexene
    (d) 2-methyl-3-ethyl-1-pentene
49. (a) 1-pentyne   (b) 2-pentyne
    (c) 3-methyl-1-butyne
    (d) 2-methyl-5-ethyl-3-heptyne
51. (a) Cyclopropane   (b) Cyclooctane
    (c) Cyclobutane   (d) Cyclohexyne
    (e) Methylcyclopentane
    (f) 4-methyl-1-ethylcycloheptane
    (g) 3-methyl-1-propylcyclohexane
56. Yes
57. $R—C\!=\!C—(CH_2)_x—C\!=\!C—R$
        $|$ $|$ $|$ $|$
        H H H H
58. Structural isomers
59. 2,5-dimethyl-3-heptene
60. alkyne, C_nH_{2n-2}
61. $CH_3—C\!=\!CH_2$
              $|$
             CH_3

62.

CH_3

63. $CH_3-CH_2-C=CH_2$
                       $|$
                      CH_2
                       $|$
                      CH_3

64. 4-methyl-4-ethyl-1-hexene
65. 1-octene, 2-octene, 3-octene, 4-octene

66.

CHAPTER 19

3. (a) alcohol   (b) ether   (c) aldehyde
    (d) carboxylic acid
7. (a) nonpolar   (b) polar
9. (a) propanol   (b) 1-hexanol
13. (a) $H_3C-CH_2-CH_2-CH-CH_2-CH_3$
                                $|$
                               OH
    (b) $H_3C-CH_2-CH_2-CH-OH$
    (c) $H_3C-CH_2-CH_2-CH_2-CH_2$
                                    $|$
                          $CH-CH-CH_3$
                                    $|$
                                   OH
    (d) CH_3OH
17. (a) butyl ethyl ether   (b) ethyl methyl ether
19. (a) $CH_3-O-CH_2CH_2CH_2CH_3$
    (b) CH_3-O-CH_3
25. (a) hexanal   (b) propanal
27. (a) $CH_3CH_2CH_2CH_2-CH$ (b) CH_3CH
                                $\|$ $\|$
                                O O
33. (a) 3-pentanone   (b) 2-hexanone
    (c) cyclohexanone
35. (a) $CH_3CH_2CH_2-C-CH_2CH_2CH_2CH_3$
                            $\|$
                            O
    (b) $CH_3-C-CH_2CH_2CH_3$
              $\|$
              O
41. (a) heptanoic acid
    (b) formic acid (or methanoic acid)
    (c) butanoic acid
43. (a) $CH_3CH_2CH_2CH_2CH_2CH_2C-OH$
                                        $\|$
                                        O

    (b) CH_3CH_2C-OH
              $\|$
              O
49. (a) ethyl hexanoate   (b) methyl ethanoate
51. (a) $CH_3CH_2CH_2-C-OCH_2CH_3$
                          $\|$
                          O
    (b) $CH_3-C-OCH_2CH_2CH_2CH_3$
              $\|$
              O
55. (a) $CH_3CH_2-C-OCH_2CH_2CH_3$
                    $\|$
                    O
    (b) $CH_3-C-OCH_3$
              $\|$
              O
57. $CH_3CH_2CH_2CH_2-C-OH + HO-CH_2CH_3$
                            $\|$
                            O
61. (a) tertiary   (b) secondary   (c) secondary
    (d) primary
63. (a) trimethylamine   (b) cyclohexylamine
    (c) cyclopentylethylamine
65. (a) CH_3CH_2-NH
                      $|$
                     CH_2CH_3
    (b) $CH_3CH_2CH_2-NH$
                        $|$
                       CH_2CH_3
    (c) $CH_3-CH-CH_2CH_2CH_3$
              $|$
             NH_2
    (d) $CH_3CH_2-CH-CH_2CH_3$
                    $|$
                   NH_2
72. 1(C), 2(B), 3(D), 4(G), 5(A), 6(E), 7(F)

73.

74. (a) amine   (b) ether   (c) carboxylic acid
     (d) ester

75. $CH_3 - CH_2 - \underset{\underset{OH}{|}}{CH} - CH_3$

76. $CH_3 - CH_2 - \underset{\underset{NH_2}{|}}{CH} - CH_2 - CH_2 - CH_3$

77. $CH_3 - CH_2 - CH_2 - CH_2 - O - CH_2 - CH_2$
                                                  $|$
                                          $CH_2 - CH$

78. Heptanoic acid

79. $CH_3 - CH_2 - \underset{\underset{O}{\|}}{C} - H$

80. $CH_3 - CH_2 - CH_2 - \underset{\underset{O}{\|}}{C} - CH_3$

81. $CH_3 - CH_2 - CH_2 - \underset{\underset{O}{\|}}{C} - OH$

82. (a) ethyl formate   (b) methyl formate

CHAPTER 20

7. Monosaccharides, disaccharides and polysaccharides

21. Those *with* fatty acids in their molecules and those *without* fatty acids in their molecules

35. $CH_2 = \underset{\underset{CH_3}{|}}{C} - CH = CH_2$

37. Vitamins A, E, and K

43. $NH_2 - \underset{\underset{R}{|}}{CH} - \underset{\underset{OH}{|}}{C} - OH$

65. (a) vitamin A, eggs and milk

(b) vitamin B_{12}, liver   (c) vitamin C, fruits
(d) vitamin B_2 or niacin, most foods, especially liver and yeast

67. They are enzymes.

68. manganese, iron, cobalt, copper, zinc, iodine

69. (a) amino acid   (b) glucose
     (c) glycerol and fatty acids

71. carbohydrates, proteins, and lipids

72. Strong alkalies hydrolyze the ester linkages of fats.

CUMULATIVE REVIEW/CHAPTERS 17–20

1. Alpha particles are helium nuclei. They carry a double positive charge. Beta particles are identical to an electron in mass and size. They have a negative electrical charge. Gamma rays are made up of energy. They have no mass and carry no electrical charge.

2. $^{13}_{7}N$

3. $^{214}_{84}Po$

4. (a) $^{210}_{83}Bi \longrightarrow \, ^{210}_{84}Po + \, ^{0}_{-1}e$
     (b) 1.56 g

5. 25 mg

6. $^{249}_{98}Cf + \, ^{18}_{8}O \longrightarrow \, ^{263}_{106}Unh + 4\, ^{1}_{0}n$

7. $^{263}_{106}Unh \longrightarrow \, ^{259}_{104}Unq + \, ^{4}_{2}He$

8. $^{209}_{83}Bi + \, ^{58}_{26}Fe \longrightarrow \, ^{266}_{109}Une + \, ^{1}_{0}n$

9. $^{266}_{109}Une \longrightarrow \, ^{262}_{107}Uns + \, ^{4}_{2}He$

10. True

11. The balanced equations are:
     (a) $Pb + SO_4^{2-} \longrightarrow PbSO_4 + 2e^{1-}$
     (b) $PbO_2 + 4H^{1+} + SO_4^{2-} + 2e^{1-} \longrightarrow$
                              $PbSO_4 + 2H_2O$

12. An electrolytic cell uses electricity to produce a chemical reaction. A voltaic cell produces electrical energy from a chemical reaction.

13. False

14. A gaseous air pollutant is composed of gas molecules, whereas a particulate is a tiny particle of dust, smoke, fume, or aerosol.

15. Sulfur in coal is oxidized to sulfur dioxide,

which eventually becomes sulfuric acid. The sulfuric acid remains as a mist in the atmosphere and can mix with moisture in clouds to produce acid rain.

16. If PVC burns it can produce hydrogen chloride gas, which is poisonous.

17. Three methods of hazardous waste disposal are incineration, landfilling, and reacting the hazardous waste with chemicals to form harmless products.

18. PCBs can leach out of plastics and end up free in the environment where they can enter the food chain.

19. Smog is a mixture of smoke and fog. Photochemical smog is smog that results from atmospheric reactions in which sunlight is present.

20. False

21. ·C·

22. False

23. *Alkane:* (1) methane, CH_4; (2) ethane, C_2H_6; (3) propane, C_3H_8; (4) butane, C_4H_{10}; (5) pentane, C_5H_{12}.
Alkene: (1) ethene, C_2H_4; (2) propene, C_3H_6; (3) butene, C_4H_8; (4) pentene, C_5H_{10}; (5) hexene, C_6H_{12}.
Alkyne: (1) ethyne, C_2H_2; (2) propyne, C_3H_4; (3) butyne, C_4H_6; (4) pentyne, C_5H_8; (5) hexyne, C_6H_{10}.

24. (a) —C—C=C—C—C—C—

(b)
—C—C=C—C—C—C— with CH₃, CH₃ CH₃

25. Isomers

26. —C_6H_{13} hexyl
—C_7H_{15} heptyl
—C_8H_{17} octyl
—C_9H_{19} nonyl
—$C_{10}H_{21}$ decyl

27. Cyclic hydrocarbons can have single, double, or triple bonds. Aromatic hydrocarbons have alternating single and double bonds.

28. When the compounds in a series differ by the same structural group (like CH_2) we say these compounds are members of a homologous series. For example, methane, ethane, propane are all members of the alkane series.

29. Cyclobutane

30. One, two or three pairs of shared electrons are possible for covalently bonded compounds.

31. C_4H_7I

(1) —C=C—C—C— with I

(2) —C=C—C—C— with I

(3) —C=C—C—C— with I

(4) —C=C—C—C— with I

(5) —C—C=C—C— with I

(6) —C—C=C—C— with I

(7) C—C=C with C and I

(8) I—C—C=C with C

(9) C—C=C—I with C

(10) C—C—I with C—C

32. A functional group is an atom or group of atoms that determines the specific chemical properties of a class of organic compounds.

33. (a) —C—C—C—C—C—C—   2-hexanol with OH

(b)
$$-\underset{|}{C}-\underset{|}{C}-\underset{|}{\underset{OH}{C}}-\underset{|}{C}-\underset{|}{C}-\underset{|}{C}-\underset{|}{C}-\underset{|}{C}-$$

3-octanol

34. CH_3CH_2OH: short R group is polar
$CH_3(CH_2)_{10}CH_2OH$: long R group is nonpolar

35. CH_3CH_2—O—$(CH_2)_2$—CH_3: ethyl propyl ether

36. cyclohexanone

37. CH_3—$\underset{\underset{O}{\|}}{C}$—$(CH_2)_5CH_3$

38. (a) secondary (b) tertiary (c) secondary

39. CH_3CH_2—$\underset{\underset{O}{\|}}{C}$—$OH + H$—$O$—$(CH_2)_3$—$CH_3$

40. CH_3—$\underset{\underset{O}{\|}}{C}$—$O$—$CH_2CH_3$ ethyl ethanoate

41. $CH_3CH_2CH_2$—$\underset{\underset{O}{\|}}{C}$—$H$ butanal

42. CH_3—$\underset{\underset{O}{\|}}{C}$—$(CH_2)_2CH_3$ methyl propyl ketone

43. Metabolism is the process by which the food we eat is transformed in the body through a series of complex chemical reactions to provide material for cellular function and energy to power the body. Anabolism is the process in which organic materials that have been absorbed by the cells in the body serve as building blocks for the synthesis of more complex organic substances. Catabolism is the breaking down of food molecules into smaller molecules, usually accompanied by a release of energy.

44. Monosaccharides are carbohydrates containing one sugar group in the molecule. Disaccharides are carbohydrates containing two sugar groups in the molecule. Polysaccharides are carbohydrates containing many sugar groups in the molecule.

45. Lipids are classified into two main types: those with fatty acids in their molecules and those without fatty acids in their molecules.

46. Fats are formed by an esterification reaction between a fatty acid and the alcohol glycerol.

47. Waxes are formed when fatty acids and high molecular weight alcohols other than glycerol undergo esterification. Steroids are lipids that do not contain fatty acids in their molecules. They are composed of an interlocking structure containing four rings. Terpenes are lipids that do not contain fatty acids in their molecules. They are made up of molecules of isoprene which link together to form large molecules.

48. The general formula for an amino acid is

amino group H_2N—$\underset{\underset{H}{|}}{C}$—$\underset{\underset{O}{\|}}{\underset{|}{\overset{R}{C}}}$—$OH$ R group side chain / carboxyl group

49. An essential amino acid is an amino acid that cannot be synthesized by the human body.

50. Water soluble vitamins are, in general, polar molecules. Fat soluble vitamins are soluble in lipids.

SUPPLEMENT A

1. (a) 17 (b) -15 (c) 15 (d) 5
2. (a) 1 (b) 5 (c) -5 (d) -1
3. (a) 33 (b) -17 (c) 17 (d) -33
4. (a) 11 (b) 99 (c) -99 (d) -11
5. (a) $\frac{1}{16}$ (b) $\frac{1}{4}$ (c) 512 (d) $\frac{1}{32}$
6. (a) $\frac{6}{35}$ (b) $\frac{18}{25}$ (c) $\frac{1}{4}$ (d) $\frac{25}{49}$
7. (a) $(4)^{-7}$ (b) $(4)^5$ (c) $(6)^2$ (d) $(5)^1$
8. (a) $(4)^3$ (b) $(4)^8$ (c) $(a)^1$ (d) $(b)^2$
9. (a) 625 (b) 1000 (c) 99 (d) $\frac{1}{512}$
  (e) 16
10. (a) $(5)^5$ (b) $(12)^{12}$ (c) $(10)^0 = 1$
  (d) $(4)^{-10}$
11. (a) $(4)^2$ (b) $(12)^5$ (c) $(20)^{-7}$
  (d) $(20)^{-1}$
12. (a) $(4)^9$ (b) x (c) $(y)^7$ (d) p
  (e) $(10)^8$
13. (a) 64 (b) $\frac{1}{81}$ (c) $\frac{1}{64}$ (d) 64
14. (a) $(10)^3$ (b) $(2)^{-5}$ (c) $(3)^{-3}$ (d) $(6)^2$
15. (a) $(3)^2$ (b) $(2)^3$ (c) $(a)^{15}$ (d) $(b)^{-13}$
16. 48 inches
17. $w = A/1$
18. 6 yards
19. 48 eggs
20. (a) 6 hours (b) 2 hours
21. 16 pints
22. 1.47 ft/sec

23. 6 feet
24. 180 inches
25. 300 feet
26. 180 inches
27. 24 dozen
28. 1440 minutes, 86,400 seconds
29. 10 gallons
30. 15,000 mi/hr
31. (a) 149.683   (b) 141.41
32. (a) 26.525   (b) 103.71   (c) 106.14
    (d) 147.78
33. (a) 29.32   (b) 38.5628   (c) 27.0657
34. (a) 2.04   (b) 5
35. (a) 31.20   (b) 46.15   (c) 39.80
    (d) 75.38
36. (a) $a = 4$ (b) $y = \frac{13}{40} = 0.33$ (c) $z = 3$
37. (a) $x = \frac{3}{2}$ (b) $y = 1$
38. (a) $P = \frac{nRT}{V}$ (b) $n = \frac{PV}{RT}$ (c) $T = \frac{PV}{nR}$

    (d) $\frac{n}{V} = \frac{P}{RT}$

39. (a) $t = 16$ (b) $y = 7$ (c) $x = 8$
    (d) $k = -8$
40. (a) $b = \frac{A}{cd}$ (b) $c = \frac{A}{bd}$ (c) $\frac{A}{b} = cd$
41. (a) $x = 20$ (b) $y = 27$ (c) $t = 2$
    (d) $k = 4$
42. (a) $k = 1$ (b) $h = 8$ (c) $f = 16$
    (d) $y = 4$
43. (a) $x = 8, y = 36$ (b) $h = 42$
    (c) $a = 1.5, b = 2.5, c = 3.5$
    (d) $d = 12.5$
44. (a) $m = DV$ (b) $V = \frac{m}{D}$
45. 5 g/mL
46. LA = 33.3%, LT = 16.7%, HT = 50.0%
47. 800 to Europe, 600 to Asia, 200 to Alaska, 400 to Miami
48. (a) 66   (b) 17   (c) 6   (d) 25
49. (a) 65   (b) 3   (c) 1   (d) 12.5
50. (a) 25   (b) 1000   (c) 0.00375
    (d) 12.5

ANSWERS TO PRACTICE EXERCISES

2.1 301 in^3
2.2 3500 cm
2.3 0.350 g
2.4 1800 cm
2.5 66 cm
2.6 (a) 5.5 cm   (b) 0.055 m
     (c) 0.000055 km or 5.5×10^{-5} km
2.7 (a) 4400 g   (b) 44000 dg   (c) 440000 cg
     (d) 4400000 mg or 4.4×10^6 mg
2.8 13.8 ft by 4.10 ft
2.9 143 L
2.10 2.7 g/cm^3
2.11 8.0 g/cm^3
2.12 5.50 g/cm^3 liquid A, 2.00 g/cm^3 liquid B. Liquid A is more dense.
2.13 $V = 1333$ cm^3
2.14 27
2.15 53.44
2.16 6.3 cm^2
2.17 10
2.18 82
2.19 (a) 2   (b) 4   (c) 6   (d) 4   (e) 2
      (f) 5
2.20 (a) 4.2×10^{-3} (b) 5.60×10^4
      (c) 6.023×10^6 (d) 1.23×10^{-3}
2.21 118°C

2.22 302°F

3.1 (a) 6.00 ft   (b) $12\overline{0}$ in.
3.2 (a) 10.0 moles   (b) 32.0 g
3.3 (a) 0.5000 mole   (b) 0.100 mole
     (c) 20.0 mole   (d) 0.000100 mole or 1.00×10^{-4} mole
3.4 (a) 975 g   (b) 16.0 g   (c) $12\overline{0}$ g
3.5 (a) 9.03×10^{24} (b) 1.20×10^{23}
     (c) 1.81×10^{24}
3.6 H_2O
3.7 Cr_2S_3
3.8 ZnN_2O_4 or $Zn(NO_2)_2$
3.9 (a) 106   (b) 130
3.10 (a) 5.0   (b) 0.0010
3.11 (a) 3.0×10^{24} (b) 6.0×10^{20}
3.12 $C_9H_8O_4$
3.13 32.8% K, 67.2% Br
3.14 (a) 85.7% C, 14.3% H
      (b) 39.3% Na, 60.7% Cl
      (c) 92.3% C, 7.7% H
3.15 75.0 g

4.1 (a) 14 p, 14 e, 14 n   (b) 15 p, 15 e, 16 n
     (c) 3 p, 3 e, 3 n
4.2 Average weight = 6.92

4.3 $^{100}Y = 20.0\%$, $^{110}Y = 80.0\%$

4.4 909,200

7.1 (a) H:N:H   (b) H:C̈l̈:   (c) :Ö:S̈
      H   :Ö:

7.2 H—O←S→O—H
              $\|$
              O

7.3 (a) covalent   (b) ionic   (c) ionic
(d) covalent

7.4 (a) 84% ionic, 16% covalent   (b) 63% ionic, 37% covalent   (c) 55% ionic, 45% covalent   (d) 59% ionic, 41% covalent

7.5 CS_2 is most covalent, H_2S is next, then $AsCl_3$ and CO_2 tie for least covalent.

7.6 KF

7.7 MgO

7.8 SrS

7.9 $K_2Cr_2O_7$

7.10 $Al_2(CO_3)_3$

7.11 CuBr

7.12 FeO

7.13 Cu_3N

7.14 +3

7.15 (a) +4   (b) +5

7.16 (a) iron(III) bromide   (b) zinc nitride

7.17 (a) Ag_2SO_4 (b) Zn_3N_2 (c) $MgCl_2$
(d) $NaClO_3$

7.18 (a) nonpolar   (b) polar   (c) polar

7.19 (a) BH_3 bonds are polar, molecule is polar
(b) Cl_2O bonds are polar, molecule is polar
(c) CO_2 bonds are polar, molecule is nonpolar

8.1 $2CuO \longrightarrow 2Cu + O_2$

8.2 $Zn + H_2SO_4 \longrightarrow ZnSO_4 + H_2$

8.3 (a) $2Al + 3Cl_2 \longrightarrow 2AlCl_3$
(b) $2Na + 2H_2O \longrightarrow 2NaOH + H_2$
(c) $2KNO_3 \longrightarrow 2KNO_2 + O_2$
(d) $2HNO_3 + Ba(OH)_2 \longrightarrow$
                          $Ba(NO_3)_2 + 2H_2O$

8.4 (a) $4K + O_2 \longrightarrow 2K_2O$
(b) $2Ca + O_2 \longrightarrow 2CaO$
(c) $H_2 + Br_2 \longrightarrow 2HBr$

8.5 (a) $Sr(OH)_2 \longrightarrow SrO + H_2O$
(b) $SrCO_3 \longrightarrow SrO + CO_2$
(c) $2KClO_3 \longrightarrow 2KCl + 3O_2$
(d) $2KCl \longrightarrow 2K + Cl_2$

8.6 (a) $2Li + 2H_2O \longrightarrow 2LiOH + H_2$
(b) $2Al + 3H_2SO_4 \longrightarrow Al_2(SO_4)_3 + 3H_2$
(c) $3Mg + 2Al(NO_3)_3 \longrightarrow 3Mg(NO_3)_2 + 2Al$

8.7 (a) $2BiCl_3 + 3H_2S \longrightarrow Bi_2S_3 + 6HCl$
(b) $2Fe(OH)_3 + 3H_2SO_4 \longrightarrow$
                          $Fe_2(SO_4)_3 + 6H_2O$

8.8 (a) No reaction   (b) Reaction does occur

8.9 (a) Copper is oxidized; silver ion is reduced.
(b) Sodium is oxidized; chlorine is reduced.
(c) Bromide ion is oxidized; chlorine is reduced.
(d) Zinc is oxidized; sulfur is reduced.

8.10 (a) Copper is the reducing agent; silver nitrate is the oxidizing agent.
(b) Sodium is the reducing agent; chlorine is the oxidizing agent.
(c) Sodim bromide is the reducing agent; chlorine is the oxidizing agent.

9.1 2.43 g Mg, 1.60 g O

9.2 0.8999 g Al, 7.099 g $Al(NO_3)_3$, 3.269 g Zn

9.3 49.07 g

10.1 $4\bar{0},000$ cal or $4\bar{0}$ kcal

10.2 12,500 cal or 12.5 kcal

10.3 5380 cal or 5.38 kcal

10.4 2680 cal or 2.68 kcal

10.5 $C_2H_6(g) + 3\frac{1}{2}O_2(g) \longrightarrow$
                $2CO_2(g) + 3H_2O(l) + 372$ kcal

10.6 $2C_6H_{14}(l) + 19O_2(g) \longrightarrow$
              $12CO_2(g) + 14H_2O(l) + 198\bar{0}$ kcal

10.7 204.9 kcal heat are produced

10.8 $PCl_5(g) \longrightarrow PCl_3(g) + Cl_2(g) - 22.2$ kcal

10.9 $\Delta H = -75.3$ kcal

10.10 $\Delta H = -135.4$ kcal

10.11 $\Delta H = -216.6$ kcal

10.12 $\Delta H = -6.1$ kcal

10.13 $\Delta H = -74.4$ kcal

10.14 ΔH_f of $H_2SO_4(l) = -217$ kcal

10.15 ΔH_f of $C_6H_{12}O_6(s) = -304.4$ kcal

10.16 ΔH_f of $CH_4(g) = -17.9$ kcal

11.1 $25\bar{0}$ mL

11.2 $5\bar{0}\bar{0}$ mL

11.3 6080 torr

11.4 1875 mL

11.5 $t_i = 17.0°C$

11.6 $V_f = 5.15$ L

11.7 $t_f = 12.0°C$

11.8 $V_f = 2.50$ L

11.9 $V_f = 228$ mL

11.10 $V_f = 187$ mL

11.11 $V_f = 4.70$ L

11.12 20.0 L of HF gas

11.13 0.0200 mole

11.14 4.48 L

11.15 36 L

11.16 0.0217 mole

11.17 M.W. = 410

14.1 (a) Salt is the solute and water is the solvent.
(b) Sulfur trioxide is the solute and water is the solvent.
(c) Water is the solute and isopropyl alcohol is the solvent.
(d) Carbon dioxide is the solute and water is the solvent.

14.2 (a) No   (b) Yes   (c) Yes

14.3 (a) 44.4%   (b) 30.0%

14.4 (a) 20.0%   (b) 55.6%

14.5 (a) 4.5%   (b) 20.00%

14.6 (a) 0.200M (b) 0.0164M (c) 1.07M

14.7 37 g

14.8 50.0 mL

14.9 2.00N

14.10 0.444N and 0.444M

14.11 12.0 g

14.12 2.00 L of the 6.00N HCl solution are needed.

14.13 16.7 mL of the 12.0N H$_3$PO$_4$ solution are needed.

14.14 (a) 1   (b) 0.04   (c) 0.16

14.15 (a) 100.83°C   (b) 100.02°C

14.16 (a) −2.98°C   (b) −0.0744°C

15.1 (a) pH = 5   (b) pH = 10   (c) pH = 1

15.2 (a) pH = 4   (b) pH = 3   (c) pH = 9
(d) pH = 3   (e) pH = 13

15.3 0.33N

15.4 0.086N

15.5 0.409N

15.6 0.102N

16.1 $K = \dfrac{[COCl_2]}{[CO][Cl_2]}$

16.2 $K = 2\overline{0}$

16.3 [glucose-6-phosphate] = 2.0

16.4 1.20 moles

16.5 (a) left   (b) right   (c) left   (d) right

16.6 (a) right   (b) left

16.7 (a) no change   (b) left

16.8 (a) left   (b) left   (c) right   (d) left

16.9 $K_a = 4.5 \times 10^{-4}$

16.10 [H^{1+}] = [citrate ion] = 2.9 × 10^{-2}M
[citric acid] = 0.97M

16.11 [NH$_4^{1+}$] = [OH^{1-}] = 4.2 × 10^{-3}M
[NH$_3$] = 0.995M

16.12 (a) $K_{sp} = [Pb^{2+}][SO_4^{2-}]$
(b) $K_{sp} = [Ag^{1+}]^2[CrO_4^{2-}]$
(c) $K_{sp} = [Ag^{1+}][I^{1-}]$

16.13 0.033 g/L

16.14 The equilibrium will shift to the left, causing a decrease in the H^{1+} concentration, with a resultant increase in pH. A condition known as respiratory alkalosis results.

17.1 0.39 g

18.1 (a) CH$_3$—CH$_2$—CH$_2$—CH$_3$

(b) CH$_3$—CH$_2$—CH—CH$_2$—CH$_3$
$\qquad\qquad\qquad$ |
$\qquad\qquad\qquad$ CH$_3$

18.2 (1) C—C—C—C—C—C—C

(2) C—C—C—C—C—C
$\qquad\qquad\qquad$ |
$\qquad\qquad\qquad$ C

(3) C—C—C—C—C—C
$\qquad\qquad\quad$ |
$\qquad\qquad\quad$ C
$\qquad\qquad\quad$ C

(4) C—C—C—C—C
$\qquad\qquad\quad$ |
$\qquad\qquad\quad$ C

(5) C—C—C—C—C
$\qquad\quad$ |$\qquad$ |
$\qquad\quad$ C$\qquad$

(6) C—C—C—C—C
$\qquad\quad\;$ |$\;$ |
$\qquad\quad\;$ C$\;$ C

(7) C—C—C—C—C
$\qquad\;$ |$\qquad$ |
$\qquad\;$ C$\qquad$ C

(8) C—C—C—C—C
$\qquad\quad\;$ |
$\qquad\quad\;$ C
$\qquad\quad\;$ C

(9) C—C—C—C
$\qquad$ |$\;$ |
$\qquad$ C$\;$ C

18.3 (a) hexane   (b) butane   (c) octane

18.4 (a) methylhexane   (b) methylbutane
(c) *n*-octane

18.5 (a) 3-methylhexane   (b) 2-methylbutane
(c) *n*-octane

18.6 3-ethyl-5-methylheptane

18.7 2,4-dimethylhexane

18.8 (*a*) 2,3-dimethylnonane
(*b*) 2,3,5-trimethyl-4-ethylnonane
(*c*) 2,3,4,5-tetramethylhexane
(*d*) 3,4-diethylheptane
(*e*) 4-ethyl-2,3,5,6-tetramethyloctane

18.9 (*a*) 1-pentene   (*b*) 1-hexene
(*c*) 3-heptene
(*d*) 3-methyl-1-hexene   (*e*) 1-hexyne
(*f*) 3-hexyne   (*g*) 4,5-dimethyl-1-hexyne
(*h*) 3,4-dimethyl-1-pentyne

18.10 (*a*) cyclobutane   (*b*) cyclopentane

18.11 (*a*) cycloheptene   (*b*) cycloheptyne
(*c*) cyclopentene   (*d*) cyclopropene

19.1 (*a*) polar   (*b*) nonpolar

19.2 (*a*) propyl alcohol or propanol
(*b*) cyclopentyl alcohol or cyclopentanol

19.3 (*a*) dipropyl ether   (*b*) ethyl methyl ether

19.4 (*a*) propionaldehyde or propanal
(*b*) heptylaldehyde or heptanal

19.5 (*a*) diethyl ketone or 3-pentanone
(*b*) butyl ethyl ketone or 3-heptanone
(*c*) ethyl methyl ketone or butanone

19.6 (*a*) propionic acid or propanoic acid
(*b*) valeric acid or pentanoic acid

19.7 $CH_3CH_2CH_2CH_2 - \overset{\overset{\textstyle O}{\|}}{C} - OCH_3 + H_2O$

19.8 (*a*) methyl propionate or methyl propanoate
(*b*) propyl butyrate or propyl butanoate

19.9 (*a*) tertiary   (*b*) secondary   (*c*) primary

19.10 (*a*) ethyl amine or amino ethane
(*b*) trimethyl amine
(*c*) ethyl methyl amine

20.1 Glucose is found in blood, fructose is found in honey, and galactose is found as part of lactose, agar, and pectin.

20.2 Sucrose is found in cane sugar, lactose is found in milk, and maltose is obtained from the breakdown of starch, dextrin, and glycogen.

20.3 Fats are lipids that contain fatty acids. Steroids and terpenes are lipids that don't contain fatty acids.

20.4 $CH_2 = \overset{\overset{\textstyle CH_3}{|}}{C} - CH = CH_2$

20.5 Twenty

20.6 Eight essential amino acids

20.7 Vegetables and cereals are examples of incomplete protein food.

20.8 Blood proteins, catalytic proteins, contractile proteins, digestive proteins, hormonal proteins, and structural proteins

20.9 A, E, D, and K are the fat-soluble vitamins.

20.10 Vitamin C, folic acid, niacin, and the B vitamins are all water-soluble.

20.11 Macronutrients are minerals we need in large amounts; micronutrients are minerals we need in trace amounts.

Index